Insights into the Relationships between Humanities and Sciences

Liliana Mammino
Editor

Insights into the Relationships between Humanities and Sciences

A Working Outlook

Springer

Editor
Liliana Mammino
University of Venda
Thohoyandou, South Africa

ISBN 978-3-031-97899-9 ISBN 978-3-031-97900-2 (eBook)
https://doi.org/10.1007/978-3-031-97900-2

This Springer imprint is published by the registered company Springer Nature Switzerland AG
The registered company address is: Gewerbestrasse 11, 6330 Cham, Switzerland

In memory of Prof. Giovanni Bonfanti,
my chemistry teacher at Liceo Ginnasio A. Canova, Treviso, Italy,
who not only taught me to love chemistry,
but also taught me to see philosophy in chemistry,
to relate language and chemistry,
and to view the challenges of chemistry education as integral part of chemistry.

Preface

How was this book born?

The first idea started from the realization that students' inadequate language mastery is often the major single cause for poor understanding and performance in chemistry courses, at all levels. It gradually also developed into a tentative response to other phenomena that appear to be getting more widely spread and more uncomfortable with time. It happens at various levels and in various contexts that some colleagues from the sciences appear to 'look down' at colleagues from the humanities, and the latter often appear to take this attitude as a fact of life. And I cannot avoid a feeling of deep hopelessness when persons in various-levels decision-making positions state, or take for granted, that language mastery is not-so-much-needed for students selecting science or technology areas. Those persons genuinely want to promote the preparation of good scientists and engineers for the future, but they are not aware that, by offering 'discounts' on the students' language mastery, they are actually denying them a tool that is essential for the attainment of deep understanding in the sciences.

What appeared increasingly evident was the absence of what could be called 'working conceptions' about the nature, importance, and benefits of cross-disciplinary collaborations between humanities and sciences. The wish to try and 'do something' grew as new observations accumulated. A book cannot bring dramatic changes, but can disseminate information and stimulate reflections that might, in turn, lead to some positive changes. In view of these objectives, what was needed was not a new treatise discussing the theoretical aspects of the relationships between humanities and sciences, but a book presenting concrete examples of constructive, mutually beneficial relationships between the two domains, and concrete documentation of the negative consequences of inadequate language mastery and inadequate epistemological awareness for science learning and science practice, and doing so in a way that is accessible to readers from all areas. This is what this book attempts to do.

Like in any beginning, initial enthusiasm made me plan a high number of different chapters, envisaging a lot of attractive themes. Had all of them materialised, the book would have largely exceeded 1000 pages. The search for authors and co-authors, however, also highlighted an unanticipated challenge—a frequent lack of real possibility to 'talk to each other' between experts in different domains, stemming from not sufficiently *knowing* each other (knowing the nature and significance of each other's fields) and, consequently, not knowing how to talk to each other. Some themes that I cherished could not be devoted individual chapters because of this reason. This actually identified an additional objective for this book and for future studies into the relationships between humanities and sciences: promoting the awareness of the vastness of themes we can and need to talk to each other about and investigate through shared approaches to build working interaction routes.

Technology and engineering are currently grouped together with the sciences into the STEM (science, technology, engineering, and mathematics) keyword—one of those 'magical' keywords that seem to embed the essence of modern society. It is not easy to find technology experts available to talk about humanities in association with their professional expertise; I was lucky because my brother (who is a structural engineer) shares the same humanities-background built in pre-university instruction and accepted to contribute some chapters; this has enabled the incorporation of implementation- and technology-oriented perspectives into the overall scope.

Different chapters focus on different themes, trying to cover both general-type aspects relevant to promote underpinning perceptions of the nature of the two domains and their interactions and the presentation of individual components of the interactions, considering both their documented or potential benefits and the impacts of their inadequacies or absence.

Chapter 1 is introductory. It outlines the reasons why it is important to consider the relationships between humanities and sciences and how each of these two domains can enhance the other. In order to be convincing, it utilises selected concrete examples to attract specific attention to extant undesirable effects stemming from underestimation or phasing out of humanities components. It also outlines historical excerpts, to recall that humanities and sciences were born together and they developed together until comparatively recent (2-3 centuries ago), when a 'parting of the ways' started, and ended up becoming nearly dramatic in our times. Historical excerpts are also present in a number of other chapters, when relevant to the themes considered in those chapters, to ensure that the assertions about the importance of specific interactions in our time are not perceived as something coming 'out of the blue sky', but as something deeply embedded in the nature and development of both the sciences and the humanities.

Chapter 2 focuses on the relationships between science and culture, considering culture in its broadest meaning, and highlights the importance of potentially useful complementarities in the contributions from different domains, and the envisaged benefits for human activities and progress. Chapter 3 explores areas of complementarity between the two domains and shows that both domains would benefit from considering the insights of the other's approaches.

Chapters 4–7 deal with different aspects concerning language. Given the importance of language for the development and communication of thought and, therefore, for the development of each discipline, whether humanities-related or science-related, it was considered important to highlight several diverse perspectives for its roles, substantiated by an adequate number of concrete examples. Considerations on language are also included in some other chapters, when they can bring additional insights.

Chapter 8 attempts to outline the bases of an epistemology of engineering—a novel investigation area, which has been growing in the last decades and is clearly cross-disciplinary between the two domains—and to relate them to engineering practice. Chapter 9 presents a case study in which different disciplines from the two domains contributed to the recovery of an ancient mine, to transform it into an underground museum. Chapters 10–12 consider relationships between specific areas in the two domains: art and chemistry (10), art and the images from the physics of a specific category of molecular structures (11), and literature and science (12).

Chapter 13 considers relevant interfaces between science research and history research, as well as the educational role of the history of science in history teaching. Chapters 14–16 focus on constructive synergies between humanities and sciences within educational contexts and perspectives. Chapter 17 recalls the motivations for the merger of the two previous international bodies, respectively for the sciences and for the human and social sciences, into the current unified body—a merger that underlines the importance and expected benefits of synergies between the two domains.

Chapter 18 has the role of concluding chapter. It reiterates the main objective of the book as the fostering of working conceptions about the relationship between the two domains, outlines the fuzzy-set character of the possible definitions of the two domains, summarises some themes that could not have individual chapters devoted to them because of non-availability of authors willing to take the challenges of writing them, and presents take-home messages from the individual chapters and from the book as a whole.

Overall, the book emphasises that the exploration and identification of interaction-routes, interfaces and overlap areas between humanities and sciences is important both from a theoretical point of view and from a practically oriented one. The former enables better insight into human intellectual endeavours, the latter facilitates a variety of applications, including major ones closely related to education. The extensive attention to practically oriented interactions, their modes and their benefits, is likely the major novelty of the book.

Thohoyandou, South Africa Liliana Mammino

Contents

About the Editor

Liliana Mammino obtained a degree in Chemistry from the University of Pisa (Italy, her home country) and a Ph.D. in Chemistry from Moscow State University (Russia). She is currently an Emeritus Professor at the University of Venda (South Africa), where she has been working since 1997, teaching the physical chemistry courses at all levels, the first-year general chemistry course and the third-year process technology course, and establishing computational chemistry research *de novo*. Her earlier work experiences include chemistry teaching at the National University of Somalia (1974–1975), University of Zambia (1988–1992), and National University of Lesotho (1993–1996). Her research interests comprise theoretical chemistry, with main focus on the computational study of biologically active molecules, and chemistry education, with main focus on conceptual understanding and the roles played by language and visualization, and on education for sustainable development, with particular attention to green chemistry education. She has published extensively in these areas. She represents Associate member of the Interdivisional Committee on Green Chemistry for Sustainable Development of IUPAC. In 2013, she was one of the recipients of the IUPAC Award for Distinguished Women in Chemistry and Chemical Engineering. Her pre-university education entailed extensive humanities education, and she has maintained this interest throughout, above all in relation to the use of language and to the history of science, including science aspects in traditional practices. In 2016, she received an award from the African Language Association of Southern Africa (ALASA) for "the most outstanding article in sociolinguistics in 2015". In 2024, she was awarded the Chemistry Education medal by the South African Chemical Institute.

Chapter 1
Relationships Between Humanities and Sciences: A Discourse with Immense Exploration Potentialities

Liliana Mammino

> The most beautiful experience we can have is the mysterious. It is the fundamental emotion which stands at the cradle of true art and true science.
> Albert Einstein

Abstract Humanities and sciences can be viewed as the two domains of human intellectual activities. A dichotomy between them has developed in the last centuries and the gap continues expanding. Several calls have been made to halt and possibly try and reverse the 'moving apart' process. This requires the generation of working conceptions of the relationships and synergies between the two domains, together with the awareness of their implementation significance. The present chapter highlights the undesirable impacts of the mutual isolation of the two domains, as they constitute fundamental motivations for efforts aimed at overcoming it. Given the current frequent tendency to somewhat 'downgrade' the humanities and ascribe a sort of 'superior status' to science and technology, the chapter opts to highlight the undesirable impacts that the impoverishment of humanities-based abilities—such as language mastery and epistemological awareness—brings on science learning, on the very capability to 'do science', and on the possibility of safe decisions about the application of new technologies made available by new science discoveries. The discourse is substantiated by a number of concrete examples whose terms are easily evident. It is concluded that building synergies between the two domains is crucial for the future of all human intellectual activities and for development in general.

Keywords Non-existence of a crisp boundary between humanities and sciences · Bridging the dichotomy between humanities and sciences · Education as a cross-area field · Epistemological aspects in the sciences · Humanities and sciences through history · Language and science-generation · Language and science

---

L. Mammino (✉)
Faculty of Science, Engineering and Agriculture, University of Venda, Thohoyandou, South Africa
e-mail: mammino@univen.ac.za; sasdestria1@gmail.com

L. Mammino (ed.), *Insights into the Relationships between Humanities and Sciences*,
https://doi.org/10.1007/978-3-031-97900-2_1

learning · Working conceptions of the relationships between humanities and sciences

1.1 Introduction

This chapter has an introductory role, presenting the motivations for specific attention to the relationships between humanities and sciences, as well as some of the crucial concepts and discourses, which will be developed more in detail in the next chapters. The major *leitmotiv* emphasises the importance of generating working conceptions about the relationships and synergies between the two domains, where 'working' refers to a variety of components: the ability to identify relationships within existing works and endeavours, including all what pertains to the history of human development; the ability to recognise when and how contributions from the other domain are relevant, or even essential, for a specific purpose or activity; and the ability to build relationships when they are absent or their presence is below the actual needs for them to be beneficial.

Humanities and sciences can rightly be considered the two major domains of human intellectual (and, often, creative) activities; even more so if the meaning of each term is extended to include the applied or practically-oriented activities related to, or rooted in, each domain. The best term to express their mutual relationships is likely the term *Ometeca*, an Aztec word that means "two in one"—two domains (or perceived domains) of the same unit—human intellectual activities. It was the term characterising the vision of the *Working Conferences on the Relationships between Humanities and Sciences* series, which continued for many years and was discontinued with the Covid-19 pandemic.

Within the currently widely-spread model entailing a de facto compartmentalization of knowledge, the two domains are often perceived as separate and scarcely interacting, or non-interacting at all. On the other hand, not only total separation is impossible, but impoverishment in the extent and quality of their interactions is bound to bring about corresponding impoverishment in the development patterns of either domain [1]. This suggests the opportunity of practically-oriented explorations of the relationships between the two domains, with the following broad-range purposes:

- Highlighting the extent of their interactions, the way they developed through history, and the significance they have in our times.
- Searching for possible options and routes to enhance and optimize their interactions, by promoting working conceptions of the possible relationships.
- Designing educational approaches that may enable students develop a deeper and more complete perception of the interactions and their benefits.

The exploration itself can follow various pathways and pursue various objectives. For instance, a pedagogically-oriented exploration, searching for ways to underline the existence and importance of relationships within formal education contexts, envisaged a classification-based route as an effective option: identifying what pertains to which domain becomes a way to attract students' attention on their relationships [1]. On the basis of common perceptions about the two domains, it would seem easy to ascribe chemistry, physics or biology to the science domain, and literature or the arts to the humanities domain. But there are areas that cannot be ascribed only to one domain, even at the most-basic-level approximation. Architecture merges engineering-based components and art-based components. Education refers to all the areas, and requires specific knowledge (which, for science subjects, is the knowledge of the given science) and humanities-based knowledge in relation to how learners actually learn, how to design teaching approaches, how to ensure communication in both directions (from teacher to learner and from learner to teacher), and to themes such as ethics or epistemology, that are relevant for science practice. Language is needed in all areas, to develop ideas and knowledge and to communicate them: thus, a humanities-based instrument (language) is equally essential within the humanities and within the sciences. Even the investigation methods may often overlap; for instance, geology is a science, but its processes, taking millions of years, cannot be reproduced in a laboratory; therefore, its research method is similar to that of history: discovering evidence (in the rock formations, through the shapes and types of layers, through fossils, etc.) and interpreting it. In other words, there are no stringently identifiable (or 'crisp') boundaries between the two domains or—in terms borrowed from mathematics—the two domains can be viewed as fuzzy sets [2]. This issue is analysed more in detail in Sect. 18.2 of Chapter 18; it is opted to analyse it in the last chapter because then its understanding can be facilitated by the awareness that is progressively built through the other chapters.

Recent decades have witnessed a progressive decrease in the consideration for humanities, observable in many contexts. While the 'usefulness' of the sciences is apparent to many, from decision makers to academics to the public, the roles of the humanities are often not perceived in an equally clear way. This may lead to different considerations for the careers chosen by young persons, such as the perception that someone choosing science is 'smarter' than someone choosing humanities. It may also lead to different funding choices [3], with the funding for the humanities often tending to lower below what could be considered 'survival needs'. It has also led to a sharp decrease in the extent of humanities components in pre-university education, with worrying impacts that are already becoming evident.

A number of worrying impacts are described and analysed in the next sections. Two concrete examples are presented at this stage, to start stressing since now how the reduction of the acquired humanities background below what could be considered a 'safety level' may have unwelcome implications for very practical aspects such as the communication of instructions or the evaluation of research project proposals. The first example refers to the description of a simple procedure. Some time ago, the author (who will from now on be simply denoted as LM for the sake of conciseness; pronouns she/her/hers) needed to quickly learn how to sign a soft copy of a PDF document and asked some colleagues for advice. The two 'extreme' (poorest and best) received answers are reported below. Both colleagues knew how to perform the operation; their answers differ by the extent of their language-mastery. Answer 1 can be understood only by someone who already knows how to proceed, because it does not provide an implementable operational procedure. Answer 2 provides all the necessary details to perform the operation; its elaborated description testifies to good language-mastery and ensures efficient communication.

1. *You can use trial pdf editor perhaps.*
2. *Open the PDF using Acrobat. Nowadays it usually opens by default in your browser, which you don't want, so make sure you use "open with" and select Acrobat. In the grey bar above, you should see a "fountain pen" icon. Click on it, it should then give you directions for inserting the signature.*

The second illustrative example shows how adequate philosophical and epistemological foundations may become essential for something as crucial in the academic world as is the evaluation of a research proposal. In response to a call offering substantial funding, a researcher A submits a proposal concerning the computational study of biologically active molecules and a researcher B submits a proposal concerning the study of certain aspects of fluid motion; a senior administrator C is not expert in either field and consults an expectedly expert D; D sees that both projects involve some form of modelling, decides that this is a reason for merging the proposals and actually merges them; the unavoidable outcome is that the merged proposal does not have any chance of being considered for funding. The decision taken by D stems from inadequate familiarity with the epistemological aspects in relation to the nature of the two individual projects. 'Modelling" is something that is done in many areas, and in different ways for different areas. In the case in question, the study of molecules is the study of discrete microscopic entities and pertains to quantum chemistry, whereas the study of fluids pertains to the 'physics of continuum' and, therefore, to classical physics; the two studies have nothing in common and the modelling approaches are entirely different. An analogy that might prompt a simpler perception of the problem could refer to the study of animals: let us suppose that A prepares a proposal concerning the study of crocodiles, B prepares a proposal concerning the study of zebras, both A and B mention that these are animals with four legs, and D merges them into a proposed study of something that is partly crocodile and partly zebra. Obviously, nothing like this would ever happen for proposals concerning animals, because everybody is familiar with them; but themes concerning physics (or other areas with strong mathematics components) require philosophical/epistemological background for one to be in a position to evaluate whether they can be merged or not. Thus, the inadequacy of a humanities-based ability relevant for the sciences determined the rejection of projects that could have had chances of approval if they had not been merged.

All this points to the importance of expanding the humanities components in pre-university instruction, at least to what could be considered a 'safety level', still largely to be identified and determined, but intended as the minimum level that enables development of thoughts (reasoning), communication efficiency, and also the recognition of categories that are relevant for assessments and decision making.

The landscape of the appreciation of the roles of humanities for the general preparation of a person, and for a better understanding of the sciences, knows also refreshing moments, like R. Shraim's note with the telling title "How philosophy is making me a better scientist" [4] and statements like

> *Philosophy has expanded my critical and creative thinking. Philosophical arguments often lead to imaginative edge cases and a dive into hypotheticals, which I frequently find creatively stimulating.*

It would be important to offer opportunities for advantages of this type to all young people, whichever the career that they will later chose, on entering university or on opting for other activities. Recalling and emphasising the main advantages is also a major objective of the present chapter.

The next section considers a quick 'journey' through history to recall how the current dichotomy between the two domains has not been part of historical developments until comparatively recent times; it also highlights representative ideas, quests or features that better exemplify the actual integration of perspectives from the two domains in the epochs considered. Section 1.3 reverts to our times, and analyses aspects for which inadequate presence of humanities-based components or expertise determines drawbacks for research, education or other components of science practice.

1.2 A Quick Journey Through History—Before or Beyond the Dichotomy of the Two Domains

The separation between humanities and sciences is comparatively recent. It was not present in ancient civilizations. The priests of ancient Egypt were custodians of religion and also custodians of the secret knowledge about substances and their transformations—the knowledge that later engendered alchemy, the 'ancestor' of chemistry. The ancient Egyptian goddess Seshat (Photo 1.1) was the goddess of writing (which she was believed to have invented), wisdom, and knowledge; on this basis, she also became the goddess of architecture, building, astronomy, astrology, mathematics, accounting, measurements and surveying [5]. She also had the title *Mistress of the House of Books*, because her priests oversaw the library in which the scrolls with important knowledge were preserved; her priests were also trained to understand the powers of alchemy [5]. In ancient Greece, a group of nine goddesses (the Muses, Photo 1.2, [6]) were inspirators and protectors of arts and sciences; the same Musa could inspire both an art and a science; for instance, Thalia inspired comedy, but also architecture and, by extension, geometry because of its role for architecture.

Ancient Greek philosophers investigated a vast range of questions, from the principles of the world and the nature of matter, motion and change, to the study of living beings and to the questions typical of human beings and their communities (ethics, i.e., the issue of right and wrong; the search for the optimal form of government; the nature of wisdom and knowledge). The etymological meaning of 'philosopher' is 'friend of wisdom', i.e., a person who pursues knowledge in all the areas and searches for optimal answers for the aspects that concern human beings. Both humanities and

Photo 1.1 The goddess Seshat (first on the left) and the god Horus (to her right). Temple of Kom Ombo. *Source* https://www.flickr.com/photos/sonofgroucho/974504276

Photo 1.2 Muses sarcophagus, Louvre [6]. The lower row represents the nine Muses and their symbols. From left to right (giving the name of each Muse and, in parentheses, the arts or sciences she inspired, and her symbol; the first mentioned symbol is the one represented in the figure; others are listed when referred to other arts or sciences ascribed to that Muse): Clio (history; scrolls, books), Thalia (comedy, but also architecture, and geometry for its role in architecture; comic mask), Erato (lyric poetry, but also erotic poetry; lyre), Euterpe (music, and also lyric poetry; two flutes), Polymnia (sacred poetry, but also dance and eloquence, agriculture, and geometry; pensive face, veil, grapes); Calliope (epic poetry; writing tablet, stylus, lyre), Terpsichore (dance; cithara), Urania (astronomy; celestial globe, compass), Melpomene (tragedy; tragic mask, a blade)

sciences were part of this search. Many terms categorising human intellectual activities were born within this search: the term *physics* means 'nature' and it referred to the study of nature; the term *metaphysics* referred to what might be 'beyond nature'; the term *politics* referred to the modes of governing the *polis*, the independent state-cities of Greece; the term *democracy* (literally, power of the people) referred to the political system that developed in Athens; and so on.

Some of the early (pre-Socratic) philosophers tried to identify the main principles of which the world consists; only one principle was hypothesised initially: water (Thales, ≈624–≈548 BC), air (Anaximenes, 586–526 BC), fire (Heraclitus, ≈535–≈475 BC), the indefinite or *apeiron* (Anaximander, 610–546 BC); then Empedocles (494–434 BC) proposed the theory that all substances are made of the four elements water, air, earth and fire—a theory that remained dominant for many centuries. Leucippus (V century BC) and his disciple Democritus (460–370 BC) inferred that matter must consist of particles (indivisible entities that they called *atoms*) on the basis of the observation that substances can mix [7], correctly understanding that their mixing could not be possible if they had continuous structure. Pythagoras (≈570–≈490 BC) and his school investigated mathematics and linked it with music and with astronomy; they believed that mathematics can explain everything in nature, and that all properties and causes can be expressed through numbers—an idea that will give many fruits across centuries and in modern science. Socrates (≈470–399 BC) focused mostly on ethical issues. Plato (≈428–≈348 BC) investigated ethics, politics (search for the best form of government), and the nature of knowledge; he also hypothesised that the ultimate constituents of matter are mathematical entities—the regular solids [8]. Aristotle (384–322 BC) investigated metaphysics, ethics, logic, the nature and characteristics of motion, and carried out accurate observations of animals and plants. Epicurus (341–270) and the philosophers of his school explored the routes that could enable a life free from anxiety and fears; they also adopted and somewhat modified Democritus' atomistic theory. Titus Lucretius Carus (99/94–55/50 BC), one of the greatest Latin poets, utilised poetry to explain their atomistic theory in *De Rerum Natura* (on the nature of things) [9].

After the Christian religion became the dominant ideology, and after the fall of the Roman empire, the type of philosophical and scientific research that had flourished in ancient Greece remained dormant in Europe for some centuries, while some of its components were investigated and expanded by Arab scholars. Many ancient writings were lost, other were preserved in monasteries, carefully copied by monks. A re-awakening of creative and investigative intellectual activities in Europe started with a new interest in the works of ancient Greek and Latin authors (philosophers, poets and writers), searching for the original texts and viewing them as inspirations for new ideas. This period, called *Humanism*, engendered the *Renaissance*, when arts and sciences bloomed together into new explorations. The discovery of Lucretius' poem in a monastery (1417) brought back the atomistic model into European academia (Photo 1.3, [9, 10]): the concept of the particulate nature of matter—although not in agreement with Aristotle's views and the way the Catholic Church had turned them into its official philosophy—became the key interpretative model in physics, alchemy and chemistry. The study of the method through which knowledge can be attained was a common research ground for philosophy and science. Leonardo Da Vinci (1452–1519) considers observation and direct experience of the reality in the world around us as the mother of knowledge; the term he uses, "esperienza", combines the *observation* concept and the *experimentation* concept. Among the philosophers developing reflections on the method, Giacomo Zabarella (1533–1589) considered logic as an expedient instrument both for the arts and for the sciences, and included inductive reasoning among the routes for the search for knowledge. Galileo Galilei (1564–1642) was the founder of modern physics; he introduced the idea that mathematics is the fundamental tool to describe and understand physical reality, and this has characterised physics in all its subsequent developments.

Renaissance artists and scientists could simultaneously be expert in fields across the two domains. Leonardo was a painter, an architect, a scientist, an engineer, a naturalist, and wrote about his observations of nature and about the approaches to pursue knowledge. Artists studied nature and the anatomy of the human body to depict them in an informed way. Painters prepared their dyes in their own workshops, often searching for new combinations of materials to get new colour hues; in this way, they were also alchemists or early chemists. Artists were also fascinated by the works of science, and often represented them. The laboratory of the alchemists, with the dominant role of the furnace and a number of non-common-use containers, was particularly attractive, and was the object of various paintings (e.g., Photos 1.4, 1.5, [11, 12]). Alchemy was perceived as bordering on magic, and depictions of a witch's den often had a furnace as its central 'piece' (e.g., Photo 1.6, [12]). Alchemists considered their art as an art under the protection of the most 'versatile' god of ancient Greece, Hermes (Photo 1.7, [14]); the term 'hermetic', denoting the tight shutting needed for chemicals' containers, derives from this assumed protection. Following the tradition of ancient Egypt's priests, alchemists considered that the knowledge about substances was risky and should be reserved to the initiated, to prevent misuses; therefore, they used allegoric names and images to denote substances or to describe processes; the creation of allegories required a good deal of imagination; these images were also attractive to artists; the bird on the head of the alchemist in Photo 1.4, and the images in the right hand side of Photo 1.7, are part of the allegoric representations used by alchemists.

Having expertise in both domains remained a characteristic of outstanding philosophers and scientists through the subsequent centuries. Immanuel Kant (1724–1804) introduced new outlooks on major philosophical issues [13], but was also interested in the nature of matter and supported Ruđer Josip Bošković's (1711–1787) theory envisaging atoms as mathematical entities. In the XX century, Albert Einstein (1789–1955) was a theoretical physicist, but also wrote extensively about philosophy; Bertrand Russell (1872–1970) was both a mathematician and a philosopher. New theories in the sciences prompted intensive philosophical debates; for instance, quantum physics prompted major debates because of the way in which it describes the world of atoms and molecules (including the uncertainty principle, which restricts our possibility of knowing with simultaneous high precision physical quantities of specific pairs, thus impacting on the views about knowledge). New

Photo 1.3 One of the early editions of Lucretius' poem *De rerum natura* (on the nature of things). [10]

applications that proved particularly destructive prompted deep philosophical and ethical reflections—a major example being the atomic bomb.

It may also be recalled that the fact that scientists used to have broad humanities background has often turned their writings, whether on science or on philosophy, into works with literature value [16]. For instance, works like those by Leonardo da Vinci and Galileo Galilei are part of Italian literature, besides being part of the history of science.

Photo 1.4 Depiction of Alchemy in Cesare Ripa's iconography. [11]

Photo 1.5 An alchemist with his family, working at his furnace, in an engraving by J.C. Bentley after A. van Ostade, 1661 [12]. Besides the furnaces, photo 1.5 shows containers corresponding to what would be termed glassware in modern chemistry, and photo 1.6 shows a fume-hood above the furnace

Photo 1.6 A witch and a devil making a nail with which to make a boy lame. Woodcut, 1720 [12]

1.3 The Need for Synergies and for 'Talking to Each Other'

1.3.1 Focusing on the Impacts of Inadequate Humanities-Based Skills on Science Practice and Education

Section 1.1 included two examples of impacts on very practical aspects, stemming from inadequacies in humanities-based skills like language-mastery and epistemological awareness. The next subsections highlight major impacts of these same inadequacies on science practice, science education and learning, normal life, or professional communications and initiatives, and illustrate them through selected examples. Some of the examples derive from LM's direct experience. Concrete examples constitute the most effective bases for reflections on their causes and on perspective remedies.

It is imperative to stress that the fact that this section highlights impacts on the quality of 'doing science' does in no way imply that the humanities could be viewed as somehow ancillary to the sciences. The two domains are viewed as equally important for human development and activities. The option of highlighting the impacts of

Photo 1.7 Hermes Trismegistos (from Stolcius, *Viridarium Chymicum*, 1624)

inadequacies in humanities-based expertise on science practice and science education has two major motivations:

- It is expected that showing how science-pertaining activities are affected is more likely to stimulate reflections also from those who are currently favourable to the continuous downgrading of the humanities in research and in education, as witnessed in recent decades.
- It is part of LM's direct experience as someone professionally pertaining to the science domain (a chemist), having keen interest in education, and having extensive humanities background from pre-university instruction (as common in the Italian educational system at the time when she was young, and having attended Classical Lyceum, the humanities-oriented secondary school).

1.3.2 When Language Mastery Decreases Steadily…

Language is the most fundamental communication tool and is also the most essential instrument of thought [15–17]. It is the only tool capable of conveying relationships between different pieces of information and, therefore, it is the only tool that enables both the development and the communication of individual thoughts and trains of mutually-linked thoughts. A decrease in language abilities implies a decrease in learning and thinking abilities. This is being verified from early age onward. A study that verifies it for pre-school age children [18] reports that:

> *Inadequate language development hampers the preschooler's readiness-to-learn in the formal teaching situation, because they have not been taught higher-order thinking skills via early-mediated language experiences.*

The roles of language are so fundamental that they are given specific and more detailed attention in some of the next chapters. This subsection focuses mainly on the concerns about the impoverishment of the language-mastery level among the young generations, and its already observed and further predictable undesirable impacts. The impoverishment comprises a variety of aspects: decrease in the number of known and used words; decrease in the ability to use the resources offered by the grammar and syntax of a language to express details and nuances, and consequent inability to build complex discourses; and the impact that this unavoidably has on the generation of thoughts. These effects can be effectively expressed through the words of a writer [19]:

> *… our understanding and use of language shapes us. Language is the way we express every idea, every thought. Sloppy thinking leads to a sloppy use of language. But the opposite is also true: imprecise language causes imprecise thinking.*

The recent impoverishment is extensively documented. The decrease in the number of words known by youth is a telling symptom [20]: for instance, for the Italian context, a 1976 study by De Mauro showed that average gymnasium (the first two years of Classical Lyceum) students knew about 1600 words; in 1996, they knew 600–700 words; now-a-days they likely would not go beyond 300 words; the seriousness of this situation is easily understood by recalling Heidegger's views that our thoughts are limited to the words that we know [20].

Clavé [21, 22] summarises facts and concerns in a clear-cut way, directly linking language impoverishment to IQ decrease and impoverishment of thought generation: "the poorer the language, the less thought exists" [21]. The text is so information-dense that it is not easy to summarise or review it. A central analysis considers that diminished knowledge of words, and of language resources in general, results in diminished ability to express emotions and to develop thoughts; in turn, the inability to express emotions is one of the sources of violence. Diminished use of words and of language resources (e.g., the tenses and modes of verbs) also jeopardise the possibility of complex reasoning. The removal of the ability to express time sequences, or to express the difference between what might have happened and what actually happened, hinders critical thinking. On the basis of all these observations and considerations, Clavé concludes that those who want to remove complexity from language are "the undertakers of the human mind". Some illustrations of the effects of diminished language mastery on thinking and communications abilities are offered in [23], which focuses on the importance of language mastery for sustainability-related explorations and communications.

All the details mentioned in [21] have impacts also for science communication, even those that may appear marginal from a superficial outlook. For instance, the lack

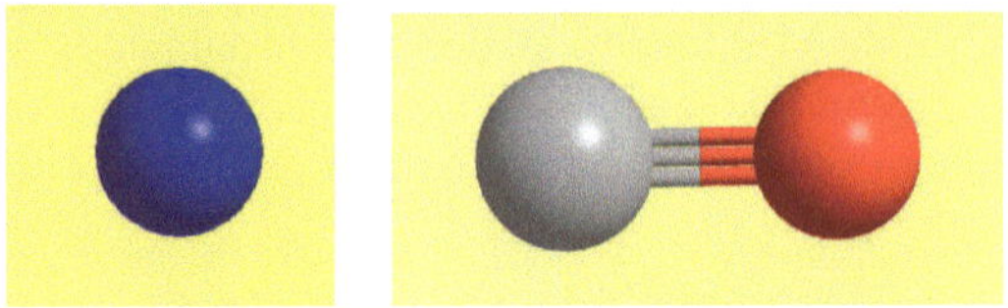

Co
a cobalt atom

CO
a carbon monoxide molecule consisting of one carbon atom (grey) and one oxygen atom (red)

Fig. 1.1 Illustration of the importance of distinguishing uppercase and lowercase letters in chemical symbols and formulas

of attention to capitalization engenders problems and errors in the use of symbols in mathematics, physics and chemistry; Fig. 1.1 shows an example in which correct capitalization is essential to distinguish between cobalt and carbon monoxide; errors of this type appear rather frequently in students' writings, when students have not been trained to pay adequate attention to how they write individual words.

Clavé's text [21] has been taken up by other authors, in different countries, to stress the importance of language and language mastery and the concerns for their impoverishment. It has been translated into other languages, including Italian [24]. It has also been used as conceptual prelude to analyses of the undesirable effects resulting from the impoverishment of language mastery in the Italian context [25], from cultural decay and the parallel increase of violence to the lowering of educational standard and to the high presence of functional illiteracy (being able to read, but not to understand the meaning of what one reads), affecting about 47% of the adult population according to PIAAC-OCSE report (PIAAC being the *Programme for the International Assessment of Adult Competencies*).

Several works expressing concern for the impoverishment of language also highlight the risks associated with the decrease in complex thinking abilities. For instance, [26] emphasises that many young people are increasingly unable to articulate complex concepts: they do not know the words to express them, or how to build sentences that actually convey the wanted meaning (an ability that requires good knowledge of grammar and syntax, and of all the resources of the language being used). A quick review of studies on these problems [27] recalls observations like the fact that many university students are unable to write or analyse even simple sentences, or the fact that radical changes in the educational approaches, like the decreased exposure of young people to poetry and literature (which had traditionally been cornerstones in pre-university instruction), are among the main responsibles for the problem. The same review ([27]) recalls the close connections between language and thought, as thought would be impossible without language, which enables us to express it through words; it stresses that the loss of the ability to express the details of complexity affects thinking abilities; it reflects on the current limitations for which we often lack the words to understand the impact of new technologies and their products on humankind and the environment; it also emphasises that the mother tongue is the language through which we think.

A concern expressed in some of the just mentioned works and taken up by many others (e.g., [28]) refers to the fast disappearance of the use of the conjunctive verbal mode from French, Italian and some other languages. The conjunctive mode has a fundamental role in the formulation of hypotheses, enabling the distinction between what is certain (expressed through the indicative mode) and what may be possible but is not certain, or would be impossible. Losing the ability to identify and express whether something is a possibility or is part of reality hampers the possibility of making hypotheses, which, in turn, hampers the possibility of a type of thought that is fundamental in the interpretation of observations and the generation of models. On a more general scale, it removes hypotheses and doubts from our mind set, thus limiting thinking abilities in general [28]. A number of the works cited in the last paragraphs also recall that G. Orwell, when depicting a system where authorities are determined to systematically curb the possibility of independent thought by citizens in his novel "1984" [29], includes the reduction of words in the language as a tool to reduce thinking abilities: when a word is removed, the concept that it expresses disappears from people's minds.

All what has been mentioned in the previous paragraphs leads to clear inferences. The deterioration of language mastery entails a decrease in the possibility of complex reasoning and its expression. This, in turn, affects the development of scientific thought, which is unavoidably complex because it needs to relate a plurality of facts together and to engender new thoughts (including interpretations and models) from the identified relationships. The generation of new ideas in the sciences requires complex thinking, and this requires language mastery. Actual and known requirements of the current world situation require complex thinking up to the point that systems-thinking (the ability to relate different pieces of information to build a comprehensive picture) is increasingly considered crucial in the education for sustainable development [30]. Systems-thinking has logical thinking as a pre-requisite, and logical thinking requires language mastery as a prerequisite [31].

Ensuring that young people acquire adequate language mastery is tantamount to equipping them with a fundamental tool for all aspects of their future professional activities. Inadequate language mastery may affect professional interactions in a number of ways. The observation that some researchers are able to obtain results but not to write about them for publication is painful. From more basically practical points of view, language errors may decrease the reliability of messages that are sent out for intended professional purposes. For instance, a message meant to advertise language-editing services and containing language errors defeats its purpose by documenting the non-reliability of the advertised service. An invitation to a conference, containing language errors, makes the reader think it twice before deciding that it may be worthy selecting that specific conference among the many that are continuously proposed. Thus, for example, an invitation containing a sentence like "This can be an important meeting for you because it is able to shape your field in this way which helps in brightening your future" or starting with a sentence like "To highlight the disorder related to the XXX conference series are going to organize the [number] International Conference on XXX" makes the reader wonder whether those who wrote them did take care to check whether their wording and grammar actually conveyed an identifiable meaning. Adequate language mastery is a necessity not only for the development of new thoughts, but for the effectiveness of any type of professional interactions.

A variety of other issues concerning language mastery and inherently entailing integration between humanities and science perspectives—such as the importance of the preservation of languages' resources, the importance of the mother tongue, and the inherent multidisciplinary nature of languages—are analysed in [23].

1.3.3 When Philosophical and Epistemological Backgrounds Fade Away…

Section 1.1 included an example of risks for very practical and crucial issues like the evaluation of a project proposal, engendered by inadequate familiarity with basic epistemological aspects of scientific discourses. Examples of the impacts of inadequate familiarity by students (and not only) are offered in Chapter 4, as inherent components of a number of incorrect wordings. The problem has actually a sort of subsequent-loops' character: it is practically impossible to guide students to adequate awareness of epistemological aspects if their language mastery is not adequate and, in turn, inadequate awareness is responsible for many cases of incorrect wording. The two inadequacies combine, for instance, in the random use of the terms expressing a basic classification of operations in relation to what is done and to the nature of their objects (whether the objects belong to physical reality or to our interpretation models), like the verbs:

- *to observe*—usually related to qualitative studies of objects or phenomena pertaining to nature, or qualitative diagnoses on performing experiments;
- *to determine*—usually related both to the verification of something pertaining to nature, or to the measurement of some quantity, or to finding some values through calculations using measured values as starting values;
- *to calculate*—usually related to mathematical operations leading to values of quantities that have not been measured directly, starting from values obtained through measurements;
- *to define*—usually associated with the introduction of a new concept and its operational characterisation [32].

Errors in the use of these verbs, or in the interpretation of questions utilising them, attest both inadequate language mastery and inadequate familiarity with the scientific method (since the nature of operations is part of the method). On the other hand, in current curricula, there is no space devoted to the analysis of words and their usage in science communication. At pre-university level, language education within humanities courses does not include the 'language of science' (although, in some contexts, it may include a variety of other 'languages', from cinema to sport to adverts). Science teachers are often not aware themselves of the role of language aspects, or not adequately prepared to explain it. In this way, students' attention is never attracted to the communication-effectiveness roles of expression rigour, or to the need that the messages conveyed by sentences should be consistent with all the aspects of the objects or phenomena concerned, including the method-related ones. In many cases, science learning is regrettably reduced to the memorisation of some definitions that will be repeated passively and to the memorisation of some algorithms that will be applied just as passively (up to the point that many students are unable to solve problems requiring the design of algorithms that are not exactly the same as the memorised ones, even when the "design' entails only small modifications of the memorised ones). Even the practical component of science courses, which traditionally requires a report, is losing its role of stimulating and expressing reflections, when students are provided with pre-set forms in which to enter some values or some simple words. Altogether, not asking students to express science is tantamount to not asking them to thoroughly think about the concerned topics: language abilities and conceptual understanding get simultaneously undermined, and the epistemological aspects disappear.

Observing, having curiosity and setting questions, searching for answers, verifying the information attained, are the essence of the scientific approach, known as the scientific method. Science courses are expected to take the most active role in fostering this attitude of the mind. Yet, too often, the scientific method is hastily presented as a rigid set of steps (nearly like a recipe) that students memorise and repeat on request, without any discussion leading to insights of its richness and complexity. In this way, the very nature of science is obscured. Students cannot develop the perception of science as a continuous search that, at each step, is able to reflect on the validity-extent and on the limitations of every new 'item' added to its body, whether experimental or theoretical. Inadequate philosophy literacy (resulting from inadequate presence of the humanities in pre-university instruction) prevents the very understanding of the significance of the question about the nature of knowledge and of the debate that led to the formulation of a method.

The frequently random use of the 'scientific' adjective is by itself a testimonial of the impact of not devoting adequate attention to epistemological aspects at educational level. For instance, recommendations like "The methodology section should be scientific" or "Results should be provided in a scientific way" do not convey a clear meaning. Is there a 'non-scientific' way of writing a methodology section of an article, or of presenting results? Statements of this type—or any statement containing the adjective 'scientific' with reference to direct requests—should also define the intended meaning of the adjective with reference to the item concerned. The unspecific use of this and other method-related terms is passively transmitted from teachers to learners in a loop that can be interrupted only if teachers start discussing their meaning in the class and relate them to basic epistemology; then, their students will not use them passively when they become professionals. Furthermore, the random use of the 'scientific' adjective may result in inherently abhorrent concepts, because of the unavoidable perception that it means "proven by science"; an example is the "scientific racism" neologism, where 'scientific' is coupled with a term ('racism') that has no scientific basis.

Inadequate epistemological awareness may surface at any level and in any context. The following example pertains to the study of molecules, i.e., to an advanced area of modern chemistry, and comes from a source that should have legitimately been expected to be reliable:

> *Molecular orbitals (MOs) are mathematical artefacts created only to construct the wave functions, but they have no physical existence. Consequently, MOs do not control any chemical thought.*

The first sentence contains a quantum chemistry error, because molecular orbitals (MOs) are actually the wavefunctions themselves, not some mathematical contraptions used to somehow build them. The second part of this sentence and the second sentence indicate inadequate epistemological cognizance. Mathematics is a description tool in the physical sciences. No mathematical entity has physical existence. And no mathematical entity can *control* human thought (since chemists are human beings, *chemical thought* is a form of human thought). As description tools, mathematical entities provide information that we use to better understand the characteristics of physical items. This is the essence of the mathematical modelling of physical objects. Mathematics is a powerful tool to generate sufficiently reliable descriptions from starting information, when an adequate model and corresponding algorithm have been developed. The scientist needs to provide correct starting information, select the appropriate tool (algorithm) for the specific investigation, perform the associated calculations (often with the help of high-performance computational resources) and interpret the results, i.e., relate the terms of the obtained mathematical description to the characteristics of the physical object concerned.

More examples specifically related to learning difficulties and students' uncertainties are included in other sections of this chapter or in other chapters. The next paragraphs consider examples of impacts of inadequate familiarity with epistemological aspects in a variety of contexts or situations, from science education to other types of happenings somehow related to science practice.

Epistemological views may greatly influence educational approaches and the consequent choice of expressions by the teacher. For instance, some lecturers choose to speak in terms of 'possibility' when presenting options that are still in an exploration stage (like some cross-disciplinary explorations), or when dealing with predictions of possible phenomena; others consider that all terms related to a possibility notion should be avoided because "students need certainties". When engaged in discussions of this type, LM insists on the importance of teaching students that science is investigation, not certainties, and that this needs to be communicated both through direct statements and through wordings that take into account the possibility concept, whenever it has a role in the given discourse.

Keener attention to epistemological aspects could have prevented some educational 'headaches', like those associated with the concepts of "relative atomic mass" and "relative molecular mass". Many students struggle with them, and there is a good deal of literature about the difficulties they encounter (e.g., [33–35]). On the other hand, the very adjective 'relative' is misleading. All measurements are 'relative', because they all imply comparison with a specific amount of a certain quantity, which is conventionally taken as unit for that quantity and is also conventionally defined. In this regard, the atomic mass and molecular mass expressed in atomic mass units (a.m.u.) are not different from any other measurement. The simplest approach would be that of explaining that we can use two different types of units to express atomic or molecular masses, the SI unit (kilogram) and the a.m.u., and that the a.m.u. is more convenient because the mass is expressed by easily manageable and comparable numbers; after this introduction, one simply needs to provide the definition of a.m.u. and then use it throughout [36]. In terms of the heritage that Lavoisier has endowed chemistry with, it can be said that the use of the adjective 'relative' is unnecessary; and unnecessary items need to be avoided, because they add complications to models without bringing any benefit.

Close attention to epistemological aspects could play fundamental roles in science education research, to enhance its ability to provide information valuable for the design of approaches effectively targeting students' diagnosed learning difficulties. However, it also happens that some frameworks become 'fashionable'; then, they end up being used as sorts of multi-cell cages, into which interpretations try to fit the observations. In other words, those frameworks become sorts of dogmas, to which reality gets somehow adapted, even if this was surely not the purpose for which they were initially proposed. For instance, the Johnston triangle [37], which was proposed as a reference to identify or assess three categories of chemical knowledge or mastery (macroscopic, microscopic and symbolic), is often treated as expressing three levels of reality, what would be epistemologically incorrect. The macroscopic level and the microscopic level pertain to what we could (at least, as first approximation) call 'physical reality': substances and their molecules would behave in the same way, whether we (humans) observe them or not; they are independent of us (at least, by neglecting the implications of sophisticated quantum mechanics with reference to the microscopic world). The symbolic representation pertains only to our descriptions and not to physical reality. The distinction between physical reality and our descriptions should be a crucial component in the preparation of science teachers, because it is a fundamental epistemological aspect that students need to acquire and internalise.

As Leonardo da Vinci stated five centuries ago, observation is the mother of knowledge. On the other hand, tendencies sometimes appears, which turn the prestige of leading researchers into a reason to use their conceptions as something that is not object of discussion (although initiating dogmas was surely not in the intentions of the leading researchers who proposed hypotheses as parts of their research, not as absolute truths to be always believed). When considering the development of abilities as young people grow (or any other aspect of human behaviour), statements may refer to a majority of cases, but they cannot be assumed to be surely true for each and every individual. A simple example from LM's direct observations may illustrate the concept. It is common belief that children are not capable of abstraction before they reach at least the beginning of their teen years. Then, one happens to observe a 3-year old child who, looking at a stained window in a church, enthusiastically explains that if one takes certain figures that he specifies (triangles), rotates some of them and puts them close to the others in a certain way, one obtains another of the figures present in the same window (rhombus). The child is happy to have made a 'discovery'. The 'discovery' clearly implies abstraction. Are the adults witnessing this allowed to feel happy for the child's newly emerging ability, or should they say that this is impossible because it is not in agreement with some statements that have somehow been turned into dogmas? The latter attitude may regrettably surface in schools, and it happens that children are not provided answers to questions involving abstraction abilities, because their teachers do not believe that they are capable of them; in some extreme cases, parents are told to tell their children to avoid such questions, what is tantamount to hampering the development of those children's thinking abilities and general curiosity. The awareness that no statement referred to human beings is an absolute truth would endow teachers with better possibilities to fully understand their students and provide them with specifically-tailored guidance.

Better familiarity with epistemological bases would also be functional to the quality of questionnaires and other types of surveys that researchers receive from a variety of sources and for a variety of reasons. This may become serious when the results of such surveys are used as indicators for important decisions, including establishing criteria for the financing of projects. For instance, LM was included in a survey about how researchers choose their references when writing papers; the survey asked a number of questions on why she had chosen to cite a specific paper (specified), including a question of how that paper has influenced the selection of her own research topics; on the other hand, the selection of that specific paper was unsuitable for those questions, as it was a paper simply providing a value useful for certain calculations; this means that the reference provided a tool; when one uses that tool, one cites the source paper, but this does not mean that that tool influences the choice of the research theme. In another case, where the sender stated that the concerned survey was carried out on behalf of a funding agency, the question was in terms of "A majority of your citations come from some journals (e.g., WWW, XXX, YYY, ZZZ) that you rarely cite their articles in your papers. Do you consider these journals irrelevant to your works or unimportant in your field?". Besides the fact that the presence of a grammar error (the simultaneous presence of "that" and "their" in the same clause with reference to the same item) immediately lowers the consideration that one could give to the message, it is difficult to understand the logic of the question. The underpinning discourse appears to be that, if an article in a certain journal cites a paper written by A, then A should cite articles appearing in that journal, and that if A does not do it, it means that A does not have a good opinion of that journal. This discourse—besides being void of internal logic—completely ignores that, when writing a paper, one chooses the references on the basis of their themes (themes that are relevant for the paper being written) and of the information they provide, not on the basis of the journals or the authors. In other words, the question shows no familiarity with the meaning and practice of scientific research. The fact that it may be linked to the establishment of funding criteria is worrying.

1.3.4 Learning to 'Talk to Each Other' Again

As already mentioned, humanities and sciences have been together since ancient times. Even after different science areas developed their separate identities, scientists continued having extensive humanities background—knowledge of philosophy, literature, history, linguistics and, quite often, also Latin and ancient Greek. This background had roles in the practice of science, favouring creativity and the generation of new thoughts. The human brain (like any other organ of the human body) needs exercise for optimal functioning, and philosophy, literature, linguistics, etc. provided quality exercise grounds.

Recent times have been experiencing a continuous increase in the dichotomy between the two domains, also in association with the progressive narrowing of the deeply specialised scope of the various sectors of individual disciplines, stemming from the fast increase of the amount of knowledge in each sector. Many scientists do not know enough about humanities to appreciate their roles for human intellectual activities. The previous sections have outlined some of the impacts of this divergence, through the consideration of concrete examples. The seriousness of the impacts points to the need of somehow bridging the current gap. Appeals to this purpose have started appearing (e.g., [2]). Bridging the gap requires learning to know each other and to recognise and appreciate the benefits that each area derives from the others. In the current context, in which the sciences are increasingly perceived or treated as dominant, it is important that scientists learn enough about humanities to realise how humanities-based skills and knowledge are vital for the practice and the very existence of science.

Talking to each other is also important for many issues that are crucial for further human development, including the huge ethical issues that should be given careful consideration for decisions on large-scale applications of new discoveries, or the challenges inherent in pursuing sustainability. Several chapters present areas and routes of possible interfaces and collaborations. Sections 18.4 and 18.5 in Chapter 18 outline interfaces and routes that are not included in the other chapters, but are in line with the scope of this book.

1.4 Discussion and Conclusions

1.4.1 Considerations and Inferences

The previous sections have considered aspects that can highlight the importance of the relationships between humanities and sciences. Brief historical outlines emphasise how these two domains have been integrated since ancient times and through many centuries. The undesirable impacts engendered by inadequate humanities-based skills, such as language mastery or epistemological awareness, have been given particular attention because they heavily affect the development of thinking abilities and the efficacy of learning. These impacts alone are sufficient indicators of the urgent need of measures aimed at ensuring the development of these skills in the young generation, through suitable approaches at all levels of instruction. If these skills are not developed adequately, the efficiency of learning is bound to decrease below 'safety levels', affecting the acquisition of knowledge in all the areas. In other words, the impoverishment of humanities components in pre-university education, apparently meant to favour the acquisition of science-related knowledge, does actually prevent the development of the thinking abilities (logical reasoning, abstract reasoning, critical reasoning) that are essential for conceptual understanding in the sciences, even at basic level. Furthermore, by affecting the possibility of generating and expressing thoughts, it affects the very level of intelligence that a person can develop, with negative impacts on the life of individuals and on society.

Calls for the recognition of the roles of humanities are appearing at an increasing rate, as the risks connected with their possible 'demise' become more apparent. Peters makes an "appeal on behalf of Arts, Humanities and Social Sciences" [2], stating that "There must not be a false division between arts and sciences" and recalling some examples, including "Ethics belongs as much in engineering and banking and medicine, as it does in philosophy and law". Gleiser clearly states that "To solve modern problems, we must integrate the sciences and the humanities and think across these traditionally disparate disciplines" [38]. Leading thinkers support the need to bridge the gap between the sciences and the humanities [39], and often stress the importance of ethics for a non-destructive management of the new technologies.

Highlighting the variety of interactions and overlaps between humanities and sciences in a way that can be easily understandable also for those who are not yet familiar with them can contribute to the building of bridges between the two domains. Doing it in a way that emphasises the practical significance and benefits is the purpose of this chapter and of the entire book.

1.4.2 Exploration Routes Through the Next Chapters

Overall, the book emphasises that the exploration and identification of interaction-routes, interfaces and overlap areas between humanities and sciences is important both from a theoretical point of view and from a practically-oriented one. The former enables better insights into human intellectual endeavours, the latter proposes or facilitates a variety of applications (practically-oriented interactions), including major ones closely related to education. The attention to the 'working conceptions' about the nature, importance and benefits of cross-disciplinary collaborations between humanities and sciences is dominant, because it appears to be largely inadequate in current times. Therefore, the book presents concrete examples of constructive, mutually beneficial relationships between the two domains, and concrete documentation of the negative consequences of inadequate language mastery and inadequate epistemological awareness for science learning and science practice. Historical excerpts are also present when it is expedient to recall that humanities and sciences were born together and they developed together until comparatively recent (2–3 centuries ago) times, when a 'parting of the ways' started, and ended up becoming nearly dramatic in our times; the excerpts also aim at ensuring that the assertions about the importance of specific interactions in our time are not perceived as something 'out of the blue sky', but as something deeply embedded in the nature and development of both the sciences and the humanities.

- The present chapter is introductory. It outlines the reasons why it is important to consider the relationships between humanities and sciences and how each of these two domains can enhance the other; it also outlines the most common undesirable effects stemming from underestimation or phasing out of humanities components.
- Chapter 2 focuses on the relationships between science and culture, considering culture in its broadest meaning, and highlights the importance of potentially useful complementarities in the contributions from different domains, and the envisaged benefits for human activities and progress.
- Chapter 3 explores areas of complementarity between the two domains and utilises a historical perspective to show that both domains can benefit from considering the insights of the other's approach.
- Chapters 4–7 deal with different aspects concerning language. Given the importance of language for the development and communication of thought and, therefore, for the development of each discipline, whether humanities-related or science-related, it was considered important to highlight several diverse perspectives for its roles, substantiated by an adequate number of concrete examples. Chapter 4 underlines the importance of language and conceptual rigour, and their inextricability, for effective science learning;
- Chapter 5 expands the scope of Chapter 4 to include the role of language in engineering and technological areas in general, with their specific requirements, first of all, the need that language conveys sufficiently clear information to ensure correct implementations.
- Chapter 6 addresses the issue of the roles of various softwares in relation to the language of texts, and concludes that it requires deep language mastery and deep subject knowledge from the humans to ensure the quality and truthfulness of the conveyed information.
- Chapter 7 offers a historical perspective of the establishing of scientific terminology, highlighting the crucial role of extensive interplays between humanities and sciences.
- Chapter 8 outlines the bases of an epistemology of engineering—an emerging investigation area, which has been growing in the last two decades and is clearly cross-disciplinary between the two domains—and relates them to engineering practice.
- Chapter 9 presents a case study in which different disciplines from the two domains contributed to the design of the recovery of an ancient mine, to transform it into an underground museum. It is a rather rare case in which the humanities side contributes information relevant to technical design.

- Chapters 10–12 consider relationships between specific areas in the two domains. Chapter 10 offers an outline of the relationships between chemistry and art—a thread of constructive synergies between humanities and sciences across many centuries.
- Chapter 11 shows how the geometrical molecular patterns of purely chemistry entities—like metal organic frameworks—can suggest similarities (or similarity perceptions) with some forms or art.
- Chapter 12 considers cases of significant convergence between science and literature, through the scientific interests and conceptions of some Italian poets, including Dante Alighieri—the poet who set the bases of the Italian language—and highlights how science components have been embedded in their poetry becoming integral parts of it.
- Chapter 13 considers relevant interfaces between science research and history research, as well as the educational role of the history of science in history teaching.
- Chapters 14–16 focus on different pathways through which humanities-based components become crucial in chemistry education.
- Chapter 17 recalls the motivations for the merger of the two previous international bodies, respectively for the sciences and for the human and social sciences, into the current unified body—a merger that underlines the importance and expected benefits of synergies between the two domains.
- Chapter 18 has the role of concluding chapter. It reiterates the main objective of the book as the fostering of working conceptions of the relationship between the two domains, outlines the fuzzy-set character of the possible definitions of the two domains, summarises selected themes that could not have individual chapters devoted to them because of non-availability of authors willing to take the challenges of writing them, and presents take-home messages from the individual chapters and from the book as a whole.

Overall, the themes of these chapters offer a broad-scope overview of existing working relationships between humanities and sciences, thus also highlighting potentialities and routes for the search of other working relationships.

References

1. L. Mammino, Some general-character reflections on the relationships between humanities and sciences. Ometeca **9**, 156–178 (2005)
2. A. Zadeh, Fuzzy sets. Inf. Control. **8**(3), 338–353 (1965). https://doi.org/10.1016/S0019-9958(65)90241-X
3. J. Peters, The last invention: An appeal on behalf of Arts, Humanities and Social Sciences in an increasingly career-driven education world. *NSR (No Shelf Required) Editorial* (June 3, 2021)
4. R. Shraim, How philosophy is making me a better scientist. Nature (2021). https://doi.org/10.1038/d41586-021-01103-x
5. Wikipedia. Accessed 19–07–2022.
6. Wikimedia Commons. Accessed 19–07–2022.

7. H. Diels, *I Presocratici: Testimonianze e Frammenti* (Laterza, Bari, 1969).
8. Platone, *Timeo*, in G. Reale (Ed.) (Bompiani, Milan, 2000)
9. Lucrezio, *La natura delle cose*, in G. Milanese (Ed.) (Mondadori, Milan, 1992).
10. Lucretii Cari De rerum natura, Ferran d'Aragó, duc de Calàbria, 1488–1550, ant. Pos, Monestir de Sant Miquel del Reis (València, Comunitat Valenciana), ant. Pos, https://weblioteca.uv.es/cgi/view.pl?sesion=2022071818273522690&source=uv_ms_0506&div=7. Accessed 19–07–2022.
11. Raffigurazione dell'Alchimia nell'Iconologia di Cesare Ripa, tomo I, pag. 69, di Iconologia Del Cavaliere Cesare Ripa Perugino Notabilmente accresciuta d'Immagini, di Annotazioni, e di Fatti Dall'Abate Cesare Orlandi, Perugia, Costantini, 1764. https://upload.wikimedia.org/wikipedia/commons/6/6c/Alchimia_nell%27Iconologia_di_Cesare_Ripa_%281764%29.png. Accessed 19-07-2022
12. An alchemist with his family in their dim dwelling, working a bellows at his furnace. Engraving by J.C. Bentley after A. van Ostade, 1661. https://upload.wikimedia.org/wikipedia/commons/3/3e/An_alchemist_with_his_family_in_their_dim_dwelling%2C_working_Wellcome_V0025535.jpg Accessed 19-07-2022
13. Witchcraft: a witch and a devil making a nail with which to make a boy lame. Woodcut, 1720. https://wellcomecollection.org/works/jtrd42ta/items
14. Hermes Trismegistos. Stolcius, *Viridarium Chymicun*, 1624. https://upload.wikimedia.org/wikipedia/commons/2/25/Hermes_Trismegistos%2C_from_Stolcius%2C_Viridarium_Chymicun%2C_1624_Wellcome_M0011829.jpg. Accessed 19–07–2022
15. E. Kant, *Monadologia Physica,* 1756.
16. L. Mammino, Science writing and beautiful writing. Ometeca **11**, 104–127 (2007)
17. J. Bruner, Language as an instrument of thought, in *Problems of Language and Learning*. ed. by A. Davies (Edinburg University Press, Edinburg, 1975)
18. E. Asoulin, Language as an instrument of thought. Glossa J. Gen. Linguist. **1**(1), 1–23 (2016). https://doi.org/10.5334/gjgl.34
19. G. Dupre, What would it mean for natural language to be the language of thought? Linguist. Philos. **44**, 773–812 (2021). https://doi.org/10.1007/s10988-020-09304-9
20. H. Naudé, E. Pretorius, J. Viljoen, The impact of impoverished language development on Preschoolers' readiness-to-learn during the foundation phase. Early Child Dev. Care **173**(2), 271–291 (2003). https://doi.org/10.1080/03004430303098
21. D.R. Koontz, *Whispers* (Headline Book Publishing, London, 1991), p.476
22. https://www.giannipreviidi.net/impoverimento-del-linguaggio-un-serio-problema
23. Christophe Clavé, Baisse du QI, appauvrissement du langage et ruine de la pensée, Agefi (2019)
24. C. Clavé, Plus le langage est pauvre, moins la pensée existe, https://www.breizh-info.com/2022/02/19/180125/plus-le-langage-est-pauvre-moins-la-pensee-existe-christophe-clave-contre-les-fossoyeurs-de-lesprit-humain. Accessed 20–07–2022
25. L. Mammino, Language and sustainability, in L. Mammino (ed.), *Sustainability—A Multidisciplinary Endeavour* (Elsevier, In Press)
26. L'impoverimento del linguaggio influisce sul QI della popolazione, https://www.gdo-notizie.it › node. Accessed 20–07–2022
27. Trilussa, Un popolo di ignoranti, *La Voce del Serchio*, 23–01–2022
28. A. Bettero, La lingua italiana è in pericolo?, Messaggero di Sant'Antonio, 12–09–2018
29. E. Tomaselli, Senza parole. L'impoverimento del linguaggio nell'era moderna, http://www.freakoutmagazine.it/09-03-2019/focus-on/101117/senza-parole/. Accessed 20–07–2022
30. L'italiano: una lingua in via di estinzione, *Redazione Quotidianoweb, Costume e società*, 30 July 2022. Accessed 30–07–2022
31. G. Orwell, *1984* (Secker & Warburg, London, 1949)
32. P.G. Mahaffy, A. Krief, H. Hopf, S.A. Matlin, G. Mehta, Reorienting Chemistry Education through Systems Thinking. Nat. Rev. Chem. **2**(4), 126 (2018)
33. L. Mammino, Roles of Systems Thinking within Green Chemistry Education: Reflections from Identified Challenges in a Disadvantaged Context. J. Chem. Educ. **96**, 2881–2887 (2019)

34. L. Mammino, Il ruolo delle definizioni nell'insegnamento delle scienze. Scuola e Citta' **1**, 25–37 (2000)
35. S.C. Fang, C. Hart, D. Clarke, Unpacking the Meaning of Mole Concept for Secondary School Teachers and Students. J. Chem. Educ. **91**, 351–356 (2014)
36. S.C. Fang, C. Hart, D. Clarke, Identifying the critical components for a conceptual understanding of the mole in secondary science classrooms. J. Res. Sci. Teach. **53**(2), 181–214 (2016)
37. S.A. Malcolm, E. Mavhunga, M. Rollnick, A lesson on teaching the mole concept conceptually: a learning study, in *Book of Proceedings: 26th Conference of the Southern African Association for Research in Mathematics*. ed. by L. Webb (SAARMSTE, University of Gaborone, Botswana, Science and Technology Education, 2018), pp.134–149
38. L. Mammino, *Chimica Viva* (D'Anna, Florence, 1993)
39. A. Johnstone, Macro- and micro-chemistry. Sch. Sci. Rev. **64**, 377–379 (1982)
40. M. Gleiser, Bridging the intellectual divide, *Dec 1, 2021*, https://www.the-scientist.com/reading-frames/opinion-bridging-the-intellectual-divide-69434
41. M. Gleiser (Ed.), *Great Minds Don't Think Alike* (Columbia University Press, 2021).

Chapter 2
Science and Culture

Philip H. Spies

> "For constantly I felt I was moving among two groups – comparable in intelligence, identical in race, not grossly different in social origin, earning about the same incomes, who had almost ceased to communicate at all, who in intellectual, moral and psychological climate had so little in common that instead of going from Burlington House or South Kensington to Chelsea, one might have crossed an ocean"
> CP Snow, 1959, Cambridge University

Abstract The interactions between science and culture can be perceived from three vantage points: from the impact of a dominant culture on science, from developments in science which may create different science cultures that affects communicability between scientific traditions (for example between the humanities and the natural sciences, and between disciplines within the sciences), and at the level of application in problem solving. Nevertheless, although different science cultures evolved their own ways of looking at problems and doing research, all of them are today guided by the same universal principles of being objective, measured, creative in observation, rigorous in practice, rational in contemplation, honest in reporting, transparent in procedure and open to peer-evaluation. Specialisation was a powerful driver behind 300 years of scientific progress through a systematically narrower focus on an aspect of a phenomenon. This produced a depth of knowledge, but it is ill-suited to provide the breadth of understanding of the pervasive systemic consequences when such knowledge is applied. Unless scholars from the different disciplines are able (and willing) to contribute and listen attentively to conflicting ideas and theories they may experience "cognitive dissonance" when faced with ideas that differ strongly from their own. It is from time to time necessary to re-evaluate the culture that shapes the beliefs and behaviour of society, and especially to reflect on the reigning ideas of a specific culture, including on what it considers to be the nature of "progress"—especially regarding the benefits and costs to society over the longer term. A new consciousness and appreciation of the potentially useful complementarities in the

P. H. Spies (✉)
Institute for Futures Research, Stellenbosch University, Stellenbosch, South Africa
e-mail: phspies@iafrica.com

L. Mammino (ed.), *Insights into the Relationships between Humanities and Sciences*,
https://doi.org/10.1007/978-3-031-97900-2_2

contributions of different scholastic traditions are needed. To discover new knowledge about the constituent elements, or workings, of matter and the living environment is necessary, but it is not sufficient to use this new knowledge wisely: a new way of thinking about the workings of complex living systems are also needed. Therefore, the contributions of both divergent and analytical thinkers, of humanists and scientists, of West and East and of modern and traditional thinking should be considered—if only to break the mould of the culturally-shaped mind. To cross the cognitive divide between the humanities and the sciences, multidisciplinary problem-solving through dialogue, heuristic learning and soft systems thinking is proposed.

2.1 Introduction

In his 1959 Rede Lecture at Cambridge University CP Snow, a chemist-turned-novelist, observed that British humanists and scientists cannot understand or communicate with one another. He warned that this was not just an intellectual loss, but something that threatens the ability of modern civilisation to address the most pressing world problems—much of these self-created [1].

He mooted the idea of a unifying "scientific culture" with a common set of values and common standards for research in humanities and natural sciences.

Fourteen years earlier, in 1945, another chemist and then president of Harvard University, James Conant, commissioned a study titled "General education in a free society". The commission's report was published in 1950. It proposed that all students should receive a "holistic" education that covers studies in the humanities and the natural sciences [2]. Snow's observation bears testimony to the fact that the Conant commission proposal was easier said than done: the proposal was not just a call for a change in curriculum but also for a fundamental change in scholastic culture.

What is "holistic education" and why could it be relevant for a new, more inclusive, scientific culture? Holism is a concept that JC Smuts developed in his 1926 book "Holism and Evolution" [3], which states that the properties and behaviour of a "whole" cannot be explained from research into its component parts alone, because the "whole" determines how the parts behave. Ideally holistic education should broaden and deepen human consciousness (or "awareness") of holistic interdependencies between the contributions of scientists and humanists (or between physics and meta-physics). It is assumed that a "balanced" education that incorporates learning in both the humanities and natural sciences can bridge the dualism in learning culture. This is not necessarily the case when the emphasis is only on multi-disciplinary education. The desired outcome of a holistic education should be one of building an understanding of, and mutual respect for, the necessary and unique contributions of scholars from different scholastic backgrounds to finding solutions for the vexed problems the world is facing today.

Nevertheless, if there exists at the scholastic level a lack of understanding between scientists and humanists, it is much less the case in the real world: i.e., at the level of application where finding solutions to practical problems are more important than abstract discourses. Since the 1940's the concerns of Snow and the Harvard commission were (incidentally) addressed on a very practical level by new developments in operations- and systems-research (OSR) where researchers and planners from different academic backgrounds worked together in problem solving teams. A key aspect of OSR was its focus on interdependency between the elements and components of a large system (or a "whole"). With the rapid progress in computing-technology it became possible for OSR researchers to develop intricate computer models that simulated these interdependencies and studied their outcomes. The ground-breaking 1972 report "Limits to Growth" of the Club of Rome [4] is a well-known application of systems modelling which illustrated the surprising outcomes of studying integrated feedback-loop processes.

However, it is only possible to analyse and model simple and complicated systems: i.e., systems that are not complex. Some natural systems and most human systems are very complex [5]. Complex systems are open systems that consist of a myriad of elements (and components of elements) that interact in a myriad feedback-loops, i.e., in a nonlinear manner. These interactions in complex systems are ongoing and evolving. The activities of elements and components of elements can loop back onto themselves, and then again and again onto the others and again onto themselves. It is therefore necessary to think and act differently when faced with complexity. The need is for a trans-disciplinary approach to inquiry—which goes beyond multi-disciplinary co-operation which is the basis of OSR.

Although not generally accepted (or known) as such, a procedure to research complex systems emerged out of the work of a number of scholars working in the field of "soft" (human) systems-thinking between the 1970's and 1980's [6]. Soft systems-thinking offered new insights into the differences and interdependencies between human systems and physical systems. Most important, it links the classical humanist's contemplative approach with the analytical approach of natural scientists.

2.2 Interactions Between Science and Culture

Science is the human practice of systematic inquiry to discover, observe, describe, explain, predict, and control aspects of human interest. The aim of science is to gain new knowledge, better insight into, and better understanding of, phenomena. The way science is practiced is influenced by the reigning culture of a society which may change with time passing. Culture is a multi-faceted and implicit "social policy" that guides individual behaviour towards what is generally considered to be right and proper behaviour within a community, including a science community. It shapes, influences, and directs human behaviour through language, concepts, world views, codes, values, images, metaphors, myths, aesthetic appreciation, communal practices, and rituals. A culture evolves over time out of the many responses of the members of a community to various aspects of collective interest when faced with change. It then becomes the defining characteristic which represents the essence of a community, including its attitude towards science.

Modern science evolved during the sixteenth century as a standard practice for gaining new knowledge through objective observation, objective perception, and rational thinking. This produced a new world view and social order in Europe during the second half of the seventeenth century, known as "The Enlightenment". The Enlightenment was a cultural phenomenon that removed the remaining shackles of medieval dogmatism. It created benign conditions for innovative thinking, socio-economic development, industrialisation, and, in general, a positive predisposition towards "modernisation": i.e., a process of innovativeness, industrialisation and urbanisation.

Culture displays resilience over the short term but can change over the long term through social learning, and through disruptive events, such as wars and pestilence. Ongoing interactions between scientific progress and cultural change are co-evolutionary processes that continuously transform the nature of a society, its scientific practices and the innovations that emerge from them. This was also the main reason for the emergence of humanism in Europe during the fourteenth century. West and East were not isolated worlds, and by the end of the so-called "Dark Ages" in the tenth century European trade with the East brought it into close contact with the Islamic scientific culture of that period. This inculcated Arabic, Indian and Hellenistic mathematical and scientific knowledge into the church-dominated feudal order of medieval Europe. A supportive factor was the emergence of cities as independent political entities which were culturally opposed to the reigning feudal order of the period. During the fourteenth century the supremacy of the church-dominated feudal order of medieval Europe was challenged by a powerful, more free-thinking, urban mercantile class in Italy, especially in Florence. Trade between European cities and the East was also a conduit for the bubonic plague which killed almost a third of Europe's population during the fourteenth Century and caused great social and political upheavals which even questioned the efficacy of the Roman Church and its religious practices.

Durant describes the period during which classical humanism emerged in Italy as follows: "*It was under the Medici, or in their day, that the humanists captivated the mind of Italy, turned it away from religion to philosophy, from the heaven to the earth, and revealed to an astonishing generation the riches of pagan thought and art*" [7]. Guided by the culture and teachings of ancient Rome and Greece the emerging European humanist culture propagated the idea that people are masters of their destiny—as so elegantly expressed in Shakespeare's Macbeth: "*What a piece of work is a man, how noble in reason, how infinite in faculties, in form and moving how express and admirable, in action how like an angel, in apprehension how like a god!*" [8].

The cultural revolution during that period is today known as the "Renaissance": the French for "rebirth", so named in 1855 by the French historian Jules Michelet. He referred to the period of approximately 200 years between the late fourteenth century to late sixteenth century when a church dominated feudal culture was transformed into a secular culture noted by the hegemony of urban political institutions, urban commercial economy and a lay patronage of education, arts, and music. The study of history (with contributions of authors such as Leonardo Bruni and Niccolo Machiavelli) became a branch of literature rather than of theology. Cultural humanism migrated to the rest of Europe during the sixteenth century and systematically infected the established church-dominated feudal order with new knowledge and new ideas. An iconic invention of this period was the Gutenberg press, named after its inventor Johann Gutenberg, which caused the first "information revolution" by the mass-printing of books and scientific papers in the national languages of Europe. This resulted in a rapid diffusion of new ideas and scientific theories throughout Europe. It became the driver behind the transformation of cultures, and thus of an ever-changing world order. An insistent process of co-evolution between science, technology and culture produced today's noetic world order which is in all respects radically different from the old medieval European world which spawned it (see Fig. 2.1).

Why did these developments occur in the West and not in the East, which was at the time in many respects—socially, economically, and technologically—far more advanced than Medieval Europe? In the cities of the Byzantine Empire, as well as in the Middle East and in China, the population was innovative and exerted a variety of trades and activities. The Ming Dynasty of the fourteenth Century was sophisticated and technologically advanced: China was a world leader in the use of gunpowder-based weapons, shipbuilding, navigation, and in the production of material that require technological knowledge. Earlier, printing had appeared already in the eleventh century which enabled the rapid diffusion of many books, Chinese

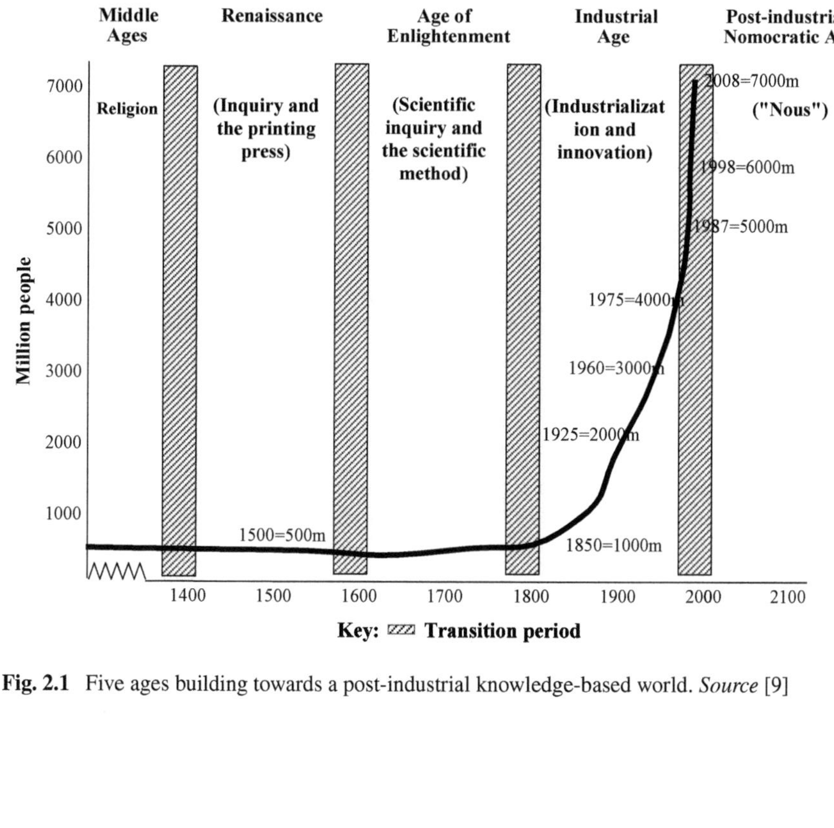

Fig. 2.1 Five ages building towards a post-industrial knowledge-based world. *Source* [9]

cities were bigger and wealthier than the ones of medieval Europe, civil and war technologies were more advanced [10].

Considering today's very rapid progress in modern science and technology in countries such as China, South Korea, and Japan, one should therefore reflect on possible reasons why the East did not pre-empt the European Renaissance: why it was a Western and not an Eastern phenomenon. One possible reason is that Eastern and Asian urban agglomerations were mostly centres of political administration: organs of the state or of an empire. The ruling class was landowners, with their feudal priorities. In European Cities, the bourgeoisie attained political power, and became autonomous state-like bodies, in opposition to the feudal agricultural world where power was centred on land ownership.

One can also not ignore the possibility that culture, and politics (and perhaps colonial politics in later years) could have been earlier detractors. David Landes proposes that totalitarianism, the closing of borders and the lack of a free market may have been important inhibiters to Chinese progress in the fourteenth century [11]. Daniel Memmi argues that Eastern culture was probably inimical to the rise of modern science [12]. He defines "Western science" as the scientific traditions that evolved out of the early (pre-medieval) Christian consciousness and worldview, and from Greek philosophy which later formed the basis of humanism. He sees obstructions to the early development of a unique Eastern science culture in a syncretism of Confucian, Taoist, and Buddhist cultural traditions.

Memmi makes nevertheless the observation that Eastern culture which emphasises an organic universe that is holistic, interactive, and dynamic—with an emphasis on ongoing change rather than stable structures—may provide a better philosophical context for, and understanding of, today's global problems. Especially the externalities (spill-over effects) of rapid economic and industrial growth created complex global problems that cannot be completely understood, or resolved, by means of analytical methods. To face problems such as over-population (and its associated socio-political implications), climate-change, inequalities in wealth, over- and unwise-exploitation of natural resources and environmental pollution a more holistic approach to inquiry is needed. The research culture of modern science is "atomistic", meaning it follows a route of analysis to discover the "building-blocks" of reality [13].

Table 2.1 illustrates how different scholastic cultures may produce divergent perspectives and approaches to inquiry which can either broaden or confuse the insights gained during inquiry, depending how it is applied. For example, classical humanist scholars may interpret today's environmental pollution largely as a (metaphysical) ethical or moral issue (which is one dimension) whereas modern science may see it as a physical or ecological issue that can be managed with better analysis and improved technology to control spill-over effects (which is another dimension). Because of different worldviews they may see and define a problem differently and may even differ in terms of their choice of research focus and the desirable outcomes of human intervention. This divergence in positions and outcomes of inquiry can be a source of inter-scholarly conflict or it can improve insight and understanding about complex problematic situations (as a whole) and their possible solutions. It

Table 2.1 Contrasting approach to inquiry in classical humanities and modern science

Dimension of inquiry	Classical humanities	Modern science
Ontology	Reality is fluid and elusive and it only exists because of people's claims of its existence	Reality is determined by facts that are waiting to be discovered
Epistemology	Knowledge is subjective—i.e., constructed subjectively by people and society –therefore interactive discursive inquiry is the route to truth	Knowledge is factual. My senses can help me to discover the objective world—therefore inquiry requires observation and empirical experimental research
Teleology	The ultimate purpose of inquiry is to build contextual understanding of reality because reality is too fluid, elusive, and complex to be precisely discovered, defined and modelled	The ultimate purpose of inquiry is to discover facts, and through the facts create useful knowledge in the form of theories, laws, and models of reality

all depends on whether scholars with different cultural traditions are willing to learn from each other and, especially, are willing to work together on real world problems.

To conclude: culture affects how scholars practice science and how society appreciates and uses the fruits of science. It is therefore from time to time necessary to re-evaluate the culture that shapes the beliefs and behaviour of society, and especially to reflect on the reigning ideas of a specific culture, including on what it considers to be the nature of "progress" —especially regarding the benefits and costs to society over the longer term. A new consciousness and appreciation of the potentially useful complementarities in the contributions of different scholastic traditions are needed. To discover new knowledge about the constituent elements, or workings, of matter and the living environment is necessary, but it is not sufficient to use this new knowledge wisely: a new way of thinking about the workings of complex living systems are also needed. Therefore, the contributions of both divergent and analytical thinkers, of humanists and scientists, of West and East and of modern and traditional thinking should be considered—if only to break the mould of the culturally-shaped mind.

2.3 Consciousness: The Mould of the Mind

Consciousness is a state of being aware of, and responsive to, external, internal, and mental stimuli. Koch (2018) defines consciousness as: "…everything you experience. It is the tune stuck in your head, the sweetness of chocolate mousse, the throbbing pain of a toothache, the fierce love for your child and the bitter knowledge that eventually all feelings will end" [14]. Consciousness covers different levels of awareness, stretching from the most mundane, physical, and basic human interests (such as sensory experiences and awareness of threats) to the deepest levels of awareness in terms of knowledge, insight, understanding and spiritual essence [15]. This discussion covers the deeper or "inner" level of consciousness, which is influenced by the dominant culture of a society, and which provides the scope of what is considered to be "relevant" thinking.

Thinking is influenced by three interacting human traits, namely rationality, emotionality, and culture [16]. Of the three traits, culture is the most clandestine effector of thinking: it is ever present, but seldom observed as such in the process and results of thinking. The reason is that cultural imperatives mould the inner consciousness of scholars and over the long term this determines the way they observe, interpret, and appreciate what is observed [17]. This determinism is a product of the paradigms and repetitive practices a culture promotes. Repetition creates habits, and habits are addictive—even in as far they influence a person's interest in a phenomenon, and ability to observe and perceive what is seen. A settled pattern of thinking and behaviour affects a person's consciousness.

In 1969 the late Edward de Bono published his seminal work on *"Mechanism of mind"* in which he explains how the human brain becomes the human mind [18]. He defines the brain as a complex thinking and knowledge system in which knowledge is processed according to the nature of the system. The brain's function is to perceive, accept and utilise information, and the way this happens is called "'thinking". Thinking is moulded by a person's memory. Memory is the ability of the mind to store and recall information that was previously acquired. It is processed through three fundamental processing stages, namely storing, encoding, and retrieval. Storing refers to the process of placing newly acquired information into memory, which is modified in the brain for easier storage. Encoding (such as the use of symbols or images) this information makes the process of retrieval easier when it needs to be recalled and then brought into conscious thinking. Retrieval is triggered when a person is faced with a problem or event and needs to act.

In de Bono's discussion of memory traces and memory surfaces he describes a memory as something that is left behind after it happened and does not completely un-happen. He differentiates between short term and long-term memories. Short term memories are those that are the results of incidental (or utility) occurrences that are soon "washed away" after it served its purpose (for example prior figures in a series of calculations). Long term memories are like "ruts on a memory surface" (he uses the metaphor of hot ink-drops falling in a bowl of jelly) [19] which ultimately mould a style of observation and perception in a scholar's mind. "Ruts" can be caused by the habits and repetitive practices within a particular scientific culture which then influence the way researchers observe, appreciate, and act upon, a subject of study.

One can only surmise about the inter-personal conflict-potential of divergent "memory ruts" when scholars' from different professional backgrounds are faced with a similar phenomenon—even to the extent of how they define the purpose of their inquiry. This may explain CP Snow's observation that British humanists and scientists cannot understand or communicate with one another because, metaphorically, they live in indifferent cognitive worlds and are "unconscious" of scientific traditions other than their own.

2.4 Breaking the Mould of the Mind Through Heuristic Learning

If scholars implicitly accept that there is only one universal and legitimate (or "true") scientific culture, they will have to put some effort into defining that culture. Disciplines (and professions) create the cultures that suit their fields of specialisation the best: for example, the images, metaphors, concepts, and procedures common to a profession or specialisation. Specialisation was a powerful driver behind 300 years of scientific progress through a systematically narrower focus on an aspect of a phenomenon. This produced a depth of knowledge, but it is ill-suited to provide the breadth of understanding of the pervasive systemic consequences when such knowledge is applied. Unless scholars from the different disciplines are able (and willing) to contribute and listen attentively to conflicting ideas and theories they may experience "cognitive dissonance" when faced with ideas that differ strongly from their own [20].

Cognitive dissonance theory is based on the work of Festinger [21] which can be summarised as follows:

- People are highly sensitive to inconsistencies between their moulded understanding of a situation and how others see the situation
- Their experience of such inconsistencies may cause cognitive dissonance (an extreme form of stressed confusion) which will then prod them to either totally (and sometimes aggressively) reject the other interpretation, or perhaps to search for a way to resolve their problem of understanding it;
- The people who are "searchers" present an opportunity for inter-cultural discourses and heuristic learning designs that can free the minds of scholars to understand and respect other scientific cultures and other ways of thinking (Fig. 2.2).

Heuristic learning is a repetitive-feedback search and re-discovery process, guided by a rule of imminent sufficiency and ongoing reconfiguration of insights thus gained. The primary condition that supports the learning process is an outcome of the condition itself, namely creative stress to resolve the contradictions between the need for new ways of thinking and acting, and the moulded culture that obstructs such thinking and acting. The purpose of heuristic learning is to systematically soften the cognitive (rational), affective (emotional) and conative (will-power) barriers to communication. It is a process of repetitive contextualisation where the culture, thinking, assumptions, paradigms, fears, and divergent worldviews of scholars are reviewed, as well as the reasons and rationale behind the differences [22].

Ulrich describes heuristics as a way of probing the unknown through a questioning process that includes three dimensions [23]:

- Probing is done by means of problem-relevant questions and problem-relevant knowledge;
- It is important to learn how to probe in order to prepare people for the task of critically reflecting upon generally accepted points of view;

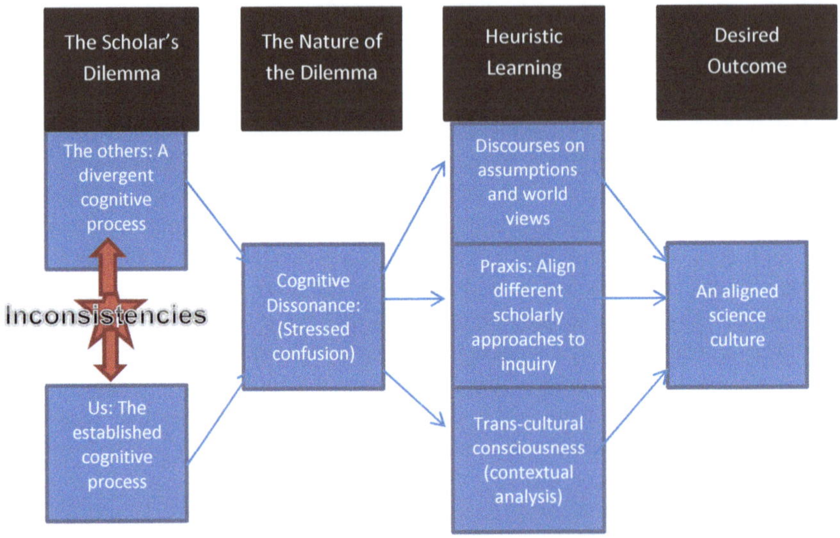

Fig. 2.2 A heuristic design for removing cognitive dissonance by changing the mould

- It is important to learn how to discover deception when faced with commonly accepted ideas, perspectives, concepts, metaphors and procedures.

  The following sequence is a typical example of such a probing exercise:

- Probing the traditions, images, metaphors, concepts, and procedures underlying different science cultures.
- Probing points of departure, i.e., assumptions and unquestioned beliefs:

  a. What are the assumptions and beliefs underlying your position, on the issue discussed?
  b. What could change if your assumptions and beliefs are proven wrong?

- Probing the rationale, or reasons behind a scholar's position on an issue:

  a. What is the problem, or the nature of the situation, as you see it?
  b. What are the main causal factors or drivers that produced the situation as defined by you?
  c. What evidence is there to support you position?

- Probing viewpoints and perspectives:

  a. Who benefits from the current situation?
  b. Who ought to benefit from the situation?
  c. If you have full control over the situation, what will you do and why will you do it?

- Probing implications and consequences

a. What are the implications of your proposal in terms of actions, financing and research needs—and how will you manage the implications?
b. What are the desirable outcomes of your proposal and why is it desirable?

Heuristic learning is also an integral part of "participatory action research" (PAR) and of soft systems methodologies (SSM). Important contributors to the PAR-methodology are Kurt Lewin's social action research and emancipatory philosophy [24] and Argyris and Schön who worked on organisational learning concepts [25]. They applied heuristic learning when selecting a research focus, clarifying theories, identifying research scope, collating information, reporting results and re-evaluating results in more rounds of inquiry. Important contributions to SSM include the work of scholars such as Ulrich and Ackoff [26], Gharajedaghi [27], West Churchman [28] and Checkland [29]. SSM engages researchers, planners and stakeholders in learning cycles to discover, *inter alia*, their own thinking and the thinking of others, to discover the nature of a problem, and to discover the general conditions (or circumstances) that produced it.

To conclude, breaking the mould of culturally determined thinking requires a mind-shift, and a transformative heuristic-learning process can help to achieve it. The perspectives and modes of thinking of scholars from different scholastic backgrounds are incrementally exchanged in this process in support of a rational, solutions-oriented, scholarly dialogue. Such a process is best managed by means of a focused inquiry into phenomena which reveals the differences and complementarities between different scholastic cultures.

2.5 The Evolution of Science: From Contemplation to Measurement

Although different science cultures evolved their own ways of looking at problems and doing research, all of them are today guided by the same universal principles of being objective, measured and creative in observation, rigorous in practice, rational in contemplation, honest in reporting, transparent in procedure and open to peer-evaluation. The journey to develop these principles started with the introduction of classical humanism into the European scholastic theatre.

Classical humanism was to a certain extent a "pre-science". Although it was the intellectual force behind the Renaissance the earliest proponents of humanism practised it only as an imitation of the classical culture of ancient Rome and Greece. This imitation of classical culture by the humanists involved a radically new way of studying the classics. The humanist contextualised classical works, studying the culture that originated them, especially from the linguistic point of view. This in-context study, largely linguistic, was radically different from the study of the classics as it was done in the Middle Ages. During the Middle Ages, the classics were read and interpreted in terms of medieval culture, not through the culture in which classical texts had originated. This approach led the humanists to analyse classical texts for what they originally intended to mean. Their essential contribution to the development of science was that it opened the minds of scholars through disciplined contextual inquiry.

Under the influence and inspiration of the classics, European humanist scholars developed a new rhetoric, new ways of thinking, new moral and civic perspectives and new values for human behaviour. This created an "enlightened world" in the seventeenth and eighteenth centuries with a strong focus on natural philosophy—Isaac Newton (who called himself a "natural philosopher) was one of the leading lights in this regard. "Enlightened" scholars opened new avenues for, and ways of, inquiry that ultimately set the ground rules for scientific research. The humanities were not left standing in this process but co-evolved with the transformation when "natural philosophy" became "natural science' in the nineteenth century. Today it is a generally accepted principle of modern humanities that knowledge of the world is derived from observation and experimentation, and that rational analysis and scientific inquiry are the best procedures for deriving this knowledge [30].

Modern humanities are noted by their interest in all things that effect human nature and the state of the world and are therefore well-placed to make meaningful contributions in discourses about the future welfare (and sanity) of humanity. It includes, amongst others, religious studies, social rise and fall theories, ancient and modern languages, poetry, the arts, history, and philosophy. As the core of a liberal arts education, the humanities investigate literature, the distant past, culture, and human values. Each discipline relies on its own unique interpretation methods. For example, historians use the historical method, which can be further divided into branches like cultural history, quantitative history, and oral history [31]. The most important contribution of studies in the humanities lies not in analytical research findings ("knowledge") but in contemplations and explanations that improve insight into, and understanding of, human behaviour and of the general conditions that transform societies. In contrast modern scientific research is more focused on analysis, collating facts, the framing of hypotheses and developing theories that may explain phenomena. The scholastic objective and practice of modern humanism may differ from modern science, but it is nevertheless not "unscientific" because of this.

Modern science includes the natural and social sciences. In academia social sciences are often organisationally positioned under the humanities but in research methodology and research culture it is more closely related to the formalities and practices of the natural sciences. The fields of interest of social sciences cover social change and development, the patterns of behaviour within societies—including the shaping of relationships among individuals within societies. The early beginnings of social science emerged during the enlightenment when new developments in natural philosophy changed the way scholars understood the meaning and requirements of formal inquiry into the human condition. It evolved out of the work of moral philosophers who tried to understand the social and economic upheavals that followed on political turmoil and the industrial revolution between the eighteenth and nineteenth centuries. By the turn of the twentieth century social sciences consisted of several formal fields of scientific study (called "disciplines") each with its own systematic knowledge-base and prescriptive practice for conducting research and constructing theories. Today the social sciences include scholarly disciplines such as anthropology, archaeology, economics, geography, demographics, law, linguistics, politics, psychology, and sociology.

The natural sciences include the physical sciences, earth sciences and life sciences. Greek natural philosophy (for example Aristotle's contributions in biology), were the earliest form of Western science. The Greeks used observation, logic, reason, and discourse to try to understand why certain events around them were happening. For ancient Christian civilizations, the unknown was explained by religion. In the twelfth century these works were translated from Greek and Arabic into Latin, and when these Latin scripts were later translated into the vernacular of different nations new ideas spread rapidly over Europe. A different scientific culture emerged when scholars started to use objective measurement to discover the mysteries of nature [32]. Metaphorically-speaking, they stopped having endless discussions on the number of teeth in a horse's mouth: they walked up to the horse, opened its mouth, counted its teeth, and noted it down.

Objective research was at first just a branch of natural philosophy but became the tree itself when scholars continued, and intensified, their efforts to understand, describe and predict (or explain) the nature and behaviour of natural phenomena. Close observation, data collation, experimentation and the formulation of hypotheses and theories were enculturated as the distinctive characteristic of good practice in science. The rules and procedures for research that were developed since the eighteenth Century are today the guiding standard for judging the appropriateness, correctness, and value of all research-findings—with adaptations to suit the particular conditions of inquiry in the humanities and social sciences.

2.6 Homo Deus: "A God that Limps" [33]

Scientific progress transformed western culture, changed the beliefs and perceptions of people, and revolutionised the world's political and economic order. One of the consequences of this new "Newtonian" world is that it convinced people that "truth" can be discovered: that science can discover how the world works and that humans can use and shape the conditions of their life according to their will and for their own benefit. This belief was expressed by the Prince Consort of Queen Victoria in 1851 at the opening of the Great Exhibition in London: "*Nobody who has paid any attention to the peculiar features of our present era will doubt for a moment that we are living in a period of most wonderful transition, which tends rapidly to accomplish that great goal to which all history points*" [34].

Considering this belief in the efficacy of science, an important question comes to mind when it involves the social sphere and not only the physical sphere of life: how can one fully discover ultimate "truth" in a complex homo-sapiens-driven world system which seems to be in a continuous state of change (i.e., in a state of "becoming")? It is, simply put, a very complex and moving target whose elements can be influenced by both looping systemic interactions and unpredictable circumstances.

The solution of science was, of course, to isolate the object of study: the Newtonian world was assumed to be a closed system: with no reference to an environment that can impact on it [35]. This world worked like a hermetically sealed clock: a mechanism that operates with a regularity that is exclusively dictated by its internal structure and the causal laws of nature. The implicit assumption is that science needs only to determine the structure and causal laws and then use this knowledge to explain what is observed. To determine results with objective certainty, a regulated (definable) environment was created, so-called "laboratory conditions" that specify the applicable scope of the findings. For example, the Law of Freely Falling Bodies determines the distance a body would fall in a vacuum (an environmental specification). In social sciences, and especially in economics, the research conditions and assumptions are specified a priori and the concept *ceteris paribus* (all other things being equal) was introduced to specify the relevant scope of application of research findings. Clearly "all other things" rarely remain the same in real life.

Another important aspect of this paradigm is an implicit assumption in the efficacy of reductionism, namely that the whole truth can be discovered through researching the constituent parts (the "stuff" or "atoms") of the truth. In the natural sciences this seems to be a reasonable assumption because of high levels of structural determinism, but it spilled over into the social sciences, which is then more questionable. So-called "social atomism" sees the individual as the "atom" of society and therefore the only true object of concern and analysis. The founder of economics, the moral philosopher Adam Smith, declared in 1776 that it is not from the benevolence of the butcher, the brewer, or the baker that we expect our dinner, but from their regard to their own interest [36]. Today social atomism underlies the free-market ideology in economics with its perception that a society is simply the integral of all the interactions between self-centred individuals. Accordingly, the self-sufficiency and moral autonomy of individuals (or companies) should be the primary concern of human enterprise. The emphasis is on individual freedoms and rights: it downplays reciprocities, duties, and connections between individuals because it is assumed that it will sort itself out in a process of what the nineteenth century French economist Leon Walras called "*tâtonnement*". "*Tâtonnement*" means that everyone in a community simultaneously calculates the pros and cons of his or her actions in terms of his or her own self-interest [37]. It assumes a balance of power between small independent players supported by a benevolent natural and social environment—in practice a false assumption.

Today this process is called "self-organization" by researchers working in the field of complexity theory [38]. Self-organization is a complex system's response to random, unintegrated, interactions between its constituent elements: often unintended spill-over consequences of self-centric individual (or "atomistic") actions. A belief in welfare-growth through social atomism is therefore based on a false premise. It is proven to be false in practice by the environmental consequences of uninhibited, innovation-driven, gravitational industrialisation and the growing global inequalities in wealth and power it spawned—to such an extent that it has become a serious threat to human existence on earth. The devil's name is sub-optimisation, i.e., to focus so much on your particular interest, that you become unconscious of the circumstances influencing your particular interest [39].

The outcomes of social atomism are clearly not benign and the kind of long-term threats it poses to humanity are due to the complexity and unpredictability of self-organising world systems. Only a better understanding of the interactions and interdependencies between the whole and its parts, and between the parts of the whole can hope to remedy this situation. There is a need for a holistic approach to inquiry that acknowledges the existence of interdependencies between people, their creations, and the environment. A holistic approach to inquiry is based on the recognition of interdependent systemic interactions between the elements of an encapsulating system, defined as a "whole",

2.7 Holism as a Synoptic Vision for a Metascience

"Metascience" is the science of studying science. The concept is used here as an approach to study the fundamental nature of how people perceive "reality" when looking at the real reality: the relationship between subjective and objective observations, between feelings and "facts", between being today (what is) and becoming, tomorrow (what can be). This also covers the question of "correctness" when a specific approach to inquiry is followed. For example: is analytical reductionism in scientific inquiry misdirecting the quest for truth—as one may conclude from the previous discussion on social atomism? If analytical reductionism is an obstruction to truth, how can one then explain the remarkable achievements in science and technology—with the associated benefits to society—over the past 300 years? And then also, how can one explain the negative impact of the same scientific and technological progress on the global scene? Something is incomplete, something is missing, and that something is a "whole" approach to inquiry.

Holism is often perceived as an approach that follows a rounded and all-over pattern of inquiry (i.e., it includes "everything"), but in fact it is much more than that. Smuts defines holism as "an attempt at synthesis, an attempt at bringing together many currents of thought—a theory of the whole" [40]. The underlying principle of this "theory of the whole" is that there is no physical matter, just arrangements where the whole and its parts are reciprocally determined: transformed and shaped by each other. Although Smuts rejected the idea of a system in his formulation of holism one should read this against the background of how the concept "system" was perceived during the time he developed his theory of holism in the 1920's. It was then perceived as an organisation of interdependent elements, causally bonded and static in design (for example a machine, a human body, or a business organisation). Smuts' holism was a dynamic, concept of "creative evolution": of change and renewal.

A second aspect of holism is its integrative and dynamic approach to interpreting the nature of mind and matter. As stated in his book Holism and evolution: "*Among the great gaps in knowledge, those that separate the phenomena of matter, life and mind still remain unabridged*" [41] Smuts intended to build a bridge between the phenomena and claimed that the three concepts constitute one and the same whole of human experience and therefore cannot be utterly unlike each other. Not only do they co-exist in humans, but they appear to give rise to each other in a creative process of co-evolution. Through a process of co-evolution, the whole becomes more than the sum of its parts. He sees human personality as the highest (or "encapsulating") whole. Wholes are amalgams which have an internal structure of elements, a role (or function), a way of functioning (or process) and a character or "personality" which differentiates them from mechanical structures, The concept "structure" does not only relate to physical structures (such as the human body) but also applies to metaphysical structures such as culture, mental (or memory) structures, attributes, values, and ideals. The structure of a whole is a living and interactive design, for example, a person's thoughts and actions are shaped by his or her personality (his or her whole) while, in return, personality shapes thoughts and actions: *"The whole and the parts reciprocally determined, formed and shaped each other"* [42].

A third aspect of holism is its emphasis on patterns of interdependencies as the essence of matter, time, and space. In recognition of Einstein's relativity theory Smuts redefined matter according to which energy, time and space become flexible and interactive concepts. Matter becomes "arrangements" (interdependencies) and accordingly it is not "stuff" that makes up the world. Underlying, and encapsulating, these interdependencies are a hierarchy of interdependent "wholes", interlinked through the "fields" that form part of wholes. A "field" is the realm of influence of a whole. For example, in physics one may refer to a gravitational field and in psychology it is the intricate set of symbolic interactions and forces which influence a person's interactions with others and the environment. A field is not something apart but a continuation of a whole that extends beyond its observable outline and exemplifies a whole through behaviour, interests, and activity. For example, part of a person's "whole" can be identified as a physical body noted by its functional design (arms, legs, head, and torso) and its interacting and interdependent internal organs that support this functional design. The field is the influence that radiates outward through a person's language, personality, character, and values. It is the metaphysical dimension of a person as a "whole", which radiates outwards and interlinks with the fields of other wholes. Personality, character, and values direct and shape a person's interactions with others and provide systemic coherence (governance) in behaviour. A person's personality, character and values have a certain power over his or her development, actions, and interactions with others. When inquiry follows the route of reductionist science it is difficult to quantify values such as truth, goodness, and beauty (aesthetics), but when you look at it from the vantage point of fields and patterns of interdependencies it becomes part of the common nature of things: part of the real reality, not the perceived reality.

The world is perceived as a system of systems consisting of a multitude interacting wholes that can be variably ranked according to the purpose of an inquiry or in terms of some or other vested interest. For example, an educational system can be ranked in different ways as an education system, or juvenile security system ("baby-sitting" system) or as an employment system or as an architectural design system. Educators, parents, employees, and architects may accordingly differ in terms of their perception of a functional (and thus systemic) hierarchy of priorities.

There are also from a second vantage point a series of interacting wholes ranked according to their level of systemic complexity, starting with matter (as defined earlier) as the simplest of wholes. From this lowest physical ranking it rises to plants, to animals, to humans (*Homo sapiens),* then on towards the metaphysical dimensions of personality and the spiritual world. Then, perhaps, the encapsulating living whole of a global ecosystem as described in James Lovelock's Gaia Hypothesis [43]**:** interacting with, influencing, and being influenced by all the world's wholes—including the respective fields of the wholes. A third pattern of wholes may focus on human organisation ranked from the smallest (less complex) forms such as a family to business organisations, societies, communities, nations, and international associations. In all these organisations there are fusions between fields (regulatory and cultural) through which the lesser forms and the larger forms interact and influence each other in a process of co-evolution.

Smuts' theory of holism is not easy to comprehend, and even more difficult to apply in practice. Smuts calls it a "synoptic vision" that brings better insight and understanding regarding a new view on reality—perhaps the real reality—that scientific inquiry should face. As such it is a new science of studying science in which natural science and humanities are on the same playing field. The basic rules of this game are:

- Thinking must be done in terms of interdependent wholes, inclusive of the fields which are integral parts of the wholes;
- The dominance of arrangements (and not "stuff") from the simplest whole to the most complex whole;
- The acceptance of change and co-evolution as a permanent reality in the real world where the assumption of permanence is just that: an assumption;
- Causality resonates through interactive loops—i.e., outcomes are "co-produced" not "caused";
- Hierarchy in the arrangements between, and within, a system of wholes is relative to the focus of an inquiry;
- Physical and metaphysical are integral parts of wholes and should be handled as such in scientific inquiry.

This suggests that as far as modern science and modern humanities are concerned, there is no need for a new science culture other than a need for a culture of cooperation between the two science cultures. Holism highlights this need but does not provide any guidance to this effect: it is a theoretical concept and not a methodology. For an answer to this problem, it would be necessary to look at how holism as a concept can also be transformed into a method of inquiry. Multi-disciplinary procedures (as, for example, applied in the professions) and trans-disciplinary soft systems thinking provide some answers in this respect.

2.8 Professional Studies and Soft Systems Thinking as Holistic Practices

Professional studies—such as those in medical schools, engineering schools, business schools and agricultural schools—accommodate multiple disciplines in their curricula to support the technical and practical requirements of their professions. For example, medical students have as part of their training selections of chemistry, physics, botany, zoology, mathematics, statistics physiology, genetics, microbiology, technology, radiology, psychology, sociology, and ethics (philosophy*)*. In their specialised medical coursework aspects of these disciplines are then foundationally integrated as one "whole". The scope (frame of interest) of professional work is defined by the focus of a profession but with time this scope may change when professionals discover new "wholes" that may impact on the quality of their work. When a general practitioner treats a patient, the functioning of the patient's body with his/her emotions and character will be the "whole" of such a diagnosis. When the doctor wants to prescribe a longer-term solution to the patient's problem considerations of interactions of different" wholes" within a larger "whole" such as the general living conditions and lifestyle and world-view of the patient should form part of a diagnosis and prescription. In community healthcare this "whole" of health practice may expand further into issues covering (amongst others) the quality of the natural environment, social welfare, community development, and economic development.

Professionals are keenly aware of the interdependencies between various specialised disciplines that inform their field of practice and will readily use this knowledge to diagnose, plan or decide on prescriptions. Within the scope of its application a profession is a semblance of CP Snow's "unifying scientific culture" because it applies the different contributions from the humanities, the social sciences, and the natural sciences in practice. This application is an essential part of a professional culture—such as a medical culture, an engineering culture, a law culture, and a farming culture (agriculture). It is supported by good communication within the professions regarding the correct application of the specialised knowledge of various disciplines. However, between the professions one may again face the proverbial 'Tower of Babel" of poor inter-professional communication. Snow's comments are therefore also applicable to the situation between different professions: "*they have in intellectual, moral and psychological climate so little in common*". This requires not only a multidisciplinary approach in professional knowledge development but also a transdisciplinary approach for a better insight and understanding of interdependencies within the world system of wholes.

What is a system? Although Smuts rejected the concept of a system in his treatise on holism, developments in soft systems studies since the 1970's seem to echo his idea of interacting wholes—including the fields of the wholes—and patterns of interdependencies between them. For example, one of the most renowned systems thinkers the late RL Ackoff defined a system as a whole that consists of two or more elements ("sub-wholes"), where the behaviour of each element has an effect on the behaviour of the whole, where the behaviour of the elements and their effect on the whole are interdependent; and where, however subgroups of elements are formed, each has an effect on the behaviour of the whole and none an independent effect on it [44]. To research, specify, and scope these wholes and their interactions with other wholes are the first steps of soft systems inquiry. The concepts "open system" and "environment" are applied to allow for interactions with wholes from outside the field of a particular whole (such as a profession). Open systems research accepts variable laboratory conditions: i.e., interactions with the environment of the system (or whole) of interest are integral parts of the research. This is obviously a very complex problem situation.

Since the 1980's scholars within the community of soft systems researchers developed procedures to scope and map such complex problem situations to understand their nature and the underlying reasons for their emergence. This was a departure from the very precise systems dynamics modelling procedure that was developed by a research group at the Massachusetts Institute of Technology under the leadership of Jay W Forrester [45]. Forrester's team focused on the structure and dynamics of extremely complicated systems. Scholars studying soft systems behaviour were interested in the dynamics of such systems, but they were also interested in the behaviour of "multi-minded" human systems—and how this differs from the mechanistic or organismic systems that Smuts had in mind in his rejection of systems in his treatise on holism (see Table 2.2).

The mechanistic worldview perceives a machine-world consisting of separate entities that are passive and purposeless. Together these separate entities work like a machine, guided by an external force ("God" or a "leader") and regulated internally by natural laws that determine change in a cause-and-effect process. Certain knowledge can be acquired through taking the "machine" apart—i.e., through reductionism—and the whole can be explained by adding the separate parts together. This is the classic Newtonian assumption.

The organismic worldview sees a world-system resembling an animal with functional organs. Evolutionary change in this world is determined by natural laws (e.g., "genetic codes") and the survival of the fittest: for example, culture is perceived as the "DNA" of human organisations. Knowledge acquisition is reductionist-analytical and sense making is deterministic—i.e., in terms of cause and effect modelling or theory development. The governance of organismic systems is cybernetically- controlled through loop-feedback processes [46]. Cybernetics is the study of communication and automatic control systems.

In contrast, the soft systems worldview is multidimensional, multi-minded and sees a diverse world of dynamically interacting wholes—including complex interactions between physical, living and social systems. It recognizes that such dynamically complex wholes cannot be aptly defined, and not precisely structured and precisely measured. This is not only because it consists of a multitude of interacting wholes, but also because it is facing a moving target. The objective of soft systems inquiry is therefore not to analyze but to observe and to interpret what is observed. In observation and interpretation, the idea of a single perspective, or of a single issue, is rejected in favour of gaining insights from scholars and professionals with different scholastic

Table 2.2 Comparative mechanistic, organismic, and soft systems cultures in human organisation

Underlying assumptions	Machine metaphor	Animal metaphor	Community metaphor
Cultural dictates	Mechanistic Systems: Newton (Henry Ford)	Organismic Systems: William Harvey (Stafford Beer)	Holistic Soft Systems: JC Smuts (RL Ackoff)
World view	"Homo Deus" "control over": people are the cogs and shafts of the machine that operates with a regularity, dictated by its internal structure, and governed by an external brain (owner, dictator, etc.) which is not part of the machine	Body, head, torso, limbs and organs controlled by a brain (top management) [46]; People are functional members of a structured whole and hierarchical (top-down) subservient to an established order	A whole of wholes, multi-minded, multi-dimensional and purposeful: People are self-thinking, innovative, interdependently linked into a process of ongoing renewal and co-evolution
Behaviour	Driven by external forces, rules and laws, functionally controlled with predictable outcomes	Driven by hierarchy and rules legitimised by its contribution to the body's survival or development: cybernetic-controlled	Driven by own volition and an alignment of perceptions, purpose insight and understanding through learning and co-evolution
Ideals	Good submitter: Optimal (effective and efficient) functioning within the fixed structure	Good team player, adherence to top management ideals, their rules of governance and culture of the system:	Good "surfer": Harmony with and within the "wholes"—to increase the ability and desire to serve self and others
Means to achieve ideals	Enforcement through penalties and rewards	Executive top-down control and management by objectives	Interactive social learning: management of information, knowledge, insight and understanding

Notes Isaac Newton (1642–1727) as a symbol for pure mechanics, William Harvey (1578–1657) for physiology and Jan Smuts (1870–1950) for holism. Henry Ford applied a mechanistic managerial model in the manufacturing of his motorcars, Stafford Beer (1926–2002) was an academic who used the human body as a model for his systems theory of management, and RL Ackoff (1919–2009) used holistic soft systems thinking in his interactive design of organisational development control in auto-control systems, such as the nervous system and brain of animals, and mechanical–electrical auto-control systems in machines [47]. Organismic systems are "uni-minded", meaning there is a single controller (the brain in animals, top-management in organisations or the president of a country) that governs their behaviour.

backgrounds in the sciences and humanities. "Soft systems" include both the physical and metaphysical dimensions of interacting wholes: i.e., "wholes" inclusive of their associated interactive bonding "fields", as defined by Smuts. The sciences and humanities are perceived to be integral parts of a unitary process of inquiry.

Soft systems studies confront problematic situations, not problems. Problematic situations are clusters of "problem wholes" that interact with each other to produce problem-symptoms in society—such as poverty, social decay, and environmental decay. Problem-symptoms cannot be solved by direct intervention (say, by handing out money) because it will reverberate within the problematic situation and produce unexpected outcomes that may even worsen the original situation. It is nevertheless possible to manage them by manipulating the conditions that underpins their existence and continuance, and by means of "strange attractors" —such as inspiring leadership and a strong shared vision in human systems [48].

The good that comes from scientific progress and the bad that it may cause is therefore an interdependent reality that needs to be researched as one whole. Moreover, the associated long-term outcomes of scientific and technological progress are the systemic emergences of human behaviour in reaction to this progress. The challenge is to develop sufficient insight and understanding of such complex, human-based, systemic processes. The following frameworks for application in soft systems studies illustrate a few ways on how to approach this challenge.

2.9 Conclusion: Finding a Common Ground for the Sciences and Humanities Through Soft Systems Thinking

It should now be clear that research into complex social systems requires a holistic approach to inquiry in which the sciences and humanities have complementary roles to play. It is within the processes of complex human systems where most of the vexed global problems are produced, and it is therefore also where the solutions for these problems should be researched. Physical substances and non-complex (albeit intricate) systems—such as machine systems and even some organic systems—can be researched and explained through analysis and modelling. This is not possible with complex social systems that are open evolving systems and, moreover, that display multi-dimensionality (i.e., evolutionary multidimensionality) in the transformation of their purpose, structure, and process over time [49]. This presents opportunities for lateral thinking and creative inquiry.

Evolutionary multidimensionality is a process of self-driven multi-faceted transformation which can produce surprising results over the long term, both good and bad. The fruits of scientific progress and technological innovation are typical examples of this dichotomy: extreme levels of wealth are countered by extreme levels of poverty in the world; the amazing benefits of industrialisation are countered by its externalities of pollution and over-exploitation. A simple example of long term evolutionary multi-dimensionality is how early human spirituality (cosmic consciousness) was transformed into human culture which produced human organisation based (*inter* alia) on belief systems, such as the various Christian religious traditions, which influenced the development of various institutions and social governance systems (including organised Roman Christianity and the emergence of Islam) during the first millennium, which in turn affected the development of educational and science systems during the early second millennium—which in turn looped back to influence culture, and social and political change.

The work of Darwin on evolutionary theory tells a similar story about multidimensional natural evolution where environmental changes, survival processes, natural selection and geographic diversity helped to shape the world's organic diversity. The development of the electronic computer in the 1940's was built for a very specific purpose but became a catalyst for socio-economic processes which produced a radically different information society by the turn of the twenty first century: it changed not only the way humans calculate and model in their research and businesses but also their culture, their values, their way of life and their way of observing and thinking.

Evolutionary multidimensionality is a holistic process, and a full understanding of this process requires both objective research and subjective contemplation. In order to explain this, the proposals of Dana Meadows [50] and Sohail Inayatullah [51], on how to integrate the objective (or tangible) and subjective (or intangible) dimensions of inquiry into one framework are briefly reviewed,

Meadows discussed ten leverage points over three layers of intervention in human systems, starting with ("hard") highly objective procedures to ("soft") highly subjective procedures. The first layer is concerned with quantitative management, research and analysis covering financial indicators (such as taxes and subsidies) and the management of human and physical resources, and technology. The second layer covers regulating processes (negative and positive feedback loops) through information management, rules, cultural change, and policies. The third layer promotes self-organisation through managing the functions and goals of the system, through managing institutional values and through breaking the moulds of the minds of people (as was discussed under heuristic learning).

Figure 2.3 presents a summary of Inayatullah's CLA-methodology that proposes an interactive integration of "hard" and "soft" procedures over four layers of inquiry which involves both the sciences and the humanities. It is one seamless procedure of research and discovery, interacting between focused research and broad contemplation, between empirical research and heuristic learning (loosely, the two modes of inquiry that are typical of the sciences and humanities).

CLA"s value lies with creating transformative spaces for learning. The first level, the litany, consists of the assembled body of practical and scientific "facts". These are the products of analytical/quantitative research and/or of objective experiences that over time (through repetitive use) became part of a general knowledgebase of researchers, scholars, professionals and decisionmakers. The second level is the (holistic) social systems level where the information in the litany is scoped and contextualized within a systems framework to explain or question the assembled facts. The third level is an inquiry into worldviews through scholastic dialogues.

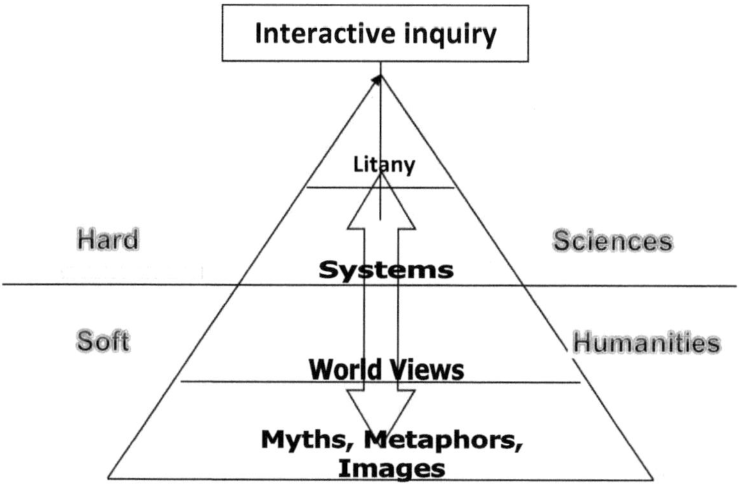

Fig. 2.3 Inayatullah's Causal Layered Analysis (CLA)—integrating contributions from the sciences and the humanities during inquiry

The objective is to surface deeper, unconsciously held beliefs and assumptions. The fourth level is an inquiry into the emotive dimensions of of the subject of research. CLA functions on an interactive process of heuristic learning that moves up and down these layers of analysis in order to discover and note different ways of knowing, a common focus, a standardised procedure for inter-scholarly dialogue and also the concepts that support such a dialogue.

Through the CLA framework of inquiry, the "politics" of scholarly interactions become part of one interactive research process in which different points of view and personal self-interests are presented to the participants within an open space for objective discourse. It is unlikely that all of these will be disclosed in one round of interactions but over time different scholarly traditions can, through CLA's heuristic learning process, arrive at a *modus vivendi* for better cooperation. While scholarly interactions between scientists and humanists may initially remain problematic at the conceptual level, on the applied level interests may converge on a need to find sustainable solutions to global problems. In other words, CP Snow's scholarly dualism can be cured on a practical level, and through practical application ultimately also on a conceptual level. In other words, once scientists and humanists learn to appreciate each other's potential to contribute meaningfully while working together to solve global problems, they will also learn to communicate better with each other at a conceptual level.

References

1. C.P. Snow, *Two cultures and the scientific revolution. The Rede Lecture* (Cambridge University Press, Cambridge, 1959)
2. Report of the Harvard Committee, *General Education in a Free Society* (Harvard University Press, Cambridge, 1950). https://archive.org/details/generaleducation032440mbp/page/n5/mode/2up
3. J.C. Smuts, *Holism and Evolution* (Macmillan Publ, New York, 1926)
4. D. H. Meadows, L. Meadows Dennis, R. Jørgen, W.W. Behrens III, *Limits to Growth* (Potomac Associates, Bethesda, 1972)
5. P. Cilliers, *Complexity and postmodernism* (Routledge, New York, 1998), pp.2–7
6. P. Checkland, J. Scholes, Soft systems" is a concept coined by Peter Checkland of Lancaster University in the 1970's to differentiate human systems from "hard" (physical) systems, in *Soft Systems Methodology in Action* (Wiley, New York, 1990)
7. W. Durant, *The Renaissance* (Simon and Schuster, New York, 1953), p.77
8. W. Shakespeare, *Illustrated Shakespeare. The Complete Works, Annotated,. Hamlet, Act II, Scene 2* (Greenwich House Inc, New York, 1986), p.1879
9. P.H. Spies, *Millennial megatrends: Forces shaping the 21st Century*, in Keynote address to the Annual Conference of the International Association of Technological University Libraries (IATUL) (1998). https://docs.lib.purdue.edu/iatul/1998/papers/33/
10. D. Landes, *The Wealth and Poverty of Nations* (Little Brown & Company, London, 1998) pp. 36–59, 336–339
11. D. Landes, Op cit.: 45–59 (1998)
12. D. Memmi, The relevance for science of Western and Eastern cultures. AI Soc. **34**(3), 1 (2019)
13. I. Stone, Read, for example, an exposition of Darwin's quest, in *The Origin: A Biographical Novel of Charles Darwin* (Cassell Ltd., London, 1980)
14. C. Koch, What Is Consciousness? *Scientific American* (2018). https://www.scientificamerican.com/article/what-is-consciousness/
15. R. Barrett, *The Seven Levels of Personal Consciousness*. https://www.valuescentre.com/wp-content/uploads/PDF_Resources/Barrett_Model_Articles/Seven_Levels_of_Personal_Consciousness.pdf
16. J. Gharajedaghi, *Systems Thinking: Managing Chaos and Complexity* (Butterworth Heinemann, Boston, 1999), pp.33–36
17. A. De Geus, *See for example the discussion of de Geus on this topic, in The Living Company* (Harvard Business School Press, Boston, 1997), pp.31–34
18. E. De Bono, *Mechanism of Mind* (Pelican Books Ltd., Middlesex, 1969)
19. E. De Bono, *PO: Beyond Yes or No* (Pelican Books Ltd, Middlesex, 1972), pp.53–55
20. E. Harmon-Jones, C. Harmon-Jones, Cognitive dissonance theory after 50 Years of development. Zeitschrift für Sozialpsychologie **38**(1), 7–16 (2007)
21. L. Festinger, *A Theory of Cognitive Dissonance* (Row & Peterson, Evanston, 1957)
22. W. Ulrich, *Critical Heuristics of Social Planning* (Wiley, New York, 1994), pp.372–417
23. W. Ulrich, Op cit: 19–23 (1994)
24. K. Lewin, *1948 Resolving Social Conflicts; Selected Papers on Group Dynamics* (Harper & Row, New York, 1948)
25. C. Argyris, D.A. Schön, *Organizational Learning: A Theory of Action Perspective* (Addison-Wesley Pub. Co., Reading, 1978)
26. R.L. Ackoff, *Creating the Corporate Future* (Wiley, New York, 1981)
27. J.J. Gharajedaghi, *Systems Thinking: Managing Chaos and Complexity* (Butterworth Heinemann, Boston, 1999)
28. C.W. Churchman, *The Systems Approach and Its Enemies* (Basic Books, New York, 1979)
29. P.B. Checkland, *Systems Thinking, Systems Practice* (Wiley, Chichester, 1981)
30. American Humanist Association, *Humanism and Its Aspirations: Humanist Manifesto III* (2013). https://americanhumanist.org/what-is-humanism/manifesto3/

31. G. Carlton, *What are the Humanities?* (2020). https://www.bestcolleges.com/blog/what-is-humanities/
32. F. Bacon, *Novum Organum* (P. F. Collier & Sons, New York, 1902)
33. C. Norman, This is in reference to the smithing Greek god Hephaestus who served as the blacksmith of the gods, and was worshipped in the manufacturing and industrial centres of Greece, in *The God that Limps: Science and Technology in the Eighties* (WW Norton, New York)
34. G. Vickers, *Freedom in a Rocking Boat: Changing Values in an Unstable Society* (Penguin Books, Harmondsworth, 1970), p.17
35. This is the implication of using the word "cosmos" to describe the universe.
36. A. Smith, *An Inquiry Into the Nature and Causes of the Wealth of Nations* (W. Strahan, London, 1776)
37. L. Walras, *Elements of Pure Economics; The Theory of Social Wealth* (Taylor & Francis, Milton Park, Oxfordshire, 1874)
38. I. Prigogine, I. Stengers, *Order Out of Chaos: Man's New Dialogue with Nature* (Bantam Books, New York, 1984)
39. J. Jaros, M.M.E. Dodd, The name of the devil is sub-optimization. World Futures. **44**(1), 33–70 (1995)
40. J.C. Smuts, *Greater South Africa: Plans for a Better World*, p. 132 (Truth Legion Publisher, Johannesburg, 1940)
41. J.C. Smuts, *Op cit*: 2 (1926)
42. J.C. Smuts, *Op cit*: 137 (1940)
43. J. Lovelock, *The Revenge of Gaia: Why the Earth Is Fighting Back—and How We Can Still Save Humanity* (Allen Lane, Santa Barbara, 2007)
44. R. L. Ackoff, In Spies, PH 2009; *Russell L Ackoff: Educator, Management Consultant, Systems Thinker.* Stellenbosch, Institute for Futures Research. Strategy Insights. Vol 7 No 3. December 2009: 3
45. For a history of how, and why, Forrester developed this methodology read: Forrester. Jay W.1989. The *Beginning of System Dynamics.* International Meeting of the System Dynamics Society Stuttgart, Germany July 13, 1989. https://web.mit.edu/sysdyn/sd-intro/D-4165-1.pdf
46. S. Beer, *Brain of the Firm* (Wiley, New Jersey, USA, 1972)
47. N. Wiener, *Cybernetics or Control and Communication in the Animal and the Machine (Second edition)* (MIT Press, Cambridge, Massachusetts, 1962)
48. D.L. Gilstrap, Strange Attractors and Human Interaction: Leading Complex Organizations through the Use of Metaphors. Complic. Int. J. Complex. Educ. **2**(1), 55–69 (2005).
49. J. Gharajedaghi, Op cit: 29–55 (1999)
50. D.H. Meadows, Places to in a system: Strategic levers for managing change in human systems. *Whole Earth Winter 1997.* https://donellameadows.org/archives/leverage-points-places-to-intervene-in-a-system/
51. S. Inayatullah, *Causal Layered Analysis (CLA) Reader Theory and Case Studies of an Integrative and Transformative Methodology* (Tamkang University Press, Tamsui, Taipei, 2004)

Chapter 3
Natural Philosophy and the Humanities

Gary Patterson

Abstract Humans have been observing the physical world and trying to make sense of it for many millennia. Eventually this activity was organized into a coherent practice known as Natural Philosophy. Humans have also observed one another and tried to understand human culture. One definition of The Humanities is the study of human creativity. A persistent notion in academic circles is that Natural Philosophy and The Humanities are from two different cultures and that there can be no useful dialogue between them. The goal of this chapter is to examine the nature of Natural Philosophy and The Humanities and to demonstrate by examples and discussion that both areas of human scholarly activity would benefit from considering the insights of the other approach. A selection of academic sub-disciplines from The Humanities will be discussed in detail, and areas of complementarity with Natural Philosophy will be explored. The author has made a career of both Natural Philosophy and The Humanities and will present a series of biographies of other historical figures that followed this path.

Keywords Natural philosophy · Humanities · Critical realism · Mental realism · Human meaning

3.1 Introduction

Modern humans, *homo sapiens,* have lived on the Earth for more than 100,000 years. For most of this time they have been wanderers; they hunted and they gathered. But they also observed the world around them and left physical evidence that they had existed. The cave paintings at Lascaux [1] are vivid reminders that visual art was a part of human culture by 30,000 BCE. They are also evidence that humans were beginning to think about more than food and safety. One of the things that distinguish

G. Patterson (✉)
Carnegie Mellon University, Pittsburgh, USA
e-mail: gp9a@andrew.cmu.edu

Science History Institute, Philadelphia, USA

L. Mammino (ed.), *Insights into the Relationships between Humanities and Sciences*,
https://doi.org/10.1007/978-3-031-97900-2_3

modern humans from other animals is the creativity that produces **meaning** through art [2].

We live in a world of sounds and humans can vocalize over more than four octaves. Singing is as human as it can get. While the melody of solo songs may inspire the singer and the audience, two or more humans can make harmony. Human vocalization also enables sophisticated sounds intended to communicate meaning more complicated than danger or food. The invention of spoken human language is one of the most important periods in human history. However, there is very little agreement on this subject in the academic communities that study it [3]. One of the goals for The Humanities is to develop a method for forging consensus on issues that matter for all of humanity.

In addition to the world of sensory perception, humans have a "life of the mind." They can remember the past, think about the present, and envision the future. Such a function does not require words, it can be carried out with images and remembered sounds, but it is greatly aided by formal language. Humans can also invent stories about the past, the present and the future. Ancient mythology played an important role in the evolution of human culture [4]. The full range of "human stories" is both rich and complicated. It will take all the power of The Humanities to harness this material for the benefit of present and future humanity.

As human culture evolved, some individuals were allowed to devote their time to higher level activities. They both invented and guarded the "tales of the tribe." They mediated between the harsh realities of human existence and the palliative measures that were discovered or invented. These individuals were called names like shaman, priest and wise man. One of the most important aspects of the history of human culture is the emergence of religion. While academic disciplines are agnostic with regard to the existence of a "spiritual world," the formal study of religion considers all aspects that can be either observed directly or inferred from discovered artifacts [5]. Another class of humans devoted their "free time" to the study of the physical world. They invented the wheel and the lever. They developed the technology that allowed humans to flourish. The human activity now known as natural philosophy is very old, just as religion is ancient [6].

Another step in human cultural evolution was the invention of written language. The understanding of human formal languages continues in the present. In addition to local dialects that characterize local communities, more universal languages were adopted to facilitate trade and allow sophisticated international communication. The Greek language became a highly dispersed communication medium after the age of Alexander. Greek is also an important language today. Latin followed the Roman troops that ruled the Mediterranean world. Latin remains an important part of some human sub-communities today. In the present, English has become the worldwide *lingua franca.*

The study of written documents is central to The Humanities. It is also an essential part of the practice of Natural Philosophy. Both communities have developed highly technical vocabularies that are opaque to the general public. Academic sub-communities, even within science or the humanities, also often cannot talk to one another. In the seventeenth century there was a strong movement to create a universal language, other than Latin, that would allow communication between all "educated" humans. John Amos Comenius (1592–1670) was a leader in this effort. (Comenius was a Moravian Bishop who promoted an irenic approach to both natural philosophy and religion. A good introduction to his work is *The Great Didactic of John Amos Comenius* [7]) One of the goals of this chapter is to encourage all scientists and humanists to seek real communication, both verbal and written. (While the term scientist was invented in the nineteenth century and is more familiar today, I will continue to use the more specific term, natural philosophy, to refer to the process employed by every real scientist and hopefully all humanists as well.)

While a focus on texts is important, there are many other aspects of human life and culture that transcend this modality. Humans have been creating solid figures for a very long time. Some of them are very large, like the Sphinx in Egypt or the Celtic stone circles in England. Even larger objects, such as the pyramids in Egypt, challenge our understanding of the creativity of ancient civilizations. In the twenty-first century there is a tendency to view "all things modern" as better than all that went before. One of the goals of The Humanities is to gain an understanding of human culture over time and geography that includes the whole story of humankind. In order to do this, a multidisciplinary approach is needed. It is time to "take a fresh look" at the acquired data. The modern practice of natural philosophy includes the creation of teams of scientists with different expertise. One of the goals of this chapter is to encourage humanists to adopt this "team oriented" approach.

In this chapter the formal system known as Natural Philosophy will be considered and it will be related to some of the academic communities known as The Humanities. This collection of academic disciplines is now quite extensive. One of the persistent notions of twentieth century thought was that the academic disciplines known as Science constituted a separate world from The Humanities, the Two Cultures position discussed by Snow [8]. Baron Charles Percy Snow (1905–1980) was both a physical chemist and a novelist, as well as an essayist. He served in many parts of twentieth century Britain, including industry, government and the academic world. He strongly encouraged all educated humans to be literate in both worlds, so that the cross-fertilization of the two communities would benefit humankind. One of the goals of this chapter is to demonstrate that progressive scholarly activity requires the full power of both Natural Philosophy and The Humanities to achieve understanding of human reality. The next section will outline the nature of Natural Philosophy, as practiced by scientists such as Sir John F.W. Herschel (1792–1871) [9]. Herschel was the son of the famous astronomer, William Herschel (1738–1822), and completed the map of the night sky as viewed from both hemispheres of the Earth. He was a Fellow of the Geological Society of London, as well as a serious chemist [10]. He published the excellent *Preliminary Discourse on the Study of Natural Philosophy* (1851). It summarizes all the principles still needed to understand this approach to physical reality.

3.2 Natural Philosophy

Humans have been observing the world in which they live as long as there has been *homo sapiens.* A critical point in human evolution was the ***permanent recording of deliberate observations***. One field that exemplifies this activity is astronomy. Humans have been observing the sky for a long time. Careful recording revealed that many observations are periodic in their time evolution [11]. Once the night sky had been carefully observed, ancient astronomers organized their data into arbitrary patterns, the signs of the Zodiac, that helped other humans to both enjoy the night sky and to navigate on the Earth below. They did not know what the stars were, but they were able to observe them and follow their locations. The concept of location requires the existence of the mathematical field of geometry. The concept of time requires the mathematical field of arithmetic. The modern practice of Natural Philosophy employs all the power of mathematics.

The curriculum of the Medieval University included the seven Liberal Arts. They were divided into a *Trivium* of Grammar, Dialectic and Rhetoric and a *Quadrivium* of Arithmetic, Geometry, Astronomy and Music. From the perspective of this chapter, it is easy to see that the Quadrivium included the activity we are denoting Natural Philosophy. Harvard University was founded in 1636 to provide an excellent education for the sons of Puritan worthies. It included Natural Philosophy as part of the Quadrivium [12]. By the eighteenth century there was an endowed Professorship in Natural Philosophy at Harvard. The greatest Harvard Professor of Natural Philosophy in the nineteenth century, Josiah Parsons Cooke (1827–1894), was fully committed to this approach and held the Erving Chair of Chemistry, one of the Natural Sciences [13]. In addition to his contributions to the study of the atomic weights of the elements, Cooke was a humanist who considered the importance of the interactions between religion and chemistry [14]. He devoted an entire chapter to the "necessary limitations of scientific and religious thought. The Dean of American historians of science, I. Bernard Cohen (1914–2003) of Harvard, called Cooke "the first university chemist to do truly distinguished work in the field of chemistry." [15]

The discipline of Natural Philosophy is founded on the practice of observation. The full collection of both actual and potential observations is denoted Nature. To be included in this set, an observation should be reproducible and should be verifiable by any trained observer. The process of training someone to be a keen observer of Nature is rigorous and challenging, and usually requires a teacher who is practiced in the art. While good books on Natural Philosophy do exist, and will shortly be discussed, it is important to note that direct human-to-human communication is essential to become truly proficient.

In order to be of value in the prosecution of Natural Philosophy, the observations need to be recorded. One of the most famous products of natural philosophy is the book, *Micrographia* (1665), by Robert Hooke (1635–1703).

Robert Hooke became one of the leading members of the Royal Society of London and helped to challenge other natural philosophers, including Newton. He has been called *The Forgotten Genius* [16]. Careful observations made through his microscope were drawn by the architecturally trained Hooke and published for the benefit of his Royal Society friends. He was an accurate reporter! Hooke also made careful observations through his telescope. One of the most important tools of the natural philosopher is the visual representation of actual phenomena. Humans need all the forms of communication to help make sense of physical reality.

Robert Boyle (1627–1691) formed a research team that included Robert Hooke [17]. They made precise measurements of the pressure of a gas as a function of its volume. Another of the primary activities of natural philosophy is the process of ***measurement*** [18]. Without a quantitative measurement, complete with valid limits of uncertainty, it is impossible to create a mathematical theory of Nature. Not all observable phenomena lend themselves to quantitative evaluation, but the goal of all Natural Philosophy is to discover and measure our world.

Once a valid set of observations has been collected, natural philosophers try to discern ways of organizing the data. This requires both mathematical skill and carefully constrained creativity. If the data can be represented by an actual mathematical formula, one of the goals of natural philosophy has been attained. One way to proceed is to try and identify the independent, measurable, quantities that constrain the observed quantities. One of the tenets of Natural Philosophy is that the physical world can be understood in terms of a set of measurable properties. When successful, the observed variable can be expressed in terms of other measurable quantities in a closed form. This is called a ***phenomenological theory***. Good examples of such successes are Thermodynamics, Hydrodynamics, Maxwell's Equations of Electricity and Magnetism, and Classical Optics. Such subjects are still taught as part of a scientific education, at least to physicists and chemists.

While there was a sense in the nineteenth century that it was time to "declare victory," no adventuresome natural philosopher was willing to abandon the goal of understanding the **macroscopic behavior** of the observable world in terms of the **microscopic processes** that determined the macroscopic behavior. James Clerk Maxwell (1831–1879) also was one of the developers of the Kinetic Theory of Gases [19]. This paradigm proposes that both the Thermodynamic and Hydrodynamic properties of gases can be understood in terms of the microscopic particles that make up the gas. He derived the distribution of particle velocities as a function of particle mass and a new variable: ***Temperature (T)***. His collaborator in this endeavor was Ludwig Boltzmann (1844–1906). In spite of the fact that Maxwell and Boltzmann were correct, many leading academic physicists resisted their theory. This led Boltzmann to commit suicide. Alas, all natural philosophers are human.

The Kinetic Theory of Gases is one of the most productive paradigms in science, but it is neither the beginning nor the end of the story [20]. The phenomenological pressure function for a gas, P(n,V,T), is quite complex and includes a special condition, known as the gas–liquid critical point, where the two phases are identical (the variable n is the number of moles of gas, and V is the volume) (a material phase is a homogeneous region at equilibrium with its external constraints. The liquid phase is an example). The shape of the pressure function implies that the gas particles interact with one another according to a complex potential. (A physical potential is the derivative of the energy of the system as a function of the geometric arrangement of the particles.) Modern gas–liquid theory is a triumph of Natural Philosophy [21]. The observed macroscopic properties can be calculated entirely in terms of microscopic, measurable quantities. The microscopic world is shown to be both real and understandable, at least to a point. One of the key concepts in Natural Philosophy is that the job of the natural philosopher is **never done**.

In an attempt to bring meaning to complex phenomena, some historical natural philosophers invented fictive substances or interactions that could account, at least imaginatively, for the observations. Other clever natural philosophers invented mathematical descriptions of the phenomenon that were untethered from any actual measurements. But the goal of "pure" Natural Philosophy is to produce understanding of a particular observation in terms of other particular observations. The spirit of Natural Philosophy is rooted in **measurable reality**. This implies that a philosophical stance of material realism is required. Another aspect of Natural Philosophy is its acknowledgement of the **partial nature of its conclusions**. In the end, there is no such thing as "Scientific Truth." At best, natural philosophers can claim that their theories explain all known observations, at least within the limits of measurement error. (All actual measurements are uncertain. Precise measurements directly acknowledge this fact.)

While there are well-defined steps in the protocols of Natural Philosophy, the actual trajectory of any actual investigation is uncertain. It takes considerable imagination to determine that an important artifact has been observed. It takes even more creativity to plan a search for related phenomena. As data are collected, it takes massive focus to recognize the "signal" in the "noise." Only near the end of the journey does the "big picture" become more than a "shadow in the mist." Once enough tentative understanding is obtained, the search for the other phenomena that influence the particular phenomena of interest can be initiated. Considerable intuition is required to make progress at this stage. After many trials and false leads, eventually a phenomenological theory may emerge from the data. But it is not automatic. A thorough knowledge and understanding of mathematics is required at this point. Many useless mathematical theories have been created at this point in the investigation. The form of the equation may contain false characteristics, especially discontinuities. Many hapless scientists have been lured to their professional death by focusing on mathematical artifacts. Some "theorists" prefer to ignore the need for a sound phenomenological theory and immediately concoct some form of fictive theory involving non-existent quantities that can neither be observed nor measured. Natural Philosophy requires both audacious creativity and strong discipline. Even then, there is no guarantee that Nature will give up its "secrets." Only the faith that the physical world is both comprehensible and representable by mathematical expressions drives a natural philosopher on to success. Nevertheless, **failure** is the normal outcome of this exercise. Natural Philosophy is hard, and there is no promise that a search will succeed.

The practice of Natural Philosophy requires a commitment to physical reality that exceeds other human desires for recognition. All known theories have "limits of validity." One essential part of the programme of natural philosophy is to determine experimentally and theoretically what these limits are. Unless these commitments are honored, humans might be tempted to treat some theory as absolute. The willingness to allow natural philosophy to devolve into ideology needs to be resisted. There is an explicitly **ethical** dimension to Natural Philosophy.

All natural philosophers are humans. Humans doubt! The fear of failure often prevents the scientist from risking her reputation by proposing explanations at an early stage, so that it can be tested and either confirmed or refuted. Real Natural Philosophy is brutal and most humans are not psychologically capable of pursuing it to the end. But, when a real success occurs, the ecstasy is unimaginable.

This essay is a personal account of more than 60 years of practice as a Chemical Physicist. It has been informed by dozens of books on the history and philosophy of science, but the current form is the creation of the author.

3.3 The Humanities

The human activity known as **The Humanities** grew out of a specifically academic milieu: the Latin term was *studia humanitatis*. This involved the "education befitting a cultivated man." Before entering the University, students were expected to master Latin. Even in the North American wilderness, students could not enter Harvard unless they were Latinate and male.

In the twenty-first century, The Humanities are "academic disciplines that study aspects of human society and culture." One of the goals of this chapter is to explore the use of concepts from Natural Philosophy to guide and strengthen The Humanities, and to consider the importance of the human factor in all areas of science and technology. While the two cultures of C.P. Snow do actually exist, there is no essential reason that it must be so. A current arbitrary list of communities that fall under the umbrella of The Humanities includes: Anthropology, Archaeology, Classics, History, Linguistics, Literature, Philosophy, Religion, Performing Arts and Visual Arts. In this chapter I will address some of these communities and review how they would benefit from including a natural philosophical approach along with more traditional protocols. I will also assess how the insights that exist in these communities would benefit the prosecution and performance of the sciences. Perhaps it is time to wed the two cultures to produce an informed scholarship based on verifiable axioms and proven protocols.

One place on Earth that exemplifies this approach is the Science History Institute in Philadelphia [22]. In addition to the finest chemical library in the world, scholars address both historical and current issues. Scientists and Humanists collaborate to enlighten historical studies and to inform current policy. Most current human problems are too complex for naïve approaches based on ideology or ignorance. Now is the time for groups of scholars to join forces for the benefit of humanity.

One of the most important concepts of The Humanities is the existence of human meaning: How does the human world in which we live make sense? It is an article of faith that it does make sense, but if the world does not make sense, there is no point to The Humanities at all. Meaning implies **relationships** between observable aspects of human culture. The earliest evidences of such attempts at meaning took the form of visual art. In a more sophisticated form we observe hieroglyphics (the solution to the problem of the Rosetta Stone, a device used to translate between hieroglyphics, Demotic Egyptian and Greek script, was found by Natural Philosophers of the nineteenth century [23]). Visual storytelling remains an important part of human culture.

Unlike a rock, humans are constantly changing. (Rocks do change too, but on a different time scale.) One form of change is denoted **emotion**. Vibrant human stories include explicit reference to the emotional states of the actors. One of the goals of The Humanities is to develop a model for human emotion and for the interactions of humans with one another. One of the purposes of emotion is to produce action. A folksy expression that incorporates this reality is: "Without emotion there ain't much motion." Trying to understand observed human actions without taking into account the full state of the actors is confusing. But external observations of humans are opaque, unless some theory of humanity exists. One form of analysis takes advantage of non-verbal cues such as body language to interpret actions. Human verbal acts can often be used to obscure the true nature of the physical or emotional act. Precision in the analysis of human behavior does not exist, but mindful experience can produce some insight. This is one of the great challenges of The Humanities. Much of what makes us human is "hidden."

Humans are a social species. We tend to associate with other humans. The sum total of human interactions we denote as **culture**. Observation of human communities, both historically and at present, reveals extreme diversity. The ability of humans to adopt a very wide range of group behaviors is part of our evolutionary legacy. One of the great challenges of The Humanities is to deal with this cultural heterogeneity. Different groups of humans respond to other cultures in various ways. One approach is cultural isolation. This produces idiosyncratic outcomes such as unique dialects and unusual practices. When two such cultures are thrust into interaction, the usual outcome is **conflict**. Is there another way forward?! One of the goals of this chapter is to recommend both appreciation of other cultures and to encourage multicultural thinking that takes the best practices of historical and present communities and adapts them for the future. While knowledge of other cultures is necessary, it does not automatically lead to productive outcomes. Something else is necessary for human flourishing.

Just as a natural philosopher must respect Nature, a humanist must respect Humanity. Unless the scholarly community is willing to value all humans, no real progress will be made in The Humanities. Human culture is exceedingly complex and only by patient observation and testing will some measure of insight be gained. Provisional paradigms are needed to guide further research. Now is a good time to encourage all scholars to be open to the future.

The essay on the humanities is also a personal statement, rather than a scholarly discussion of the many ways that humanists have behaved in the past. This set of stances is not shared by all academic Humanists. It is also colored by an American provenance and another 60 years of practice. The author takes full responsibility for these positions.

3.3.1 Homo Sapiens *and the Humanities*

Physical science is the study of the material world. One of the presuppositions of this discipline is that the outcome of any observation is independent of time, location and observer. While there are always limits of error associated with any measurement, the belief of a natural philosopher is that further observations will **converge** on a reliable and verifiable result. The Humanities are the study of humans and their culture. This fact immediately creates a dilemma: no two actual humans are the same. Even "identical twins" have unique developmental histories, both biologically and socially. And each actual human constantly changes. How can a scholarly enterprise be carried out under such circumstances?!

In the twenty-first century, the legal identification of any particular person is determined by their DNA profile. A monumental survey of the history and geography of human genes has been carried out [24]. One of the clearest outcomes of this work is that every aspect of the species *homo sapiens* can be characterized by a **distribution** of manifestations. Every actual human is unique, biologically. Broad generalizations usually fail, when subjected to careful analysis. Statistics is one of the most important tools of Natural Philosophy. A focus on the mean of the distribution obscures the potential wide variability of the quantity. Every scholarly person who deals with humans should be competent in Statistics. This does not mean that all Humanists need to be Social Scientists, but if they are going to be able to both assess and contribute to discussions involving humans, they are going to need to be able to understand statistical data and analysis.

The infants of the species *homo sapiens* are not independently viable at birth. They require years of care. They are also born with very little programming in their brain that deals with issues important to human culture. They can immediately "make noise," but it takes years to be able to talk intelligently. They must learn language from other humans. They soon acquire the ability to distinguish shapes of objects in their environment. Human infants seem to have a special skill in identifying "faces" of those closest to them. They spend years learning to distinguish between safe and dangerous objects in their world. While newborn infants are highly vulnerable animals, it takes years to become competent adults. One definition of what it means to be human is the ability to learn and adapt to the ambient environment. Each actual person must adapt to whatever situation it experiences, or they will perish. This need to adapt remains throughout a human lifetime.

While atoms are sometimes expected to remain "the same" over times of the order of 100 years, humans change on a daily basis. Even atoms change as they emit radioactive particles or undergo chemical reactions. The rarest of situations in the material universe is the "static state." One historical theme in The Humanities is the "essential" nature of groups of humans. A notorious verse in the biblical epistle of Paul to Titus states that: "The Cretians are always liars." (Titus 1:12, KJV). Natural Philosophy started to make progress when it insisted that the full set of observations be taken into account when theories that comprehended the phenomenon were invented. Human existence is both more interesting and more complicated than we can imagine, but assuming anything less renders the scholarship of limited value. The commitment to reality on the part of the natural philosopher is just as important for the humanist.

One of the most dynamic scholarly communities in the twenty-first century is the field of cognitive science. Each person arrives at adulthood with a brain wired by both genetics and experience. Most of the activity of this organ is "unconscious." But unconscious thoughts can often influence actual behaviors. The monograph *Blindspot* (Banaji, 2013) discusses how we are influenced by many unconscious attitudes created during our prior life [25]. One of the expectations of Natural Philosophy is that when a particular experiment on a material system is carried out, the result will be predictable, at least in a statistical sense. When such an expectation fails in the laboratory, scientists start to look for "hidden" variables that were not controlled for in the design of the experiment. Human behavior is extremely complicated, and naïve attempts to explain any behavior in terms of a priori causes is likely to fail. While it may not be possible to attain consensus on a "theory of humanity" in the present academic community, that does not excuse the group from the need to try and find one. In the next sections of this paper, I will address specific issues in each humanities discipline that are ripe for study and would benefit from employing both the classic tools of the humanist and the empirical stance of the natural philosopher.

3.3.2 Natural Philosophy and Anthropology

Since the founders of Anthropology viewed it as a form of science, it might be supposed that they would have automatically adopted the stance and methods of Natural Philosophy. The field has become extremely heterogeneous, with dozens of sub-communities. A standard description of the field of Anthropology is that it is "the scientific study of humans, human behavior and human societies in the past and present." At the most scientific end, the subfield of Biological (or Physical) Anthropology examines the full range of historical and current human populations in terms of their biological characteristics. When this is done, it is clear that, taken as a worldwide group, humans are continuously distributed with regard to almost every measurable characteristic.

The irremediable aspect of Anthropology is its subject: humans. When a chemist interrogates a molecule, it just does "what comes naturally." When an anthropologist interrogates a living (or dead) human, the problem of deception immediately arises. Humans are notoriously unreliable when it comes to both observation and reportage. While "eye-witness" testimony is revered in a judicial setting, it has been shown many times that untrained observers are often distracted by irrelevant factors and then invent narratives to make sense of the situation. The party game of "telephone" reveals that, even when there is no need to outright lie, humans are still bad reporters of what they have heard. With such an unreliable subject, how do cultural anthropologists sift through the "noise" to obtain accurate data? Modern anthropologists know this problem well and are committed to "advancing anthropology as the science that studies humankind in all its aspects, through archeological, biological, ethnological and linguistic research" [26]. In spite of a history that demonstrates only too well that all anthropologists are humans, the current community strives to develop reliable methods and to train scholars that seek verifiable narratives. It is an explicitly multidisciplinary community that seeks to form alliances with all relevant specialties.

Since human behavior is one of the essential issues in anthropology, collaboration with the field of behavioral psychology is obvious. Great progress has occurred since this field acknowledged that human behavior is exceedingly complex. Certain aspects of human behavior are encoded in the genetic makeup of the individual. But human behavior is not just "unconscious;" actual people can choose consciously to act in ways that make sense. Other "fixed" variables occur during gestation and early childhood, such as fetal alcohol syndrome. Irreversible changes during development can be palliated, but they cannot be ignored.

Humans are complex chemical systems. Imbalances, such as vitamin B_{12} deficiency, have marked influences on human behavior. Anthropologists have become very sensitive to the effects of nutritional imbalances on individual humans. They can also play an important role in informing current human societies of the need to use accurate information to regulate food and nutrition. Cultural anthropology has documented societies where certain dietary practices are both beneficial and harmful.

Humans are also complex biological systems. Cultural practices can either benefit or damage human physiology. Some societies demonstrate a large occurrence of heart disease, or kidney disease, or lung disease. The current state of the scientific and medical field of human anatomy and physiology is very advanced. Partnering with cultural anthropologists is a natural marriage.

Humans form individual associations; we are not, by nature, a solitary species. The anthropology of the family has always been an important area [27]. Some current societies have seriously eroded both the relevance and the importance of the extended family. Anthropology can play an important role by elucidating the role played by the family in the history and flourishing of humanity. The issue is not intrinsically an ethical or religious one; how have families contributed either to the problems of humanity or to its survival and progress? Once the range of demonstrated outcomes has been obtained, the way forward can be formulated in practical terms. I sense a serious interest in the present anthropology community in making the field a clear benefit to all of humanity.

3.3.3 Natural Philosophy and Archaeology

The roots of modern archaeology are related to the Natural History movement of the seventeenth and eighteenth centuries [28]. In addition to the fossils and current arcana, eager foragers identified artifacts that were created by humans. A modern definition of archaeology is "the study of human activity through the recovery and analysis of material culture."

Since Archaeology is associated with antiquarians like John Aubrey (1626–1697) and the Royal Society of London, it is no surprise that it strives to achieve the protocols of Natural Philosophy. While amateurs muddle important sites, and looters steal important artifacts, archaeologists "protect, preserve, collect and analyze material evidence of human activity." [29] The professional discipline has adopted the latest technology to assist in this effort, including sophisticated computer tools and drones to conduct aerial surveys.

The need for this discipline, if any, is to make sure the rest of the Humanities are aware of the current state of archaeology. No accurate picture of human culture is possible without the results of archaeology. Multidisciplinary teams are the way forward for the Humanities, and whenever material culture is an important part of the project, a professional archaeologist is essential.

While the methods and tools of archaeology were developed to examine ancient sites, a real opportunity for the present is "urban archaeology." Careful surveying, excavating, curating and analyzing are required to create true understanding of any human site.

Another area where professional archaeologists can be of help to current human culture is in the curation and preservation of artifacts of national or world significance. Many current nations are now interested in the material artifacts created by their culture, and they "want them back." Archaeologists can help to put such issues in their proper context and to mediate resolutions of conflicts. The sub-discipline of "legal archaeology" seems like a natural area for growth.

All the work of extracting and analyzing artifacts is not the final goal of archaeology. Understanding how these artifacts help define the human story, both locally and globally, is the desired outcome. Once the story has been created, all of humanity needs to know. Especially since much of archaeological work is now funded with public money, it is important to produce "public archaeology." Lively museums, rather than dusty repositories or private hoards are the goal.

Current leaders of the worldwide community of professional archaeologists have largely adopted the suggestions and ideas contained in this brief essay. A good example of this is the book, *The Archaeological Process* (1999) by Ian Hodder, the Professor of Archaeology at Stanford University [30]. Archaeology is hard, and technical experts are required. But only teams of skilled workers can achieve the level of field work needed to produce truly useful artifacts. In order to provide the necessary context for any artifact, the full power of the Humanities is required: history, philosophy, religion, anthropology, etc. Understanding the creation of meaning (hermeneutics) is essential as the archaeologist works with sites in the field. The degree of well-informed creativity required by archaeology is very high. The current state of the actual Earth is so complicated, both by physical evolution and by human activity, that naïve approaches are certain to fail. Hodder recommends being realistic about archaeological complexity, but optimistic about the ability of great archaeologists to surmount the difficulties. He discusses in great detail the types of mental processes used in archaeology. All the power of human thought is required to succeed. Rigid approaches usually fail.

One of the most important insights into the scientific method is the need to be trained to observe. Untrained observers often see nothing at all. This issue is critical in archaeology. Hodder also discusses the importance of scale. Modern archaeology uses drone surveys to gain a "bird's-eye view" of the site. But, this important scale can both reveal and obscure important details. Some archaeologists focus on the finest products of the sieving process. The fact that they are studying a human site is ignored. Sophisticated chemical analysis often reveals dramatic details that upset the whole theoretical picture envisioned for the site. Archaeological work is "never done." This perspective is central to all of the Humanities, and leaders like Hodder encourage the kind of thinking that should be adopted in the "world of humans."

Hodder summarizes his approach to archaeology in terms of four key themes: reflexivity, contextuality, interactivity and multivocality. By reflexivity he means the kind of mindful research that is constantly assessing the current state of the project. All good science works like this, and archaeology can be viewed as a science in this sense. The goal of archaeology is not just a blind discovery and cataloguing of artifacts. In order to create a compelling story, all objects need to be placed in context; geographically, culturally and chronologically. Archaeology is not carried out in a scholarly silo. It is important to be aware of and to put into practice the full range of other scholarly work, both scientific and humanistic. And, archeology exists in a global context. Different scholars from different cultures, both human and technical, may read the data differently and tell the story in another way.

The time has come for archaeology to truly "come of age." It should be integrated into the grand project of The Humanities: to tell the human story in all its richness and variety. It should be open to all the help it can get and receptive to the best ideas created by others. Leaders like Ian Hodder are practicing this kind of archaeology.

3.3.4 Natural Philosophy and Human History

Within the context of this chapter, the treatment of History will be restricted to the human story. The goal of such history is the creation of an accurate and meaningful story of the evolution of humans and human societies over the past aeons. The task begins with the collection and analysis of "**primary sources**" of information. They could be manuscripts, archaeological artifacts or other evidences of human activity.

One of the largest tasks is to envision the context for the evolution of the story. Human stories take place in particular places, both physically and socially. Human activities range from the basics of animal existence, (food, clothing and shelter), to highly sophisticated programs of technological or political adventure. One concept that is useful in the practice of History is called the **hermeneutical spiral**. As data is collected, the historian begins to envision how it was that humans could have engaged in such activities. These partial pictures are then used to probe both past and future discoveries. The historian's job is never really done. Only provisional histories are ever actually possible.

One issue involves an incomplete data set. Most of the past has vanished from human view. How can we be sure the small set of observations we make in the present are **representative** of the actual past? Even given the reliable but limited information, many different stories could be told. How do we know that our story is within the bounds of reality and accuracy? The basic stance of the historian needs to be audacious humility.

Natural philosophers face all these issues with material artifacts. One potential difference is the ability to probe the material world in the present to try and replicate artifacts from the past. One of the most exciting areas of current Natural History is **replication** [31]. Scholars with expertise in both science and history are solving the secrets of the past by replicating them in the present.

The practice of Natural Philosophy often leads to a set of **axioms**. These are then treated as absolute, and conclusions are drawn, as if it was mathematics, like geometry. The actual highly contingent nature of these conclusions is hidden from later readers, and the human story that went in to the adoption of the axioms is obscured. The missing element is a commitment to Natural Philosophy as an explicitly human activity. While all humans are highly flawed, and hence should proceed cautiously and humbly, we have no other choice. All historians are human.

While it used to be said that all texts are created by humans, this is no longer the case. Nevertheless, no text can be accepted uncritically, without further corroboration. The history of Alchemy is replete with texts. There are even extant artifacts, such as "alchemical gold." (No, they are not Au.) Antimony cups are prized museum artifacts. But determining the actual author of these texts has provided one of the most challenging aspects of scholarly work. One of the triumphs of this enterprise is the analysis of the *Summa perfectionis* [32]. This thirteenth century alchemical text was traditionally attributed to one of the most famous names in Arabic alchemy, Jabir ibn Hayyan or Geber. The attribution of alchemical texts to either actual, but ancient, authors, or to mythical characters, such as Hermes Trismegistus, was very common. Since much of alchemy itself was fictional, including a bogus author is hardly the worst of its sins. Newman was able to show, by careful reading of hundreds of alchemical texts in many different forms of Medieval Latin, that the actual creator of the *Summa* was Paul of Taranto, a monk in Assisi. The really hard work of discovering the provenance of a text is essential to accurate humanities scholarship.

Along with establishing the authorship of a text, no text arrives magically. There is always a human context for the creation of a text. Who was the intended audience, if any? Where was it written? When was it written? What other documents are part of its literary context? This is more hard work. The notion that the Humanities are "easy" was debunked long ago. But how many modern students realize the incredible challenge associated with the practice of The Humanities?

The marriage of Natural Philosophy and The Humanities has been achieved in current practice. One of the best examples is the recent book, *Newton the Alchemist: Science, Enigma, and the Quest for Nature's "Secret Fire"* [33]. Newman is one of the principals in the "Newton Project."[34] A large corpus of verified Newtonian writings (often from manuscripts now collected in major libraries) have been digitized and made available to the scholarly community. This allowed Newman to use all the power of digital humanities to analyze the texts. Newton was one of the best laboratory chemists of the late 17th and early eighteenth centuries. Newman and his collaborator Principe have reproduced many of the experiments gleaned from Newton's actual laboratory notebooks. Isaac Newton was a voracious reader and made notes of the books he read. Newman has also read these works and used the full alchemical corpus to place Newton's work in its full literary context. The resulting book is a "tour de force." It sets a standard for the History of Science that should inspire a new generation of scholars.

3.3.5 Natural Philosophy and Linguistics

The Linguistics Department at Carnegie Mellon University features research in **computational linguistics**. Words are well-adapted to the practice of digital Humanities. While the discipline that studies human language, spoken, written and acted, is a member in good standing in the Humanities, it has embraced Natural Philosophy as a valuable approach.

One of the most dynamic departments at AT&T Bell Laboratories during its final years (1970–1984) was the Speech and Communication Research Department. They employed computer tools to both synthesize speech and to analyze human vocalizations. One of the tools in this work was the concept of a **phoneme**, a set of unique vocalizations that can be found in all human audible communication.

The current state of the academic field of Linguistics is highly heterogeneous. While sophisticated tools are employed and rigorous criteria are applied, human language is still a creation of actual communities of humans. This introduces considerable variability in any "universal" axioms.

One of the goals of serious philosophy is to create an "exact" language that allows mathematical precision in the statement and discussion of propositions. Kevin Kelly, the Director of the Center for Formal Epistemology at Carnegie Mellon University has made striking progress in this quest [35]. The "holy grail" of a universal language was very much a part of the seventeenth century world of Francis Bacon (1561–1626) and Samuel Hartlib (1600–1662). But even Natural Philosophy is always carried out by humans and the "telltale heart" cannot be hidden.

Humans also communicate using symbols and actions. The formal study of this human activity is called **Semiotics** [36]. Many forms of communication are unconscious, such as "body language." There is considerable scope for studying the actual vocalizations that transcend the written version of the speech act. Humans are both more complicated and more interesting than algebra. A careful balance between digital technology and human analysis promises great progress in this area.

3.3.6 Philosophy

Modern Philosophy encompasses an enormous range of human activities [37]. While it has roots in antiquity, and owes an incalculable debt to a worldwide group of previous philosophers, the work of this discipline is even more needed today. One of my favorite books is by the Greek historian Diogenes Laertius (3rd Century): *Lives and Opinions of the Eminent Philosophers.* Like Gaul, he divided philosophy into three parts: natural, ethical and metaphysical. Ethical philosophy is hard. It is rooted in the deepest parts of human experience and requires sifting through millennia of human thought and behavior [38].

An even deeper quandary is associated with the field of metaphysics. While Natural Philosophy deals with the material world, metaphysics is concerned with nonmaterial issues: existence, causation, and mind. In the last century, many philosophers of mind asserted that human cognition was an epiphenomenon with no verifiable reality. One of the most positive developments in the twenty-first century is the acknowledgement that, while human thoughts may not correlate with anything happening in the external world at the time, they are real phenomena and must be analyzed as such. This stance of "mental realism" offers humans the chance to understand humanity at a deep level [39].

The existence of the mental world is what makes us human. Research in neuroscience has demonstrated that our chemistry and our physiology are influenced by our mental states [40]. This implies that philosophers should engage with questions related to human "**will**," the influence of thoughts on actions.

Another human emotional response is called "**awe**." It can occur from many forms of sensory input, but a real view of the night sky, undisturbed by adventitious light pollution, often produces awe. The sight of the Grand Canyon in Arizona or a Giant Redwood tree in California can also produce emotions that can be "felt." The stance of emotional realism is also an important part of The Humanities.

3.3.7 Natural Philosophy and Religion

One of the earliest types of evidence of human activity was burial of the dead. One of the things that "make us human" is the ability to envision our own demise and the respect shown to those who have died. Independent of the existence of the "spiritual world," including the existence of a divine being or beings, human religious communities do exist. And the academic study of this human activity constitutes one of the enduring fields of The Humanities.

Long before formalized religions were organized and codified, families revered their ancestors. Family traditions and values were attached to previous patriarchs or matriarchs and practiced in the present. One of the enduring themes in Religious Studies is the importance of the family for human thriving. Divine beings were envisioned as Fathers and Mothers. Family values like Brotherhood were extended to larger groups, such as the clan or tribe. The existence of families and family groups introduced social boundaries that protected the tribe. Eventually, certain families were entrusted with keeping and enforcing the ethos of the clan.

The principles of Natural Philosophy emphasize observing actual human activities. Religious practices are extensively documented, both historically and at present. One of the hardest issues with regard to religious data is verification. The religious scholar must observe a religious community for a very long time in order to sift the wheat from the chaff. Nevertheless, some scholars have devoted their lives to trying to catalog and verify the range of human religious activities. The greatest religion scholar of the twentieth century was Huston Smith (1919–2016) [41]. His *The World's Religions* (2009) is the gold standard for an irenic presentation of the current international religious world.

One of the central purposes of The Humanities is to provide a metanarrative for humanity that includes meaning and purpose. It is no surprise that Huston Smith is also the author of *Beyond the Post-modern Mind* (1998) [42]. The academic assault on meaning and purpose is antithetical to everything he devoted his life to. One of the central purposes of Natural Philosophy is to provide a metanarrative for the physical world that illuminates its nature and history. One of the greatest twentieth century natural philosophers was John Polkinghorne. In addition to outstanding work in particle physics (*The Particle Play*(1979)), he is the author of many books exploring the relationship between Science and Religion. He received the Templeton Prize in 2002 for his leadership in this field. Another friend from this world is John Hedley Brooke from Oxford University, the former Andreas Idreos Professor of Science and Religion. One of the outcomes from their work is a programme for the irenic interactions of Science and Religion. This same approach could apply to all the Humanities. It starts with the stance of Critical Realism: observable worlds should be accepted as real, but claims of full understanding are unreasonable, since reality is complex and humans are finite. For example, neuroscience improved dramatically when it accepted human thought as real.

3.4 Examples of Natural Philosophers Who Were also Humanists

In 2007 the author had the privilege of giving a series of lectures in China. The philosopher, Melville Stewart, was invited to bring a group of scholars to present papers on the dialogue between science and religion [43] as part of an academic course at Shandong University in Jinan. Similar events took place at Peking, Fudan, Wuhan and Tsinghua Universities. My topic was Natural Philosophy and Religion. Five examples of the type of thinking that benefitted from and gave back to both philosophy and religion were presented.

The first subject was the third century philosopher and theologian, Origen of Alexandria. He was one of the most educated men of his time and lived and worked in the intellectual capital of the Roman Empire. He sought to create a coherent picture of all of reality and his masterpiece, *On First Principles,* is still worth reading today. It is one of the central documents of Neo-Platonism. There are two great themes in this work: Unity and Diversity. He created a metanarrative that included a baroque collection of connected beings, the "Great Chain of Being," but he stressed that the system as a whole was unified. Every being in the world mattered. But there were many different kinds. His system included a vision of ultimate bliss, even though the present is often dominated by evil. Each human person is on a journey, and must be both aware of their environment and ready to adapt to the challenges presented in real time.

While Origen earned his living as a *grammateus,* (a teacher of Rhetoric) he was in constant demand as a mediator. He was especially good at listening to both sides, creating a common space where discussion could take place, and shepherding the negotiations so that both parties recognized the best way forward. This produced a high level of jealousy in his religious superiors. Because Origen was aware of many forms of human thought, he could envision many possible outcomes. He was the ultimate Humanist. One of my own goals as a Humanist is to remain open to my colleagues in the History community so that the best story can be told, and the other stories can be appreciated for their insights, not just rejected as "written by others."

The second figure was Isaac Newton (1643–1727). While most people know he had something to do with Calculus and Classical Mechanics, very few are aware of the breadth of his scholarship. His masterpiece, *Philosophiae Naturalis Principia Mathematica* (1687), set a standard for Natural Philosophy. Interestingly, he was able to derive Boyle's Law of Gases using a fictive model of the gaseous state and an inverse-square law potential of interaction. This illustrates both the creativity needed to carry out Natural Philosophy and the underdetermination of all inductive schemes. Even Newton needed to remain humble.

Newton was interested in all aspects of observable reality, as well as all the tools of scholarship. The subject that consumed more of his time than anything was Alchemy [33]. In order to pursue this activity, he needed to become a Humanist; he needed to be able to read arcane texts in ancient Latin. In fact, Newton learned to read many ancient and modern languages and was one of the most learned men of his era. But just reading was not sufficient to determine the meaning of these texts. Newton became the finest laboratory chemist in seventeenth century England. (His monetary reward for this was to be elevated to Master of the Mint!) Almost all alchemical texts only hinted at the secret of the philosopher's stone. Only by laborious trials could the actual recipe be determined. But one of the goals of Natural Philosophy is to challenge received wisdom and to establish the limits of validity of even the best notions.

One of the best-kept secrets about Isaac Newton was his devotion to religious studies [44]. Even as he was able to proceed from the astronomical observations to a theory of planetary motion, he tried to determine what religious facts existed, and to develop an understanding of human religious history. He developed tools to evaluate ancient texts and to detect frauds. He envisioned alternative scenarios for the evolution of both Judaism and Christianity. His extant writings on these topics are extensive and are housed in Jerusalem. An extract of one of his clearest statements about religion is contained in his Commentary on II Kings [45].

> "To celebrate God for his eternity, immensity, omnisciency, and omnicompetence is indeed very pious and the duty of every creature to do it according to their capacity, but yet this part of God's glory as it almost transcends the comprehension of man so it springs not from the freedom of God's will but the necessity of his nature."

Although God is not "visible" to the naked eyes of humans, Newton believed that He could be discerned by the effects He creates in the observable world. A natural philosopher is always on the lookout for unexplained phenomena, and is open to any coherent explanation that can be articulated and employed to understand a wider range of our world.

The nineteenth century was the great age of Natural Philosophy in England. Michael Faraday (1791–1867) was one of the best, even though he never went to college and did not know the calculus [46]. He is most famous for his studies in electricity and magnetism, but he was officially Professor of Chemistry at the Royal Institution in London. As a natural philosopher, he had a knack for envisioning the physical world in geometric terms. He is most responsible for the entire field of physics that invokes the presence and effectiveness of the concept of **"fields."** (A field is a three-dimensional representation of the potential of interaction between objects.) Such fields can be visualized, as in the behavior of iron fillings around a bar magnet, which follow the magnetic field. Faraday was lionized throughout the scientific world for both his devotion to verifiable phenomena and for his warm personal relationships with all fellow workers in the fields of his endeavour. He had a voluminous correspondence, which has been lovingly curated [47].

While Faraday was one of the greatest scientists of all time, I revere him most for his humanity. A biography that is especially sensitive to this side of his life and character is *Michael Faraday: Sandemanian and Scientist* [48]. Faraday was fierce in his exploration of Nature, but gentle with his interactions with other scientists. He sought to create universally validated phenomena, and encouraged others to probe the limits of validity of any of his ideas. Even though he was subjected to outrageous prejudice by Humphry Davy, his boss at the Royal Institution, including trying to blackball him from election to the Royal Society, he remained calm and open. Hermann von Helmholtz (1821–1894), the great German Physicist and Natural Philosopher, described him as a man of "perfect simplicity, modesty and undimmed purity of character."

The perfect colleague, in so many ways, was James Clerk Maxwell (1831–1879). Like Newton, Maxwell was a brilliant mathematician. Like Faraday, he was a consummate experimentalist. During his time in London at Kings College, he partnered with Faraday to produce the "Maxwell Equations of Electricity and Magnetism." This is one of the high points of classical physics [49]. But Maxwell was more than just a Natural Philosopher. He was accepted as one of the leading lights in the Anglican Church and convened with religious scholars at Cambridge, like Westcott and Hort. (Brooke Foss Westcott (1825–1901) and Fenton John Anthony Hort (1828–1892) were the authors of *The New Testament in the Original Greek*

(1881). Upon Maxwell's early death due to the same cancer that killed his mother, Westcott eulogized:

> "It was impossible to think of him whom they had so lately lost, to whom the first charge of the Cavendish Laboratory had been committed, Prof. Clerk Maxwell, and to recollect his genius and spirit, his subtle and profound thought, his tender and humble reverence, without being sure that the close connection between Physics and Theology which was consecrated by the past was still a living reality among them. That was an omen for the future. He felt, as probably all present felt, that he owed a debt of gratitude to him, both for his researches, and for the pregnant words in which he gathered up their lessons." [50]

A highly visible scientist in nineteenth century England was under pressure to expostulate on many areas outside his expertise. For James Clerk Maxwell, not much was "outside." But he resisted calls for him to "take sides" on many of the controversies in his day. He rested on two verities: physical reality will progressively reveal its secrets, and humans should adopt a provisional stance with respect to all of their opinions. In the scientific arena, Maxwell led the effort to create the kinetic theory of gases, which transcended the merely phenomenological description of the equations of state. In the Humanities, Maxwell championed the role of creativity in the understanding of the role of emotions and thoughts in the life of actual humans. If he had lived, he would have also revolutionized psychology.

The greatest of all the nineteenth century classical physicists was Lord Rayleigh, John William Strutt (1842–1919) [51]. He was educated at Trinity College, Cambridge and had the great humanities scholar, Joseph Barber Lightfoot(1828–1889), as his tutor. Like Maxwell, he was the Senior Wrangler in mathematics. (the highest honor at Cambridge in mathematics) Rayleigh never made a distinction between his natural philosophy and his more general efforts in the humanities. One of Rayleigh's triumphs would be viewed as a failure in some circles. He was the finest classical mathematical physicist of the nineteenth century and calculated the classical optical emission spectrum from a heated "black body." It was flagrantly in disagreement with the emerging measurements of this phenomenon. Rather than just abandon the theory, he promoted the notion that it was the underlying assumptions of classical mechanics that were at fault. Not only was Rayleigh correct, but the knowledge that Rayleigh's calculations were a completely accurate conclusion based on classical theory helped to drive the evolution of thought to quantum physics. One of Lord Rayleigh's best friends was William Thomson (1824–1907), Lord Kelvin. They made a great team and served on many governmental panels. Rayleigh was able to abstract the best of any idea, and Kelvin was able to intuitively detect the errors in any proposition. Between them they created the magnificent edifice of Classical Thermodynamics. But they also cared about the human side of both science and life in general. Their advice to government ministers showed a depth of human understanding. These examples of natural philosophers who were also serious humanists demonstrate that there need be no "great gulf" between these communities.

3.5 Concluding Reflections

Natural Philosophy is a coherent and effective approach to the study of the material world. It is grounded in careful and quantitative observation of physical reality. It arose in the context of the Royal Society of London and its desire to produce an understanding of the world that would benefit society. It can be applied to any form of scholarship that includes observable phenomena and an underlying coherent system of interactions. The benefits of Natural Philosophy have been proven many times, especially in the advanced technologies of the twenty-first century.

However, most of the current problems in world culture transcend the merely technological. Human and societal factors dominate the solution to these issues. The Humanities seek to observe and understand humans and human culture. In so far as it is able to accurately and coherently provide an understanding of our world, it is time for The Humanities to join together to help solve real world problems. So-called applied humanities are both possible and imperative, if the world is going to flourish in the twenty-first century and beyond. Multidisciplinary teams of Natural Philosophers and Humanists are needed to solve human problems. The goal of this chapter is to encourage scholars to join this effort for the benefit of humanity.

While the realities of academic scholarship and the funding of research cannot be ignored, there is a crying need for scholars who will take a more synoptic view of human reality and join with other scholars to solve real problems. Since very few people can be true experts in every skill, the need for teams of scholars is obvious when the goal is to solve problems. In the industrial world, engineers are workers who "solve problems." The primary commitment of a humanities scholar needs to be to the worldwide community of humans. Our future depends on it.

References

1. C. Eshleman, *Juniper Fuse: Upper Paleolithic Imagination and the Construction of the Underworld* (Wesleyan University Press, Middletown, 2003)
2. A.E. McGrath, *Surprised by Meaning* (John Knox Press, Louisville, 2011)
3. S. Pinker, *The Language Instinct* (William Morrow and Co., New York, 1994)
4. M. Eliade, *Mythes, reves et mysteres* (Gallimard, Paris, 1972)
5. H. Smith, *Why Religion Matters: The Fate of the Human Spirit in an Age of Disbelief* (HarperOne, New York, 2001)
6. O. Neugebauer, *The Exact Sciences in Antiquity* (Brown University Press, Providence, 1970)
7. M.W. Keatinge, *The Great Didactic of John Amos Comenius* (A&C Black Ltd., London, 1923)
8. C.P. Snow, *The Two Cultures and the Scientific Revolution* (Cambridge University Press, New York, 1959)
9. J.F.W. Herschel, *A Preliminary Discourse on the Study of Natural Philosophy* (Longmans and Co., London, 1851)
10. R. Brashear, G.D. Patterson, Josiah Parsons Cooke, the Natural Philosophy of Sir John F.W. Herschel and the Rational Chemistry of the Elements, in *150 Years of the Periodic Table*. ed. by C.J. Giunta et al. (Springer, Switzerland, 2021)

11. J. Silk, *The Infinite Cosmos: Questions from the Frontieres of Cosmology* (Oxford University Press, Oxford, 2006)
12. S.E. Morison, *Harvard College in the Seventeenth Century* (Harvard University Press, Cambridge, 1936)
13. J.P. Cooke, *First Principles of Chemical Philosophy* (Sever, Francis & Co., Boston, 1870)
14. J.P. Cooke, *Religion and Chemistry* (Charles Scribner's Sons, New York, 1880)
15. I.B. Cohen, Some reflections on the state of science in America during the nineteenth century. Proc. Nat. Acad. Sci. USA **445**, 666–677 (1959)
16. S. Inwood, *The Forgotten Genius: The Biography of Robert Hooke (1635–1703)* (CMacAdam Cage, San Francisco, 2003)
17. M.B. Hall, *Robert Boyle on Natural Philosophy* (Greenwood Press, Westport, 1965)
18. K. Ferguson, *Measuring the Universe: Our Historic Quest to Chart the Horizons of Space and Time* (Walker & Co., New York, 1999)
19. B. Mahon, *The Man Who Changed Everything: The Life of James Clerk Maxwell* (Wiley, Chichester, 2003)
20. J. Jeans, *An Introduction to the Kinetic theory of Gases* (Cambridge University Press, Cambridge, England, 1940)
21. J.P. Hansen, I.R. McDonald, *Theory of Simple Liquids* (Academic Press, London, 1976)
22. www.sciencehistory.org
23. En.wikipedia.org/wiki/Rosetta_Stone
24. L.L. Cavalli-Storza, *The History and Geography of Human Genes* (Princeton University Press, Princeton, 1994)
25. M.R. Banaji, A.G. Greenwald, *Blindspot: Hidden Biases of Good People* (Delacorte Press, New York, 2013)
26. American Anthropological Association, americananthro.org (2020)
27. K.R. Allen, A.C. Henderson, *Family Theories: Foundations, and Applications* (Wiley Blackwell, Chichester, 2017)
28. B.M. Fagan, N. Durrani, *A Brief History of Archaeology* (Routledge, London, 2016)
29. Society for American Archaeology, saa.org (2020)
30. I. Hodder, *The Archaeological Process* (Blackwell Publishers, Oxford, 1999)
31. L. Principe, *The Secrets of Alchemy* (University of Chicago Press, Chicago, 2013)
32. W.R. Newman, *The Summa Perfectionis of Pseudo-Geber* (Brill, Leiden, 1991)
33. W.R. Newman, *Newton the Alchemist: Science, Enigma, and the Quest for Nature's "Secret Fire"* (Princeton University Press, Princeton, 2019)
34. www.newtonproject.ox.ac.uk
35. K.T. Kelly, *The Logic of Reliable Inquiry* (Oxford University Press, Oxford, 1996)
36. U. Eco, *Semiotics and the Philosophy of Language* (Indiana University Press, Bloomington, 1984)
37. American Philosophical Association, apaonline.org (2020)
38. A. MacIntyre, *A Short History of Ethics* (University of Notre Dame Press, Notre Dame, 1998)
39. P. Churchland, *Touching a Nerve: Our Brains* (Our Selves. W.W. Norton & Co., New York, 2013)
40. M.S. Gazzaniga, R.B. Ivry, G.R. Mangun, M.S. Steven, *Cognitive Neuroscience: The Biology of the Mind* (W.W. Norton & Co., New York, 2009)
41. H. Smith, *The World's Religions* (HarperOne, New York, 2009)
42. H. Smith, *Beyond the Post-modern Mind* (Quest Books, Wheaton, IL USA, 1989)
43. M.Y. Stewart, *Science and Religion in Dialogue*, vol. 2 (Wiley, Chichester, 2010)
44. F. Manuel, *The Religion of Isaac Newton* (The Clarendon Press, Oxford, 1974)
45. I. Newton, (undated) Yahuda Manuscript 21, Jerusalem.
46. M. Faraday, in Crookes, W. (ed) *On the Various Forces of Nature and their Relations to Each Other* (Chatto and Windus Press, London, 1884).
47. B. Jones, *The Life and Letters of Faraday* (Longmans, Green and Co., London, 1870)
48. G. Cantor, *Michael Faraday: Sandemanian and Scientist* (Macmillan, London, 2016)

49. L. Campbell, W. Garnett, *The Life of James Clerk Maxwell* (Macmillan and Co., London, 1884)
50. B.F. Westcott, Eulogy of James Clerk Maxwell. The Christian Life **8**, 68 (1882)
51. R.J. Strutt, *Life of John William Strutt, Third Baron Rayleigh* (Edward Arnold and Co., London, 1924)

Chapter 4
Language Mastery, Conceptual Rigour and Science Learning

Liliana Mammino

Abstract The crucial role of language mastery for science learning has been gaining increasing recognition, up to the point of being often viewed as the possible single major factor determining the learning quality and efficiency. The present chapter focuses on the role of language mastery for chemistry learning, offering a broad spectrum of information from direct experience at the tertiary level to document the impact of inadequacies in language-mastery extent and depth. This analysis also provides indications on the language-mastery sophistication level needed for successful chemistry learning. Considerations on the importance of mother tongue instruction are added because of the essential role of the mother tongue for clearer conceptual understanding and for building language-mastery, including the ability to express acquired knowledge through other languages. The envisaged 'way forward' to overcome diagnosed challenges entails the integration of language components into chemistry teaching in such a way that language analysis (analysis of the relationships between chemistry concepts and the way in which they are expressed) becomes an explicit tool for the clarification of chemistry concepts and for active learning.

Keywords Action research · Active learning · Chemistry learning · Epistemological access · Language mastery · Language of instruction · Logic mastery · Science learning · Visualization literacy

4.1 Introduction

Language is the most fundamental communication tool for human beings. Its study and codification pertain to the humanities domain, but its roles pertain to all human practical and intellectual activities. It is also the most fundamental instrument of thought [1–3]. The teaching and learning process relies on communication: the more effective the communication in both directions (from teacher to learner and from

L. Mammino (✉)
Faculty of Science, Engineering and Agriculture, University of Venda, Thohoyandou, South Africa
e-mail: mammino@univen.ac.za; sasdestria1@gmail.com

L. Mammino (ed.), *Insights into the Relationships between Humanities and Sciences*,
https://doi.org/10.1007/978-3-031-97900-2_4

learner to teacher), the more effective the learners' acquisition of knowledge can be. The communication cannot be passive: it involves active reflection, with the learner reflecting on each new piece of information and the teacher reflecting on how to best respond to any response from the learner. Within this type of activity-framework, language has the role of essential instrument of thought, the tool through which reflections are realised. Since this chapter focuses on the teaching and learning process, the idea that language is the instrument of thought is taken as a functional hypothesis, without any need to go more deeply into the details of analyses about this role. The fact that science education is the focus of attention further supports this option, because language is the instrument for the generation of new scientific information and models, as pointed out by a long history of philosophical analyses (Chapter 5).

The role of language mastery for science learning has been gaining increasing recognition in recent decades, up to the point that it is often viewed as the single most important factor determining its quality and efficacy. Investigations of this role span across a broad range of perspectives, from comprehensive points of view (e.g., [4–8]) to focus on specific disciplines (e.g., [9–16] focusing on mathematics learning [17, 18], focusing on physics learning [19–22], focusing on chemistry learning). Recommendations aimed at improving science learning often advocate explicit attention to language and propose a broad variety of options, from the holistic approach of "teaching science as a language" [6, 23, 24] or integrating science teaching and language teaching [25, 26] to the analysis of specific aspects such as the interconnections between language and symbolic forms of communication [27] and to the provision of practical guides relating the details of concepts to the details of the way in which they are expressed [28].

Inadequate language-mastery affects the pillars of the learning activities: understanding when reading science books [29, 30] or when listening during classroom activities, and being able to express concepts [5], i.e., to express acquired knowledge, both verbally and in writing. These abilities could be comprehensively called 'functional literacy', as understanding sentences or building meaningful ones goes beyond the bare ability to read or write individual words. The issue is clearly highlighted by Arias [31]: "I have often observed a student reading a sentence, yet, even though they understood all the words in the sentence they have no idea what they are reading". The literacy concept has been increasingly emphasised by educators in the last decades. Norris and Phyllis [32, 33] recommend the use of reading and writing as literacy tools—where literacy is meant "in its fundamental sense" i.e., the ability to read and understand, and to write something that can be understood by others. In this way, typically language-related exercises such as reading [34, 35] or writing [36–43] become tools for effective science learning. It is interesting to note that writing has been considered an important tool for chemistry learning for more than two decades, and its use has been explored for various courses and levels [44–59]. The idea that writing facilitates understanding is actually more than 2000 years old; Cicero, in the first century BC, already stated that "it is doubted whether a man ever brings his faculties to bear with their full force on a subject until he writes upon it" [60].

The recommendation advocating specific attention to literacy is particularly relevant for underprivileged contexts, where the number of science students who have not yet acquired the literacy level which enables effective science learning may be considerable even at the tertiary level [25]; but it is becoming increasingly crucial also for other contexts, in relation to the observed decrease in youngsters' language mastery and general literacy and the worrying increase of functional illiteracy (e.g., [61–63] for the Italian context). Interactions on students' difficulties with colleagues from countries that would not be considered disadvantaged have highlighted the appearance of language and expression errors that only a couple of decades ago were present only in disadvantaged contexts; this confirms that greater attention to students' language mastery is important and urgent in increasingly more contexts.

This chapter focuses on the role of language mastery in science learning, largely utilising documentation from the teaching and learning of areas of chemistry involving considerable proportions of conceptual understanding, such as general chemistry and the various branches of physical chemistry. Besides highlighting the crucial function of language mastery in the understanding and internalisation processes, it also highlights how careful investigation of language related difficulties experienced by students within chemistry courses is critical for the design of pedagogical options in which emphasis on language features provides powerful tools to enhance the efficacy of explanations and clarifications of chemistry concepts. It also shows the huge adaptability of such options, capable of being finely tuned to the specific needs of each group of students.

It is considered that the inferences derived from language-related investigations within chemistry courses can be easily extended to other science areas, because of the characteristics of chemistry as a science. By its nature, chemistry offers an ideal ground for the investigation of language-related aspects in science teaching and learning [64], because it comprises all the major crucial communication and reflection components, as it involves:

- extensive description components, from the description of the phenomena presented in learning materials (which learners need to understand) to the detailed descriptions of direct experimental observations (which learners need to compose on their own);
- extensive theoretical components, which require sophisticated use of language to understand their frameworks, to recognise how and why they were developed, and to identify inferences relevant for problem solving and other implementation aspects;
- extensive presence of mathematics, which requires the ability to relate physical aspects and their mathematical descriptions through mental processes involving logical and abstract thinking, whose quality, in turn, depends on language mastery;
- broad use of visualization, which requires prior acquisition of the ability to utilise it as a communication tool. The level to which such ability develops is largely determined by the level of language mastery [65, 66];
- the most articulated and sophisticated system of symbols available in the sciences to denote physical objects (atoms and molecules), whose understanding depends

on language and requires the ability to mutually interconvert information through symbols and information through words.

In view of this, findings and inferences attained through investigations of language-related aspects in the teaching and learning of chemistry may be easily extended to other science areas.

4.2 Documentation Sources and Analysis Approaches

Most of the information about the impact of language-mastery inadequacies on chemistry learning presented in this chapter derives from a systematic investigation of students' language-related challenges [67, 68] conducted at the University of Venda (UNIVEN, South Africa) through the 1997–2018 years, within courses entailing substantial conceptual understanding components, namely, first year general chemistry, undergraduate and postgraduate physical chemistry, and process technology. Their enrolments have gradually increased from $\approx$300 to > 500 for the general chemistry course, from $\approx$50 to > 100 for the second year physical chemistry course (chemical thermodynamics), from $\approx$30 to > 60 for the third year physical chemistry course (electrochemistry, chemical kinetics, surface chemistry), from $\approx$5 to > 20 for the postgraduate physical chemistry course (quantum chemistry) and from $\approx$8 to > 20 for the process technology course (a third year course). The time span of the studies and the enrolments of the courses endorse the systematic character of the investigation.

UNIVEN is located in a rural area and mostly serves an underprivileged student population. Like in most South Africa and Sub-Saharan Africa contexts, students learn through a second language (English), what adds challenges inherent in the use of a language different from the mother tongue to the language-related challenges experienced by students who use their mother tongue as medium of instruction. This entails a broader variety of challenges than might be identified in mother-tongue-instruction contexts.

The investigation was an integral component of the teaching approach, aimed at identifying the details of learning problems in order to design options to try and address them at classroom level. Such an approach is typical of action research [69–71]; a diagram highlighting the action-research recursion pattern for the integration of educational research and the design of in-class approaches is included in [71]. The initial information for each new loop was derived from the analysis of students' answers in their routine written works (tests, laboratory reports); these have been viewed as the most reliable sources because they correspond to maximum students' efforts to produce their best in view of assessment [72]. Information surfacing within in-class interactions was also given extensive consideration because the interactions enable more complete understanding of students' conceptions. In-class discussions utilised error analysis as an explanation tool, which is particularly apt to engage students' attention and particularly suitable to integrate the consideration of language aspects and conceptual aspects [72–74]. Errors from other sources (including some teaching materials or some internet sites) were also added when expedient to highlight relevant aspects of the concepts of interest. The details of how errors can be turned into pedagogical resources are outlined in [75]. The analysis integrated the conceptual (chemistry) point of view and the language point of view (integration made possible by the author's humanities background-training).

For most errors, it is impossible to untangle the language and the conceptual components or to ascribe individual weights to each of them [74]. Errors regarding the simplest and more basic language aspects (one might say, literacy aspects) have dominant language component; they comprise features such as the sound-meaning correspondence of individual words, whose impact is particularly substantial in a second language context [76]; the combination of words with specific functions [77]; the use of prepositions [78]; the use of verbal forms, and the use of logical connectives. In many cases, the analysis may be quite complex, entailing a variety of perspectives. For instance, a statement like "spontaneous reactions are negative" can be viewed from several angles: as an incorrect combination of noun and adjective (language point of view); as an outcome of the tendency to oversimplify sentences, omitting key words (the omitted key term being the 'free energy change'—the quantity that is negative); as an indication that the student is not aware of the actual meaning of what he/she writes (also in relation to inadequate proof-reading abilities); as an indication that the student does not distinguish between qualities pertaining to physical quantities (such as the possibility of taking negative values) and qualities pertaining to objects or to processes; or as an indication that the student does not actually know that the quantity to be considered is the free energy change accompanying the reaction. When this error is made by more than one student, it entails discussion in the class. Then, all the just-listed points of view are considered, starting from the question whether it is possible for a reaction to be negative, and developing the rest of the discussion according to student's responses, but ensuring that all the aspects are included. The comments to the examples reported in Sect. 4.3 offer further illustration of how the analysis is performed and highlight the main types of features that are brought to students' attention. Examples showing interaction pathways are included in [21].

Complex sentences often present errors with regard to the logic of the discourse or to crucial epistemological aspects, such as cause-effect relationships [79], condition-consequence relationships, or the distinction between key categories like physical quantities and their changes [80], systems and processes [81], numbers and values [82], 'general' and 'particular' [83]. Studies focusing on specific chemistry concepts also highlight the impact of inadequate language-mastery on their understanding and general handling by students (e.g., the Hess law [84], the Hess law and the state-function nature of enthalpy [85], the rate law [86], what happens during a chemical reaction [87], the spontaneity concept [75, 88–90]).

Throughout the investigation, a clear distinction is made between language and terminology [21, 75]. Terminology is a set of names (technical terms) important within each discipline, because they denote the objects and phenomena of interest in that discipline, but it is not the knowledge-carrier or information-carrier. Language in its fundamental meaning is the knowledge-carrier, through sentences and ensembles of sentences. Sentences are constructed through the normal components of common language (nouns, verbs, adjectives, prepositions, conjunctions, etc.) and according to its grammar rules. Technical terms are embedded into frames made of these components. Incorrect use of common language components results in incorrect scientific communication. Thus, what could be called the 'language of science' (the way of expressing things in the sciences) inherently depends on the correct use of common language [21, 90]. This motivates the option of discussing language in its fundamental meaning when interacting with students whose language mastery is limited, within a science course.

The largest space in this chapter (Sect. 4.3) is devoted to the impact of learners' inadequate language mastery on the quality and efficiency of chemistry learning. This information is crucial for the design of classroom interventions. It is also crucial to substantiate the essential roles of humanities-based expertise (language mastery, familiarity with basic epistemological aspects) for the effectiveness of science learning. Two other key issues are also given specific attention to make the panoramic more complete:

- The issue of the language of instruction (mother tongue or second language; Sect. 4.4). It is particularly important in Sub-Saharan Africa contexts, where instruction is mostly done through a second language (the language of the former colonial power). The different levels of language mastery enabled by the two options largely determine the efficacy of science learning.
- The search for approaches that can help learners develop adequate language mastery (Sect. 4.5). This is becoming an increasingly urgent necessity. The integration of language training into science teaching, in a form for which language analysis is utilised as an effective tool for the explanation and/or clarification of science concepts, is viewed as a promising option [25].

As already mentioned, errors have primarily been used as resources for in-class clarifications, through their analysis and discussion. They have also been 'stored' as documentation of the impact of language-mastery inadequacies on chemistry learning. It was considered important to build extensive documentation because of the frequent misconception that 'language is not so important in the sciences', which surfaces through a variety of routes in a variety of contexts, from explicit statements to lower demands on language-mastery level for students enrolling in science faculties. This misconception can only be counteracted by documenting the negative impacts of inadequate language-mastery on science learning. This is also the purpose of the present chapter: using the analysis of an adequate number of concrete examples, with adequate diversity in the nature of the errors involved, to demonstrate that language-mastery inadequacies affect the quality, clarity and completeness of conceptual understanding.

On storing them, errors were grouped according to their nature, as this classification makes them ready-for-use as documentation. No specification of the source of individual errors was considered relevant, and this applies also to errors from other sources (communications from colleagues in other contexts, teaching materials, internet, etc.), which were stored together with those from students, because of their usefulness for classroom work. Recurrent errors were 'stored' together, reporting each statement individually, to provide indications about their recurrence; the errors selected as documentation in this chapter correspond to errors with considerable or high recurrence.

4.3 When Learners' Language Mastery is Inadequate

The analysis of the impact of inadequate language-mastery on the extent, clarity and quality of chemistry learning convincingly highlights the role of language-mastery in chemistry learning. This section examines the major aspects of such impact:

- the role of language as the fundamental communication tool and the impact of diminished communication efficacy when language mastery is inadequate;
- the impact of inadequate language mastery on conceptual understanding and overall learner's performance;
- the intimate connections between language and logic, and the impact of inadequate language mastery on logical thinking, on the expression of logical relationships, and on the ability to build a logically sound discourse;
- the fundamental role of language for the development of literacy in, and mastery of, other communication tools relevant for science learning (first of all, visualization).

4.3.1 Impact on Communication Extent and Efficacy

Language is the fundamental communication tool in the teaching and learning process. The most direct communication—what we call *teaching*—implies that the teacher talks to the learners or—in the more effective approaches involving interactions—teacher and learners talk to each other. Another fundamental type of communication is realised through the resource materials, e.g., through books. Both talking (or listening) and reading imply the use of language; thus, their communication effectiveness depends on the learners' language-mastery.

The science concepts that are objects of learning at the tertiary level may have considerable degrees of complexity. Correspondingly, language-mastery needs to have a parallel level of sophistication. When this level is not present, the effectiveness of the learning process may decrease severely. Language mastery inadequacies affect the understanding component, because students may not fully grasp the meaning of what is said in the classroom or in a textbook; then, concepts remain unclear, or misunderstandings of the literal meaning of sentences and texts generate misconceptions. They also affect the expression level, because students find it difficult to express concepts in a way that responds to the nature and meaning of the given issue, or, in more extreme cases, find it difficult to build sentences that convey a meaning.

A teacher comes into contact only with the latter component, i.e., with what students express. What students express comprises the compound effects of understanding inadequacies and expression difficulties. It is impossible to untangle these two components, because the teacher can only see the overall outcome. From the analysis of students' works (complemented by interviews in selected cases), it appears reasonable to assume that, in many cases, students manage to express less than what they might have at least partially understood or at least intuited. It even appears possible to hypothesise that the two components may not sum up linearly, but enhance each other at least in some cases, although their untanglability prevents realistic evaluations of each of them. This brings a considerable degree of uncertainty into assessment, as language errors may actually overweight conceptual errors, but the outcome (what a student writes) appears wholly as conceptual errors. Clerk and Rutherford [91] consider language inadequacies as a "confounding variable" for diagnosing misconceptions. The same holds for assessment in general and, since this "confounding variable" is difficult to remove, it may affect the assessment reliability [74, 92]. In the rest of this text, the outcome (what students express) is considered as evidence of what students understand; although this is a simplification that may not correspond to reality in several cases, it is determined by the impossibility of untangling the language and conceptual components.

4.3.2 Interconnections of Language-Related Difficulties and Conceptual Difficulties

Incorrect sentences and statements constitute the main data for an investigation on the impact of language-mastery inadequacies on science learning. Representative examples are reported in this section as illustrations of frequently-encountered difficulties. The consideration of an adequate number of examples is viewed as the most effective way to document the relationships between sentence-wording and the level and correctness to which chemistry concepts are understood. The examples are gathered into subsections considering key issues, which are specified in the subsections' titles; the titles refer to the category of language-related challenge to which the examples of the given subsection refer.

The examples are numbered progressively to facilitate references throughout the text; references to them are made by writing ex-1, ex-2, ... to indicate example 1, example 2, ... Because of their roles as documentation, the sentences are here reported in their original formulations; on the other hand, only the part of each sample relevant to the discussion of the specified error is reported. Short comments are added to identify the imprecisions or errors, and to highlight the main threads for in class-discussions utilising the analysis of error and integrating conceptual and language aspects in order to clarify chemistry concepts. Some comments also consider the teacher's uncertainties or possible diverse reflection-paths in front of certain errors, in view of both in-class clarifications and of assessment. The word or words that are object of the comments are underlined when this is expedient to facilitate their immediate identification. The comments focus on language aspects because this is the focus of the current work; chemistry concepts are discussed only to the extent that their discussion is functional to the analysis of the language error. Since not all the examples are from students' works (although they have all been utilised as resources for in-class clarifications), they are just identified by "ex", without any reference to their source; the terms 'student', 'students' are used with reference to in-class explanations or discussions, or to assessment-pertaining reflections.

4.3.2.1 The Sound-Meaning Correspondence for Individual Words

The perception of the sound-meaning correspondence is largely natural within the mother tongue, but it may be challenging within a second language, where words with sounds perceived as similar may be interchanged on expressing concepts [76]. In some cases, the substitution does not affect the communication within a learning environment because a teacher can guess the intended word; on the other hand, these errors also require discussion, because students need to acquire the ability to communicate to other persons, who will not be as trained to guess intended meanings as a teacher does. For instance, if a student writes *veisa-veisa* in place of *vice versa*, it is possible to consider it a spelling mistake on assessment, and to consider the concept conveyed by the sentence. Even in cases such as ex-1, where *side* is written in place of *sign*, or ex-2, where *constance* is written in place of *constant*, it is still possible to give the student the benefit of doubt and assume that the correct meaning was intended (although there will be a certain grading difference with respect to answers using the correct words).

1. *The entropy change of the system and the surrounding are equal with opposite side.*
2. *R is the gas constance.*

Similarly, the frequent random use of *the*, *they*, *there* and *then*, which are perceived as similarly sounding and often interchanged, or simplified to *the* (ex-3) or, more rarely, to *that* (ex-4), although incorrect from the grammar and meaning point of view and requiring in-class discussion, can still be included among the cases for which it is possible to guess what the student intended.

3. *The materials that are going in and going out of the system must be balanced so that the won't be more accumulation of material within the system.*
4. *Ionic compounds can conduct electric current in the electrolytic solution when that are positive and negative ions.*

In other cases, the interchange does raise concern, because it is impossible to evaluate whether the student knows the meaning pertaining to that context. For instance, the correct term in ex-5 would be *reactor*. Since the reactor is a unit (a physical object, part of a bigger structure) and a reaction is a process, the confusion between the two terms raises the question whether the student knows the difference between the two concepts (knowledge that is definitely required in a process technology course). In a case like ex-6, it remains unsure whether the student knows the concept, although it is likely that *constract* (which appears repeatedly in the same work and is, therefore, not an occasional typo) is written in place of *contract*.

5. *The prototype chemical plant contains the preparation unit, the reaction , and the separation unit.*
6. *Compressibility is the ability to constract.*

Not distinguishing between words that may be perceived as having similar sounds may lead to serious errors, as in ex-7, which is an answer to the question "Explain the mechanism of heat transfer by convection". The two words (*convection* and *conversion*) are both technical terms within the same course and they are likely perceived as similarly sounding. The answer refers to the *conversion* concept instead of the *convection* one. The fact that the question specified "heat transfer" (thus providing a context for the term) suggests lack of familiarity with the convection concept.

7. *Conversion is given by this expression: conversion* $= (F_{i\,out} - F_{i\,in})/F_{i\,in}$.

4.3.2.2 The Meaning of Individual Words

Poor knowledge of the meaning of individual words may seriously affect understanding and performance. It is particularly evident for words that have a method-related meaning. Students often do not clearly distinguish the meanings of key words used in questions, such as *describe*, *explain*, *derive*, and the like, and this affects their answers. The same is true for relevant terms utilised in certain types of answers. For instance, the verb *assume* has a key role in the explanation of some problem-solving procedures (e.g., in chemical thermodynamics). Students often use it with reference to information directly provided in the question, or to facts that are common knowledge, as in ex-8, (the melting point of water at standard pressure is something that is known, not something that needs to be hypothesised):

8. *I assume that the melting point of water is 0 °C.*

Similarly, the word *relationship* is encountered often in explanations, including the description of diagrams, and it may be added even where it does not have a role. For instance, in ex-9 *the relationship* is not needed and the sentence would provide correct basic explanation without it; moreover, *relationship* should be followed by explicit specification of the two items that get related, more often introduced by the preposition *between*.

9. *The heating curve is a graph showing the relationship how a substance phase changes from solid to liquid and from liquid to gas when heat is supplied.*

Inadequate awareness of the importance of grammar results in the neglect of several grammar rules, including features as fundamental as the distinction between the singular and plural forms of nouns. In several cases, the chemical meaning is not affected, but in other cases this type of error engenders doubts about the student's familiarity with the topic. In ex-10 (besides the fact that the whole answer is unfocused), the use of the singular form *isotherm* suggests lack of familiarity with the isotherms diagram, which comprises several curves (isotherms) corresponding to different temperatures and highlighting different behaviours; on the other hand, the error may simply be caused by inadequate attention to grammar, not by the conviction that the diagram contains only one curve; only if the answer is accompanied by a drawing of the diagram it is possible for the teacher to know whether the student is actually familiar with it:

10. *Compressibility factor is the factor that affect the* <u>*isotherm*</u> *of ideal gas and real gases.*

4.3.2.3 The Association of Words in View of the Meaning to Be Conveyed

The selection of individual words and the combination of words into pairs or triplets having key roles in the communication of information (such as subject-verb, subject verb-object, adjective-noun etc.) are basic aspects of language literacy. They are also crucial for science communication. Incorrect pairing testifies to incorrect understanding of at least some important features of the concepts concerned. For instance, the frequent cases in which the verb *occur* is referred to a subject denoting a system (a physical object, as in ex-11) indicates inadequate awareness of the distinction between systems and processes, and of what pertains to each of the two categories [81]. Similar indications stem from the use of the verb *undergo* with reference to a process (ex-12), because it can only be referred to a system (i.e., a system must be its subject).

11. *I will assume that* <u>*benzene is occurring*</u> *at constant pressure.*
12. *Reversible process is a* <u>*process that undergoes*</u> *isothermal or adiabatic expansion or compression.*

Cases in which the selected verb is not suitable for the subject in questions are very frequent and may be associated with incomplete understanding of the concept or context, or with inadequate ability to analyse a sentence and verify the actual meaning it conveys, or with both. Ex-13 uses an incorrect verb to connect the subject and a grammatical object (*different entities*) that has no role for the given concept (the correct verb should be *are*, with *different entities* becoming part of the overall verbal component of the clause).

13. *Waves and particles* <u>*have*</u> *different entities.*

When complex sentences (sentences consisting of more than one clause) are involved, inadequacies in the ability to analyse a sentence may play significant or major roles. For instance, in ex-14, *it* is grammatically referred to *number*, and a number cannot *be generated or consumed*. In ex-15, *impossible to solve* is referred to *repulsion term*, and individual terms of an equation are not *solved*. In ex-16, *the pipe* is the grammatical subject of *to have*, conveying the message that the pipe is moving (instead of the liquid flowing inside it).

14. *The number of moles is not conserved because it can be generated or consumed within the system by the chemical reaction.*
15. *It is impossible to solve the schrödinger equation for the helium atom exactly because the repulsion term r_{12} is impossible to solve using separation of variables.*
16. *The pipe will be designed to have a low velocity of motion.*

Sentences containing the verb *accompany* (typically used with reference to changes in thermodynamic quantities during a process) often show uncertainty about what accompanies what. For instance, in ex-17, *accompanying* is referred to an object (liquid benzene), while it should be referred to a process (the heating of liquid benzene).

17. *Consider the entropy accompanying the liquid benzene from 278.56 K to 353.1 K.*

Ex-18 entails the combination of inadequate language mastery and inadequate familiarity with the nature of models: *quantum chemical* qualifies an approach that we use to study phenomena at the molecular level and, therefore, it should qualify *study*, not *reaction mechanism* (the reaction mechanisms pertains to the reaction, i.e., to what we can consider as physical reality; the reaction occurs with its own mechanism whether we study it or not; and the mechanism can be studied with different approaches).

18. *Consider the study of the quantum chemical reaction mechanism of methane and oxygen.*

Another frequent problem, likely associated with expression oversimplifications instigated by inadequate language mastery, is the omission of key terms. In ex-17, the omitted key term is the name of the process (heating). The most commonly omitted term is the noun in an adjective-noun pair: e.g., *function* or *free energy* is omitted after *Gibbs* in ex-19 and *potential* is omitted after *chemical* in ex-20 and after *electrochemical* in ex-21; since the noun conveys the principal information in the pair, these sentences fail to convey a literal meaning.

19. *Chemical potential is the molar gibbs of a pure substance.*
20. *When the solute is dissolved in the solvent the chemical of the solvent is lowered.*
21. *Consider the electrochemical of all species considered.*

Ex-22 is an extreme case in which what should have been a considerably longer sentence (the electrochemical potential of Cu is equal to its chemical potential) is reduced to an adjective (*electrochemical*) and an equality ($\overline{\mu}_{\mathrm{Cu}} = \mu_{\mathrm{Cu}}$), the nature of whose terms is not specified within the sentence.

22. *The electrochemical of* $\overline{\mu}_{\mathrm{Cu}} = \mu_{\mathrm{Cu}}$ *because they are not charged.*

If the verb is omitted, the sentence cannot convey any meaning. In ex-23, there is no main verb, but only the modal verb *can*, which, alone, does not convey a meaning.

23. $\Delta U_{mix} = 0$ *can always with each other in any phenomena.*

4.3.2.4 Connecting Individual Words to Build Clauses

Connections between words within individual clauses are expressed by prepositions. Incorrect uses of prepositions result in incorrect expression of concepts, frequently associated with inadequate understanding [78]. A typical example concerns the distinction between intramolecular and intermolecular bonds, respectively associated with the use of the prepositions *in* and *between* (Fig. 4.1). Ex-24 shows a case in which *in* is used in place of *between* with reference to intermolecular interactions. It is possible to check students' actual understanding by asking them to draw stick-and-balls models of the molecular systems considered in the answers. An independent verification is obtained by inserting incorrect entries such as "the bonds between the O atom and each of the H atoms in a water molecule are hydrogen bonds" in multiple-choice questions. Both exercises show that the confusion frequently concerns the conceptual level. On the other hand, it also appears that students do not realize the importance of clearly distinguishing the meanings of the two prepositions, or they view it as a "language issue" (expecting that the teacher should not take it into account on assessment) and not as a chemistry issue (the distinction between two types of chemical bonds); in other cases, they often do not have a real perception of the distinction between the meanings of the two prepositions.

24. *Real gases behave differently because of the intermolecular interactions in molecules.*

In ex-25, the main problem is the omission of the preposition *of* after *free energy*.

25. *The molar free energy is the free energy formation of pentane.*

The preposition *of* often specifies what pertains to what. When the concept is not clear, the specification may become absurd, as in ex-26 (where the correct wording would have been *these properties of solutions*; from a chemistry point of view, properties are not objects that can be dissolved in a solvent) and ex-27 (a liquid cannot melt, because the liquid phase is the outcome of the melting process).

26. *The magnitude of colligative properties depends on the number of particles of solute present. The solution of these properties depend on the concentration of solute molecules or ions, but not upon the identity of the solute.*

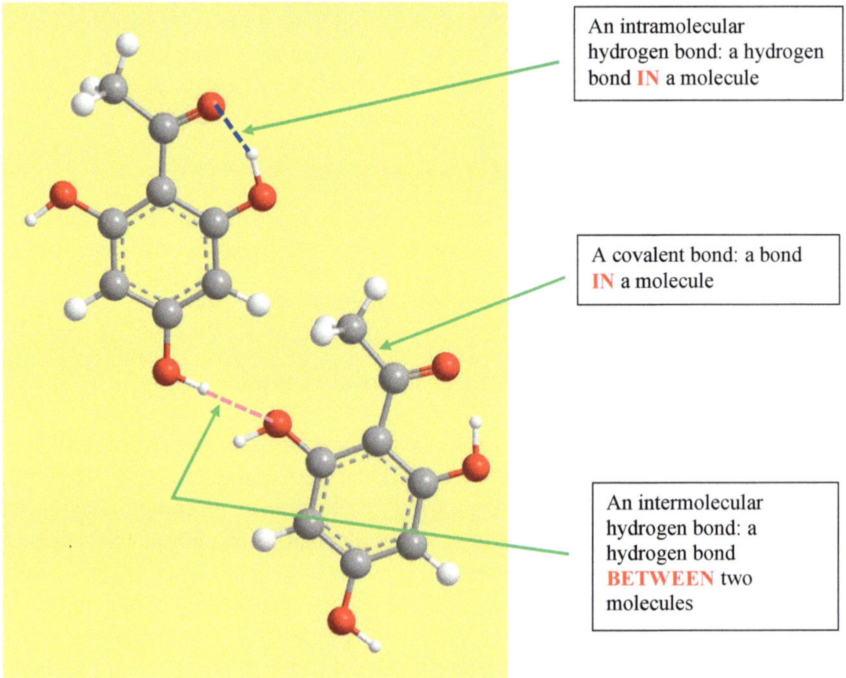

Fig. 4.1 How prepositions identify types of chemical bonds

27. *Consider the melting of liquid methanol at 278.56 K.*

In ex-28, the preposition *from* is omitted before *353 K*—an error encountered frequently when the *from....to....* form would be the correct one.

28. *Consider the heating of gaseous benzene 353 K to 393 K.*

Other errors concerning the *from....to....* form may neglect that the *from* implies a direction which is determined by the rest of the sentence. For instance, the presence of *heating* in ex-29 establishes that the lower temperature value should come after *from* and the higher temperature value after *to.*

29. *Stage V: heating of gaseous benzene from 353.1 K to 283 K*

The preposition *than*, having a key role in the expression of comparisons, is often used without being preceded by a word or form identifying the first term of the comparison or informing that a comparison is considered. For instance, in ex-30, *time* should be preceded by longer"

30. *High mass of ice took time to melt than smaller mass of ice*

4.3.2.5 Logic Aspects in Sentences and Discourses

Logical thinking is essential in the sciences. It is necessary for the analysis and interpretation of observations and in all the mental processes involved in the development of models and theories or in problem-solving. Logic is expressed through language and language has its own internal logic. This engenders intimate connections between language and logic, making their mastery-acquisitions interdependent and mutually enhancing. This subsection considers logical relationships and logical frameworks from the point of view of their expression through language.

A basic logical principle excludes internal contradictions. This closely relates to conceptual understanding. The presence of internal contradictions in a sentence is often a symptom of insufficient understanding, but it may also be related to insufficient ability to verify the meaning actually conveyed by the selected wording. Ex-31 and ex-32 simultaneously contain the *heating* concept and the *isothermal* or *constant temperature* concept, which is contradictory. Similarly, ex-33 states that the temperature is constant and gives an equation involving a temperature change ($T_2 - T_1$). Ex-29 contains a similar contradiction, as it uses the verb *heating* and then considers an initial temperature (*353 K*) higher than the final temperature (*283 K*).

31. *The benzene is heated isothermally, so we will use the equation...*
32. *We are heating benzene. Assume all processes occur at constant temperature.*
33. *Since temperature is constant,* $W = n\,C_p\,(T_2 - T_1)$.

Connectives between clauses within a sentence build the logical framework of the sentence, by expressing the relationships between the pieces of information conveyed by each clause. The appropriate use of connectives and the ability to build each clause correctly are essential for a sentence to convey a meaning. Some key categories of logical and epistemological relationships having fundamental roles in the sciences are considered here: time-related relationships, cause-effect relationships, and hypotheses-inference or condition-consequence relationships.

Time-related relationships involve the ability to describe series of events or operations in their appropriate time sequence. The difficulties in this regard (e.g., those mentioned in [12]) are better analysed from the combined points of view of inadequacies in language-mastery and inadequacies in narrative abilities [12]. Typical examples include incorrect sequences of operations in the description of experimental procedures [12].

The **cause-effect relationship** has a fundamental role in the interpretation of experimental information and in all the theoretical frameworks deriving from it. From a conceptual point of view, it is essential to distinguish and clearly identify 'what causes what'; once this is identified, it is essential to communicate it in an understandable way. Students' writings often show incorrect identification of the cause, or even reversal between cause and effect [79]. Interviews mostly reveal that errors in this regard are not purely language-related, but express incorrect understanding; on the other hand, the close relationships between language and logic imply that inadequate language-mastery development is often associated with inadequate logic-mastery development [8]. In ex-34 (pertaining to a description of the photoelectric effect), the cause is not identified correctly, as the wording conveys the message that no current flows because a potential difference is applied (actually, no current flows because the circuit is open). Ex-35 does not identify the correct effect: if the thickness of a barrier is infinite (identified as a cause, because it is introduced by *since* in the sentence), the particle will not go beyond the barrier according to both classical physics and quantum mechanics.

34. *The potential difference is applied, consequently there is no current flow.*
35. *In case (iii) since the thickness of barrier is infinite according to classical physics the particle will go beyond the barrier.*

Ex-36 reverses cause and effect: actually, the enthalpy of elements is taken as zero by definition (third principle of thermodynamics) and, as a consequence, the enthalpy of compounds is equal to their enthalpy of formation.

36. *Since the enthalpy of compound is equal to the enthalpy of the formation, the enthalpy of the elements in standard conditions is zero.*

The difficulties in identifying correct causes often results in failure to answer questions of the 'Explain why…'. type. Ex-37 (meant to explain why $\Delta U_{mix} = 0$ for the mixing of ideal gases) only expresses the meaning of $\Delta U_{mix} = 0$ through words, but does not indicate the cause.

37. $\Delta U_{mix} = 0$ *because the internal energy of the gases does not change.*

Ex-38 responds to a question asking whether the moles of propane are conserved or not during its combustion; the absence of the verb in the main clause does not allow the reader to know whether the answer is YES or NO; the part following the *because* conjunction does not identify a cause (propane is the reactant in the considered reaction; a combustion reaction does not involve a simultaneous reverse reaction; therefore, propane can only be consumed, not produced).

38. *The number of moles of propane because it can be produced or consumed by the chemical reaction within the system.*

Hypothesis-inference or condition-consequence relationships require complex sentences consisting of a clause introduced by *if*, stating the hypothesis or the condition, and a principal clause stating the consequence. The situation may have different nature, according to whether condition and consequence are realistic (pertain to reality), express a possibility, or express something that does not or did not happen (non-reality cases). In the last two cases, the verbal mode is different (conjunctive). The expression of reality cases is often reasonable in students' answers (e.g., ex-39).

39. *If the energy is enough to make the electron leave the electron will leave the metal.*

Non-reality cases appear to be particularly difficult, both because of lack of familiarity with the use of the conjunctive mode and because of the perceived complexity of a discourse of the type "if A were true, then B should also be true; but, since B is found not to be true, then A is not true". A discourse of this type is relevant in chemistry learning, because it is important in the interface between experimental observations and their interpretation leading to theoretical models. A typical example [93] concerns the inference from the results of Rutherford's experiment, which can be expressed with the wording 'If Thomson's model had been true, all the α particles would have behaved in the same way. Since they did not behave in the same way, Thomson's model cannot be true'. It may be useful to highlight the correspondence of the individual statements in this description (right-hand column) with the general-type statement for non-reality (left-hand column):

• If A were true	If Thomson model were true
• then B should also be true	then all the α particles should behave in the same way
• but, since B is found not to be true	but since not all the α particles behaved in the same way
• then A is not true	then Thomson model is not true

The difficulties in grasping the logic of this reasoning and the difficulties in constructing complex sentences may lead to oversimplified expressions suggesting the co-existence of both behaviours (ex-40). It may also happen that reading difficulties (difficulties in understanding the full meaning of a text) give space to the perception (or even the conviction) that both behaviours exist in some parallel way.

40. *In Thomson's model, all the α particles behave in the same way, in Rutherford's model they behave in different ways.*

The problem may also be enhanced by the difficulty in understanding the nature and role of models in the sciences (something whose understanding requires both adequate language-mastery sophistication and epistemological awareness). This is confirmed by other cases where different models are viewed as parallelly existing, like in ex-41, which is an oversimplification of a more complex discourse comprising the following logical steps: *i)* the consideration that Rutherford's model is not in agreement with classical electrodynamics because, according to the latter, the electron would fall on the nucleus; *ii)* the inference that, since the electron does not fall on the nucleus, a new model becomes necessary; *iii)* the confirmation that Bohr's model responded to this necessity.

41. *In Rutherford's model, the electron falls onto the nucleus. In Bohr's model the electron does not fall onto the nucleus.*

These examples (and many similar others) highlight a huge discrepancy between the complexity of the actual reasoning and the expression-oversimplification caused by language-mastery inadequacies and blurring or removing important components of the overall reasoning. What gets lost in terms of students' familiarization with the scientific approach is likely huge.

Understanding the **logic of sentences** and of ensembles of sentences (discourses) is fundamental to understand what one is reading or listening (i.e., is fundamental for functional literacy). Difficulties in this regard lead to incorrect answers because the concept does not get to be fully understood. If the meaning of a sentence or a discourse is not understood in its entirety, students may focus only on what they perceive as keywords. Ex-4 is an answer to the question "Explain why ionic compounds can conduct electric current only when they are in the molten state", but totally neglects the condition of being in the molten state. Ex-42 is the most frequent type of answer to the question "Explain how the electronic conduction and the ionic conduction connect to each other in electrochemical cells"; it appears that students focus on the keywords "electronic conduction" and "ionic conduction" without considering the complete sentence and without realising that the verb "connect" is actually the keyword in the question; then the answers solely provide a description of the two types of conduction.

42. *Electronic conduction it is involve movement of electrons that occurs in electrodes and metal wire. Ionic conduction it is involve movement of positive and negative ions. It occurs in salt bridge and electrolytic solution.*

Inadequate familiarity with the logic of sentences limits the ability to write even minimally complex sentences (sentences with at least two non-paratactic clauses), as is the case of the cause-effect or condition-consequence relationships. It also leads to oversimplified, incomplete answers even when the answer should just provide a straightforward description, like in ex-43 and ex-44, which constitute two answers (the entire answers provided) to a question asking to describe laminar flow.

43. *Laminar flow ==> adjacent layers of fluid move other without mixing in a direction.*
44. *Laminar flow ==> smooth layers.*

4.3.2.6 Language Mastery and the Development of Other Communication and Learning Tools

Language plays fundamental roles in the development of the literacy and mastery of other communication tools relevant for science learning [65, 66, 94]. The most important for chemistry learning is visualization. In general, the importance of visualization in science learning relates to its ability to provide external representations that facilitate understanding [95] and to contribute to the generation of mental images, which, in turn, are important for conceptual understanding [96]). Within chemistry courses, visualization plays crucial roles in the familiarisation with the invisible word of atoms and molecules: any textbook section and any in-class explanation about molecules contain images.

Chemistry students' works often highlight interdependence between visual literacy and language mastery [65, 66]. Although the consideration of illustrative examples would be important to substantiate the interdependence, reporting imagery examples would considerably increase the space demands of this chapter. It is therefore opted to consider only three examples (other examples can be found in [66]). Ex-45 refers to the prototype chemical process. Students are asked to draw a diagram and to explain it. The pronoun *some* is grammatically referred to *reacted materials*, making them the logical subject of *are recycled*, whereas the ones to be recycled are the unreacted materials (the reacted materials have formed the desired products).

45. *Purification and separation ==> in this unit reacted materials are purified and <u>some</u> are recycled back to the reactor.*

Ex-46 refers to a comparison between the general diagram of the prototype chemical process and the corresponding diagram for a concrete case (the production of ammonia). It refers to the diagrams drawn in the answer: the diagram for the former included a preparation unit, whereas the diagram for the latter did not, starting with a mixer unit without any inlets and only an outlet. Both the diagrams and the sentence highlight inadequate understanding of the details of the process, and a problem that relates to science and to logic, as a unit cannot have only outlets without inlets (if nothing enters the mixer, nothing can come out of it).

46. *The prototype chemical process starts with a preparation unit while ammonia flowsheet starts with mixer unit.*

Questions simultaneously asking to draw a diagram and to describe it are meant to stimulate students to relate the two descriptions (through visualization and through words), thus favouring the familiarization with both [66]. In the case of comparatively simple diagrams, students often memorise the diagram, but fail to describe it [66]. A typical example is offered by the phase diagrams of pure substances. Ex-47 is a description accompanying a correct phase diagram. It is difficult to guess whether the student understands the diagram, but is not able to describe it through sentences (an informative description would require a considerable number of sentences), or whether the diagram has been memorised passively, without understanding.

47. *Shows the changing of liquid, solid, gas phase which in the illustration are according to the way of the curve.*

4.3.3 The Importance of Mother Tongue Instruction

The issue of the language of instruction (mother tongue or second language) is particularly crucial in sub-Saharan Africa, where the language of instruction is usually the language of the former colonial power. The information and documentation deriving from the sub-Saharan Africa experience may also be relevant for other contexts, such as contexts where there is ongoing debate about which language (mother tongue or English) to choose for general instruction, or contexts where it has been opted to teach one or more science subjects through English in secondary schools, with the aim of improving students' mastery of English or because it is considered the best choice in classes with significant proportions of non-native speakers, often also having different mother tongues.

Studies of the situation in Sub-Saharan Africa amply document that second language instruction has a huge impact on students' ability to learn, enormously increasing understanding difficulties [8, 65, 94, 97–103]. Rubanza [104] clearly highlights the communication failure stemming from the poor mastery of the second language by the learners and, in several cases, also by the teachers. Furthermore, the fact that students do not sufficiently master the theory of the mother tongue is a major factor responsible for inadequate development of logical thinking abilities [94, 102] as well as for the poor mastery of the second language [105]. Learning the theory of the mother tongue entails learning to relate all the expression-components to the resources of the mother tongue, from the expression of individual logical relationships to the expression of any type of details and nuances; when, then, a person learns or uses a second language, that person can utilise the awareness of how language resources express the details relevant to a given information ensemble and search for corresponding resources in the other language. The results of the previously-mentioned studies provide clear indications of the importance of a good knowledge of the mother tongue and of mother tongue instruction for science learning, including chemistry learning. Therefore, these indications need to be taken into account on decision-making about language policies [106].

The author's experience of having learnt through her mother tongue throughout her education (up to M.Sc. level included) and having worked for more than 30 years with students who study through a second language (in three different Southern Africa countries) has enabled direct experience of the enormous differences between the two options. Chemistry is often not easy to understand even through the mother tongue. The second language vastly increases the difficulties, above all in learners' initial encounters with it. Interactions with some teachers teaching chemistry through English in Italian schools (e.g., within CLIL projects) highlighted further complications. They explain that, in order to enable learners to understand through a non-familiar language, they utilise simple, mostly one-clause sentences [107]. This deprives learners of the contact with logical frameworks, resulting in a sort of 'chemistry in pills'. The learner is deprived of the possibility of coming into contact with the world of science, with its approaches and its inquiries, its questions and its investigations, the formulation of hypotheses and the process aimed at their verification. No humanities-born aspect can be incorporated, because expression abilities are not sufficiently sophisticated in a second language to support the sophistication level of a discourse going beyond simple lists of facts, expressed with the simplest possible terms and through the simplest, shortest sentences. The 'chemistry in pills' option thus entails a serious risk of not communicating enough chemistry to build a new generation of scientists who can also discuss and analyse theories and create new ones, not only solve standard-type problems using memorised basic algorithms. It may be hypothesised that the very proposition of using a language different from the mother tongue in students' first approach to science concepts, in countries where mother tongue instruction is the common practice, can only be linked to inadequate awareness of the role of language in science, both at understanding level and at creative level. This inadequate awareness, in turn, results from the dichotomy between humanities and sciences being brought to extreme limits, so that the whole issue of the role of language as a tool for communicating and generating science information is ignored. It also appears that the suggestion of teaching science through a language different from the mother tongue has a common ground with the suggestion claiming the development of skills, and not the acquisition of knowledge, as the objective of science learning (Sect. 18.4.7). As long as understanding is considered a fundamental objective, the tools facilitating understanding are given paramount importance; no option that could decrease the extent of understanding, or the comfort with which it is achieved, would be considered viable; the possibility of fully using all the resources of a language would not be denied to students. The suggestion of second-language instruction when students approach new topics and concepts is possible only within a perspective that does not consider understanding as a crucial objective.

Although the extent of second-language instruction in Sub-Saharan Africa enables the identification of the most extensive documentation on its impacts on learning, including science learning, the issue of the language of instruction has also been investigated by scholars from other contexts, and has also led to the recognition of the importance of mother tongue instruction. Few excerpts are here recalled for illustration purposes. Noormohamadi [108] writes.

> *Mother tongue... is essential for learning as a part of intellectual ability. ... It ... has a central role in education that demands cognitive development. ...*
>
> *Studies show that children who come to school with a solid foundation in their mother tongue develop stronger literacy abilities. ...*
>
> *Incomplete first language skills often make learning other languages difficult.*

Nishanthi [109] stresses that the problem mostly affects developing countries.

> *Many children across the developing world are learning very little in school, a reality that can be linked to teaching that is in a language they do not fully understand. It is a practice that leads to limited or non-existent learning and acquisition of knowledge and skills, alienating experiences, and high drop-out and repetition rates.*

The importance of mother tongue instruction is also recognised by UNESCO [110].

> *Research shows that education in the mother tongue is a key factor for inclusion and quality learning, and it also improves learning outcomes and academic performance.*

Finally, it may be mentioned that the importance that young people learn to express science also through English, because of its current role as *lingua franca*, cannot be considered a reason to teach science through English. The objectives of a science course in secondary schools should remain exclusively those of acquainting learners with the specific science, not that of strengthening learners' mastery of a foreign language. It becomes important to utilise different approaches to the two fundamental components of learning: understanding new information, and expressing the acquired knowledge [94, 97, 111]. Understanding new concepts needs to be as complete and deep as possible, and this can be effectively pursued only through the mother tongue.

Only after concepts are clear, can students express them correctly; then, they may learn to express them through the mother tongue within the science courses, and through other languages within other forms and time-frames that are made available (including the foreign language courses).

4.4 Building Learners' Language-Mastery Through Chemistry Teaching and Chemistry Knowledge Through Language Teaching

In an ideal situation, language mastery should be gradually and efficiently developed during primary and secondary instruction and, therefore, it should not be a concern at the tertiary level. However, in recent decades, this has become less and less true. In South Africa, the standard of pre-university instruction is frequently unsatisfactory [e.g., 104, 112–114] leaving learners largely unprepared for the needs of tertiary education; language-mastery inadequacies are the major factor, both by themselves and for the drawbacks they determine. Similar problems are increasingly appearing also in contexts where language education used to have a strong tradition, like Italy [62–64]. Universities in many countries have started initiatives to improve the language and literacy skills of incoming students (e.g., [114]). Several Italian universities have started offering Italian language courses for entering students, to enhance their language skills, also in view of writing in science and technology (e.g., [115, 116]). Some South African universities have established Writing Centres, to offer support to students for the language challenges they experience (e.g., [117–119]).

UNIVEN offers an English Communication Skills course, compulsory for all entering students and meant to develop their language abilities. However, the still widely spread misconception that language is "not so important" in the sciences influences science students' perceptions and attitudes. They go through the course because it is compulsory, but they often do not put adequate efforts to actually improve their language skills. The fact that the language-proficiency entry-requirements are lower for students registering for science degrees than for the other students does not contribute to enhance science students' awareness of the importance of language mastery.

It becomes important to convince science students of the roles of language in the sciences by stressing the interconnections between language and conceptual aspects when learning or communicating science, or even when thinking about it. The integration of language training into chemistry teaching, in a form for which language analysis is utilised as an effective tool for explanations and/or clarifications of chemistry concepts, can be a promising option. It has been applied at UNIVEN within the courses taught by the author [120], resulting in significant clarifications at least for those students whose language mastery had a minimum level. The selected interactive teaching approach enabled sets of questions guiding students through active and coordinated analysis of language and chemistry features, until the desired clarifications were attained.

The strategies for integratively building language mastery through chemistry teaching and chemistry knowledge through language analysis can be further developed to become the key tool for relevant courses. It could also be optimal for a chemistry course specifically meant to bridge secondary and tertiary instruction [25]. In this case, students would be taught to read chemistry texts and write about chemistry concepts, to foster the development of language mastery in relation to chemistry, and errors would be analysed to stress the connections between language errors and chemistry errors. The concrete details of the approaches can be designed on the basis of careful investigation of students' answers aimed at identifying the language aspects that affect the understanding of concepts, or their expression, and linking them to the relevant chemistry meanings, so that both aspects are illustrated within classroom discussions. The investigation approach outlined in Sect. 4.2 has proved functional. It enables continuous interplay between investigation results and teaching modes [70], with the terms of the interplay sufficiently flexible for the teaching modes to tune to the needs of each specific group of students (needs that may change from topic to topic, from student group to student group and from year to year).

The greatest challenge for a large-scale implementation of the integration of language teaching into chemistry courses is the fact that it requires sophisticated language knowledge from the chemistry/science teacher—a type of knowledge that incorporates the identification of conceptual and epistemological relationships between the details of clauses and sentences and the details of the concerned scientific information. Developing this ability should become part of the training of science teachers at all levels. This is tantamount to incorporating humanities-based components into the preparation of science teachers to the purpose of enhancing the efficacy of science teaching and learning.

4.5 Discussion and Conclusions

The examples considered in the previous sections are not exhaustive. Thousands of examples have been collected, and a high number of categories of language-related difficulties affecting chemistry learning can be identified. However, the considered examples are sufficiently representative to underline the importance of language mastery for chemistry learning, and the serious impacts on the clarity and completeness of conceptual understanding when it is not adequately developed.

Although the importance of language mastery for science learning is gaining increasing recognition among educators, this awareness has not yet reached relevant implementation levels. For instance, it should be important to set high language mastery demands for incoming students registering for science degrees, both to ensure the possibility of effective learning and to promote students' awareness of the importance of language in the sciences.

The implementation level also entails attention to rigour from the teacher as a basis to request it from students [120]. Insisting on language rigour is not part of a pedantic attitude, but the utilisation of an important pedagogical tool [121–123]. Proving the impact of inadequate rigour on problem solving may be the most convincing way to show that language rigour is not "just semantics" (as several people still dismiss it), but has very practical implications. For instance, students tend not to distinguish between solute and solution—for example, between 'NaCl' and 'NaCl solution'—when talking or writing. Year after year, second year lab reports show a high proportion of students who calculate the molarity of an NaCl solution by dividing the mass of the solution by the molar mass of NaCl (the molarity is referred to the amount of NaCl dissolved per volume of solution, not to the overall mass of the solution). Only insisting on rigorous wording (for which 'NaCl' and "NaCl solution' are not allowed to be used interchangeably) can effectively prevent errors of this type, by building the mental perception that the two things are different. Regrettably, similar lack of precisions is frequent in media, e.g., in statements like ex-48, where the correct term would be "sodium chloride" or "table salt" (adding pure sodium into a can or a bottle containing a sauce would cause an explosion); examples like this also highlight the importance of basic chemical literacy for the public.

48. *Most canned and bottled tomato sauces are preserved with sodium.*

As mentioned previously, UNIVEN student population is largely underprivileged. This also means that language-related difficulties are likely to be more frequent and more extensive than in more advantaged contexts. On the other hand, incorrect use of language and misunderstanding of words or sentences are diagnosed with increasing frequency also in other contexts, including contexts with mother tongue instruction (and some of the examples considered in Sect. 4.3 are from other sources, including media and teaching materials). An escalating drop in the literacy levels of the young generations appears to be a phenomenon affecting several contexts (Sect. 1.3.2 in Chapter 1). Exchanges of views and experiences highlight the appearance of analogous errors in Europe and the Americas, or in South African universities with a good proportion of native speakers of the language of instruction. Some of the examples reported in [62] are comparable with some of the imprecisions or errors diagnosed at UNIVEN. This also implies that the need for effective addressing measures is fast-growing in an increasingly higher number of contexts.

The integration of language training into chemistry courses (or other science courses) can offer the most permanent addressing option, as it can span throughout a student's learning career, thus enabling better development of the language skills needed for chemistry learning while, at the same time, providing valuable opportunities to analyse and clarify chemistry concepts. Such a route entails continuous interplay between the analysis of students' works and the design of approaches to address identified problems taking into account all their features. Both steps involve investigation: investigation of language and conceptual components in the analysis of students' answers, and investigation of students' responses during classroom discussions. They thus require adequate expertise in both areas from the teacher's side.

The impossibility of untangling the language and the conceptual components of most errors, or to ascribe individual weights to each of them [72], is an indication of the impossibility of untangling humanities-based and science-based components in science teaching, learning and communication. The integration of language training into science courses requires simultaneous humanities-based and science-based expertise.

References

1. J. Bruner, Language as an instrument of thought, in *Problems of Language and Learning*. ed. by A. Davies (Edinburg University Press, Edinburg, 1975)
2. E. Asoulin, Language as an instrument of thought. Glossa J. Gen. Linguist. **1**(1), 1–23 (2016). https://doi.org/10.5334/gjgl.34
3. G. Dupre, What would it mean for natural language to be the language of thought? Linguist. Philos. **44**, 773–812 (2021). https://doi.org/10.1007/s10988-020-09304-9
4. A.H. Munby, Some implications of language in science education. Sci. Educ. **60**(1), 115–124 (1976)
5. S. Muralidhar, The role of language in science education: Some reflections from Fiji. Res. Sci. Educ. **21**, 253–262 (1991)
6. C.R. Sutton, *Words, Science and Learning* (Open University Press, Buckingham, 1992)
7. J. Wellington, J. Osborne, *Language and Literacy in Science Education* (Open University Press, Buckingham & Philadelphia, 2001)
8. L. Mammino, The essential role of language mastering in science and technology education. Int. J. Educ. Inform. Technol. **3**(4), 139–148 (2010)
9. R.A. Zepp, Relationships between mathematics achievements and various English language proficiencies. Educ. Stud. Mathem. **12**, 59–70 (1981)
10. S.J. Howie, Language and other background factors affecting secondary school students' performance in mathematics in South Africa. African J. Res. Math. Sci. Technol. Educ. **7**, 1–20 (2003)
11. M.J. Schleppegrell, The linguistic challenges of mathematics teaching and learning: a research review. Read. Writ. Q. **23**(2), 139–159 (2007). https://doi.org/10.1080/10573560601158461
12. J.N. Moschkovich, Language(s) and learning mathematics: Resources, challenges, and issues for research, in *Language and Mathematics Education: Multiple Perspectives and Directions for Research*. ed. by J.N. Moschkovich (IAP Information Age Publishing, 2010), pp.1–28
13. L. Jourdain, S. Sharma, Language challenges in mathematics education: a literature review. Waikato J. Educ. **21**(2), 43–56 (2016)
14. A. Msimanga, P. Denley, N. Gumede, The pedagogical role of language in science teaching and learning in South Africa: a review of research 1990–2015. Afr. J. Res. Mathem. Sci & Technol. Educ. (2017). https://doi.org/10.1080/18117295.2017.1367874()
15. A.A. Essien, The role of language in the teaching and learning of early grade mathematics: an 11-year account of research in Kenya, Malawi and South Africa. Afr. J. Res. Mathem. Sci. Technol. Educ. **22**(1), 48–59 (2018). https://doi.org/10.1080/18117295.2018.1434453
16. K. Erath, J. Ingram, J. Moschkovich, S. Prediger, Designing and enacting instruction that enhances language for mathematics learning: a review of the state of development and research. ZDM—Mathem. Educ. **53**, 245–262 (2021). https://doi.org/10.1007/s11858-020-01213-2
17. D.T. Brooks, *The role of Language in Learning Physics*. PhD dissertation (State University of New Jersey, USA, 2006)
18. N. Semeon, E. Mutekwe, Perceptions about the use of language in physical science classrooms: A discourse analysis. South Afr. J. Educ **41**(1), 1–11 (2021). https://doi.org/10.15700/saje.v41n1a1781

19. S. Markic, J. Broggy, P. Childs, How to deal with linguistic issues in chemistry classes, in *Teaching Chemistry—A studybook: A Practical Guide for Student Teachers, Teacher Trainees and Teachers*. ed. by I. Eilks, A. Hofstein (Sense Publishers, Boston, 2013), pp.127–152
20. L. Mammino, Importance of language mastery and mother tongue instruction in chemistry learning, in *The Role of Language in Teaching and Learning Science and Mathematics*. ed. by Z. Desai, M. Qorro, B. Brock-Utne (African Minds, Somerset West, 2013), pp.33–56
21. L. Mammino, Clarifying chemistry concepts through language analysis. Lumat Spec. Iss. ECRICE **3**(4), 482–500 (2015)
22. S. Markic, P.E. Childs, Language and the teaching and learning of chemistry. Chem. Educ. Res. Pract. **17**, 434–438 (2016)
23. C. Sutton, Los profesores de ciencias como profesores de lenguaje. Enseñanza de las Ciencias **21**(1), 21–25 (2003)
24. B.A. Brown, K. Ryoo, Teaching science as a language: A 'content-first' approach to science teaching. J. Res. Sci. Teach. **45**(5), 529–553 (2008)
25. L. Mammino, Focused language training as a major key for bridging the gap between secondary and tertiary instruction, in *ISTE International Conference Proceedings*. ed. by D. Mogari, A. Mji, U.I. Ogbonnaya (UNISA Press, Pretoria, 2012), pp.278–290
26. L. Mammino, The first year general chemistry course: great challenges and great potentialities, in *ISTE International Conference Proceedings*. ed. by D. Mogari, U.I. Ogbonnaya, K. Padayachee (UNISA Press, South Africa, 2014), pp.40–48
27. A.A. Lahore, Lenguaje literal y conotado en la enseñanza de las ciencias. Enseñanza de las Ciencias **11**(1), 59–62 (1993)
28. L. Mammino, *Il linguaggio e la scienza* (Societá Editrice Internazionale, Turin, 1995)
29. G.G. Mallinson, H.E. Sturm, L.M. Mallinson, The reading difficulty of textbooks for high school chemistry. J. Chem. Educ. **29**, 629–630 (1952)
30. Z. Fang, The language demands of science reading in middle school. Int. J. Sci. Educ. **28**(5), 491–520 (2006)
31. D. Arias, *Action Research Proposal: The Effect of Conceptual Change and Literacy Strategies on Students in High School Science Classes* (California State University, Northridge, 2007)
32. S.P. Norris, L.M. Phillips, How literacy in its fundamental sense is central to scientific literacy. Sci. Educ. **87**, 224–240 (2003)
33. S.P. Norris, L.M. Phillips, Reading as inquiry, in *Teaching Scientific Inquiry: Recommendations for Research and Implementation*. ed. by R.A. Duschl, R.E. Grandy (Sense, Rotterdam, 2008), pp.233–262
34. F. Davies, T. Greene, *Reading for Learning in the Sciences* (Oliver and Boyd, Edinburgh, 1984)
35. A. Horning, Reading across the curriculum as the key to success. *Across the Disciplines*, 4. http://www.wac.colostate.edu/edatd/articles/horning2007.
36. M.A.K. Halliday, Language and the order of nature, in *Writing Science: Literacy and Discoursive Power*. ed. by M.A.K. Halliday, J.R. Martin (The Palmer Press, London, 1993), p.118
37. R. Moore, Writing to learn biology. J. Coll. Sci. Teach. **1**, 289–295 (1994)
38. B. Hand, V. Prain, Writing for learning in science: a model for use within classrooms. Austral. Sci. Teach J. **42**(3), 23–27 (1993)
39. B. Hand, V. Prain, C. Lawrence, L.D. Yore, A writing in science framework designed to enhance science literacy. Int. J. Sci. Educ. **21**(10), 1021–1035 (1996)
40. L.P. Rivard, S.B. Straw, The effect of talk and writing in learning science: an exploratory study. Sci. Educ. **84**, 566–593 (2000)
41. L. Mason, P. Boscolo, Writing and conceptual changes. What changes? Instruct. Sci. **28**, 199–226 (2000)
42. L. Mason, Introducing talk and writing for conceptual change: a classroom study. Learn. Instr. **11**, 305–329 (2001)
43. A.P. Cabral, J. Tavares, Reading and writing skills in higher education: lecturers' opinions and perceptions, in *Proceedings of the European Conference on Educational Research* (University of Lisbon, September, 11 – 14, 2002).

44. H. Beall, In-class writing in general chemistry: a tool for increasing comprehension and communication. J. Chem. Educ. **68**(1), 148–149 (1991)
45. H. Beall, Literature reading and out-of-class essay writing in general chemistry. J. Chem. Educ. **70**(1), 10–11 (1993)
46. H. Beall, Probing student misconceptions in thermodynamics with in-class writing. J. Chem. Educ. **71**(12), 1056–1057 (1994)
47. M.M. Cooper, Writing—an approach for large enrolment chemistry courses. J. Chem. Educ. **70**(6), 476–477 (1993)
48. J.W. Wilson, Writing to learn in an organic chemistry course. J. Chem. Educ. **71**(12), 1019–1022 (1994)
49. W.G. Holliday, L. Yore, D. Alvermann, The reading-science learning-writing connection: breakthroughs, barriers, and promises. J. Res. Sci. Teach. **31**(9), 877–893 (1994). https://doi.org/10.1002/tea.3660310905
50. E.A. Castro, Escribir sobre química para aprender química. Panam. Newslett. **8**(4), 8–9 (1995)
51. F.M. Rossi, Writing in an advanced undergraduate chemistry course: An assignment exploring the development of scientific ideas. J. Chem. Educ. **74**(4), 395–396 (1997)
52. S. Kovac, Writing in chemistry: an important learning tool. J. Chem. Educ. **76**(10), 1399–1402 (1999)
53. R.J. Whelan, R.N. Zare, Teaching effective communication in a writing-intensive analytical chemistry course. J. Chem. Educ. **80**(8), 904–905 (2003)
54. B. Dawidowitz, The chemistry writing project at UCT: from teaching communication skills to writing as a tool for learning. Afr. J. Res. Math. Sci. Technol. Educ. **8**(2), 127–139 (2004)
55. L. Ablin, Student perception of the benefits of a learner-based writing assignment in organic chemistry. J. Chem. Educ. **85**(2), 237–239 (2008)
56. J.A. Reynolds, C. Thaiss, W. Katkin, R.J. Thompson, Writing-to-learn in undergraduate science education: a community-based conceptually driven approach. CBE—Life Sci. Educ. **11**(1), 17–25 (2012)
57. L. Mammino, Teacher-students interactions: the roles of in-class written questions, in *Chemistry Education and Sustainability in the Global Age*. ed. by C. Mei-Hung (Springer, Dordrecht, 2013), pp.35–48
58. Science Writing: A Tool for Learning Science and Developing Language, 2015, *Institute for Inquiry, Exploratorium.*
59. S. Graham, S.A. Kiuhara, M. MacKay, The effects of writing on learning in science, social studies, and mathematics: a meta-analysis. Rev. Educ. Res. **90**(2), 179–226 (2020). https://doi.org/10.3102/0034654320914744
60. One of the quotations appearing at the end of the outputs of calculations performed with the Gaussian software
61. A. Baricco, I. Barbari, *Saggio sulla mutazione* (Fandango, Rome, 2006)
62. P. Mastrocola, *Togliamo il disturbo. Saggio sulla libertà di non studiare.* Parma: Guanda, 2011.
63. L. Aterini, È l'analfabetismo funzionale «la più grande emergenza dell'Italia», *Greenreport.it,* 7 September 2020
64. L. Mammino, La chimica come area ideale per l'educazione al linguaggio della scienza. *Edichem '99: Chemistry in the Perspective of the New Century*, Bari, 1999, 212−230
65. L. Mammino, Interplay, interfaces and interdependence between visual literacy and language mastering, as highlighted by chemistry students' works. *Anuario Latinoamericano de Educación Química (ALDEQ)*, 2010, XXV, 16−20
66. L. Mammino, The interplay between language and visualization: the role of the teacher, in *Science Teachers' Use of Visual Representations.* ed. by B. Eilam, J. Gilbert (Springer, Dordrecht, 2014), pp.195–225
67. L. Mammino, Science students and the language problem: Suggestions for a systematic approach. Zimbabwe J. Educ. Res. **10**(3), 189–209 (1998)
68. L. Mammino, Language-related difficulties in science learning. I. Motivations and approaches for a systematic study. J. Educ. Stud. **4**(1), 36–41 (2005)

69. R. Levin, Frontiers in group dynamics. II. Channels of group life: social planning and action research. Hum. Rel. **1**, 143–159 (1947)
70. L. Mammino, Educational components in the supervision of chemistry postgraduate students: experiences and reflection. Phys. Sci. Rev. **5**, 1 (2021). https://doi.org/10.1515/psr-2020-0116
71. L. Mammino, Chemistry teaching and chemical education research: 30-year experience in integration pathways. In Mammino L, Apotheker J. (Eds), *Research in Chemistry Education - Proceedings of the 2nd ACRICE Conference* 143–160 (Springer, 2021)
72. L. Mammino, Empleo del análisis de errores para aclarar conceptos de química general. Enseñanza de las Ciencias **20**(1), 167–173 (2002)
73. L. Mammino, L'analisi degli errori come strumento didattico. Nuova Secondaria **10**, 75–76 (1996)
74. A. Love, L. Mammino, Using the analysis of errors to improve students' expression in the sciences. Zimbabwe J. Educ. Res. **9**(1), 1–17 (1997)
75. L. Mammino, The spontaneity of chemical reactions: challenges with handling the concept and its implications. Phys. Sci. Reviews (2022). https://doi.org/10.1515/psr-2021-0144
76. L. Mammino, Language-related difficulties in science learning. II. The sound concept correspondence in a second language. J. Educ. Stud. **5**(2), 189–213 (2006)
77. L. Mammino, Language-related difficulties in science learning. III. Selection and combination of individual words. J. Educ. Stud. **6**(2), 199–214 (2007)
78. L. Mammino, Language-related difficulties in science learning. IV. The use of prepositions and the expression of related functions. J. Educ. Stud. **8**(4), 142–157 (2009)
79. L. Mammino, Cause and effect—a relationship whose nature involves conceptual understanding, scientific method, logic and language. ALDEQ **21**, 33–37 (2006)
80. L. Mammino, Physical quantities and their changes: Difficulties and perceptions by chemistry students. South. Afr. J. Res. Math. Sci. Technol. Educ. **5**, 29–40 (2001)
81. L. Mammino, La percepción de la distinción entre sistemas y procesos por parte de los alumnos de química. ALDEQ **15**, 125–129 (2002)
82. L. Mammino, Actitud hacia números y valores en cursos de química general. ALDEQ **16**, 5–10 (2003)
83. L. Mammino, General y particular en química. ALDEQ **14**, 25–28 (2001)
84. L. Mammino, The Hess law—chemical information from a rigorous statement and challenges experienced by students. ALDEQ **28**, 167–172 (2013)
85. L. Mammino, The Hess law and the state-function nature of enthalpy: the concept and its expression by chemistry students. ALDEQ **29**, 100–105 (2014)
86. L. Mammino, Students' understanding of the rate law, as highlighted by laboratory reports. ALDEQ **30**(1), 175–180 (2015)
87. L. Mammino, What happens during a chemical reaction? Insights into students' perceptions. ALDEQ **30**(2), 20–25 (2015)
88. L. Mammino, The spontaneity concept: an investigation of the dichotomy between learning the definition and handling the concept. ALDEQ **27**, 120–125 (2012)
89. L. Mammino, An investigation of students' difficulties in handling the spontaneity concept in electrochemistry. ALDEQ **27**, 155–160 (2012)
90. L. Mammino, *Terminology in science and technology—an overview through history and options* (Ditlou Publishers, Thohoyandou, 2006)
91. D. Clerk, M. Rutherford, Language as a confounding variable in diagnosing misconceptions, in *Proceedings of the Sixth Annual Meeting of the SAARMSE* 126–131 (UNISA Press, Pretoria, 1998).
92. E.N. Lee, M.K. Orgill, Toward equitable assessment of english language learners in general chemistry: identifying supportive features in assessment items. J. Chem. Educ. (2021). https://doi.org/10.1021/acs.jchemed.1c00370
93. L. Mammino, Method-related aspects in an introductory theoretical chemistry course. J. Molec. Struct. (Theochem) **729**, 39–45 (2005)
94. L. Mammino, The mother tongue as a fundamental key to the mastering of chemistry language, in *Chemistry as a Second Language: Chemical Education in a Globalized Society*. ed. by C. Flener, P. Kelter (American Chemical Society, Washington, 2010), pp.7–42

95. J.K. Gilbert, Visualization: an emergent interdisciplinary field of practice and enquiry in science education, in *Visualization: Theory and Practice in Science Education*. ed. by J.K. Gilbert, M. Reiner, M. Nakhlekh (Springer, Dordrecht, 2008), pp.3–24
96. R. Harre, *The Principles of Scientific Thinking* (Chicago University Press, Chicago, 1970)
97. L. Mammino, Language mastery, narrative abilities and oral expression abilities in chemistry learning: importance of mother tongue and traditional narration. South Afr. J. Afr. Lang. **5**(1), 19–27 (2015)
98. S.M. Case, The language barrier in science teaching. Teach. Educ. in New Countries **9**(1), 15–26 (1968)
99. C. Benson, *The importance of mother tongue based schooling for educational quality* (2005/ED/EFA/MRT/PI/9, 2004)
100. N. Alexander (ed.), *Mother-Tongue Based Bilingual Education in South Africa: The Dynamics of Implementation* (Salty Print, Cape Town, 2005)
101. B. Brock-Utne, R.K. Hopson (eds.), *Languages of Instruction for African Emancipation—Focus on Postcolonial Contexts and Consideration* (CASAS, Cape Town, 2005)
102. B. Brock-Utne, I. Skattum, *Languages and Education in Africa: A Comparative and Transdisciplinary Analysis* (Symposium Books, Oxford, 2009)
103. K.A. Heugh, Into the cauldron: An interplay of indigenous and globalised knowledge with strong and weak notions of literacy and language education in Ethiopia and South Africa. Lang. Matters Stud. Lang. Afr. **40**(2), 166–189 (2009)
104. Y.I. Rubanza, Competition through English. The Failure of Tanzania's Language Policy. In Prah, K.K. (Ed.) *Rehabilitating African Languages* 39–51 (CASAS, Capetown, 2002)
105. M. Qorro, Language planning and policy in Tanzania: When practice does not make perfect, in *The Role of Language in Teaching and Learning Science and Mathematics*. ed. by Z. Desai, M. Qorro, B. Brock-Utne (African Minds, Somerset West, 2013), pp.151–170
106. L. Mammino, Essential roles of language mastery for conceptual understanding and implications for science education policies, in *Human Factors of a Global Society: A System of Systems Perspective*. ed. by T. Marek, W. Karwowski, M. Frankowitz, J. Kantola, P. Zgaga (Taylor & Francis Inc., CRC Press Inc., Rosa Boca (USA), 2014), pp.835–847
107. P. Dall'Antonia, M. Rosolini, La lingua straniera come veicolo di comunicazione nella didattica della chimica, in *IV Conferenza Nazionale sull'Insegnamento della Chimica*, Assisi (Italy), 9-1-December (2004).
108. R. Noormohamadi, Mother tongue, a necessary step to intellectual development. J. Pan-Pac. Assoc. Appl. Linguist. **12**(2), 25–36 (2008)
109. R. Nishanthi, Understanding of the importance of mother tongue learning. Int. J. Trend Sci. Res. Dev. (IJTSRD) **5**(1), 77–80 (2020)
110. Why mother language-based education is essential, https://www.unesco.org/en/articles/why-mother-language-based-education-essential. Accessed 15-07-2022.
111. L. Mammino, The African context: investigating the challenges and designing for the future, in *Research in Chemistry Education—Proeeedings of the 2nd ACRICE Conference*. ed. by L. Mammino, J. Apotheker (Springer, Berlin, 2021), pp.177–189
112. V. Reddy, A. Kaniee, G. Diedericks, L. Winnaar, *Mathematics and Science Achievements at South African Schools in TIMMS 2003* (HSRC Press, Pretoria, 2006)
113. B. Fleisch, *Primary education in crisis: Why South African school children underachieve in reading and mathematics* (Juta & Co., Cape Town, South Africa, 2008)
114. N. Mtshali, N. Smillie, Many of our kids can't read or write: Study shows pathetic state of primary school education. *Sunday Time*, June 29 (2011)
115. D. Motta-Roth, Academic literacy in the South. Writing practices in a Brazilian university). *Writing Programs Worldwide. Profiles of Academic Writing in many places* 105–116 (2012).
116. M. Dota, G. Polimeni, M. Prada (Eds.), Scrivere oggi all'università, *Quaderni di Italiano Linguadue*. 15932-Articolo-47794–1–10–20210702.pdf (2019)
117. S. Demartini, Dalle tecniche di scrittura alla scrittura tecnica: scrivere nei corsi di laurea scientifici. Quaderni di Italiano Linguadue **12**(1), 3 (2020). https://doi.org/10.13130/2037-3597/13985

118. A. Archer, Investigating the effects of writing centre interventions on student writing. South Afr. J. Higher Educ. **22**(2), 2010–2016 (2008)
119. A. Archer, R. Richards (eds.), *Changing Spaces: Writing Centres and Access to Higher Education* (Sun Press, Stellenbosch, 2011)
120. S. Kane, Reading and writing in the writing centre. An integrative approach, in *Proceedings of the HELTASA (Higher Education Learning and Teaching Association of South Africa) 2012 Conference* 46–54 (Stellenbosch University, South Africa, 2012)
121. L. Mammino, Teaching physical chemistry in disadvantaged contexts: Challenges, strategies and responses, in *Chemistry Education in the ICT Age*. ed. by M. Gupta-Bhowon, S. Jhaumeer-Laulloo, H. Li Kam Wah, P. Ramasami (Springer, Dordrecht, 2009), pp.197–223
122. J.D. Bradley, M. Brand, G.C. Gerrans, Excellence and the accurate use of language, symbols and representations in chemistry, in *Widening the Scope of* Chemistry 135–144 (IUPAC Blackwell Scientific Publications, 1987)
123. L. Mammino, Precision in wording: a tool to facilitate the understanding of chemistry. CHEMEDA Austral. J. Chem. Educ. **48**, 30–38 (1998)
124. L. Mammino, Rigour as a pedagogical tool, in *The Language of Science*. ed. by S. Seepe, D. Dowling (Vyvlia Publishers, Johannesburg, 2000), pp.52–71

Chapter 5
Language and Communication in Science, Engineering and Technology

Armando Mammino and Liliana Mammino

Abstract Language mastery has a fundamental role in science, engineering and technology, both in the learning stage and throughout professional practices. The present chapter focuses mostly on the roles of language in engineering and technology, spanning through a variety of relevant aspects, from the optimisation of language usage in reports and article writing to historical synopses about crucial issues such as the relationships between language and logic and between language and knowledge. It thus comprises practically implementable outlines, practically-oriented motivations for the need of high level language mastery, and theoretical motivations through the awareness of the role of language as a crucial focus of philosophical and epistemological investigation throughout centuries.

Keywords Characteristics of a good engineering text · Concerns about language mastery lowering · Improving language mastery through self-teaching · Language and engineering · Language and science · Research on language · Logic and knowledge through history

5.1 Introduction

According to common understanding, language is a communication instrument utilised by the members of a given community. In other words, language refers to the ability—peculiar to human beings—to communicate thoughts and information by means of sounds (spoken language), symbols corresponding to sounds (written language), graphic signs (communication through various types of drawings), or mathematical symbols combined into formulas (formal-symbolic language).

A. Mammino
Consultant Structural Engineer, Treviso, Italy

L. Mammino (✉)
Faculty of Science, Engineering and Agriculture, University of Venda, Thohoyandou, South Africa
e-mail: mammino@univen.ac.za; sasdestria1@gmail.com

L. Mammino (ed.), *Insights into the Relationships between Humanities and Sciences*,
https://doi.org/10.1007/978-3-031-97900-2_5

A huge ensemble of reflections (philosophy of language) has been developed, from ancient Greece times up to current times, entailing theories and hypotheses on the nature and structure of language, the roles of the symbols it utilises (for the previously listed categories as well as other possible ones), and the communication behaviours that are recurrent within natural and artificial language systems. Such reflections have attained new perspectives in the XX century, also thanks to the contributions of new developments in disciplines such as symbolic logic, linguistics, epistemology, semeiotic, and information theory.

Besides using the common language as essential communication tool, various science and technology areas have developed artificial and/or symbolic languages consisting of digits, letters, symbols, or codes. They are utilised in a strictly conventional way on the basis of shared pre-set rules ensuring that:

- concepts are expressed in an unambiguous way;
- the formal connections of a logical system are explicit and accessible to analysis (as common, e.g., in the formalised languages of mathematics, logic, and their applications in physical sciences and derived technological areas);
- instructions for procedures and practical operations are clear and unambiguous.

The XX century has also witnessed the development of programming languages, which express instructions in a way that computers can execute them and perform the wanted operations. They implement communication between humans and a machine, thus responding to an extension of the *language* concept beyond the human-to-human realm (a different extension of the language concept to communication between humans and non-humans could be the *gesture language* between humans and animals, which the latter learn to comprehend).

As already stated, language is a communication tool which transmits information from one thinking being to another by utilising vocal, written, graphical or symbolic signs. The signs for a specific purpose are always the same within the same language (e.g., the letters corresponding to specific sounds in a written language); in this way, they act as signals recalling the same individual responses, while their combinations transmit specific meanings. For instance, each letter of the alphabet represents a specific sound; a combination of letters constructed in a way that the overall sound corresponds to a word in a given language represents that word, and its meaning is understandable to all those who share that language; a combination of words according to logical patterns builds a sentence, i.e., an item capable of transmitting information.

The current chapter focuses on the language and communication tools used in science and technology, and relates them with the humanities-pertaining roots of language. The language in application-oriented and/or technology-oriented disciplines is given major attention, and engineering (more often, structural engineering) is here selected to represent them. The chapter considers both practical aspects, such as the quality of language usage needed in technical writings and the motivations to pursue good quality, and theoretical aspects reviewed in their historical developments and providing motivations for the necessity of expression rigour and attention to logic in any writing.

For the sake of conciseness, the authors will from now on be simply denoted as AM and LM respectively (pronouns he/him/his and she/her/hers respectively).

5.2 Language in Science and Technology

5.2.1 Is Language Mastery Important for an Engineer or a Technology Expert?

The use and roles of language in science, engineering and technology communication constitute a cross-area or interface between science and technology on the one side and the humanities on the other. The cross-disciplinarity is guided by, and simultaneously realises, the requirements and modes of epistemology, i.e., the philosophy of science.

The roles of language in engineering professional practice can be conveniently illustrated by considering the kind of documentation entailed by engineering projects, according to their nature and location:

1. report with a description of a construction intervention (building, infrastructure) or a product to be manufactured and commercialised, maybe even to be patented (e.g., a prefabricated structure, a car);
2. report on the materials utilised for the item considered in (1);
3. report on the structural calculation of the same item;
4. reports on systems (e.g., electrical, hydraulic) to be built in, or on mechanical components;
5. geological, geotechnical, geomechanical and geohydrological reports;
6. hydraulic report;
7. report on environmental impact;
8. report on landscape features or impact;
9. archaeological report;
10. reports on static, technical-administrative, mechanical, electrical, and other, tests.

This list is not exhaustive, but is sufficient to illustrate the extensive writing requirements in the normal practice of engineering and other application-oriented professions.

A diffuse commonplace believes that engineers, architects and other application-oriented specialists communicate basically through images, formulas and numbers. In reality, all these graphical, ideographical and symbolic ciphers are not able to communicate anything if they are not accompanied by written texts, i.e., texts expressed through sentences. Written texts are indispensable in any document within any discipline.

It then becomes important to analyse the characteristics of specialised language in application-oriented disciplines. This language needs to have high quality, as the contents that are the communication objects must be conveyed without errors, omissions, ambiguities or lacunae. The text must primarily aim at full understandability of its contents, both in terms of the representation of the documented results and in terms of the demonstrative logical flow that has led to those results after the definition of the objectives. Logical rigour is one of the heritages that technology has acquired from science (as a sort of parental inheritance) and from which it cannot depart in any of its application, under penalty of distortions and the loss of the reliability of its production-related messages. It follows that fostering students' mastery of the language needed in engineering, up to attaining adequate levels, needs to be an integral component of their training.

Ideally, an engineer's training would comprise high standards of content and teaching at university level, followed by systematic guidance by senior experts in the first years of professional career. This was also the common practice in the Italian system up to only few decades ago: high quality demands within university courses, followed by situations in which the best construction enterprises provided practical training internally, by associating a newly-employed engineer with a more experienced one. This simultaneous transmission of theoretical knowledge and applied mastery entailed also the appropriate use of language in relation to the various components and objectives of the 'job'. Some teachers made the importance of language very explicit; for instance, Prof. Claudio Datei, in his first lecture of the Hydraulic Constructions course at the University of Padua, used to inform the students that, at exam level, an Italian language error would have been considered as an engineering error. Chapter 4 analyses this issue in relation to chemistry, showing that, in most cases, a language error is inextricably entwined with incorrect chemistry.

Rigorous correctness in science, engineering and technology communication is essential to ensure that no space is left to misunderstandings, incomplete acquisition of knowledge, and consequent implementation errors. To further emphasise the importance of training rigour and completeness, it may be recalled that errors in some areas may cost human lives. This is more easily evident for medicine (whose practice is directly associated with human health), structural engineering (which is responsible for the stability and resilience of buildings and other structures, like bridges or tunnels) and chemical engineering (which is responsible for the safety of the production processes in the chemical industry). Errors in the practice of medicine may cost the lives of patients; they may or may not be detected, because they affect one person at a time, and are not so blatantly evident to the public. Errors in structural engineering may lead to the collapse of a building or a bridge, causing a high number of causalities in one moment. Errors in chemical engineering (either in the design phase or in the process control during the functioning of a plant) may cause disasters like explosions or the release of toxic chemicals into the surrounding areas, where casualties may be in the order of thousands (more than 2000 deaths in the Bhopal disaster in India in 1984) and vast areas may become heavily polluted. Therefore, ensuring good training for students preparing for these professions is an ethical duty of the educational systems in all countries. Teachers need to keep this in mind and to inform students in a straightforward way. An anecdote may help illustrate the latter point. In one of the years in which LM was teaching the process technology course at the University of Venda, a group of students approached her to tell her that it would be important that they all pass, because if many of them failed "it would not look good for you" (it was the time when the idea that it is the teacher's fault if students do not pass a course was starting taking hold in the institution). LM answered that "I see three possibilities: (i) you work hard and you pass, and we are all happy; ii) you do not work enough, and then I will have to choose between two options: (a) you fail the course, come back next year for this same course, and work more; (b) I pass you even if you have not learnt, and in few years' time I will have to visit you in jail". The students asked the reasons for the last statement, and LM explained that process technology concerns the chemical industry; if, after graduating, they go to work in an industry without having learnt enough, they will end up causing some disasters for which they will be held responsible. LM's trust in students' basic reasonability paid off: the students got the message, worked hard, and all of them passed the course because they deserved it. This anecdote is also meant to convey the message that students (with the basic honesty inherent in most young people) are available to get messages that are conveyed in convincing terms; lecturers teaching courses related to practical applications, where errors can involve high risks, should never compromise the quality of the learning standards expected from students; they need to explain why the standards cannot be lowered, and students are likely to understand and respond constructively. Similarly, those same lecturers need to counteract suggestions based on other persons' (not involved in these types of courses) educational opinions, which would make allowance for partial learning, such as the suggestions that a lecturer should be content if students understand 50% of the course material; for these courses, the remaining 50% might entail a disaster in the making and, therefore, the suggestion is not applicable [1]. Of course, the lecturer's guidance through correct language needs to be part of the guidance to full understanding, and the analysis of professional implications stemming from language errors can be used as one of the constructive routes to facilitate students' professional development.

What has been mentioned in the previous paragraphs is sufficient to effectively dispel another aberrant cliché: the diffuse statement—often repeated also by politically-influential persons—that professionals in scientific and technological areas do not need to write well: they would thus appear to be authorised to write in 'whichever way', because the things that matter for them are 'others'. Two aspects are particularly worrying in this highly questionable statement. 'Writing well' enables a univocal interpretation, which can be viewed as universal within a given language, whereas 'writing poorly' (the frequent outcome of writing in 'whichever way') leads to situations in which the desired content gets lost in a myriad of possible aberrations; the variety of expression errors may become so broad and chaotic that its effects emulate the "Babel tower", resulting in absence of communication, i.e., defeating the objective of communicating all the details of crucial information. The idea that "other things matter" within technology communication is also worrying, if it is meant to exclude the importance of good writing. A technology expert needs to know the numbers, formulas and state-of-the-art of his/her field of expertise; furthermore, the "expertise" term also entails being a bearer of culture—a mind-set that is attained through extensive reading, which, in turn, automatically results in the ability of good language usage.

The Italian engineer and mathematician Giulio Krall (1901–1971) can be chosen as an outstanding example of this type of mind-set. He was one of the pioneers of the construction science, which he applied to innovative designs of bridges and other large structures, with the support of deep mathematics knowledge. His books (e.g., [2–6]) and articles utilise language in a highly rigorous way and highlight deep and broad knowledge background; they also took for granted the information already reported in previous literature. On the other hand, when he went to the building sites where his projects were being constructed, he could speak the workers' language and explain them not only what they were expected to do, but also why it was important to do it in the indicated way, thus stimulating their willingness to work at their best. In other words, he could modify the use of language according to the persons for whom the communication was meant: a concise and extremely precise language in works meant for specialists; a popular and even colourful language when talking to workers, managing to maintain conceptual rigour while simplifying the expression, so that even persons who had not gone beyond primary education could understand all the messages. This ability to adapt language to the communication recipient in a way simultaneously ensuring understandability and rigour is of paramount importance in science and engineering communication, whether meant for specialists, for workers, or for students.

5.2.2 Origin and Development of Language and Written language—a Quick Outline

Organised languages, corresponding to precise conventions shared within individual communities, were moulded, at the beginning of the various civilisations, through the prose of legend narrations and through the poetry of the long epic and mythological poems narrating the origins of a people; the metrics and often rhyme of the narrations served as support to their memorisation. Scientific and technical language started forming at later stages of a civilization development, when a community became sedentary, organised, and sufficiently advanced to afford theoretical investigations and reflections through the minds of its philosophers, mathematicians, astronomers, land surveyors, healing practitioners, and other professionals. By their nature, science and technology needed to be communicated, at least within the group of those practicing them and to the persons in power, who would identify interesting implementations and organise their realisation. The latter objective prompted scientists and other professionals to build a language that could communicate concepts also beyond the group of specialists who developed them, maintaining communication efficacy.

A rational and clear integration of language and symbolic representations is extant in Euclid's *Elements of Geometry* (c 300 BC; [7]): 13 books in which text, images and symbols combine to build demonstrations with high logical rigour, and the explanations expressed through words reflect the traditions of prose and poetry as they had already been present for centuries in the Greek civilisation. Similar trends appear in subsequent works. The ten books *On Architecture* by Marcus Vitruvius Pollio (c. 90—c. 20 BC, [8]) constitute a technical treatise; its style is that of the *aurea mediocritas* (the golden average) that was the average basis in Latin literature, with several passages rising above it for their descriptive and narrative efficacy.

The convergence of the qualities of prose and poetry into these texts has important roles, because each of them brought specific contributions to the scientific and technical language of those times. Prose is the mould of ordinary metaphors and expression tools, which builds the structural basis of a text. Poetry is the mould of expression tools meant to leap towards intuition and the emotional aspects of learning, coming in when pure exposition-rationality may not be sufficient to guide the reader to full understanding of a given theme. The contribution of prose to scientific and technical language is the more extensive and has dominant proportions, whereas the contribution of poetry appears localised in places where additional qualitative representation ability is relevant to ensure understanding by the reader. In any language, the metaphors, expression tools, images, style and strength of poetry are the driving ones with respect to the correspondent components of prose; a telling similitude could entail a comparison of the mechanics of sport cars with respect to the mechanics of common cars, in terms of technological advancement and continuous improvement. When the prose-poetry synergies promote enhancements in the basis prose (representing the *aurea mediocritas* of a certain community at a certain time), all the writing communication abilities attained by that community are strenghtened and enhanced, and science and technology texts also benefit from this enrichment. In other words, as far as language and written texts are concerned, multidisciplinarity between humanities and sciences involves not only prose, which, starting from literature, provides ordinary metaphors and expression tools, but also poetry, which, also starting from literature, provides special metaphors and expression tools to be used sparely, when they can improve understanding in ways that would otherwise be impossible.

As already observed, the correct and masterful use of language is fundamental in science and technology writings to ensure correct and complete communication of information and, consequently, the rational management and realization of projects with full implementation of all the communicated details. Therefore, a science or technology expert needs to acquire and master the expression resources of the language utilised, as they enable optimal communication of thoughts, information and instructions. This, in turn, entails adequate level of multidisciplinarity with the humanities, which are the univocal repositories of the knowledge and mastery of language resources and the art of communication, as well as of logic, which is essential for science and technology communication (as paradigmatically highlighted by mathematics or mathematical-physics texts). Such multidisciplinary attitudes should not be limited to particularly sensitive individuals—as in the current situation—but should become inherent components of science and engineering practicing, and possibly expand to other professional areas and to public sensitivity. The very fact that we use the word 'literature' in terms like 'scientific literature' or 'technical literature' recalls the humanities roots of expression through language.

The engineer also needs to be a 'humanist', capable of thinking in terms of the human species and its future, because his/her work affects the environment, which is the habitat of the *Homo sapiens* species. Such dual 'soul'—engineer and humanist—corresponds to optimal multidisciplinarity between humanities and technology. In turn, this multidisciplinarity is essential for the correct communication of messages intended for implementation.

Education needs to play crucial roles to ensure that multidisciplinary and interdisciplinary attitudes become a generalised acquisition, instead of being an asset possessed only by few individuals. This would require a reform of educational systems (e.g., in Europe and North America), aimed at equipping people graduating in science or technology areas with adequate language and epistemological mastery. In turn, this would require: calibrating explanations to ensure that students understand textbooks, articles and lectures; promoting reflections; and fostering writing abilities up to the levels that will be needed in professional life. The opposite process would be fatal—barbarization of the language and its usage, eventually leading to a Babel tower scenario.

The issue of education is very broad; it is a crucial object of investigation and debates in many countries, also because of the lowering of its standard witnessed in the last decades in many contexts; in particular, the role of quality teacher-students' relationships, that had been a key to ensure culture and expertise continuity in the past, is fading fast. This topic will be given specific attention in Sect. 5.2.3.6.

5.2.3 Some Considerations on the Prose of a Scientific or Technical Text

5.2.3.1 Pursuing a Text's Quality

As already mentioned, the quality of language usage in a science or engineering text needs to be high, whichever the theme of the text (theoretical, practical, or a mix of the two). The major characteristics of such texts can be summarised as follows:

(a) maximum understandability;
(b) logical sequencing consistent with the logic typical of science and engineering discourses. It corresponds to what was formerly termed "engineer's logic", derived from mathematics and verbatim applied to practical implementations;
(c) construction simplicity, tending to linearity in the text structure as well as in individual sentences;
(d) formal rigour, to be optimally managed through the resources of the language utilised.

The routes or levels through which such quality is pursued can be hierarchically summarised as follows:

1. correct spelling;
2. correct grammar;
3. correct syntax;
4. understandability by the reader;
5. establishing of a personal style by the author;
6. style enhancement, aiming at high style levels.

'Grammar' refers mostly to the organisation of words within a clause, and 'syntax' to the connections of clauses within a sentence. The difference between (5) and (6) may be compared to the difference between the style of a good student and that of a professional capable of composing artistic prose.

The characteristics are listed hierarchically because each of them is possible only if the preceding ones are implemented; in other words, a characteristic n, with $1 \leq n \leq 6$, implies that all the characteristic i, with $1 \leq i \leq n - 1$ are implemented, whereas the characteristics $n + 1 \leq i \leq 6$ are not necessarily implemented.

Besides the technical reports listed in Sect. 5.2.1, which are inherent to the professional practice and are required by tradition or by law, an engineer may wish or need to prepare:

1. papers to be published in specialised journals or presented at conferences or congresses;
2. treatises or books;
3. papers to be published in science popularising magazines;
4. papers to be published in newspapers.

The content of each of them may be mostly technical, scientific (with main focus on theoretical aspects), or multidisciplinary, with reference to a variety of other disciplines, including humanities-based ones. The papers in point (3) and (4), although both meant for the broad public, differ by their approach and style. Papers for popularising magazines have a wider scope, utilise an information route which could be viewed as a 'cognitive service', fulfil the dissemination objective, and suggest the possibility of further research by interested readers through a good bibliography. Papers meant for newspapers are short, provide only basic outlines, and need to be captivating, so as to stimulate curiosity and prompt further deepening.

Authors who have learnt to write well can do it within all these types of writing. Besides the engineers who have written treatises that became milestones in their field, like the previously mentioned Giulio Krall, or Eduardo Torroja (1899–1961), Pier Luigi Nervi (1891–1979), Sergio Musumeci (1926–1981), and many others, there have been engineers who were also poets, like Umberto Saba (1883–1957), or writers, like Achille Collas (1795–1859), Fyodor Dostoyevsky (1821–1881), Rober Musil (1880–1942), Carlo Emilio Gadda (1893–1973), Roberto Vacca (born 1927) and several others; one might recall the association envisaged by a renowned statement by Stalin, according to which a writer is an "engineer of the soul".

5.2.3.2 Practical Stages in the Preparation of a Text

The practical operations to pursue quality in a written text involve drafting, followed by refining stages and proofreading stages. Drafting the text puts the seeds of quality, through the outline on the main issues that are the object/s of the text, as well as the necessary supporting information like the conceptual background, the general state of the art to which the object/s belong, the discussion of the selected approaches, and all the other aspects that, together, build the literature review. It may be useful (above all for beginners, like postgraduate students [9]) to initially prepare a framework delineating the predictable components of the text and their logical sequence; although the framework will often be modified/upgraded during the actual writing, it helps getting started and proceeding in a viable way. Experienced authors conceive this framework mentally, as part of envisaging their approach to the text.

The refining stages may entail working on the initial draft to add content and/or parts clarifying the logical connections between different pieces of information, until the overall information is communicated fully, i.e., the entire information ensemble that the author wishes to communicate is present and understandable. After this status is attained, the next refining stages involve accurate checking of the words' selection, the correctness of individual sentences, and the logical sequence with which individual pieces of information are presented and connected to each other, as well as the internal logic of things typically depending on logic, such as mathematical proofs. This also entails ensuring that there are no repetitions of the same statement and no internal contradictions, as well as making sure that the overall treatment makes the logical thread—from the initial information (hypotheses, description of the system considered, or any other information that introduces the topic) to the conclusions—clear for the reader. In practice, the checking requires both accurate verification of the use of language and accurate verification of how technical components (e.g., mathematical proofs or descriptions of any type) are presented. It is not easy for an author to detect all the imprecisions or readability-problems in a text, because that same text is somehow in the author's mind. More than one proofreading is advisable. It is often useful for the author to carry out a final proofreading by trying and playing the role of the reader—a 'game' that becomes easier if time-constrictions allow the possibility of doing the final proofreading at some time-distance after the previous ones.

As already mentioned, the ability to analyse the relationships between the wording of a sentence and the actual details of the concept/s involved develops automatically through reading; authors who have read extensively take automatical care of these relationships also when they write. Books specifically focusing on these relationships may be useful for beginners (postgraduate students, people in early stages of professional careers), to build awareness about what requires attention. An analysis of actual errors, discussing how each error fails the correct communication of one or another relationship, is viewed as an effective option to attract attention on practical routes to match expression and actual concept/s on writing and on proofreading. Such an analysis is carried out in [10], with an extensive selection of error categories, focusing mostly on chemistry and occasionally extending to physics; a similar approach for the applied sciences, first of all engineering, could provide an effective learning tool, above all for students and early-career professionals. It is also interesting to note that, while grammar errors remain within a specific language, the conceptual and/or epistemological aspects of various language errors are detected in more than one language; e.g., LM has verified the presence of corresponding errors of this type in Italian and English chemistry writings.

5.2.3.3 Complex Sentences and Complex Thinking

The quality of text writing depends on an author's mastery of the resources of a given language and on the level of the internalised conceptual and epistemological knowledge. It cannot depend on externally-decided artificial restrictions. One of the suggested restrictions that appear rather frequently in recent decades concerns the length of sentences. Any generalization of preconceived schemes in this regard would be absurd, because the length of a sentence depends solely on the nature, complexity and articulation of the concepts to be expressed. For the same reason, a request that a text should contain only sentences of comparable length would be equally absurd.

Comparatively shorter sentences are more suitable for the representation of simple concepts, and a sequence of short sentences is suitable for the representation of a series of simple concepts. Long sentences are more suitable for the presentation of complex concepts, with a natural correspondence between the complexity of a concept and the complexity of the sentence expressing it. The expression of a complex concept requires an ensemble of principal, coordinated and subordinated clauses connected to each other in such a way that the logic of the connections is easily apparent, to ensure univocal understandability. An easy way to facilitate readability of a complex sentence is that of starting with the principal clause, so that the reader is immediately informed about the theme of the sentence, and organising the subordinated clauses in such a way that they provide the associated information in a logical sequence (i.e., each clause is preceded by the ones that condition its understandability). A complex sentence may contain one logical thread or more than one (multiply connected sentence). A normal text contains concepts of different complexity and, therefore, it unavoidably contains corresponding sentences with different complexities and lengths.

It may be added that some recommendations in favour of short sentences stem from the awareness that the observed continuous impoverishment of language mastery entails impoverishment of the ability to understand complex sentences, in turn associated with impoverished ability for complex thinking (Sect. 1.3.2, Chapter 1)—an unavoidable consequence because of the interdependence of language and thought, for which language does not only express thoughts, but is the essential instrument for their creation [11–13]. Some reviewers' comments requesting that a complex sentence is split into shorter ones are actually expressed in ways that highlight those reviewers' lack of familiarity with complex sentences and complex thinking. On the other hand, this should never be a reason to request the unavoidable loss of rigour that would result from fragmenting a complex reasoning into pieces, with unavoidable disappearing of some logical connections. The ability of complex thinking is essential for the development of science and of the theoretical bases of engineering. Therefore, the correct and constructive approach should entail the development of complex thinking abilities by students, not the abolishment of the expression of complex thinking.

A concrete example may help illustrate the epistemological relevance of complex thinking. A rigorous statement of the third law of thermodynamics is expressed through the following complex sentence [14]:

A. *If the entropy of each element in its most stable state at $T = 0$ K is taken as zero, then every substance has a positive entropy which at $T = 0$ K may become zero, and which does become zero for all perfect crystalline substances, including compounds.*

Those who prefer to avoid complex sentences simplify it to something like.

B. *The entropy of all perfect crystalline substances at $T = 0$ K is zero.*

Statement B neglects fundamental epistemological aspects, starting from the fact that it is us (humans) who make a hypothesis (ascribing value zero to the entropy of each element at $T = 0$ K, provided the element is in its most stable state at $T = 0$ K); in this way, the student or reader will take statement B as an established truth, without knowing where it comes from. From an educational point of view, it is preferable to use statement A and to help the student analyse its logic and meaning (Scheme 5.1), so that understanding is ensured and, simultaneously, the student is exposed to the complexity of thinking in physical sciences.

5.2.3.4 The Standardised Sentences Between Mathematical Equations

Mathematics is a language of its own, expressing relationships in a symbolic format, through equalities, equations, inequalities, etc. Mathematical components—including expressions of physical laws, derivations of relationships important for specific theoretical inferences or specific applications, and other roles—are present in many science and engineering texts. A physics or physical chemistry article may contain the derivation of new mathematical tools to be used for certain studies or problems. An engineering report on a project proposing an innovative solution will include the mathematical demonstration of the validity of that solution, usually starting from the theoretical bases pertaining to rational mechanics and/or the construction science. The language and organisation of these components is more standardised than those of other components of the text. As progressively-numbered formulas and equations provide the key pieces of information, the short sentences between equations have the essential role of expressing the logical connections between previous and subsequent equations, thus guiding the reader throughout the mathematical discourse.

The initial sentence of a proof can be of the type "Consider…", defining the object of the proof in question, and often also the symbols used the denote them

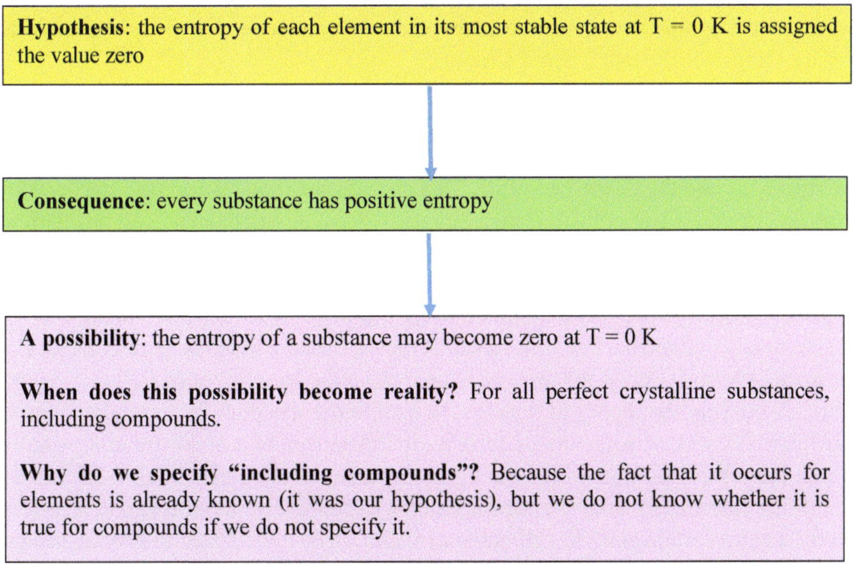

Scheme 5.1 Explanation of the logic of the statement of the third law of thermodynamics

(e.g., Consider two even integers *a* and *b*). The next steps may have formulations like "Since… then…", "Let *z* denote…", "Then…", "Therefore…", "Hence…", or somewhat longer sentences like "Use equation (3) to express … in equation (5); then…", or may introduce the operation that is realised in the next equation, acting on the previous one, such as "by differentiating with respect to…", "by integrating with respect to…", and the like. Mathematics guides itself (is bound by its internal logic) and, therefore, is not influenced by an author's style preferences or abilities; what an author can choose is the detailing of the subsequent steps to be shown; this largely depends on the target readers: fewer details can be shown in texts meant for other specialists, whereas many more may need to be shown in texts meant for students.

The sentences preceding a proof usually explain the purpose of the proof, and other information that might be relevant to understand its motivations or facilitate its reading. The sentences following a proof usually discuss its physical–mathematical significance and possible uses. In both cases, the quality of the text depends on the author's language mastery and familiarity with mathematical reasoning and with relating mathematics and physics, merging them into one discourse.

5.2.3.5 How the Definition of New Concepts Transfers into Language

The generation of new concepts unavoidably entails the generation of new words, which enter the ensemble of terms specific of each discipline. The new words need to harmonise with the pre-existing terminology, enriching it with their novel meanings while ensuring that no everlaps or ambiguities are introduced, and that the already existing terms-consistency is not altered.

The generation of new definitions in association with new concepts requires extremely rigorous use of language, to ensure that the new idea is fully conveyed to other experts; this, in turn, is essential for the new idea to be evaluated, approved and incorporated into the given discipline. This progressive language-expansion pertains to all disciplines—scientific, technical, humanistic, sociological, etc.—as an inherent component of their development as their body of knowledge grows.

Specialists may introduce new terms within their professional practice, to concisely identify concepts that are relevant within specific contexts; the new terms are obviously defined the first time in which they are used within the same context. For instance, AM introduced the term *adferendi damnum proclivitas (ADP)* in a report, while discussing the theory of damage, to denote the tendency to cause damages to third parties by certain factors or certain circumstances in a major building site, in which important and potentially dangerous operations are carried out. ADP can be quantified (measured, with the measurements expressed by numbers) and it is very useful when a certain already-estimated damage can be traced back to several causes; then the determination of the ADP proportion associated with each cause provides the percent responsibility of the corresponding individuals or bodies. The option of using the Latin language to denote it responded to the fact that the concept pertains to forensic engineering, i.e., to an area that is cross-disciplinary between engineering and law; the latter frequently utilises Latin terms, both because of the inheritance from Roman law and because of the ability of Latin to express concepts with few words and absolute meaning-uniqueness.

Another example of a new term—introduced by some environmental urbanistics specialists—is the *eterotropicity* term, which cumulatively denotes the ensemble of factors relevant for building a new city: the size and nature of the territory, its ability to provide the materials needed for the city to start and grow, and to provide the components that are needed for its inhabitants to have a common-standard life, such as water supply and other facilities. The concept is complex, but can be quantified on the basis of the quantification of the individual components that it entails.

Many other examples could be cited. Kant introduced the word *noumenon* (that had already been used by Plato with reference to ideas) to denote the 'intellegible', the 'thing-in-itself', the object as assumed to exist on its own, independently of possible interactions with us humans (opposed to the *phenomenon*—something that we can perceive with our senses) [15]. Tullio Levi Civita introduced the terms 'base' and 'roller' into the body of Rational Mechanics, to denote the trajectories of rotational centres in the relative motions of objects simultaneously subjected to rotation and translation under conditions of mutual interaction [16]. An exhaustive list of new words introduced in experts' works would occupy many volumes.

5.2.3.6 Some Considerations on the Current Historical Minimum of Language-Usage Quality

In a number of countries, spoken and written language is undergoing an unprecedented crisis regarding orthography, grammar and syntax correctness, the variety and richness of the known and utilised words, the knowledge of different meanings, the style elegance and the metaphores' relevance. The quality of language usage has been decreasing steadly through the last 50 years in all areas, humanistic, scientific and technological alike. The cause of this progressive standard-lowering can be traced to two major factors:

1. The decrease in learning demands, for which students are progressively asked to learn less and with less depth with respect to what was customary until about the mid Sixties of the XX century. This decrease became rapid after 1970, following students' protests, the so called 'war on notionism', and various misunderstandings about the significance and implications of 'modernity'. The effects have been so deep that a reversal, in terms of once again accepting that quality learning requires commitment and hard work, is so far difficult to imagine, or to view as viable.
2. The impoverishment of language usage through the mental dynamics prompted by indiscriminate and uncritical use of computers in intellectual activities, delegating to computers roles that pertain to the human mind. The computer is an excellent tool, but the quality of a text depends of the mastery of language and ideas by the human being using it [17].

Cause (1) was still quite rampant when cause (2) started to rise at the transition between the XX and the XXI centuries. The combination of the two causes brought greater endangerment to the priority and institutional role of language, i.e., to the communication of information. Written texts started becoming less and less understandable and full of errors. The frequent replacement of the word "error" by the word "typo" (or corresponding terms in other languages) brought a sort of absolution, minimizing the "error" concept; the psychological impact is obvious: an error is something that should be avoided, that denounces inadequacies of one or another type, whereas a typo is something that can be condoned. Logical consistency throughout a text becomes increasingly rare.

In addition, the inadequacies or absence of systematic training in the use of language result in differences in the languages used by different individuals because of the different types of wording poverty, imprecisions, errors, and logic inconsistencies that they contain; this further decreases mutual understandability; the communication of information correspondingly becomes fuzzy, uncertain and confusing; the ultimate extreme scenario might become similar to a Babel tower situation. The fact that the phenomenon has started extending to scientific and technical communications (articles, reports, books) is cause for futher concerns, because of the undesirable practical consequences that may ensue.

One may wonder whether this continuous downgrading process could be reversed, or whether it has already become unstoppable. At educational level, it would require political decisions that are likely to encounter strong oppositions. Requiring adequate learning progress and achievements from students within their mother tongue courses would be reasonable and would build powerful bases for the students' future, whichever the career/occupation that they will later choose. On the other hand, children are (understandably, because of their juvenile inesperience) reluctant to accept to be engaged more extensively and deeply in their learning process; too many parents are not in a position to understand the actual meaning of training and acquisition of skills, and only demand that their children are given a pass, whether it corresponds to acquisition of at least basic skills or not; and decision makers are principally concerned about making the public happy at the moment, without looking into the future. This is not true in all contexts, but it is unfortunaltely true in some contexts of the so-called "first world", and gets imitated in some other contexts.

What can a person (of whichever age) do if he/she wants to acquire the ability to write a text with a satisfactory, good, or even excellent language level, and has not received the corresponding training through formal instruction? The only viable option in such situations appears to be self-teaching. Some practical suggestions to this purpose are briefly outlined here, in a bulleted-type format (numbered for practical purposes, but with no hierarchical meaning associated with the numbers). The first set of suggestions aimes at improving one's own familiarity with good language usage:

(a) Consulting or systematically reading a grammar book for the mother tongue. This helps setting essential foundations where they have not yet been acquired.

(b) Reading extensively. Reading enables indirect but quite effective familiarisation with correct wording, gradually resulting in good internalisation. It can be viewed as a gradual acquisition of skills through an absorption process (absorption of correct and refined modes of expression and thoughts' organisation).

    It is important to choose texts with good language quality. Since good language quality cannot currently be taken for granted even for journals or other sources where quality should in principle be expected, it is crucial to make accurate selections. Language quality is not strictly associated with the *genre* of a text or a book. Of course, one can trust works that have already entered the literature realm. But there are also texts that one can read for relax and which have good language quality, including some science fiction or thriller novels; they can facilitate the familiarization with correct and good language in a basically effortless and enjoyable way.

    The option can be effective also when learning a foreign language. For instance, LM started familiarising with English modes of expression by reading the original texts of English works that she liked (e.g., G. B. Shaw's theater pieces), and has more than once recommended science fiction or thriller novels with good language quality to Italian friends wishing to strengthen their English—a way of combining usefulness (improving the mastery of another langage) with the relax associated with novel-reading.

(c) Reading texts related to one's professional field. On the one hand, this is essential to keep updated with one's field of expertise and to continuously enhance the level at which one masters it. On the other hand, it combines the enhancement of conceptual knowledge with the enhancement of mastery of its specific mode of expression.

 It has also to be noted that it is comparatively easy to spot poorly worded texts if one has sufficient conceptual bases in a given field. Regrettably, such texts are encountered with a certain frequency in scientific and technical literature. More than once, postgraduate students have shown LM articles' texts whose meaning they failed to understand, and the students proved right, because the concerned sentences did not have sufficiently correct grammar for grammar analysis to help identify their meaning, and did not even leave sufficient space for guesswork leading to a reasonable meaning. The only option was to invite students to check the reported tables of values and derive their own inferences. On the one hand, encountering texts of this type reinforces the teacher's recommendations (and, hopefully, students' conviction) about the importance of appropriate usage of language in scientific communication. On the other hand, it is sad when, for instance, one or more among the few articles available on a speficic topic happen to be published in journals accepting such sloppy language that the meaning is not clearly identifiable and the reader remains unsure about the reliability of the entire article.

 In any case, identifying sentences or texts conveying information that is not compatible with the conceptual background of a certain field, or communicating practical instructions where the poor language makes the instruction incorrect or incomplete (and, therefore, not implementable) contributes to reinforce the reader's awareness about what is not correct and to prompt the search for correct options.

(d) Using dictionaries whenever needed. One should never leave words ununderstood on reading a text. If the meaning of a word is not known, it is necessary to check it in a dictionary. If the text is professional, understanding each word is essential. If the text pertains to the things that one reads for relax (e.g., a novel), ensuring that each word is understood is part of the routes to familiarise oneself with higher levels of language mastery through reading.

The second set of suggestions refers to actually writing a text and aiming at good level of correctness, internal logic and communication efficacy. It complements and reinforces the suggestions in Sect. 5.2.3.2, targeting more specifically authors on a self-teaching process:

1. Reflecting on what one wishes to communicate and preparing a plan. The plan (e.g., as outlined in [9]) needs to contain all the components that one can predict in the initial stage.
2. Preparing a first draft. The operation of actually writing a text enables a hands-on elaboration and organisation of one's thoughts, and may entail modifications of the plan as the communication develops.

3. Reviewing the first draft and improving on it. This likely becomes a loop-type sequence of operations with subsequent drafts, each of which improves on the previous one, until one reaches a draft with which one is satisfied.
4. Complying with essential criteria during the entire process. For instance:

   (i) One should never use a word of whose meaning one is not completely sure. In case of uncertainty, it is important to consult a dictionary. Using a word without being sure of its meaning entails the risk of incorrect communication. Similarly, if one is not sure about the rigorous wording of certain concepts, one can check how other authors express analogous concepts, to get a logical-pattern model.

   (ii) One should keep a text 'clean' by using only terms pertaining to the language in which the text is written. On the one hand, this facilitates readability, because the reader who knows a certain language may not know all the randomly inserted words pertaining to another language. On the other hand, a text filled with words pertaining to another language is unavoidably sloppy, and suggests that the writer does not have adequate mastery of the language used in that text. For instance, a text written in Italian may contain those English words that have already been accepted in Italian in such a way that no Italian word has been created (such as, e.g., the word *mouse* to denote the item that we use to interact with a computer), but does not need to replace Italian words with English ones on the basis of some snobbish attitude (a wish to show that the writer knows some English words).

5. Carefully proofreading a text before submitting it to wherever it is meant to be submitted. This is an essential operation to ensure that the text is correct and conveys the desired information. It is also an operation whose challenging character increases as the language mastery decreases, up to the point that, when language mastery is poor, people are not able to perform it.
   The operation is necessary for all written texts, from tests, reports and exams at educational level to articles, reports, chapters, translations from one language to another, and the like, within professional activities. A useful option is trying to put oneself in the place of the reader, keeping in mind that, while one knows what one wishes to communicate through a given text, the reader will not know it if the text is not completely rigorous. A reader reads what is actually written in the text, not what the person who wrote it was thinking and failed to express. This is true for any reader, from a teacher assessing students' works to anybody reading texts relevant to their activities.

5.3 Some Reflections About Language

5.3.1 The Nature of Language

Given the nature of the current section, it is important to clarify a fundamental distinction between the two roles with which the *language* term is used in English (expressed through different words in many other languages): language as communication of information and language as something pertaining to a specific human community. For the sake of clarity, this section utilises the *language* term in association with the former meaning and the *tongue* terms in association with the latter (which is legitimated by the existence of common expressions such as *mother tongue*).

The *language* term indicates an ensembe of codes (werbal, written, gestual, formal, symbolic, etc.) which generate, express, store and transmit information. The meaning can be extended to communication among animals, which can be expressed through sounds, or also through other forms, like the dances with which bees communicate information about nectar-rich flowers to other bees, with the velocity, trajectory and direction of the dance having the roles of symbols univocally providing information about the location of the flower-place, its distance from the beehive, and its extension (or amount of flowers). Zoosemiotic investigates how animals communicate. Plants also communicate; the mechanisms are not yet well known; a recent study found that fungi communicate with electrical spikes that might be signs pertaining to a form of language [18]. All this stresses that the circulation of information is essential for all living forms. On the other hand, while language among animals refers to concrete objects or events of interest at a certain moment, and evolves through extremely long periods of times, human language evolves during the life of an individual or the history of a community, may refer also to abstract objects, and has a degree of complexity responding to the amount and complexity of the knowledge constituting the object of communication.

A *tongue* is a mode of communication historically established in a certain place and in a certain period as a shared asset of a given community; it is characterised by specific sounds, words, grammar rules, and written characters corresponding to individual sounds or to individual words. Then *language* becomes a particular utilization criterion, or a particular application, of a certain *tongue* to express a certain ensemble of information and make it communicable. In this way, a given discourse can be translated into different *tongues* and continue being expressed in the same *language*. For this reason, we speak of English, French, German, Italian, etc. *tongues*, whereas we speak of common *language*, scientific *language*, narrative *language*, or the *language* of mathematics, the *language* of engineering, and the like. The requirements of a given *language* remain the same whichever the *tongue* utilised, because they respond to the nature of the content—a network of concepts and the criteria connecting them. It can thus be said that *language* is invariant with respect to *tongues*. For instance, the definition of a certain concept contains the same pieces of information, connected in the same way, whichever the *tongue* utilised; a specific mathematical proof contains the same equations and the same statements between subsequent equations, whichever the *tongue* utilised. *Language* and *tongues* are mutually correlated, with *tongues* having an instrumental role with respect to *language*.

5.3.2 Communication Through Language

In its basic meaning, language is the communication of a discourse (an ensemble of connected pieces of information). Within a basic picture, the communication involves a transmitter (T) and a receiver (R), which might also be a group of receivers. In the case of verbal communication, T is the speaker and R are the listeners. In the case of written communication, T is the writer and R are the readers. The roles may also interchange, according to the type of activity. For instance, education involves a mixture of oral and written communication: oral communication from the teacher to the students within traditional teaching options, or with frequent exchanges of the T and R roles within interactive teaching options; written communication through textbooks and handouts, where students have the R roles, but also through the works students submit for assessment, where they have the T role.

The efficiency of the communication depends on three factors: the quality of T (Q(T), i.e., T's level of knowledge of the discourse (D) that is the object of communication); the quality of R (Q(R), i.e., the extent to which R's background knowledge is adequate for R to understand D); and the quality of the expression of D, Q(E), where E is the way in which T expresses D. Q(E) is different from Q(T), although it depends on it; it can be high only if Q(T) is high; on the other hand, there are several cases of highly knowledgeable Ts who do not speak or write well enough to attain high Q(E). Furthermore, T needs to have the ability to tailor E according to Q(R). For instance, within a specialised conference, T can confidently assume that R (all the persons listening to T's presentation) have adequate background knowledge to be in a position to fully understand D, and this decreases the challenges inherent in the preparation of E. On the other hand, within educational contexts, Q(R) is not homogeneous, as different students have different levels of acquired knowledge and different levels of interest in the subject. The core challenge of education is the design of E in such a way that as many students as possible can attain the highest possible level of understanding and internalisation.

Whichever the circumstances, the key to communication success is the quality of E. This depends on the quality of language, which needs to be rigorous and have the logic and organisation characteristics already outlined in previous sections, to ensure understandability, and possibly also a narrative style capable of attracting R's attention and maintaining it. In turn, these features are attained through the careful selection of individual words, so that each item is denoted in a precise and unambiguous way, the careful construction of sentences, so that they convey a clearly identifiable meaning, and the careful organisation of the sentences to build the entire discourse in a way that makes it clearly understandable.

It can be noted at this point that no hypothesis has been made in this subsection about whether D pertains to the humanities domain or to the science and technology domain. The same requirements of rigour, internal logic and understandability on the one hand, and the ability to captivate and maintain R's attention on the other, are crucial for both. This convergence can be very useful for language training, and offers a platform to promote cross-disciplinarity. The reason for chapters (like the present one) specifically focusing on language in science and technology is not related to substantial differences in the requirements of language quality in the two domains, but to the desire of counteracting the diffuse attitude failing to acknowledge the importance of language quality and language mastery for the science and technology domain.

5.3.3 Views About Language Through the History of Philosophy and Science

This subsection is meant to further emphasise the importance and roles of masterful use of language in science and technology communication through a quick tour across the reflections that scientists and philosophers have devoted to the nature and roles of language. Language has been a major object of theoretical investigation because of its fundamental role in the development and expression of thoughts and, therefore, in the development of knowledge. Scholars investigated both these aspects and the characteristics that make the communication of information truthful and complete. The present overview is necessarily fast, outlining only principal ideas related to the nature of language and its relationships with thought and knowledge. On the other hand, by pointing to the vastness of investigations and reflections on language, it aims at underlining the importance of language for any knowledge-related activity, of which science and technology are examples. This importance is bound to continuously increase as we steadily move towards a knowledge-based society, where the handling of knowledge and the ability to quickly learn new knowledge are increasingly playing dominant roles in human activities.

Reflections on language date back to ancient time. Democrit (about 460 BC) considers that language was born when humans shifted from a barbaric status to building a society (it may be interesting to recall that "barbar" in ancient Greek originally meant "someone who does not speak", or "someone who does not speak in an understandable way"); in his words [19, 20]:

> *And while humans earlier emitted sounds void of meaning and articulation, they gradually started to articulate words.*

Words result from agreements within a community and, therefore, they have a conventional nature. Humans within a community.

> *by establishing among themselves conventional expressions to denote each object, ended up creating a way, known to all of them, to denote all things.*

Given the nature of the process, it is easy to understand why different communities denoted the same objects with different words and developed different languages:

> *Since such groupings of human beings got formed in all the inhabited regions of the earth, there cannot be a language with the same sound for everybody, because each of those groups designed words in their way; this is why the characters of languages are greatly varied.*

Plato considers thought and language as one: thinking is speaking silently, is a talk of the soul with itself (*Theaetetus* and *Sophist* dialogues); if the thought is expressed to another person, it becomes a discourse (*Philebus*) [21, 22]. He also insists on the need to define the terms used in a discourse (*Sophist, Cratylus*) [23] and on the importance of the way in which words are connected, i.e., on the importance of grammar, to convey a meaning (*Cratylus, Sophist*) [22, 24, 25].

Aristotle too considered names as symbols, denoting things on the basis of common agreement within a given community. He devoted particular attention to how language conveys meaning (*De interpretatione*) [25], and how words (e.g., noun and verb) need to be combined to transmit information and to make sure that the information is true. While this analysis relates more specifically to grammar and syntax (construction of sentences), Aristotle gave major attention to logic and how it is an integral component of a discourse, thus setting the theoretical bases for the investigation of the requirements of what could generally be called 'technical discourse', i.e., discourse related to specific themes or topics.

During the Middle Ages, language was a major object of study; the arts that were grouped in the *trivium* (grammar, logic and dialectic) are all associated with language and the search for optimal ways of expressing thoughts. Scholars utilised the texts of classical times, either translating them from Greek or utilising Latin translations based on Arab translations, and also Arab commentaries (for instance, Al-Farabi commentaries on Aristotle's logic and his own compendium on logic) [24].

After the Middle Ages, scholars' interest continued with a variety of new directions. Leonardo da Vinci listed thousands of Italian words destined for specific uses; in this way, for instance, he gave a substantial contribution to the Italian lexicon of mechanics. Jacopo Zabarella published *Logica* (Venice, 1578), where he analyses the nature of logic, and concludes that logic is neither a 'science' nor an 'art', but an 'instrument of science' [26].

From the second half of the XVI century, the question about what words denote (the epistemological nature of the semantic operation of 'denoting') becomes a major focus of the studies on language, and is intertwined with the question of how humans attain knowledge. Francis Bacon (1561–1626) considers that language is inadequate and words may confuse human intellect because they may recall concepts that do not coincide with anything real, or be ambiguous by having more than one meaning, and this generates confusions in science and philosophy. He recommends the creation of a language without ambiguities, with a strictly univocal correspondence between the sign (word) and the item that it denotes, and unchangeable. On the one hand, this criterion is largely implemented in the specialised languages of highly specialised areas; on the other hand, language needs flexibility to enable the generation of new ideas, and the ability to incorporate new terms in correspondence to new ideas, or to add new meanings to existing terms [27].

René Descartes (1596–1650) views language as the main characteristic that differentiates humans from animals (*Discourse on method*, 1637). He considers that the general features of grammatical structure are common to all languages because they reflect certain fundamental properties of the human mind that are common to all humans.

John Locke (1632–1704) expressed his ideas on language in "*An essay concerning human understanding*" (1690). He considers that human acquisition of knowledge starts from experience: empirical information enables the generation of simple ideas, where words correspond to the images of actual things that we form in our minds, and the syncretism of simple ideas engenders more complex ideas. Language enables us to organise ideas into thoughts and to communicate these thoughts to other human beings; thus, language has a key role in the generation of knowledge.

Gottfried Wilhelm Leibniz (1646–1716) saw an intimate connection between the form and content of language and the operations of the mind, and considered that an accurate analysis of the signification of words could give us the best information about the operations involved in the understanding process [28]. He viewed complex concepts as composed of simpler concepts and suggested that, by identifying the simpler concepts and assigning symbols to each of them, it would be possible to build an artificial language composed of symbols (which would stand for concepts or ideas) and logical rules to connect them in a viable way; this would constitute a "universal language" capable of mirroring the processes of human reasoning [28]. This suggestion is interestingly close to the basic idea of the languages now utilised in computer programming, and he actually viewed human cognition in essentially computational terms ("everything done by our mind is a *computation*") [28]; Scheme 5.2 provides an illustrative example of how an operation is split into simple instructions in order to communicate how to perform it to a computer.

Pierre Louis Moreau de Maupertuis (1698–1759) wrote a book on the origin of language and the meaning of words (*Réflexions philosophiques sur l'origine des langues et la signification des mots*, 1740) and another on the different ways that humans have utilised to express their thoughts (1756). He considers that language gives identity to what the eye perceives, i.e., the way in which we identify objects is influenced by the way in which we denote them through language. The way in which the first humans denoted their first ideas has influenced all subsequent developments of knowledge [29].

Gian Battista Vico (1668–1744) relates the development of science and the development of human knowledge to language through human history (*La Scienza Nuova*, 1725). Étienne Bonnot de Condillac (1714–1780) considers that the progress of language and the progress of the mind are mutually dependent and that the progress of human knowledge depends on the progress of language; thus, the historical development of language, human mind and society are tightly interconnected [29]. These ideas were largely adopted also by Jeremy Bentham (1748–1832).

French Enlighteners investigated the issue of language, and some of their ideas entered the *Encyclopédie*. Nicolas Beauzée is the most known representative of the

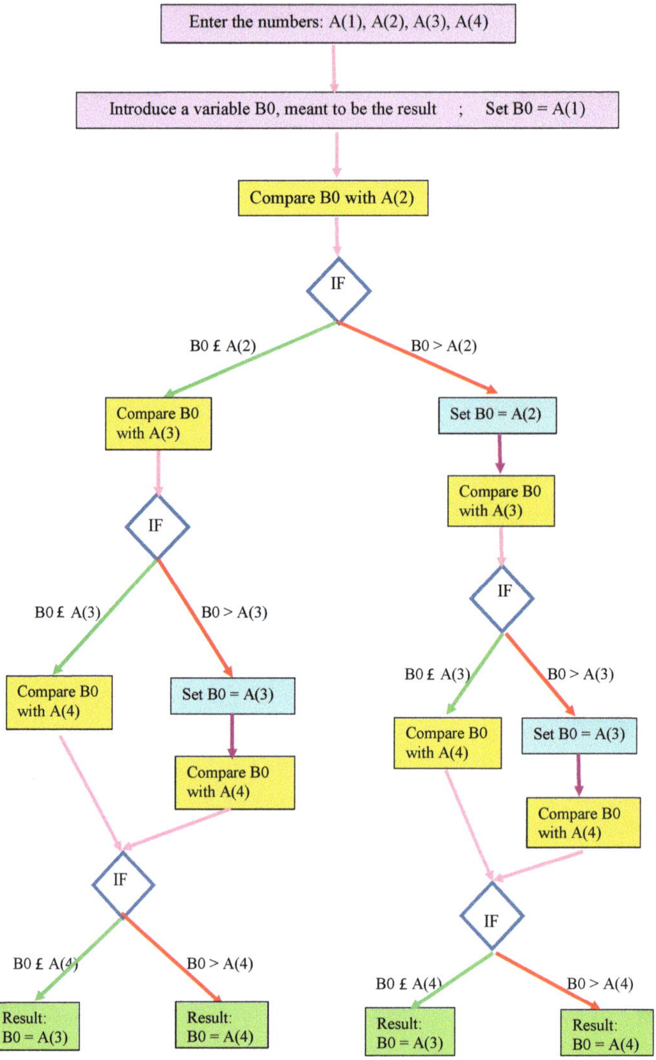

Scheme 5.2 Illustration of the simple steps in which the operation "Choose the smallest number out of four given numbers" can be subdivided. The numbers are generically denoted as A(1), A(2), A(3), A(4). This is an operation that the human mind processes immediately, because of deep internalisation of the operational components. However, it needs to be clarly split into simpler operations for a computer to perform it. The scheme follows the individual steps of the computer reasoning (in actual programming, repeated instructions would be grouped into a loop). To better follow the procedure, let us suppose that A(1) = 6, A(2) = 9, A(3) = 4 and A(4) = 8. At the end of the procedure, the value of the variable B0 will be the result, i.e., the smallest number. The first step will set B0 = 6. The second step will compare 6 and 9. Since 6 < 9, B0=6 remains valid. The next step will compare 6 with 4. Since 6 > 4, the next step will set B0 = 4. The following step will compare 4 and 8. Since 4 < 8, B0 maintains the value 4, which is the result

rationalist position, according to which all humans possess the same intellectual categories, which are inborn and follow the same logical criteria. These criteria are expressed through language and correspond to a sort of universal grammar inherent in all languages. Different languages use different terms to denote the same things, but organise them according to logical criteria responding to the categories of human thought [29]. César Chesneau du Marsais (1676–1756) also considers a universal logical grammar—a structure of the human mind which organises the deep roots of the grammar of individual languages, making allowance for a high diversity variety [30].

Wilhelm von Humboldt (1767–1835) considers that humans surround themselves with a world of sounds to the purpose of assimilating the world of objects, but "Sounds do not become words until a meaning has been put into them, and this meaning embodies the thought of a community" [30]. He views language as a system which "makes infinite use of finite means", meaning that a language can create an infinite number of sentences and corresponding meanings utilising a finite number of resources (the words and the grammar rules of the given tongue) [30]. His views emphasise both the creative power and the cultural specificity of language.

In the first decades of the XX century, the investigation of language is closely connected with the investigation of logic and—given the importance attained by mathematical logic—is also often related to the latter. Many of the philosophers studying language are also mathematicians or have a deep knowledge of mathematics. Meetings on the philosophy of science were organised by Hans Hahn, Philip Frank, Richard von Mises and Otto Neurath since 1907. In 1922, Moritz Schlick (a physicist and philosopher), with the support of Herbert Feigl and Fredrich Waismann, initiated a stable discussion group, initially called "Ernst Mach Society", and later "the Vienna circle"; other scientist joined (among which Rudolph Carnap, Kurt Gödel, and Victor Kraft) and the circle was active in the 1924–1938 years. Similar circles arose in Berlin (Reichenbach, Lewin, Köhler, von Mises, Hempel) and in Warsaw (Twardowski, Kotarbiński, Łukasiewicz, Tarski). The main interest of these philosophers and scientists was the ascertainment of the correctness of the scientific discourse. They viewed philosophy as a discipline meant to verify the procedures and methods of science, thus ascribing an epistemological character to it, and considered the analysis of language and logic as the tool to verify the truthfulness of statements [31, 32, 33]. The title of one of Alfred Ayer's books (*Language, truth and logic*, 1935) effectively summarises the core of their thought.

Since then, language has remained a central issue in philosophical investigations. A detailed analysis of the most recent views would go beyond the scope of this section, whose main objective is that of illustrating the close connections between the scientific discourse and the reflections on the roles of language throughout 25 centuries, thus also emphasising the crucial role of language in any scientific and technological discourse.

5.4 Discussion and Conclusions

The humanities domain provides expression and communication tools to the science and technology domain. Scientists and engineers need to use them expertly, consistently with the requirements that texts need to be correct and expressed in such a way as to be fully and unambiguously understandable by the reader. In addition, they need to have a comprehensive vision of their field and to know it sufficiently deeply to be able to identify and appreciate also its philosophical and epistemological characters, its conceptual foundations and its connections with other areas. In this way, they can benefit from interdisciplinary and multidisciplinary contributions that can enhance procedures and final outcomes, and simultaneously be able to write about them in a suitable way.

Engineering is implementation and production oriented. If the conceptual bases are not adequate, both the design and the implementation stages are affected. The currently observed general lowering of language mastery affects the quality and completeness of knowledge acquisition in the learning (student) period; this, in turn, affects the level and quality with which an engineer can design solutions, and the expertise with which the designed procedures are communicated to those who are tasked with their practical implementation. All these factors lead to poorer outcomes, i.e., affect the final product. It may even be impossible that viable solutions get designed for particularly challenging (non-standard) problems without deep conceptual foundations. The obvious inference is that engineers, as well as scientists, need good levels of language mastery to be in a position to acquire deep knowledge and utilise it to design, propose and supervise viable, safe, sound and cost-effective projects, and to train young people so that they will, in turn, become capable of doing the same.

References

1. L. Mammino, Roles of system thinking within green chemistry education. Reflections from identified challenges in a disadvantaged context. J. Chem. Educ. **96**(12), 2881–2887 (2019)
2. G. Krall, *Appunti di meccanica superiore* (Università degli Studi di Roma "La Sapienza", Roma, 1931).
3. G. Krall, *Appunti dalle lezioni di scienza delle costruzioni* (Pubblicazioni del GUF, Napoli, 1937)
4. M. Picone, G. Krall, C. Ferrari, *Questioni di matematica applicata, Monografie del CNR* (Zanichelli, Bologna, 1939)
5. G. Krall, R. Einaudi, *Meccanica tecnica delle vibrazioni, Monografie del CNR* (Zanichelli, Bologna, 1940)
6. G. Krall, *Appunti di meccanica e fisica matematica* (Libreria Eredi V. Verschi, Roma, 1968).
7. *Euclid's Elements of Geometry*, The Greek text of J.L. Heiberg (1883–1885) from Euclidis Elementa, edidit et Latine interpretatus est I.L. Heiberg, in aedibus B.G. Teubneri, 1883–1885 edited, and provided with a modern English translation, by Richard Fitzpatrick. https://farside.ph.utexas.edu/books/Euclid/Elements.pdf

8. Vitruvius, The Ten Books On Architecture, Translated by Morris Hicky Morgan, Cambridge Harvard University Press, 1914, https://www.chenarch.com/images/arch-texts/0000-Vitruvius-50BC-Ten-Books-of-Architecture.pdf
9. L. Mammino, Educational components in the supervision of chemistry postgraduate students: experiences and reflection. Phys. Sci. Rev. **5**, 1 (2021). https://doi.org/10.1515/psr-2020-0116
10. L. Mammino, *Il linguaggio e la scienza* (Socidta' Editrice Internazionale, Torino, 1995)
11. J. Bruner, Language as an instrument of thought, in *Problems of Language and Learning*. ed. by A. Davies (Edinburg University Press, Edinburg, 1975)
12. E. Asoulin, Language as an instrument of thought. Glossa J. Gen. Linguist. **1**(1), 1–23 (2016). https://doi.org/10.5334/gjgl.34
13. G. Dupre, What would it mean for natural language to be the language of thought? Linguist. Philos. **44**, 773–812 (2021). https://doi.org/10.1007/s10988-020-09304-9
14. P. Atkins, *Physical Chemistry*, 5th edn. (Oxford University Press, Oxford, 1994), p.140
15. E. Kant, *Critique of Pure Reason, Chapter III* (Cambridge University Press, Cambridge, 1998), pp.338–353
16. T. Levi Civita, U. Amaldi, *Lezioni di Meccanica Razionale*, Bologna: Zanichelli, p. 228
17. P. Pozzati, Proliferazione delle normative e tecnicismo, ultima lezione ufficiale del corso di Tecnica delle Costruzioni nell'anno accademico 1991–1992, Facolta' di Ingegneria, Universita' di Bologna.
18. A. Adamatzky, Language of fungi derived from their electrical spiking activity. R. Soc. Open Sci. **9**, 211926 (2022). https://doi.org/10.1098/rsos.211926
19. F. Enriques, M. Mazziotti, Le dottrine di Democrito di Abdera. Testi e commenti, Bologna: Zanichelli, 1948.
20. https://urraroblog.wordpress.com/2007/05/11/democrito-e-il-linguaggio/
21. https://www.skuola.net/filosofia-antica/platone-identita-linguaggio-pensiero.html
22. The Dialogues of Plato (428/27–348/47 BCE)
23. https://webs.ucm.es/info/diciex/gente/agf/plato/The_Dialogues_of_Plato_v0.1.pdf
24. G. Licata, Teoria dei nomi e teoria delle idee in Platone, Dialegesthai, Rivista telematica di filosofia, 2002. https://mondodomani.org/dialegesthai/. ISSN 1128–5478
25. Filosofia del linguaggio nell'antichitá e nel medioevo, https://isentieridellaragione.weebly.com/filosofia-del-linguaggio-nellantichitagrave-e-nel-medioevo.html
26. P. Fantin, Teoria del linguaggio in Aristotile: il *De interpretatione*, https://www.homolaicus.com/teorici/aristotele/aristotele3.htm
27. C. Vasoli, Jacopo Zabarella e la 'natura' della logica. Rivista Di Storia Della Filosofia (1984-) **66**(1), 1–22 (2011)
28. M. F. Suozzo, https://www.lacooltura.com/2017/07/linguaggio-bacone-contro-vaghezza/
29. M. Kulstad, C. Laurence, Leibniz's Philosophy of mind, *The Stanford Encyclopedia of Philosophy* (Winter 2020 Edition), Edward N. Zalta (ed.), https://plato.stanford.edu/archives/win2020/entries/leibniz-mind/
30. https://www.treccani.it/enciclopedia/linguaggi-e-lingue_%28Storia-della-civilt%C3%A0-europea-a-cura-di-Umberto-Eco%29/
31. https://en.wikipedia.org/wiki/Wilhelm_von_Humboldt
32. https://www.treccani.it/enciclopedia/circolo-di-vienna_%28Dizionario-di-filosofia%29/
33. http://www.storiadelleidee.it/index.php/idee-della-contemporaneita/il-circolo-di-vienna

Chapter 6
Thinking Humans Versus Non-thinking Software: The Latest Frontier Requiring Deep Language Knowledge from the Humans

Liliana Mammino

Abstract Computers provide invaluable aid for an increasingly high number of operations. However, this does not authorise human beings to decrease the amount and depth of knowledge that they need to acquire and apply: the extent and functionality of computer-provided aids depend on the extent and quality of the knowledge possessed by the humans using the computer. The present chapter analyses this requirement with regard to text-writing, considering aspects such as automatic spelling correction, computer's suggestions for different wording, and the complexity associated with the use of plagiarism and self-plagiarism detecting software. It is concluded that good language mastery and adequate epistemological bases are fundamental for the humans (authors, editors, reviewers, students, and anybody who writes texts) to be in a position to maximise the benefits of the information provided by the computer and, simultaneously, to avoid errors or lowerings of the rigour and general quality of a scientific text. Since language mastery and epistemology familiarity are humanities-based abilities, it is thus concluded that optimal usage of the writing-related computer resources depends on the presence and adequacy of humanities-related expertise. This remains true also with regard to the recently developed AI-based text-generating chatbots.

Keywords Automatic spelling corrections · Common domain wordings and concepts · Correct identification of plagiarism and self-plagiarism · Importance of epistemological bases · Importance of language mastery · Language quality in scientific texts · Reflections on AI-based large language models

---

L. Mammino (✉)
Faculty of Science, Engineering and Agriculture, University of Venda, Thohoyandou, South Africa
e-mail: sasdestria1@gmail.com; mammino@univen.ac.za

L. Mammino (ed.), *Insights into the Relationships between Humanities and Sciences*,
https://doi.org/10.1007/978-3-031-97900-2_6

6.1 Introduction

Computers have become increasingly important tools for an increasingly high number of applications. Doing lots of things without them would currently be unthinkable. We are so used to them and to all the things that they do for us that we quite often forget that they are tools: they can do all the things that we tell them to do (through programs' instructions), but they cannot think. Human beings are in charge of thinking. This has crucial implications for lots of aspects. First of all, it requires extensive and deep knowledge from the humans: knowledge of the underpinning theoretical bases of any science or technology area in which one is interested, and specific knowledge of any field where computers are utilised to process information. Some technical terms catch this situation perfectly; for instance, "computer-aided molecular design" clearly conveys the information that the computer is an aiding tool, what obviously implies that humans are in charge of the design of molecules.

This chapter is devoted to a humanities-based component whose fundamental roles in the sciences are analysed in detail in other Chapters (4 and 5): language and its uses. It is important to recall that language is a human activity; it is related to thought (Chapters 1, 4, 5), and thought is a human activity. A computer does not think, and cannot understand a text or create a new one expressing surely reliable information. All the comparisons done by a computer are automatic (mechanical).

Computers have quite convenient roles when one is writing, starting from spellchecking. On the other hand, the fact that computers do not think means that they do not 'read' in the same sense as humans can: they can compare typed words or sentences with words or sentences stored somewhere in their memory, but they do not understand the meaning of those words or sentence. This means that humans must be in charge. In order to be in charge, they must have adequate subject knowledge and language knowledge. In other words, deep subject knowledge and language knowledge are essential for humans to efficiently enjoy the aids provided by a computer when writing a text.

The next sections will substantiate this statement through an adequate number of examples—having the roles of documentation—and their analysis. The frequent recalling of the computers' inability to think does in no way diminish the usefulness of computers: it serves the purpose of underlining the humans' responsibility to acquire the knowledge that enables optimal utilisation of computer resources and prevents the risk of errors or drawbacks of other types. Major attention is given to the use of software checking a text for similarities with other texts, because inexpert handling can prompt serious damages to the quality of scientific communication.

For the sake of conciseness, the wording "the author of this chapter" (necessary to maintain impersonal style) is replaced by the LM acronym in the rest of the text (pronouns: *she*, *her*, *hers*).

6.2 The Human Mind Versus Software-Based Thought-Void Word-Checking

As already mentioned, a computer compares typed words and sentences with those stored in its memory, and makes corrections, takes some actions, or provides other types of information on the basis of such comparisons. Some categories of computer's interventions are easy to identify and analyse, and are considered in the next subsections.

6.2.1 Blessings and Pitfalls of Automatic Corrections

The computer automatically corrects many typos while one is typing (automatic spelling corrections), and this is definitely a great advantage. On the other hand, there are two pitfalls that require the author of a text to double check it by direct proofreading.

On the one hand, the computer does not correct a typo if the word resulting from the typo is also present in its dictionary. For instance, if one types *angel* in place of *angle*, the computer will not correct it, because both words exist; more than once, LM has encountered the *bond angel* term in students' typed assignments describing one or another molecule (the correct term would be *bond angle*). Hundreds of such pairs of words are encountered: *principal* and *principle*, *next* and *nest*, and many others.

On the other hand, the computer does not recognise words that are not present in its dictionary and converts them into words that it finds in that dictionary. The outcomes are often hilarious or, at least, surprising, above all when they concern technical terms. For instance, the phenomenon for which a solute molecule entering a solution gets surrounded by solvent molecules is called *solvation*; this word is rarely present in computers' dictionaries; the automatic correction converts it to *salvation* because this is the spelling embedded as correct in its stored information, and the software does not understand the meaning of the two words or their relationships with a given context (whether they are appropriate for it or not); students (and sometimes not only students) tend not to proofread after finishing typing a text, and the *salvation* concept gets ascribed to molecules. The proofreading that can detect these errors is entirely human-based (it is not incorporated in any automatized review form). In some cases, such automatic changes may enter the proofs of an article on copy-editing, and needs to be picked up on reading the final proofs; there are also cases in which they were not picked up and remained in the published article (suggesting the hilarious scenario of molecules queuing at some celestial gates).

In the early stages of the covid-19 pandemic, there was a period in which asymptomatic patients were feared as extremely dangerous; therefore, many texts, or online comments, expressed concerns about them. In several cases, the comments used the term *asymptotic* in place of *asymptomatic*; this was the outcome of automatic corrections (regrettably, it ended up having some prophetic value, as the pandemic did not then seem to come to a clearly identifiable end, it appeared to linger with a sort of asymptotic approach to a real disappearance without reaching it).

An author can easily address these problems by correcting an automatic change when it occurs for the first time, and then adding the correct term to the computer's internal dictionary.

The automatic corrections of terms borrowed from different languages can be equally surprising. For instance, an automatic correction by an English-based spellchecker transformed the well-known pasta-type *capelli d'angelo* into *cappelli d'angelo* (where *capelli* means *hair* and *cappelli* means *hats* in the original language—Italian). And *Homo sapiens* (the scientific name of the human species) may get corrected to *Home sapiens*.

The automatic comparison of terms may lead to other incredible situations. For instance, after publishing [1], LM received the following automated e-mail message:

> *The journal Building..... is currently running a Special Issue .... We think you could make an excellent contribution based on your expertise and your following paper: Computational chemistry for green design in chemistry and pharmacy: Building awareness in the classroom. SUSTAIN. CHEM. PHARM. 2020, 18, 100283.*

followed by a list of topics of the special issue, all referred to building materials and their properties. It is easy to guess that no human being had prepared or seen this message. Since a computer cannot understand the meaning of sentences, it could not distinguish the two different meanings with which the term *building* is used in the name of the journal and in that article's title; because of this, it issued the invitation. It may not be easy to program a computer to distinguish all the possible uses of the *building* term, but it would also not be necessary, because there is no actual problem involved with a message of this type being sent to, or received by, someone who has no expertise in building materials. The example is included here simply because it provides a good illustration of the meaning of the contraposition between 'thinking humans' and 'non-thinking software'.

6.2.2 When the Computer Suggests Sentence Changes…

Language knowledge is required to evaluate whether changes suggested by a computer improve a text or make it less informative, less rigorous or straightforward incorrect. For instance, after one types the sentence:

> *I am attaching the permission request form, in case you need to reproduce something from other sources*

some software requests an incorrect punctuation change:

> *I am attaching the permission request form; in case you need to reproduce something from other sources*

and the writer is the one who is expected to know that the semicolon after *form* would introduce a grammar error, because *in case you need to reproduce something from other sources* is not an independent clause, is tightly linked to the clause preceding it.

Improvement suggestions also need careful verification by the human. For instance, after the following sentence was typed:

> *The HF method considers the motions of electrons as independent, averaging the repulsion potential of one electron with all the others. In reality, the motions of the electrons are correlated...*

the software recommended what it considered a more concise form:

> *The HF method considers the motions of electrons as independent, averaging the repulsion potential of one electron with all the others. The motions of the electrons are correlated...*

The latter version removes a crucial logical connection expressing contraposition and, therefore, enabling the identification of the actual situation (the motions being correlated) with respect to the simplified vision (the motions being independent). A writer with good language knowledge ignores the software's recommendation; but a writer with inadequate language knowledge, and inadequate awareness of how the logic of language and the logic of the scientific discourse relate to each other, will follow the software's recommendation, thus losing a crucial aspect of the scientific discourse.

6.2.3 When the Computer Helps with Translation

The help that a computer can provide when one translates a text from one language to another is definitely precious, above all in terms of the rapidity with which it provides the translation of individual words or short groupings of words. On the other hand, the person who is doing the translation needs to have full understanding of the original text and good knowledge of the language into which that text is being translated. Without this knowledge, the translation risks to convey incorrect meanings, sometimes with hilarious results. The famous example of the literary translation of the English "out of sight, out of mind" with the equivalent of "blind, insane" in another language is a perfect illustration of the risks. Many others can be picked up from various sources: the lack of meaning in the translated sentence requires the reader to check the original version. For instance, on reading in an English text that Margarita de Saboya was "Portugal's last Spanish *priest*"—what is absurd in a Catholic context because the Catholic Church does not allow females to be priest—LM checked the original text and the Spanish term was *virrey*, which means *viceroy* (females were allowed to be viceroys in those times, 1634–1640).

The only possible conclusion is that the final responsibility of the correctness of the translation of a text into another language pertains fully to humans.

6.2.4 *When a Software 'Hunts' for Plagiarism*

As already anticipated, this issue is here given extensive attention because of the embedded risks for the quality of scientific communication. The major aspects are analysed in separate subsections. The analysis is not meant to 'put blame' on the software (that has the unavoidable limitations of a tool of this type), but to attract the attention of the human players (Sect. 6.2.4.5), to foster an expert use of the indications coming from the software.

6.2.4.1 Recalling the Requirements of Language Usage in a Scientific Discourse

The discussion that follows is based on the consideration of the requirements for the use of language in a scientific discourse. These aspects are extensively analysed in Chapters 4 and 5 and, therefore, only the concepts essential for the current discussion are recalled here.

The selection of individual words, their assembling into sentences, the selection of the logical connections, and the organisation of the overall discourse, need to closely match the details of the concept/s to be conveyed. This requires both good language mastery and deep understanding of those same scientific concepts.

On the other hand, the wording of many types of information in the sciences has become largely standardised. This is particularly true for the sciences embedding extensive mathematics components, like physics and theoretical/quantum chemistry. It is also true for experimental procedures, as they list sets of operations, and each of them is described through basically standard clauses: changing the wording of instructions such as "*dissolve* [value of amount] *g of* [name of substance] *into* [value of volume] *ml of* [name of solvent]", or "*heat the solution to* [value of temperature] °*C*" or "*dilute the solution to* [value of concentration] *M concentration*" and the like, from one article to another, is simply unthinkable for a scientist, and would only generate unwarranted confusions in the communication of experimental procedures.

6.2.4.2 The Nature of Computer-Based Wording Comparisons

The computer-based identification of similarities among manuscripts is based on mechanical comparisons, because this is the way in which the computer operates. The software identifies a group of words that appears in more than one text and highlights them. It does not understand their meaning, or how they relate to the context. It does not identify names of things or concepts which consist of more than one word, nor ways of saying things that are standard in the sciences. In this way, common combinations of words, or statements that need to be made in a certain way, get automatically highlighted by a similarities-checking software. For instance, a common-language wording like "from the point of view of…" gets highlighted as a similarity wherever it appears. Even the author's name gets highlighted, because it appears in more than one work.

6.2.4.3 Computer-Based Identification of, or Request for, Sources

The software may request sources, or suggest sources, for the items that it underlines. Some requests are clearly to be ignored, like the request for the source of the author's name (unless one ends up attaching a birth certificate to each submitted manuscript). Similarly, all the requests concerning common wordings (like "from the point of view of") are to be ignored.

Any source identification proposed by the software needs to be checked. Few examples can adequately illustrate these aspects.

The software appears to have a sort of inbuilt 'obsession' (if this term could be applied to a software) for self-plagiarism as the main feature it is hunting for. Therefore, it requests source-citations even for common language terms that an author has utilised in previous works, and it provides the references to those works. However, following those recommendations and utilising those references would be blatant dishonesty (of course, the software does not realise it, but human beings do). For instance, on being run through a new manuscript by LM, a plagiarism-identification software indicated a previous LM's work as the reference for the term "production of pottery"; but this term is nobody's property, it denotes a common domain concept that has been used for some thousands years and in different languages; no previous work can be considered as its origin. Similarly, the software indicated a previous LM's work as the reference for the term "the use of a second language as a medium of instruction"; but this, too, is a common-domain concept (the name of a fact, or a practice) that has been used by hundreds (more likely, thousands) of experts, before and after the work indicated by the software as the source to be acknowledged. 'Obeying' the software indication and giving the suggested chapter as reference would be dishonest.

Following the approach apparently inherent (built-in, i.e., programmed) in that software, no set of words identifying concepts could be used in more than one manuscript by the same author (actually, that software highlights even the article "the" alone in some cases). Should such indications be obeyed, a researcher would be forced to change research area after publishing an article, and this would obviously be absurd. Therefore, in all such cases, the software's recommendation must be ignored.

Another fruit of the software's 'obsession' for self-plagiarism is the failure to identify references to works not pertaining to the manuscript's author as correct references. For instance, when writing about green chemistry, it is normal to make reference to the works that set its foundations [2–4]. But the plagiarism-detection software requested LM to make reference to another work, in which she had mentioned those same references in relation to green chemistry. The extrapolation of the scenario outlined by these recommendations would lead to a situation in which, when an author makes reference to something in a second work, reference should be made also to the first work in which that reference was present; when that author makes reference to that same 'something' in a third work, reference should be made also to the second and the first work; and so on. By extrapolation of the same criterion, an author should make reference to all the works where other authors cited the reference of interest, and this would clearly lead to a realm of the absurd. It suffices to think that, for instance, if a certain work or set of works has more than 1000 citations (as normal for works setting the foundations of something new), one would have to cite all the 1000 works that have cited the seminal works. Common sense and normal practice indicate that the seminal works are the ones that need to be cited, not those that have cited them. But the software is inherently unable to make the distinction.

In other cases, the software does not recognise references that are present in a text, and recommends less focused references. For instance, in a manuscript in which LM had reported the definition of "sustainable development" and given the first source of that definition as reference [5], the software indicated "internet" as the correct reference.

Similarly to not recognising common domain terms consisting of more than one word, the software cannot recognise common domain concepts (whose recognition would obviously require thinking abilities). Therefore, it provides unsuitable references. Its capability of scanning sources where certain concepts have already appeared is further limited by the fact that it searches for possible sources solely within one language. For instance, a software of this type ascribed the concept that chemistry studies the properties of substances and what happens when two substances come into contact to a 1997 work, although the concept is common-domain in chemistry (e.g., LM had explained it in [6], dated 1993, but it had then already been common domain for many years).

6.2.4.4 Expression Rigour, Language Quality, and Standardised Statements or Descriptions

As mentioned in Sect. 6.2.3, in any scientific communication, the selection of individual words, the construction of sentences, the logical connections between different pieces of information, and the organisation of the overall discourse need to closely match the details of the concept/s involved. Furthermore, several modes of presenting specific statements or descriptions are practically standardised. This subsection provides illustrative examples; most of these examples pertain to LM's professional activity; on the other hand, the concepts that they illustrate, in terms of items that need to be expressed in rigorous and largely standardised ways, may be easily transferred to many other areas and disciplines.

There are statements for which a request to express pieces of information differently borders with the absurd. An example is offered by the sentence

A. *All the calculations were performed with Gaussian-03 [xx].*

where "*[xx]*" stands for the number of the reference to the Gaussian-03 software. If an author uses this same software to calculate 20 different molecules or sets of molecules, that author needs to write this statement in the corresponding 20 articles, including the reference to Gaussian-03. The statement gets unavoidably highlighted by the plagiarism-hunting software. On the other hand, sentence “A” is the simplest, most straightforward and correct way to communicate the information; other ways might be clumsy; furthermore, the terms “*calculations*” and the name of the software will always be present, and the automatic comparison will always highlight them. An author cannot be expected to invent 20 different sentences to communicate this information. A particularly zealous editor once asked LM to get permission if she wanted to maintain this sentence. Actually, there is nothing here for which one could ask for permission. One cannot be expected to ask for permission to calculate molecules: it is normal part of computational chemistry research. One cannot be expected to ask for permission to use a software that is licensed to one’s institution: once the software is licensed, it can be used whenever needed. One cannot be expected to change calculation software every time one starts studying a new molecule: it would be unfeasible and absurd. And one cannot ask the journals, where articles containing this statement have been published, for permission to use this statement: there are several thousands of articles making this statement. On the basis of all this, sentence “A” has to be considered as standardised wording and should never be viewed as self-plagiarism.

There are other items that might need to be repeated as part of the preliminary (introduction) information in an article. For instance, if one studies ten different molecules, or sets of molecules, pertaining to the acylphloroglucinols’ class, one must provide the pieces of information that are necessary for the reader to be in a position to understand the rest of the article (whose core content will obviously be new). This means explaining what acylphloroglucinols are and why it is interesting to study them. One of the important pieces is the information that these molecules can form an intramolecular hydrogen bond, and the specification of the donor and acceptor atoms (the atoms between which it forms). There is no way of inventing 10 rigorous, complete and very different ways of communicating this. Then, the automatic comparison will detect similarities, and a fact that is common knowledge (like the formation of hydrogen bonds between certain atoms) gets pointed out as self-plagiarism. Removing the information, or some of its crucial components, would make the rest of the article ununderstandable. Therefore, that part, although highlighted by the software, is not self-plagiarism, but essential scientific information, pertaining to common domain information.

There are similarities in the introductory parts of related articles, which pertain to what could be called ‘readers’ rights’, like using the same basic atom-numbering for molecules of the same class that are objects of separate studies, to facilitate comparisons across different studies. An example is shown in Fig. 6.1. Image (a) has been used as the first figure in all the works on acylphloroglucinols of which LM is author or co-author, because it provides information essential to understand the text and the results (molecular structures are in any case common domain, and the atom numbering is just a convention, not something that can be the subject of some copyright form). Different acylphloroglucinol molecules may differ by the natures of R, R^ and R*, but the atom numbering of their central moiety (the one shown in the figure) has been maintained the same, so that the reader can easily compare results and discussions of geometry-dependent molecular properties for different molecules (across different articles). Image (b) shows a numbering-consistency application to a more complex molecule pertaining to the same class [7].

Similarly, each article needs to state the calculation methods and justify their choice. If one uses the same levels of theory in different works, the statements will necessarily be similar, because they refer to the same tools and respond to the same concepts. The author will normally mention that the same methods have been used in the calculation of other related molecules, and provide references to the previous works (besides the references to the works that introduced those methods), thus complying with all transparency requirements. But using the same calculation methods in different works is a researcher's right and often a necessity, not self-plagiarism.

Some findings are necessarily similar in different works. For instance, if there are no imaginary frequencies, one will state this as a fact and add the inference that all the identified stationary points are true minima on the potential energy surface of the given molecule. There are not many ways of using different sentences to state this; on the other hand, every time this is confirmed for a new molecule, it is a new result that needs to be stated.

Each new article on a new acylphloroglucinol molecule, or set of molecules, will contain tables reporting relative energies, or hydrogen bond lengths, or vibrational frequencies of the O–H bonds, or other relevant molecular properties; the tables unavoidably have similar captions in different works, because they refer to the same quantities, but then the values in the tables are different for different works because they refer to different molecules. The plagiarism-hunting software highlights the captions because of their similarities (same quantities, same dimensions, same computational methods), but these similarities are unavoidable, because they are inherent components of the analysis of computational results; requesting different wordings for the captions would be tantamount to requesting that the authors renounce standard analysis and data-reporting criteria and tools.

An author may get the suggestion of writing the same concept using somewhat different wordings. This may be possible in some cases, but it too often results in precision-lowering, and precision-decrease is not acceptable. An example is offered by sentence "B":

B. *For the sake of conciseness, the calculation methods are denoted with the following acronyms in the rest of the text: ....*

and the alternative wording "C":

C. *The calculation methods are briefly denoted with the following acronyms in the rest of the text: ....*

Simple language analysis shows that "B" is more precise and complete than "C". Wording "B" states the reason why the use of acronyms is introduced ("for the sake of conciseness"), and it is well known that conciseness is highly recommended in

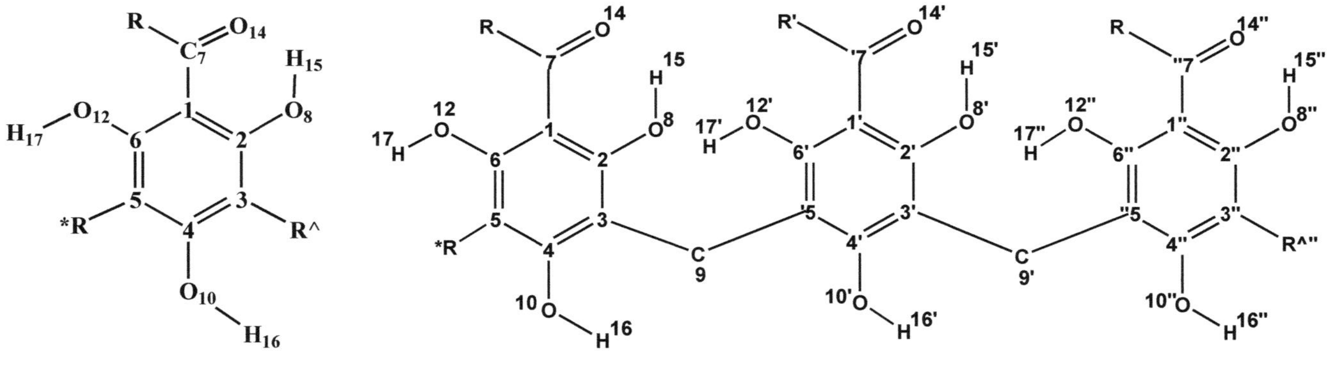

(a) (b)

Fig. 6.1 Illustration of a general-type molecular structure and how it may be recalled in different works. Image (a) shows the general structure of acylphloroglucinols and the atom numbering utilized in the works on acylphloroglucinols of which LM has been author or co-author (the first atom of R^ is given the number 9). Image (b) shows a molecular structure containing three acylphloroglucinol units, and an atom numbering of each unit consistent with that of the general structure [7]

scientific texts. Wording "C" leaves the reason unspecified: the reader might legitimately infer that the author uses the acronyms because of getting bored at writing certain long terms frequently. The adverb "briefly" is not synonym of "for the sake of conciseness". Therefore, wording "B" is to be preferred. Forcing an author to replace "B" with "C" would be tantamount to forcing that author to decrease the precision of his/her text. Furthermore, assuming that such a change would remove alleged self-plagiarism would be ludicrous. Plagiarism refers to passing somebody else's works as one's own, and self-plagiarism refers to passing one's previous work as new. There is no self-plagiarism in the provision of information that is essential for the reader to be in a position to understand a text.

An author puts a lot of efforts in trying to make an article's text as rigorous and complete as possible, and this refers not only to the entire text, but to the wording of each and every sentence, description and piece of information. Many wordings are standardised and are used over and over because they are the best (e.g., statement "B"). If an inherently thought-unable software considers optimised wordings as self-plagiarism, then the human mind (the editor, the author) must overrule diagnoses that are done by an instrument that is not capable of reading texts or understanding meanings. Under circumstances of this type, authors have all the motivations to refuse lowering the language-standard of their manuscripts because a machine finds that some groups of words are too similar to those of a previous article.

6.2.4.5 The Key Players: Editors, Reviewers and Authors

Humans—authors, reviewers and editors—play all the key roles in the preparation of a written work. Therefore, they need to have very clear ideas about the limitations of plagiarism-hunting software, to ensure that the quality of scientific communication does not get damaged by blind acceptance of the software indications. Editors need to distinguish when it is correct to request that the indications are implemented and when following those indications would result in decreased quality, absurdities or dishonesty; in order to do so, they also need to distinguish whether something highlighted by the software is actually plagiarism or self-plagiarism, or is something pertaining to normal discourses or common-domain knowledge. Authors need to be fully aware of the language requirements of scientific communication, and also of all the distinctions that are relevant for them to defend their choice of wording, sentence construction, referencing, and the like; for instance, they need to be aware of group of words pertaining to common language (like "from the point of view of"), or common domain concepts (like the previously-mentioned ones), for which no plagiarism or self-plagiarism can exist.

It is to be acknowledged that the greatest majority of editors are fully aware of the pitfalls of blind belief in the indications from the plagiarism-hunting software: when they see that the only highlighted items are common sentences and standardised (or nearly standardised) expression forms, they do not claim the presence of self-plagiarism and do not ask for changes. Only a very small proportion of editors view all the highlightings as self-plagiarism and requires the author to make changes for all of them. To further stress the comparative rarity of this phenomenon, it may be added that—so far—LM has encountered this problem only on four occasions, two of which were quickly resolved in a reasonable way (the other two required long messages to defend her selection of words and statements). Other authors have had similar experiences and had to defend the correctness and quality of their texts and wording choices.

In some cases, editors operate under the frameworks of some 'policies', requesting that authors follow certain stated criteria, usually expressed in terms of numbers or percent. In one such case in LM's experience, one of the criteria requested that all the groups of highlighted words containing three or more words needed to be changed. This would result in the impossibility of using expressions like "from the point of view of…", or "for the sake of", and many other expressions pertaining to common language and regularly highlighted by the software. Unavoidably, LM had to write a long message to demonstrate the non-reasonability of such a request.

In another case, a request stated that there must be less than 1% similarity with each previous work by the same author. A simple analysis is sufficient to show that such a request is not realistic. There are 33 lines in an A4 page with 2.54 cm margins, Times New Roman font and 1.5 spacing. This means that a 20-pages manuscript contains 660 lines. "Less than 1% similarity" with a previous work by the same author would mean "less than 6.6 lines" similarity. If one considers one line for the names of the authors (that the software usually highlights as plagiarism) and two lines for institution, address and e-mail, one is left with 3.6. lines. In addition, a couple of lines may be taken by standardised acknowledgments (e.g., to funders or to a high performance computation centre). Thus, in principle, one is left with 1.6 lines available. This is not enough to introduce the class of compounds considered, or to provide information about the calculation software (like sentence "A") and methods for a computational chemistry manuscript, or to provide essential introductory information for any theme (even the initial 6.6 lines would rarely be enough). An article pertaining to a series of related studies could never comply with this requirement, unless it suppresses all the introductory information; but suppressing it would destroy the readability of the rest of the text.

6.2.4.6 The Importance of Humanities-Based Expertise

The analyses and discussions in the previous sections clearly highlight the need that authors and editors have good language mastery and adequate epistemological background to be able to distinguish reasonable and unreasonable indications by the plagiarism-hunting software, so as to prevent getting entangled in a realm of the absurd. These types of expertise are humanities-based. They are essential for critical thinking, which is another ability that needs humanities bases to be engendered and grow. In other words, a rational usage of software-provided indications about issues concerning language requires that the humans concerned with the usage have good humanities background.

Editors need this expertise to avoid requesting that authors make changes that are not reasonable. Authors need this expertise to know whether the requests for specific changes are reasonable or not and, in the latter case, to be able to provide fully rational explanations for their choice of not implementing those requests. Actually, it should always be possible to assume that an author selects a certain wording because it is the most appropriate. Therefore, an author should always be able (and available) to explain why a certain wording is the most appropriate and should not be changed (for instance, by providing explanations along the lines with which statements "B" and "C" have been compared in Sect. 6.2.4.4).

6.2.4.7 Some Additional Considerations

In principle, the hunt for plagiarism should not be based on the hunt for words or groups of words that are repeated. It should be based on concepts, distinguishing which ones are unavoidably repeated in works that utilise the same methods (for instance, an experimental procedure or a calculation method) or refer to different items of the same class (for which it is necessary to introduce the class and its characteristics), and which ones pertain to the new information, which must constitute the largest part of an article and must actually be 'new'. However, a software cannot do this type of comparison, because it does not have the "concept" or "results" concepts.

The automatic indications produced by the software need to be inspected by a human being. Editors and reviewers should be in a position to do this efficiently. If they request that the author makes changes that would lower the quality of a text, or the conceptual rigour with which the content is presented, it is up to the author to inspect the requests, analyse their language and epistemological validity and implications, and show the results of the analysis. The language quality of a paper, or the logic of its organization, have by far greater priority than any request to make perfunctory changes for the sake of saying that some changes have been made even if those changes lower the quality of the communication. Requests for changes that lower the rigour or quality of the discourse are actually alien to the inherent requirements of scientific communication, where conceptual and expression rigour is the first priority. What needs to be new in a new article are the components that make it new: the specific object/s and objectives of the study, the results, the analysis of the results and, finally, their interpretation.

Why does it happen that unsuitable or straightforwardly non-rational requests like some of the previously-mentioned ones occur and, apparently, they are starting occurring with increasing frequency? Why does it happen (luckily, rarely) that a well-intentioned editor implements some of the indications given by the software after an author has approved the final version of a text? The answers are various, and their effects probably combine. On the one hand, the tendency to uncritically believe that the computer is always right is steadily increasing. On the other hand, language mastery and the familiarity with the history and epistemology of the sciences are steadily decreasing. These are all categories of humanities-based knowledge. In other words, the rationality with which the indications from plagiarism-hunting software are handled depends on the familiarity with humanities-based abilities (like language mastery), or abilities at the intersection between humanities and sciences (like the history and philosophy of science) by all the parties involved—editors, reviewers and authors.

Finally, it is important to recall that the use of plagiarism-hunting software was born because of cases of misconduct by some authors. This concerned only a minority of authors, but, when an unacceptable practice is detected and a certain recurrence is proved, it becomes unavoidable that prevention measures are designed and implemented. Therefore, the current need to untangle the confusions often springing from the use of plagiarism-hunting software can only be blamed on all those unknown authors whose misconduct made this use a necessity. The issue of ethics in research work is considered more in detail in chapter 18, Sect. 18.4.1.

6.3 The New AI-Based Text-Generating Chatbots and the Role of Humans' Knowledge and Language Mastery

Artificial intelligence (AI) is increasingly utilised in a variety of fields. The importance of understanding the capabilities and weaknesses of AI systems to prevent undesirable outcomes is outlined in [8]. Various AI-based *large language models* (LLM) meant to assist with writing have been developed in recent years. A particularly sophisticated LLM is the ChatGPT chatbot, which was released towards the end of 2022 (where GPT stands for "generative pretrained transformer"). It can interact with the user in a sort of conversational form and it "can produce sophisticated and seemingly intelligent writing after training on a massive data set of text" [9]. Its ability to write texts that are not easily distinguishable from those written by humans (it can "write summaries of research papers that are convincing enough to fool scientists into thinking that they were written by a human" [10]), and its fast diffusion, has created a variety of concerns, from the reliability of the information it produces to risks to individuals' privacy. A number of specialists prepared an open letter calling for a pause on the development of advanced AI systems, to give time to evaluate potentialities and risks and design suitable regulations [11, 12]. The next paragraphs will briefly recall the main concerns about research and publications and about students' learning and assessment, thus also highlighting that the inferences derived in the previous sections apply also to LLM.

Using ChatGPT for specialised research may entail the risk of introducing inaccuracies, biases and plagiarism [9]: researchers "risk being misled by false or biased information, and incorporating it into their thinking and papers" [9]. Since ChatGPT may reproduce texts without citing the original sources, researchers may risk unknowingly plagiarising other authors' texts; they may even risk giving away their own ideas, as the information that a researcher gives to ChatGPT may get incorporated into the overall reservoir of information that it uses to create its texts [9]. While ChatGPT may assist in text-writing, maintaining full reliability of the information presented in a paper depends on the author's (human being) vigilance.

Various verifications of the reliability of answers produced by ChatGPT have been carried out. When asked to summarise a systematic review previously published by the authors [9], ChatGPT "fabricated a convincing response that contained several factual errors, misrepresentations and wrong data". A study meant to check how it can understand and answer chemistry questions [13] showed that it could answer certain types of questions with sufficient accuracy, while it struggled with others, including compounds' names. Another study concerning chemistry [14] highlighted a variety of conceptual errors, including equating the speed of electromagnetic radiation with its energy and errors related to periodic properties, ionization energy, and effective nuclear charge; the conclusion was that the ChatGPT chemistry level is lower than that of the average chemistry student, and that its best utilisation in chemistry learning would be that of errors-hunting (asking students to identify and discuss the errors present in the answers by ChatGPT). ChatGPT could not even pass the exams for the 1 and 2 year general applied chemistry modules of a pharmaceutical science program [15].

Several authors highlight general-type risks, such as the risk that, because of their good mode of expression, ChatGPT outputs could "be confused with expert opinions, even though LLMs will continue to have limited ability to judge the reliability and relevance of information, in part because LLMs have not achieved natural language understanding" [16]. Such outputs might then be used to design measures or regulations, with the risks of negative impacts stemming from their implementation.

ChatGPT could also generate false or fabricated information; for instance, it can provide made-up references or sources of specific text, such as fabricated DOI or URL links [17]. Examples of fabricated information concerning environmental sciences issues are reported in [17]. Another telling example is the provision of number of cases to be used as legal precedent, provided by ChatGPT to a lawyer, which were "found to be non-existent" by the court [18]. It might appear that ChatGPT is able to fabricate references and other information which are verisimilar, i.e., they look like true but are not true. Since "it is nearly impossible to completely eliminate false or fake information… caution must be exercised when inviting AI into decision-making processes, especially for environmental problems that are closely tied to public welfare" [17].

An overview of the broad-scope risks that may derive from excessive use of ChatGPT is outlined by Yuval Noah Harari, who said that artificial intelligence has "hacked" the operating system of human civilization because of its mastery of language [19]. He states that "language is the stuff almost all human culture is made of" and illustrates it with examples such as "Human rights, for example, aren't inscribed in our DNA. Rather, they are cultural artefacts we created by telling stories and writing laws". Democracy is a language that relies on meaningful conversations, and when AI hacks language it could also destroy democracy. He also wrote that "Through its mastery of language, AI could even form intimate relationships with people and use the power of intimacy to change our opinions and worldviews", and emphasised that the establishing of regulations is extremely important, as well as the need to have an ethical framework to respond to the challenges posed by AI.

All what has been mentioned in this section shows that the same inferences that had been derived in the previous sections remain valid for the more advanced AI-based LLM. The use of ChatGPT can be beneficial as long as it is considered a tool, and humans remain fully in charge. Paradoxically, the presence of more sophisticated AI tools requires greater subject knowledge and greater language mastery from the humans who interact with them, to be able to verify the reliability of their responses and to distinguish between true and fabricated information.

Education needs to equip students with all the needed knowledge and mastery. It needs to find optimal balances between training students to use LLM as tools and ensuring that they acquire their own independent knowledge at a level that will enable them to always be in charge with the AI tools that they will use in their professional lives. The verification of students' acquired knowledge is realistic only if AI use is excluded [20–23]; this may also entail the return to traditional, AI-free forms of exam [24] and to the increase of the proportion of oral assessment.

Finally, there may also be pitfalls related to other types of unethical behaviours. An example from a recent experience outlines the following scenario: a reviewer with extremely poor language mastery is unable to conceive that an author might be in a position to write responses with a reasonable level of language mastery and accuses the author of having used ChatGPT. Cases like this need to be prevented. Editors need to be vigilant for this risk, and this also entails the need to exclude reviewers with too poor language mastery (e.g., reviewers whose comments contain too many language errors and also sentences without literal meaning). Authors who should find themselves in situations of this type need to be in a position to defend themselves; this, in turn, requires adequate language mastery to document the reviewer's inadequacies through the analysis of the errors in his/her comments.

6.4 Discussion and Conclusions

On considering the general issue of the relationships between humans and computers as far as text-writing is concerned, the most important thing is to always remember that the software is not able to think and, therefore, the human mind must be in charge. The computer cannot understand a text or create a reliably-error-free new one; it works only in terms of its stored information, through comparisons or combinations of different parts.

A text correction software cannot evaluate meanings or roles of individual words or words-groupings; it makes automatic corrections on the basis of its in-built dictionary, without having the tools to know whether that correction makes sense in the given context; the author (human mind) must check to make sure that no automatic correction is strange or destroys the meaning of a sentence.

Similarly, a plagiarism-hunting software identifies groups of words that appear in more than one text and highlights them; it does not understand their meaning, or how they relate to the context; it does not know whether they are essential components of the resources of a language, or expressions pertaining to common language, or expressions of common-domain concepts and information; the editor and the author (human minds) are the ones who need to distinguish among all these possibilities and identify the true plagiarism or true self-plagiarism cases.

Analogous considerations hold for the utilization of AI-based text-writing tools like ChatGPT. The tool is useful as long as the human mind remains fully in charge. Only a human can verify whether the information provided by the chatbot is true or fabricated. It appears that the chatbot does not distinguish between 'true' and 'verisimilar', or has not been programmed with a computer-type correspondent of an 'ethical' principle preventing the fabrication of verisimilar information; on the other hand, it might also not be possible to program something of this type, because the evaluation reference should be actual knowledge, what is totally different from a huge reservoir of texts from which to select components and utilise them to build new texts. In other words, uncritically believing that the computer is always—or nearly always—right would be highly risky even with regard to LLM.

Computers and software can be extremely useful, but operations like writing and ensuring that a written text expresses the wanted information in the best and most rigorous way pertain to the human mind. Adequate humanities-related preparation (first of all, language mastery) are essential to discriminate between valid indications and quality-lowering or absurd ones, among the automatic corrections while one is writing, and among the indications provided by a plagiarism-hunting software on a completed text; they are also essential to discriminate between realistic and fabricated information in a text built by a chatbot.

In summary, a computer can provide precious, wonderful help, but this is possible only as long as the human remains in charge. The fact that the computer can help does not mean that the human being is allowed not to know, or not to know enough. Actually, more complete and detailed knowledge is required for the human to fully benefit from the assistance provided by the computer, because the way in which its programming guides the computer's actions entails attention to every detail; therefore, the human must be in a position to critically analyse every detail. A good deal of the required knowledge is humanities-based when text-writing is involved. Decreasing the humanities foundations and the mastery of humanities-based abilities in the preparation of specialists will unavoidably decrease their ability to fully benefit from the use of computers, and to avoid the risk of unwarily utilising fabricated information.

References

1. L. Mammino, Computational chemistry for green design in chemistry and pharmacy: building awareness in the classroom. Sustain. Chem. Pharm. (2020). https://doi.org/10.1016/J.SCP.2020.100283

2. P.T. Anastas, J.C. Warner, *Green Chemistry: Theory and Practice* (Oxford University Press, New York, 1998)
3. P.T. Anastas, T. Williamson, in P.T. Anastas, T. Williamson (eds.), *Green chemistry* (American Chemical Society, Washington, 1996).
4. P. Tundo, P. Anastas, *Green Chemistry, Challenging Perspectives* (Oxford University Press, Oxford, 2000)
5. G.H. Brundtland, *Our Common Future* (Oxford University Press, Oxford, 1987)
6. L. Mammino, *Chimica Viva* (D'Anna Press, Florence, 1994)
7. L. Mammino, Correlation effects in trimeric acylphloroglucinols. Computation **9**, 121 (2021). https://doi.org/10.3390/computation9110121
8. R. Burnell, W. Schellaert, J. Burden, T.D. Ullman, F. Martinez-Plumed, J.B. Tenenbaum, D. Rutar, L.G. Cheke, J. Sohl-Dickstein, M. Mitchell, D. Kiela, M. Shanahan, E.M. Voorhees, A.G. Cohn, J.Z. Leibo, J. Hernandez-Orallo, Rethink reporting of evaluation results in AI. Science **380**(6641), 136–138 (2023). https://doi.org/10.1126/science.adf6369
9. E.A.M. van Dis, J. Bollen, R. van Rooji, W. Zuidema, C.L. Bockting, CahtGPT: five priorities for research. Nature **614**, 224–226 (2023)
10. C.A. Gao et al., Preprint at bioRxiv (2022). https://doi.org/10.1101/2022.12.23.521610
11. L. Clarke, *Alarmed tech leaders call for AI research pause.* https://www.science.org/content/article/ (11 Apr 2023)
12. L. Clarke, *Computer Science Call for AI Pause Highlights Potential Dangers.* NEWS|IN DEPTH0414NID_16964852.indd 1204/11/23 5:53 PM
13. C.M. Casto Nacimiento, A.S. Pimintel, Do large language models understand chemistry? A conversation with ChatGPT. J. Chem. Inf. Model. **63**(6), 1649–1655 (2023)
14. T.M. Clark, Investigating the use of an artificial intelligence chatbot with general chemistry exam questions. J. Chem. Educ. **100**(5), 1905–1916 (2023)
15. S. Fergus, M. Botha, M. Ostovar, Evaluating academic answers generated using ChatGPT. J. Chem. Educ. **100**(4), 1672–1675 (2023)
16. M.C. Rillig, M. Ågerstrand, M. Bi, K.A. Gould, U. Sauerland, Risks and benefits of large language models for the environment. Environ. Sci. Technol. **57**, 3464–3466 (2023)
17. J.-J. Zhu, J. Jiang, M. Yang, Z.J. Ren, ChatGPT and environmental research. Environ. Sci. Technol. (2023). https://doi.org/10.1021/acs.est.3c01818
18. J. Felton, *New York Lawyer Caught Using ChatGPT After Citing Cases That Don't Exist, IFLScience* (Wednesday, May 31, 2023) newsletters@iflscience.com
19. Tech Desk , Edited by Bijin Jose, Yuval Noah Harari paints a grim picture of the AI age, roots for safety checks, *The Indian Express*, New Delhi, June 5, 2023
20. O. Chia, Teachers v ChatGPT: Schools face new challenge in fight against plagiarism, The Straits Times, https://www.straitstimes.com/tech/teachers-v-chatgpt-schools-face-new-challenge-in-fight-against-plagiarism
21. D. R. E. Cotton, P. Cotton, R. Shipway, Chatting and Cheating. Ensuring academic integrity in the era of ChatGPT. https://doi.org/10.35542/osf.io/mrz8h
22. Top French university bans use of ChatGPT to prevent plagiarism. https://www.euronews.com/next/2023/01/28/francechatgpt-university.
23. University of Hong Kong issues interim ban on ChatGPT, AI based tools. https://news.cgtn.com/news/2023-02-19/University-ofHong-Kong-issues-interim-ban-on-ChatGPT-AI-based-tools1hxWzqgcMxy/index.html.
24. C. Cassidy, Australian universities to return to 'pen and paper' exams after students caught using AI to write essays, https://www.theguardian.com/australia-news/2023/jan/10/universities-to-return-to-pen-and-paper-exams-after-students-caught-using-ai-to-write-essays

Chapter 7
Scientific Terminology: A Long Thread of Interactions Between Humanities and Sciences

Liliana Mammino

Abstract Terminology is something typical of each discipline, as the ensemble of names of the objects and phenomena with which that discipline deals. It has developed through the history of each discipline, adding new terms every time an object, or a phenomenon, or a descriptive quality was discovered, or new models were proposed. This chapter presents an overview of the generation of scientific and technical terminology, highlighting how the new terms were often rooted in deeply humanities-based cultural backgrounds, from the use of root-words pertaining to ancient languages, meant to foster perceptions of scientificity and abstractness, to the expression of the features that the discoverer perceived as more striking, and to attention to myths and beliefs.

Keywords Epistemology of definitions · Generation of technical terms · Meaning of terminology · Nomenclatures · Technical terms and common language

7.1 Introduction: The Nature of Terminology

The *terminology* concept refers to individual disciplines. Each discipline has its own specific terminology, that is, an ensemble of terms denoting the items that are objects of study within that discipline, or the phenomena concerning those items, or properties that are not described by common-language terms. Its terminology is thus an intrinsic component of each discipline. The terms are considered its *technical terms*. The terminology terms of humanities-pertaining areas were mostly generated from root-words, perceptions or images within that area and, therefore, are not analysed here from the point of view of cross-domain (humanities and sciences) relationships. The current chapter considers the terminologies of science and technology areas,

L. Mammino (✉)
Faculty of Science, Engineering and Agriculture, University of Venda, Thohoyandou, South Africa
e-mail: sasdestria1@gmail.com; mammino@univen.ac.za

L. Mammino (ed.), *Insights into the Relationships between Humanities and Sciences*,
https://doi.org/10.1007/978-3-031-97900-2_7

because their generation often utilised root-words from mythologies, expressions of perceptions and other features that relate to the humanities domain.

In most cases, technical terms do not belong to the ensembles of familiar terms, because the corresponding objects, phenomena and properties do not belong to the fields of everyday activities and interactions; they are used nearly exclusively within the communities of specialists of each area. Few examples can easily illustrate this aspect. Terms like *boron trifluoride, nitration, electronic configuration*, are technical terms in chemistry; terms like *angular momentum, paramagnetism, conductivity*, are technical terms in physics; terms like *differential equation, series expansion, exponential dependence*, are technical terms in mathematics; terms like *cable-strayed bridge, entrenched tunnel, prestressed reinforced concrete*, are technical terms in civil engineering; terms like *anaphylactic shock, apyrexia, pyelonephritis*, are technical terms in medicine; and so on. They are rarely encountered or utilised in normal conversation or in everyday activities and interactions. Specialists in a given area may know several or many terms of a close area, but may not be familiar with the terms of a not-so-close area (for instance, physicists will know most of the terms specific to mathematics, but may not know many of the terms specific to biological sciences).

With the progress of science and technology, the amount of technical terms has increased enormously. Each science, whether pure or applied, has a high number of such terms. Learning them is part of the education and training within each discipline, and their learning is often perceived as a sort of 'initiation stage' into that discipline. As a consequence, technical terms are viewed as something belonging to the domains of individual sciences in a most exclusive way, up to the point that they may end up being perceived as knowledge carriers, although they alone cannot communicate self-consistent or identifiable information.

The fact that technical terms are not information carriers by themselves is analysed in detail in [1]. A quick illustration can be offered by any statement conveying science-based information. For example, in the statement:

A <u>molecule</u> is a stable grouping of <u>atoms</u>. Therefore, it consists of <u>nuclei</u> and <u>electrons</u>

the nouns that have been underlined are technical terms in chemistry and physics; the adjective 'stable', although not underlined, takes specific meanings with reference to the context (in this case, it means that the atoms remain together unless something intervenes to change the grouping specific of that molecule). However, no information would be conveyed by these terms alone; the information is built by the other words, which connect the technical terms and belong to common language. This is also a major reason why adequate language mastery (meaning of individual words, grammar and syntax of the language that one is using) is essential in science learning, teaching and practice; this necessity is discussed in detail in Chapter 4.

A first-approximation classification of technical terms can identify the following two broad categories:

- Terms belonging to the nomenclatures of specific sciences.
- Terms related to concepts, models or technology *items*.

They are analysed in some detail in the next sections.

7.2 Nomenclatures: systematic ways of generating names of items of a given class

Terms pertaining to nomenclatures have the role of being names of individual objects within specific classes of analogous objects. Their ensembles constitute the *nomenclature* of a given discipline. Nomenclatures are typical of sciences that deal with a large number of *items* among the objects of their studies, such as compounds in chemistry, plants in botany, or animals in zoology. The names assigned to these *items* are meant to respond to classification criteria; therefore, they are formed in a way that enables easy identification of each individual object and of the category or subcategory to which it belongs. Sometimes, they transmit information also on other features that are considered particularly relevant, provided it has been possible to design criteria for such a transmission; for example, the name of a compound in chemistry enables the identification of its formula and of the structure of its molecule.

In general, we may say that nomenclature names are constructed in such a way as to provide the maximum information-wealth compatible with comparatively simple construction criteria. For example:

- The combination of two words in the names denoting animals or plants attains the objective of transmitting information about the genus and the species. The specification of the genus is functional to an indirect transmission of all the information associated with the features that are common to that genus.
- The names of chemical compounds provide the following information (Fig. 7.1):
  - the constituting elements
  - the number of atoms of each element in the molecule (or the ratios between the numbers of oppositely charged ions for ionic compounds);
  - for organic compounds, the structure of the molecule.

Terms belonging to *nomenclatures* do not require specific individual definitions, because they are sufficient to identify the intended item on the basis of how they are built.

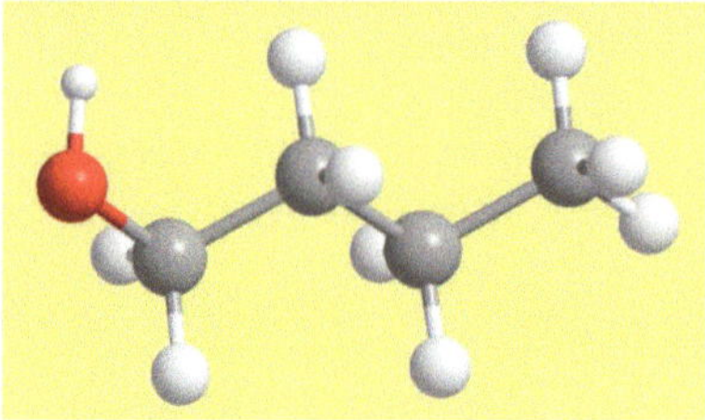

1-butanol

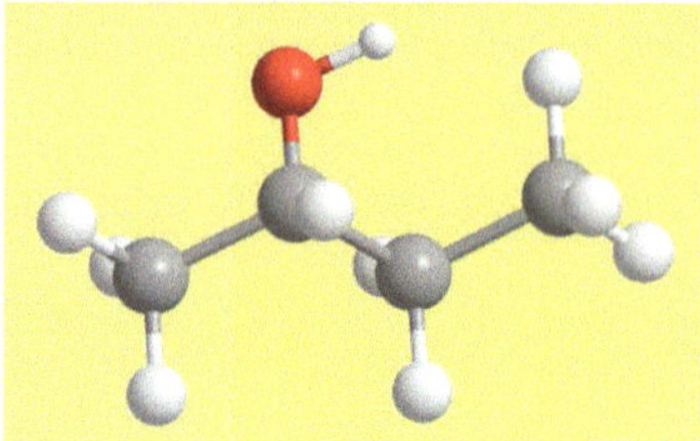

2-butanol

Fig. 7.1 Illustration of the information conveyed by the names of compounds. The 'butan' word-root informs that the compounds derive from butane, whose molecule contains 4 carbon atoms and ten hydrogen atoms. The "ol" ending informs that one of the hydrogen atoms is replaced by an OH group. The number at the beginning tells us the position of the OH group, i.e., to which carbon atom it is attached: the first one (left) or the second one (right). From this information we also know that the molecular formula of both compounds is $C_4H_{10}O$, which can be written in more detailed form as $CH_3CH_2CH_2CH_2OH$ for 1-butanol and $CH_3CH_2CHOHCH_3$ for 2-butanol

7.3 The Definition of Technical Terms not Pertaining to Nomenclatures

Terms not pertaining to nomenclatures often require definitions that enable the identification of the corresponding item. They include the following classes of terms:

- Terms denoting concepts that are parts of our interpretation models or our logical constructions. They always require definitions. They comprise nouns like, e.g., *mole, oxidation number, force, angular momentum, polynomial*, and so on. Some of them are common terms that respond to specific meanings and definitions within a given science, like *force, field, energy*, and the like.
- Terms denoting practical operations. Such terms may require a definition or be accompanied by a description. Examples may be the terms *flotation, smelting, filtration, differentiation*, and so on.
- Terms denoting objects produced by technology. When such objects are relevant in the discussion of models or in designing activities, they may be associated with individual, rigorous definitions or descriptions. Examples may be *distillation apparatus, mass spectrometer, cable-strayed bridge.*
- Adjectives denoting qualities associated with independently defined concepts. Their definition is related to that of the corresponding concept/noun. Examples may be the adjectives *acidic* or *basic*, whose meanings are associated with the definitions provided for the terms *acid* and *base* respectively, or *electromagnetic,* or *gravitational*, whose meanings are associated with the theories about electromagnetism or about gravitation. In many cases, the adjective may also play a role in an "adjective + noun" *pair* whose specific meaning may require a separate definition; pairs like *basic solution* or *electromagnetic radiation* offer illustrative examples. Verbal forms that can take the role of adjectives and express qualities are also included, like *oxidising* or *reducing* in chemistry.

- Verbs associated with processes or with operations. If they are related to our interpretation models, they may require specific definitions. Examples may be the verbs *to oxidise* and *to reduce* in chemistry, or *to differentiate* and *to integrate* in mathematics.

A definition is a complete description of the item that is being defined, and must be operational, i.e., it needs to contain all the information that is necessary for the identification of that item whenever it is encountered [2]. Because of their nature, definitions relate to the epistemology of the given science and, therefore, also to the philosophy of science. The way in which they are expressed requires rigorous use of language. In this way, the moment is which a definition is established can be viewed as a moment in which science and humanities converge.

Some definitions may constitute a sort of cluster, by being conceptually related. An example is offered by the definition of *oxidation number* in chemistry, and of the terms that relate to it. What could be called the 'parent definition' is the definition of oxidation number:

> *The oxidation number of an element is a formal charge assigned to that element on the basis of the conventional criterion that all the electrons of each bond are ascribed to the more electronegative element.*

Understanding this definition requires good language mastery and adequate epistemological awareness. The adjective "formal" before "charge" informs that the charge is not actual (the atoms in the molecule are not ions, unless one is dealing with actual ions); these formal charges are considered because they provide an easy framework to discuss the reactions in which the oxidation numbers of some elements change. The definition entails a set of mental operations that enable the identification of the oxidation number of the elements of interest; Scheme 7.1 [3] illustrates them considering the sulphuric acid molecule as an example. In practice, a number of rules derived from the definition are convenient for simple cases; it is however important to be aware that the practical rules derive from the definition, but do not replace it.

Two concepts related to the oxidation number are defined independently and identify processes:

> *Oxidation: a change whereby the oxidation number of an element increases.*
>
> *Reduction: a change whereby the oxidation number of an element decreases.*

The definition of two more concepts stems from the previous ones on the basis of normal language analysis:

> *Oxidising agent: an element that causes the oxidation of another element.*
>
> *Reducing agent: an element that causes the reduction of another element.*

The definition of redox reaction is based on the previous ones:

> *A redox reaction is a reaction in which the oxidation numbers of some elements change.*

This definition has a clear operational character. If one wants to know whether a reaction is redox, one needs to identify the oxidation numbers of all the elements among

a) Since electrons are considered, draw the Lewis structure of the molecule, representing each outer electron with a dot

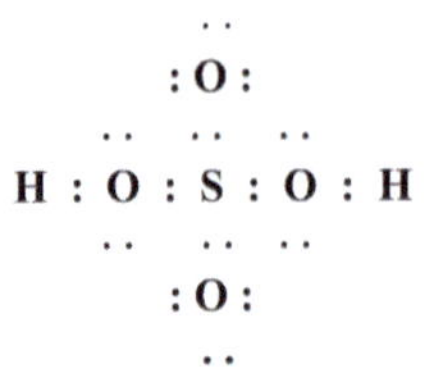

b) For each pair of bonded atoms, ascribe the electrons of the bond to the more electronegative atom

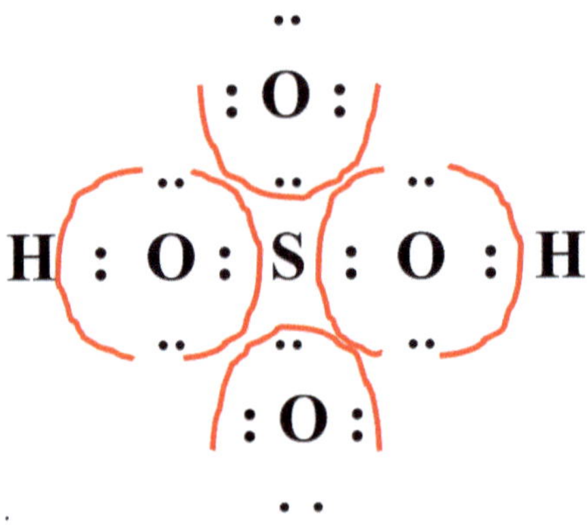

c) Count the number of outer electrons thus ascribed to each atom:
0 for H
0 for S
8 for O

d) Remember the number of outer electrons when these atoms are isolated
1 for H
6 for S
6 for O

e) Compare these numbers of electrons when the atom is in the molecule and when the atom is isolated:

- H the number of electrons is one unit smaller in the molecule. Its oxidation number in the molecule is +1
- S the number of electrons is six units smaller in the molecule. Its oxidation number in the molecule is +6
- O the number of electrons is two units greater in the molecule. Its oxidation number in the molecule is -2.

Scheme 7.1 Illustration of the operational character of the definition of oxidation number. The sulphuric acid molecule is selected as example. The operational steps corresponding to the definition are shown in detail, expressed in the form of instructions the reactants and among the products, and check whether some of them change. Scheme 7.2 highlights the way in which these definitions are related; Scheme 7.3 provides an example of the operational application of the definition of redox reaction, and examples of the other definitions of this cluster. All these Schemes (7.1, 7.2, 7.3) are meant to emphasise the importance of highlighting the operational implications of definitions by outlining approaches that can be easily used at educational level.

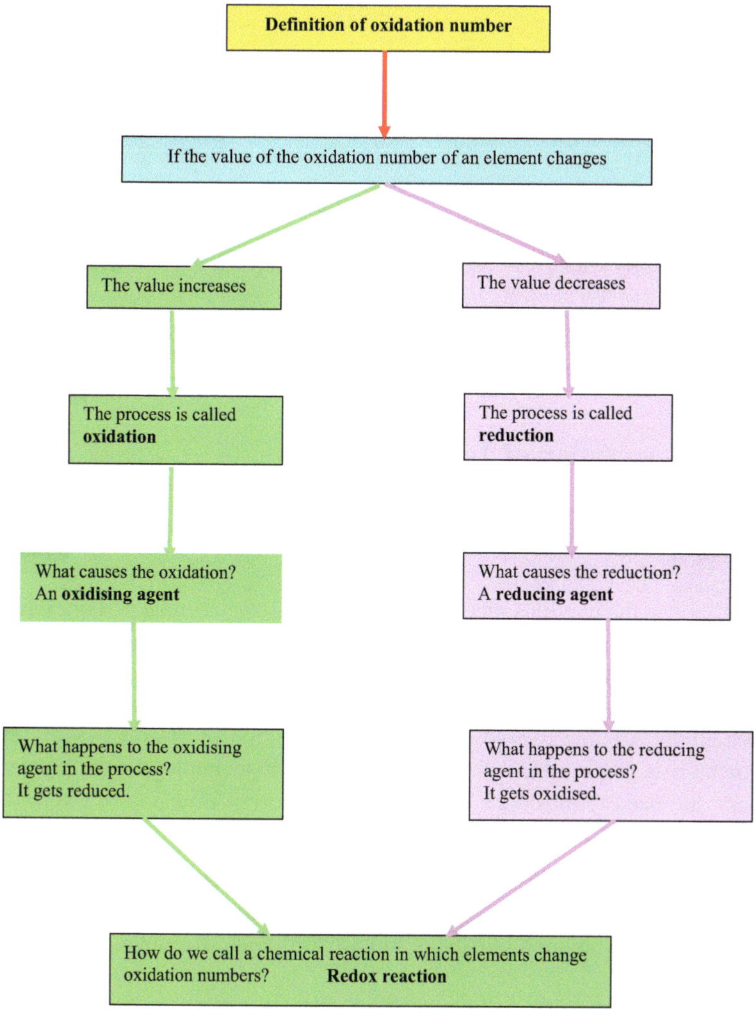

Scheme 7.2 Example of a cluster of related definitions. The cluster of definitions related to that of oxidation number is selected for illustration. The terms that are objects of definitions are written in bold

Task: Establish whether the reaction

$$H_2S + 2\,HNO_3 \rightarrow 2\,NO_2 + S + 2\,H_2O$$

is a redox reaction. In case of positive answer, identify the species that is oxidised, the species that is reduced, the oxidising agent and the reducing agent.

Procedure:

1. Identify the oxidation number of each element

$$\overset{+1\;-2}{H_2S} + 2\,\overset{+1\;+5\;-2}{HNO_3} \rightarrow 2\,\overset{+4\;-2}{NO_2} + \overset{0}{S} + 2\,\overset{+1\;-2}{H_2O}$$

2. Compare the oxidation number of the same element in the reactants and in the products:
   The oxidation numbers of S and N are different in the reactants and in the products. Therefore, this is a redox reaction.
3. Consider how the oxidation numbers change.
   - ➢ The oxidation numbers of S increases. Therefore:
     - S is oxidised.
     - By getting oxidised, it causes the reduction of N; therefore, S is the reducing agent.
   - ➢ The oxidation numbers of N decreases. Therefore:
     - N is reduced.
     - By getting reduced, it causes the oxidation of S; therefore, N is the oxidising agent.

Scheme 7.3 Example of verification of whether a given reaction is a redox reaction, and identification of the roles of the species involved, through direct application of the relevant definitions

7.4 The Generation of Technical Terms

7.4.1 When Do New Technical Terms Need to Be Generated?

As underlined in the previous section, terminology is an ensemble of names assigned to the objects of study. It is thus easy to view the development of terminology as closely associated with the development of science and technology, in the broadest meaning of these terms. Technical terms have been introduced, one by one, whenever something was discovered or a new interpretation of physical reality and its phenomena was proposed, or whenever new *items* of technology were invented. In summary, the need to introduce a new term arose every time any of the following situations occurred:

- Something new was discovered in nature. That *something* might have been an object like, e.g., a previously unknown plant, animal or substance. It might also have been a phenomenon like the pollination of flowers or the metamorphosis of insects.
- A new concept was introduced within the search for self-consistent and satisfactory explanations of observations. The introduction of new concepts might also be part of the proposition of an overall novel model, like, e.g., the introduction of the *atom* concept in Leucippus' model of matter.
- A new device or process is designed and becomes part of the technological development.

On any of these occasions, the discoverer, the philosopher, the scientist or the inventor had to answer the question "What name shall I give to this new *thing*?".

This very question highlights the fundamental difference between scientific terminology and the names that are part of common language. The latter have developed in a sort of 'spontaneous' way, through the historical development of languages in close association with human activities and societal relationships. The formers have been introduced in the moments in which the need arose, and they were invented *ad hoc* in relation to this need.

For the inventions that took place in prehistoric times, we do not have records. Therefore, we may just try to imagine, e.g., the inventor of the wheel trying to answer that question, possibly with the help of those who started using it for one or another application. The only guidelines to trace the roots of these terms are provided by etymology. On the other hand, those terms have entered common language so early that they are now part of common language in all the languages. For historical times, records can be perused, stories had been reported, and reasons can be known. Therefore, it is possible to identify options, criteria, perceptions and other features that have led, across several centuries, to the generation of the terminology of the various scientific and technical disciplines.

The consideration of how scientific terms were generated through history may by itself constitute a type of information relevant to a deeper understanding of the nature and roles of scientific terminology with respect to the body of scientific knowledge. Quite often, the new terms recorded the aspect that the discoverer considered more significant or perceived as more impressive, or the first emotional perceptions in front of the new *items*. Furthermore, the generation of new words often responded to a broad cultural/humanistic preparation, including the knowledge of other languages besides the discoverer's mother tongue.

7.4.2 *An Excursus Through Origins and Options*

The current section proposes to offer a panoramic overview of the origin of technical terms and the options followed in their generation. The selection does not aim at forming a chronologically or otherwise organized pattern, or at being exhaustive, but rather at spotlighting features and approaches, and the convergence of science and humanities that they entail.

7.4.2.1 Terms Invented in Ancient Greece

Ancient Greece is considered the cradle of European science. Reading the works of ancient Greek philosophers actually leads to a perception of continuity (from those times to our times) in the proposition of questions and in the corresponding search for answers. Philosophers were, by definition (by the meaning of the term), "friends of knowledge", without any separation between what now we call humanities and what we call sciences. Philosophers endeavoured to interpret observations of the world around them and proposed models of physical reality. Models that introduced new concepts required the *invention* of new words. The words were created utilizing the root-words and tools of the Greek language. Many of these words are still part of our terminology. Few examples are recalled in the next paragraphs.

The word *atom* was proposed by Leucippus to denote the key concept of his model of matter, i.e., that matter consists of particles [4]. He created it by utilizing Greek root-words to transmit the meaning "something that cannot be cut" and, therefore, "something that cannot be divided into smaller parts" (ἄτομος is built from the stem of the verb τέμνω = to cut, and the α preceding it, that denies its meaning). The word became a technical term within the philosophical model that was consequently called *atomistic*. The model came back to the scientists' attention after the discovery of the text of Lucretius' poem *De Rerum Natura* (On the Nature of Things) in a monastery, in 1417, and the word *atom*, generated within philosophy, entered modern science.

The regular solids (tetrahedron, hexahedron, octahedron, dodecahedron, and icosahedron) were identified by Greek philosophers and mathematicians and played a role in Plato's model of matter [5], as mathematical constituents of the four elements and the ether (tetrahedron for fire, hexahedron for earth, octahedron for air, dodecahedron for ether, and icosahedron for water). Their names are descriptive: they specify the number of faces (regular polygons: 4 triangles, 6 squares, 8 triangles, 5 pentagons and 20 triangles respectively) and have been retained in modern geometry.

The names of several sciences were also invented by Greek philosophers. The term for mathematics was introduced by the Pythagoreans in the VI century BC, from a word that indicates what is being learnt (μάθημα, mathema). The word *geometry* combines two word roots, for earth (γῆ, ge) and for measure (μέτρον, métron), referring to the practical uses of measurements of surfaces; it became the name of the science studying polygons and solids, which was systematized by Euclid in his "Elements of Geometry" (mid of the IV century BC). The word *physics* derives from the Greek word for nature (φύσις, physis) and initially denoted all the studies concerning nature; in the form "physica" became the title of the work where Aristotle described his observations and interpretations about many aspects of nature, from motion to living organisms.

7.4.2.2 Later on: Generating Technical Words from Greek and Latin Root Words

The European scientists of past centuries had extensive humanistic education, because it was considered essential to a comprehensive education and for the generation of an attitude of "love for study" or "love for knowledge". Moreover, until the XVI-XVII centuries (in some contexts, even through the XVIII century), Latin was the *lingua franca* for the communication of intellectual creations in most of Europe. This included scientific communication. Consequently, scientists had a good mastery of Latin as a working language. Most of them also knew ancient Greek. When inventing new names, they tended to use Latin root-words if they wished to remain within their working language and Greek root-words when they wished to resort to different roots. For the formation of the new term, the root-words were selected and connected in such a way as to transmit a meaning that was considered particularly significant in relation to the *item* to be named.

The tendency to utilize those root-words persisted even after Latin ceased to be the unique and unifying language of scientific communication and scientists started teaching and writing in the national languages. At this stage, the selection of Latin or Greek root-words responded mainly to the criterion for which words from a different language are likely to be more easily perceived as *abstract* and/or *scientific*, because of being different from the commonly used ones. This criterion falls within the implications of a fundamental observation expressed by Enriquez and De Santillana [6], who noted that the enormous abstraction ability attained in ancient Greece was largely the outcome of encounters and interchanges with many different cultures. The implication for terms-generation in later centuries entailed the perspective that terms belonging to languages different from the one utilized in a certain country are spontaneously perceived as 'non-common' and, therefore, special. This makes it easier to associate them with abstract and/or scientific concepts. In turn, the presence of terms to denote abstract and/or scientific items somewhat stimulates abstract thinking or scientific elaboration. In order to better understand this statement, we might ask whether, e.g., we would perceive words like *metamorphosis*, *pyrolysis* or *polynomial* as much scientific as we perceive them, if the images behind them were expressed through common-language words like "change of shape" or "separation by fire" or "something with many terms" respectively. Furthermore, a technical term formed from Greek root-words, joined together with the criteria developed by scientists, has the additional advantage of being a single word, what makes it immediately recognisable as technical term in a sentence.

It may be added that the same perception of 'scientificity' refers to the letters of different alphabets. For instance, the letters of the Greek alphabet are mostly perceived just as scientific symbols in countries using the Latin alphabet. The author of this chapter has often tried to convey the information that the Greek alphabet is first of all an alphabet used for normal writing in Greece, by writing the students' names using Greek letters and relating them to the alphabetical sound and to the way we use them as scientific symbols (this exercise is also meant to stress the fact that symbols are chosen by convention).

Most terms used in medicine were formed from ancient Greek word-root selected in such a way as to describe the meaning of the term. Thus, *apyrexia* is the time needed for a fever to disappear, or the outcome (absence of fever), from a Greek root-word meaning "to be in a fever" (in turn related to 'pyr', the word for *fire*), and the initial α that denies it (denies the presence of a fever); *pneumonia* incorporates a Greek word root that means lungs, and therefore refers to a disease of the lungs; *arthritis* combines the *arthro* word root that means joint and the *–itis* ending that means inflammation; and so on.

In the zoology and botany nomenclatures, each species is denoted by a term consisting of two words which are both formally Latin, i.e., they have the structure and endings of Latin names. Quite often, the Latin name is formed from Greek root words; in some cases, it contains root words from other languages. For instance, the terms *Panthera leo*, *Panthera tigris*, *Panthera pardus* identify the lion, the tiger and the leopard respectively; in each of them, *Panthera* identifies the common category (the genus), thus informing that the three animals belong to the same genus; the second noun (*leo*, *tigris*, *pardus*) identifies the species (lion, tiger and leopard). The terms *Plasmodium falciparum*, *Plasmodium vivax*, *Plasmodium ovale*, *Plasmodium malariae* identify four species of malaria-causing protozoa; they all belong to the *Plasmodium* genus; the second word in the names is an adjective referred to some characteristic of the parasite (*falciparum*, *vivax*, *ovale*) or a specification (*malariae*, meaning "of malaria").

The names of plants usually consist of a noun followed by an adjective, where the adjective largely refers to some characterising or impressive property of the plant. For instance, *Lamium purpureum*, *Lamium album* and *Lamium maculatum* are three species of *Lamium* whose flowers are respectively red, white and spotted (*purpureum*, *album* and *maculatum* are the Latin names of these colours); *Cardamine pentaphyllos*, *Cardamine heptaphyllos* and *Cardamine enneaphyllos* are three species of *Cardamine* respectively bearing five, seven and nine small leaves on individual stems; the adjectives are formed from the Greek root-words for the numbers five, seven and nine (*penta*, *hepta* and *ennea*), and the root word for 'leaf' (φύλλον, *phyllon*). The adjective *officinalis*/*officinale* (depending on the grammatical gender of the noun preceding it) informs that the given plant was used in traditional medicine; for instance, *Taraxacum officinale* is the common dandelion, whose medicinal properties have been known for centuries (other types of *Taraxacum* include *Taraxacum erythrospermum*, *Taraxacum platycarpum*, and *Taraxacum albidum*, where the adjectives after *Taraxacum* refer respectively to the red colour of the seed, to the shape of the seeds and to the white colour of the flowers).

It may be briefly recalled that the practice of using root-words from Latin and ancient Greek does not belong only to the past. For example, television was invented around the mid of the XX century. The term denoting it was built from a Greek root word meaning *far* and a Latin root word meaning *to see*, so that the overall meaning is "seeing from far" or "seeing things that are far away".

Scientists were in a position to use Latin and ancient Greek root-words easily because the extensive humanities background embedded in education resulted in comfortable familiarity with these languages. The converging of humanities and science in the generation of new terms to denote newly discovered objects or phenomena entails also other aspects, including the familiarity with mythologies, the attention to local cultures, and other features more closely related to imagination. The next subsection offers an overview of such images, through the consideration of representative examples.

7.4.3 *The Images Behind New Terms*

A new term was often built on the basis of an image or perception about the new *item* to be denoted. The features of the image suggested or determined the selection of the root-words. The image itself depended largely on the perceptions of the individual scientist or discoverer, and often related to the aspects or properties that impressed the scientist more deeply. A number of these images are highlighted in the following subsections through the consideration of representative examples. Many of the selected examples refer to the formation of the names of elements or the common names of some compounds, viewed as particularly illustrative as their discoveries largely span across the last six centuries.

7.4.3.1 Images Related to the Circumstances of the Discovery

In some cases, the new names were formed with reference to significant circumstances of the discovery. For example, *lithium* was given its name (from a Greek term meaning *stone*) because it was discovered in a mineral, whereas the previously known alkali metals had been discovered in plant ashes (the word *alkali* comes from an Arab root-word related to ashes). The name of *helium* was formed from a Greek term meaning *sun*, because its presence was identified in the Sun (through the analysis of spectral lines) before being discovered on Earth. The Russian name of fluorine, *ftor*, was formed from a Greek root-word meaning *destructor*, with reference to the difficulties experienced by the scientists who attempted to isolate it, including injuries (when, in 1886, Henri Moissan announced to the Paris Academy of Sciences that he had isolated pure fluorine, he had a bandaged eye).

7.4.3.2 Images Related to the Properties of the New *Items*

The names of metals known since ancient times are part of common language. Some of them reflect the property that was perceived as the most impressive. For instance, the Latin name for silver, *argentum* (from which the chemical symbol Ag) derives from a Sanskrit root-word related to the *shining* or *bright* quality.

Many names built after the discovery of a new substance recall the characteristics that had most deeply impressed the discoverer or the persons who came in contact with the substance immediately afterwards. The name *phosphorus* was formed from Greek root words meaning *bearer of light* because of its ability to emit light. The name *chromium* was formed from a Greek word meaning *colour*, because of the beautiful colours of its compounds. *Bromine* got its name from a Greek word meaning *smell*, because of its bad smell. *Chlorine* and *iodine* owe their names to their colours, from Greek root-words meaning *green* and *violet* respectively. The Latin name of mercury, *hydrargyrum* (from which the chemical symbol Hg) combined the root-words for water (Greek root) and silver (Latin root), to convey the meaning of an element shiny like silver and liquid like water.

Some names have disappeared from modern terminology, or are in the process of disappearing, but their historical memory transmits the perception of the property that was considered most impressive. The adjective *caustic* (from a Greek word related to the *burn* concept) was recurrent for compounds that caused effects analogous to burning effects; thus, *sodium hydroxide* was called *caustic soda*, *potassium hydroxide* was called *caustic potash*, and *silver nitrate* was called the *hell caustic*.

Other properties that entered into newly-formed names were related to traditional and/or industrial uses of the substance to be named. For example, the name *fluorine* was formed from a Latin term meaning *to flow*, because some fluorine compounds can lower the melting point of some minerals and were utilized for this purpose in the metallurgy industry (their addition made the mineral start 'flowing' (becoming liquid) faster). The name *manganese* was formed from a Greek root-word meaning *purifier*, because some manganese compounds were added to molten glass to remove the greenish colour commonly caused by the presence of dispersed iron.

Properties could also be related to discoveries going beyond the pure isolation of an element and expanding into new interpretations. It is the case, for example, of *hydrogen* and *oxygen*. The two names were introduced by Lavoisier. The term *hydrogen* was meant to stress the discovery (verified by Lavoisier) that water is a compound and hydrogen is one of its constituting elements; it is built from two Greek root-words and means *generator of water*. The discovery marked a milestone in the development of chemistry, because water had been considered an element since ancient times; naming an element *generator of water* stressed the novelty that water consists of more than one element (that there is an element generating it). The term *oxygen* responded to a generalisation of the observation that the acids known at that time contain it; it is built from two Greek root-words and means *generator of acids* (the realisation of the existence of acids that do not contain oxygen came at a later time). The meanings conveyed by these names were perceived as so important that some languages formed the names of these two elements using their own root-words to convey them (e.g., Wasserstoff and Sauerstoff in German (where the second root word is referred to the "matter of which something is made of" concept rather than to the 'generator' concept), водород (vodarod) and кислород (kislarod) in Russian, maintaining the 'generator' concept.

Discoveries of phenomena with totally new nature prompted new perceptions. When it was discovered that elements emit radiations that could be seen like images (sets of signals) specific of each element, and did not have material nature, the perception associated them to the idea of *ghost*, as if they were the ghosts of the elements; so, the image for each element was named with the Latin term for 'ghost', i.e., *spectrum* (plural, *spectra*). The instrument that enabled them to be seen was called *spectroscope*, and the instrument that enabled them to be recorded was called *spectrograph*, with the *spectrum* term respectively associated with the Greek root words for 'see' and for 'write'.

7.4.3.3 Images and Names from Mythologies

Mythologies are part of the tradition and culture of each country and/or context. In Europe, a humanities-based educational background entailed also good familiarity with ancient Greek and Roman mythologies. This familiarity prompted images for the formation of new names, sometimes even in connection with some features of the discovery. For example, the isolation of pure *tantalum* (previously written *tantalium*) was such a challenging task that scientists developed the perception that their efforts were analogous to those involved in Tantalus' punishment. According to Greek mythology, Tantalus had been condemned to remain near a tree full of fruits that would move further away every time he attempted to pick them, and near a stream of pure water that would move away every time he attempted to drink it. When, at last, chemists isolated the element, they denoted it with the name of the Greek mythological hero, as a permanent memory of the many frustrated attempts before the result was attained. The element that is always present together with tantalum in minerals, and from which it had been so difficult to separate, was named after Tantalus' daughter, Niobe, and is called *niobium.*

Other examples of names from Greek mythology are *titanium* (from the Titans, the Earth's giant children), *mercury* (from the Latin god corresponding to the Greek Hermes, who had been considered a sort of patron by European alchemists), *uranium* (from the Greek god Uranus, the god of the sky) and *promethium* (from the name of the hero who had given fire to mankind, in Greek mythology). The practice is still alive: e.g., *neptunium* and *plutonium*, radioactive elements discovered in the XX century, were named after the ancient Greek gods of the sea and of the underground kingdom respectively.

Mythologies other than the Greek and Latin are also represented among the names of elements. For example, *vanadium* was named after Vanadis, a Scandinavian goddess of beauty (also known as Freya), because of the beautiful colours of some of its compounds.

Mythologies include believes in spirits of various kinds. Spirits are often perceived as entities more familiar and close to Earth than gods; therefore, they become parts of traditions related to everyday life, they are included in popular interpretations of phenomena, and thus may enter into the development of legends. They can even enter the work place. For example, the element *nickel* was named after a spirit of the mines in the German mythology and legends. When miners tried to obtain copper from nickel ores (without knowing that they were not copper minerals, but contained a different element) and their attempts were unsuccessful, they thought that those ores had been bewitched by the Nickel spirit. When chemists discovered a new element in those ores, they denoted it with the name of the mines-spirit of the popular tradition. The name *cobalt* has a very similar origin, but the spirit, in this case, belonged to the Scandinavian mines' traditions.

The names of the constellations that could be seen by naked eyes in the absence of light pollution (as was the case in ancient times) responded to images prompted by their shapes, and were often associated with myths. It may also have happened that some myths were prompted by the observation of certain constellations. For instance, the shape of the *Orion* constellation was associated with the image of a hunter, with his dog (*Canis*); the smaller constellation of seven stars, which seems to be 'followed' by Orion, was viewed as a group of seven maidens pursued by the hunter, and a legend became associated to them: Orion was pursuing the seven maidens on the Earth, the maidens asked the gods for help, the gods transformed them into stars (the *Pleiades*), and Orion also moved to the sky, becoming a constellation and continuing pursuing the Pleiades. Many constellations have similarly been associated with myths. Modern astronomy does not incorporate the myths, but the identification of those constellations remains the same as the one that was born by associating the recognition of spatial relationships and geometric shapes with the artistic imagination that coupled those shapes with figures of beings or objects, and the figures, in turn, recalled myths.

The names of individual stars are also often associated with myths or perceptions. The same star may respond to different myths in different cultures, and have different common names. The examples that follow refer to the names as used in astronomy (not to the common names). For instance, the name *Aldebaran* (the red giant in the Taurus constellation) is an Arabic name meaning "the follower", because it seems to follow the Pleiades; the name *Betelgeuse* (the supergiant star in the Orion constellation) is also Arab and means 'the shoulder of the giant', or 'the hand of the giant', where the giant is the hunter Orion. The name *Antares* (the red supergiant in the Scorpio constellation) comes from Greek and means "rival of Ares", where *Ares* was the ancient Greek god of war, corresponding to the Latin *Mars*; both Antares and Mars appear red, and Antares is brighter (winning the competition). The name *Sirius* (the brightest star in the *Canis* constellations, and also the brightest star in the Northern hemisphere sky) comes from an ancient Greek word that means 'glowing'.

All the planets of the solar system have names of ancient Latin gods. The first seven were named in ancient times, as they are visible to the naked eyes. Some names are easily traceable to possible perceptions: for instance, Venus is the brightest planet, extremely beautiful when it appears immediately after sunset or immediately before dawn, and was given the name of the goddess of beauty; and the reddish colour of Mars might have suggested a perception of fierceness, prompting the association with the god of war. When Neptune and Pluto were discovered, in 1846 and 1930 respectively, they were also named after Latin gods, to continue the tradition.

The resort to mythologies is not restricted to chemistry and astronomy. In the first half of the XX century, Italian engineers were asked to avoid importing terms from foreign languages and, therefore, they had to form new words from Italian root words (including, obviously, Latin, the ancestor of modern Italian). The fantasy of the engineers (who, according to the Italian educational system up to the Sixties included, acquired also deep humanistic education) brought about a variety of specific terms, utilising ancient roots, ancient mythologies and local traditions. An example is offered by the term *trefolo ceraunico* to denote devices protecting big structures from lightening: *trefolo* is an ancient Italian word for "wire" and *ceraunico* derives from the *Cerauni*, minor gods of ancient Roman mythology who were supposed (or expected) to protect places from lightening; thus, *trefolo ceraunico* means "the wire of the Cerauni".

7.5 Presenting the Generation of Technical Terms in the Classroom

7.5.1 Highlighting the Convergence of Humanities and Sciences

Two major general features can be identified from the information presented in the previous sections.

- Scientists often resorted to their general, humanities-based knowledge in forming new words to denote objects they had discovered or invented, or concepts they had established. This highlights a spontaneous interplay between the scientific component and the humanities component of their overall cultural preparation. As a consequence, the generation of scientific terms can be viewed as a long thread, developing through history and connecting humanities and sciences in what can be considered a series of recurrent interactions and exchanges.
- The selection of the option/s for the formation of new terms was often related to an image stimulated by the new item, e.g., through the discoverer's first or main impressions. It would not be easy to separate and/or differentiate these impressions from the impressions that, for instance, an artist might have perceived in front of the same features. Therefore, the images behind the generation of the new terms can also be considered an overlap area in which the perception component (a typically *human* component) and the science component combine together to yield a shared result.

Information on the origin of technical terms in science and technology can be easily presented to students, even at junior level. In order for it to be a fruit-bearing component of educational options encompassing efforts to bridge the gap between humanities and sciences, the presentation should not be limited to a list of etymologies and other isolated bits of information, but needs to highlight the overall picture. The following components may be suitable to this purpose:

- The features of historical moments. These include the perspectives of concepts' development and of general perceptions. They may also include information about the philosophy and arts of the period concerned.
- The reasons and/or circumstances for which the considered 'item' was new for the given time.
- The features that impressed the discoverer more deeply and the possible reasons for which those features were perceived as impressive.
- How the image outlined by these features resulted in the generation of the new word.
- In the case of new concepts, how the establishing of the definition of a given concept relates to the development of models and, therefore, to the history and philosophy of science.

For example, in the case of the discovery of phosphorus, it can be underlined that it was the first time scientists came across a substance that can emit light spontaneously (on its own). Therefore, this was the feature that most impressed them. It was considered so extraordinary that phosphorus became famous, and samples of phosphorus were presented in various courts, for the entertainment of the aristocracy and the kings. It is thus obvious that this characteristic would be considered the most deserving to be 'fixed' and transmitted in the name of the new substance.

The just–outlined type of presentation includes and incorporates the history component, the science component, the method component, the perception component, the language component and, at least for particularly beautiful images, the artistic component. Such ensemble of components is by itself a link and overlap between humanities and sciences. Moreover, it has the advantage of showing these links in association with a feature that is often perceived as pertaining most exclusively to the sciences alone, i.e., the technical terms.

Two other features can also significantly contribute to highlight the interactions between the humanities and science domains. They are briefly recalled in the next two paragraphs, for the sake of completeness.

There has been substantial interaction, through history, between common language and scientific terminology. On the one hand, the images associated with the discovery of some new 'items' prompted the scientist/discoverer to borrow terms from common language. It is the case, e.g., of the terms *force*, *work* and *field*, which were incorporated into physics terminology by defining new meanings for them, which are different from the meanings they have in common language. On the other hand, scientific terms enter into common language by a sort of *osmotic* process. Some of these terms maintain their mythological origin. For instance, when we speak of a 'hermetically sealed container' we use an adverb (hermetically) related to Hermes, the god perceived as patron of alchemy (the 'ancestor' of chemistry): alchemy was the science of Hermes; alchemists needed to be able to close containers tightly; thus, a hermetically sealed container is a container suitable for use in alchemy practice, i.e., suitable for use in the science of Hermes.

In our times, the process by which scientific terms enter common language is favoured and facilitated by the media. Students can be invited to identify technical terms in articles from newspapers or from magasines and to analyse/discuss the extent to which some of them have entered or are in the process of entering common language; they can also be invited to evaluate whether and to what extent those terms have been used rigorously in the articles that they are considering.

It is also important to highlight that the need for technical terms is not restricted to the sciences, but pertains to most areas of intellectual activities. Each humanities discipline has its own terminology. Illustrative examples can be taken among the specific terms utilized in law, or in linguistics, or in the arts. In this way, the need to have a specific terminology can be presented as a frequent requirement of human intellectual activities, thus contributing to prevent the generation of perceptions viewing the two domains as isolated and having nothing in common, or the perception that one of them is somehow 'superior' to the other.

7.5.2 Opportunities for Attention to Contextual Characteristics

The selection of the technical terms to whose origin specific attention is devoted may also respond to some characteristics of the students' context, considering the expectation of aspects that may be perceived as particularly interesting. Two examples from the author's experience in Sub-Saharan Africa may serve as illustration.

On more than one occasion, the author opted to present the origin of the term *nickel* in general chemistry courses. The students' response was particularly positive because this information indirectly transmitted information about the presence of myths involving spirits and bewitching also in Europe. One of the many side-components of colonization was the accurate care by the colonizers to present their cultures as totally different, with no similarities to the indigenous ones. This was obviously meant to support claims about their 'superiority'. Aftermaths of these perceptions are still present. For instance, many Africans are surprised on hearing that the use of herbal remedies is quite common in Europe, as the colonizers presented the use of herbal remedies (common in Africa) as something pertaining to 'primitive' societies. Information about historical experiences that can be viewed as fundamentally similar has also the advantage of counteracting perceptions generated by colonialism, thus introducing an ethical-action component (correcting preconceptions not responding to reality) into science teaching.

The names of animals and plants in zoology and botany nomenclatures combine rigorous classification criteria (identification of genus and species) with name-forming ways that remain close to criteria also used in common language. In the names of plants, the second term relates to some characteristic of the plant: morphological characteristics like the size and shape of the plants, of the leaves, of the fruits, or whether the plant has medicinal properties, or the types of places it prefers. The species of the *Taraxacum* genus listed in Sect. 7.4.2.2 provide examples of a second name referring to morphological characteristics (*Taraxacum erythrospermum*, *Taraxacum platycarpum*, *Taraxacum albidum*) and to the presence of medicinal properties (*Taraxacum officinale*); another *Taraxacum* species (*Taraxacum palustre*) lives in marshes, as specified by the adjective *palustre*. Explaining the etymological meanings of some of these names can be expedient to make these names less abstract (they are perceived as highly abstract as long as their meaning is not known). In addition, analogous criteria are likely present for the formation of the common names of plants in several cultures; highlighting this aspect contributes to establish links between cultural components and scientific nomenclatures. The classification in modern botany is often not the same as in traditional names, because the identification of the genus was not based on all the characteristics that are taken into account in botany; on the other hand, the formation of names follows the same basic criterion, i.e., the idea of giving the same first name to plants that appear somehow related, and adding another term to distinguish them on the basis of characteristic that differentiate them. The examples reported in [7] can serve as illustration. In Tshivenda (one of the official languages in South Africa), the name *Munungu* denotes two plants of the *Zanthoxylum* genus: *Zanthoxylum davyi* and *Zanthoxylum lepricurii.* It also serves as parent name for the names of other plants having analogous uses in traditional medicine: Munumgudaka (meaning 'munungu of the forest'), Munungumaswi ('munungu with latex'), Munungubani ('munungu of the mopani veld'), Munungwane (a smaller type of munungu), Munungwanungwane (the smallest of the smaller types of munungu). Current studies show that these plants have similar antibacterial activities, what confirms that their grouping together in the traditional nomenclature had an expertise-based motivation. Analogous examples may be found in many cultures; their presentation to the students of the given culture may facilitate the understanding of the roots and criteria of botany nomenclature.

7.6 Conclusions

The previous sections considered major features of the generation of scientific terminology through history, from a perspective highlighting the interplays between the humanities domain and the science domain. Such generation can be viewed as a long thread of interactions between the two domains, developing through a variety of aspects, in correspondence with the circumstances and contexts of each discovery, with the discoverer's impressions and perceptions, and with the richness of options made available by deep humanities preparation.

The outline itself was developed according to a pattern that can be easily utilized as a suitable framework for the presentation of these features to students. More specific considerations on possible options for such presentation have also been included. It is considered that knowing the origin of at least some of the terms that are technical terms in the various sciences can stimulate both the awareness that terminologies are ensemble of names (that are assigned when something new 'is born' or discovered within a given science) and the perception of broader connections between a name and the time when it was introduced. It also contributes to foster the perception that science is not isolated from culture, or that any new item generates a variety of perceptions in the discoverer scientist and these, in turn, may get reflected into the names and enter the body of science.

References

1. L. Mammino, *Terminology in Science and Technology—An Overview Through History and Options* (Ditlou Publishers, Thohoyandou, 2006)
2. L. Mammino, Il ruolo delle definizioni nell'insegnamento delle scienze. Scuola e Citta' **1**, 25–37 (2000)
3. L. Mammino, *Chimica Viva* (G. D'Anna, Firenze, 1993)
4. H. Diels, *I presocratici: testimonianze e frammenti* (Laterza, Bari, 1969)
5. Platone, *Timeo*, G. Reale (Ed.) (Bompiani, Milan, 2000)

6. F. Enriquez, G. De Santillana. *Storia del pensiero scientifico. Vol. I. Il mondo antico* (Zanichelli, Bologna, Italy, 1932)
7. L. Mammino, Chemical heritage in sub-Saharan Africa and its significance for chemical education. Southern Afr. J. Folklore Stud. **33**(2), 147–158 (2012)

Chapter 8
The Epistemology of Engineering: A New Area Merging the Epistemology of Science and the Specificities of Engineering Practices

Armando Mammino and Liliana Mammino

Abstract Epistemology is a cross area between humanities and sciences, addressing the issues related to the meaning of the knowledge humans attain, and analysing the details and power of the methods through which it is attained. While the epistemology of science has been the object of investigation since ancient times (when science was part of philosophy), specific attention to the epistemology of engineering is recent. The present chapter aims at offering an overview of core issues pertaining to the epistemology of engineering, highlighting their connections with the epistemology of science while also emphasising the specificities of their connections with engineering practice—an aspect that is fundamental because the practical character of engineering determines the way in which it utilises theory, modelling and experiments in the design process.

Keywords Engineering · Epistemology of engineering · Mathematical modelling · Numerical modelling · Philosophy of engineering · Technique · Technology

A. Mammino
Consultant Structural Engineer, Treviso, Italy

L. Mammino (✉)
Faculty of Science, Engineering and Agriculture, University of Venda, Thohoyandou, South Africa
e-mail: sasdestria1@gmail.com; mammino@univen.ac.za

L. Mammino (ed.), *Insights into the Relationships between Humanities and Sciences*,
https://doi.org/10.1007/978-3-031-97900-2_8

8.1 Introduction

8.1.1 The Philosophy and Epistemology of Engineering: Disciplines in the Process of Being Constructed

Engineering and other technology areas—as the ensemble of implementation-oriented disciplines—are a direct filiation of science; they can be viewed as its deductively logical outcomes, whose developments are linked to the development of science but simultaneously take a variety of independent features. Because of the close links between science and engineering, the existence of a philosophy of science entails the possibility of a philosophy of engineering—and of technology-related areas in general—which will have important relationships with the philosophy of science and simultaneously build an independent outlook in association with the specific theoretical and practical requirements of engineering. The term 'engineering' is here selected and utilised to collectively represent all the branches of engineering and technology, i.e., all the implementation-oriented areas (applied sciences); therefore, it denotes the ensemble of applied areas deriving from physics, chemistry, natural sciences and mathematics. Structural engineering is more frequently selected as specific reference for concrete examples, because of its close connections with physics, which, in turn, has been the science more frequently associated with the philosophy of science.

While the philosophy of science has already known a long, autonomous and fruitful development, the philosophy of engineering has been taking its first steps in the last decades. As the presentation to [1] states, "Engineering has always been a part of human life but has only recently become the subject matter of systematic philosophical inquiry". Beddoes and Schimpf [2] consider it a "field in the making", and Wikipedia [3] defines it as an emerging area:

> *The philosophy of engineering is an emerging discipline that considers what engineering is, what engineers do, and how their work affects society, and thus includes aspects of ethics and aesthetics, as well as the ontology, epistemology, etc. that might be studied in, for example, the philosophy of science or the philosophy of technology.*

Actually, the importance of philosophy for engineering had already been clearly stated in 1998 [4]:

> *Philosophy has not paid sufficient attention to engineering. Nevertheless, engineers should not use this as an excuse to ignore philosophy. ... philosophy is important to engineering for at least three reasons. ... there is a tradition of engineering philosophy that is largely overlooked, even by engineers. .... philosophy, especially ethics, is necessary to help engineers deal with professional ethical problems. ... because of the inherently philosophical character of engineering, philosophy may actually function as a means to greater engineering self-understanding.*

However, most publications are quite recent. They comprise a comparatively high proportion of books (e.g., [1, 5–9], to which one might add the investigation of the philosophy of activities germane to engineering such as design [10]). It is also interesting to note that a number of articles have titles reflecting the awareness that the philosophy of engineering is still in the process of being born, either by using the preposition *toward/s* in their titles [11, 12] or explicitly using the term "construction" to indicate that the epistemology of engineering is still in the process of being constructed [13].

Epistemology is the area of philosophy that investigates the nature of knowledge; therefore, the epistemology of engineering investigates the nature of engineering knowledge [14, 15] and is a subset area of the philosophy of engineering. The philosophy and epistemology of engineering are cross-disciplinary areas [14–16], with philosophy and epistemology pertaining to the humanities domain and engineering to the science and technology domain. Doridot [12] emphasises that "*engineering and philosophy are complementary ways of approaching the same reality, distinct methodologically but not distinct in kind*". Research in the epistemology of engineering may focus on general issues [17], refer to specific areas, like system engineering [18], construction engineering [19], or information systems engineering [20], or to specific activities, like engineering design [21], control [22] or measurements [23]. Possible approaches for the infusion of epistemological information into educational practices are also analysed considering engineering education [12, 24–26] as well as architecture [27], which is an area with broad overlaps with engineering. The activities of each of the branches of civil engineering can be defined also through epistemology terms [13]. Epistemology can also be used to distinguish the critical terms of 'science', 'engineering science' and 'engineering', particularly in a learning environment [28].

8.1.2 Motivations for this Chapter

Despite the increasing interest in the relationships between philosophy and engineering, a bias viewing them as too distant from each other to have any form of conceptual interference, or interdisciplinarity, is still largely present. This chapter proposes to show that such prejudice is not sound and that, given the current complexity of applied sciences, the times are ripe to infuse engineering with a net of epistemological threads analogous to the established patterns of the philosophy of science.

The presence of relationships between engineering and the humanities can be highlighted through some easy examples. For instance, a vision viewing engineering as an arid ensemble of practical problems to be solved through numbers, formulas and drawings, would unavoidably neglect ethics, which is extremely important in engineering. Ethics is both a philosophical and a legal concept, and the very fact that it interfaces with engineering implies the presence of interfaces between philosophy and engineering. Many other interfaces—besides the one with ethics—can be identified.

The ability to analyse concepts in order to understand and utilise them in an optimal way constitutes an interface between engineering and philosophy. Familiarisation with the history and philosophical background of ideas entails familiarisation with the ways in which the key concepts, through which engineering and technology operate, were born, have evolved up to their current formulations, and have been tailored for the various practical applications and the simultaneous theoretical validation of design outcomes. Because of this, it is important that philosophy and engineering interface and mutually integrate at least along two major lines:

- A historical perspective, to highlight that engineering ideas, concepts, solutions, materials, etc. have not always been the same, but have evolved through different times and methodological phases;
- A 'pragmatic' perspective, to establish correct, appropriate and advanced connections between those ideas and concepts and the art of building things in its current advancement stage.

Within these two perspectives of the interfaces between engineering and philosophy, adequate space needs to be reserved for complexity—a feature that has become of paramount importance and all-permeating in modern engineering. All the components of a project are complex: the starting points; the problems to be solved in order to design and realise a certain work, building or manufact; the design; and also the final product. Every stage of the route from the initial ideas to the completed product involves choices that have relevant impacts on the lives of individuals, on the existential dimension of a community, and on the environment where those individuals and community live and operate and which they exploit. Without philosophy, engineering risks to become shortsighted within this challenging scenario, and prone to be randomly conditioned by politics, or by the speculative voracity of some economic and financial powers. On the other hand, philosophy without any attention to engineering would mostly develop into abstract frameworks, without taking into account the practical outomes responding to humans' desires to enhance their living conditions, or to build those systems and contexts that are considered inherent in the 'civilization' concept.

For the sake of conciseness, the authors of this chapter will from now on be simply denoted as AM and LM respectively (pronouns he/him/his and she/her/hers respectively). In the next sections, the contents specifically related to engineering reflect the conceptions developed by AM throughout his professional activity, where the design of complex projects has required the contribution of epistemology to engender viable outcomes. The content is therefore practice-based and connects philosophical outlooks with the practical requirements of engineering design, above all when the project is challenging and requires innovative (non-standardised) approaches that the engineer has to design through multisided research. A number of engineering models are mentioned, as necessary references for the discourse to be meaningful; on the other hand, their specific or detailed knowledge is not necessary for a reader to be in a position to understand the development of the qualitative discourse about the related epistemological aspects.

8.2 Some Flashlights from the History and Philosophy of Science and Applied Sciences

The consideration of key components of the historical evolution of the philosophy of science (which largely coincides with the evolution of the epistemology of science) is important because engineering derives from science and, therefore, its epistemology needs to be matched to that of science to identify agreements and differences in relation to the different natures of the two areas. Only the crucial aspects of the various philosophy-of-science theories that can be related also to engineering and applied sciences are considered in the following rapid historical overview.

8.2.1 Birth, Evolution and Decay of Theories in Science and Engineering

Karl Popper (1902–1994), after analysing the situation of his times with regard to sciences and applied sciences, predicted the fading out of the authoritarianism of the sciences and applied sciences, as a consequence of the crisis of the certainty concept (of the certainty per se) that had been the idol of XIX century positivism [29]. Popper envisaged that the new pathway of science (and, consequently, also of engineering) would tend to an infinite scope, to the study of increasingly new, more general and deeper problems, under increasingly rigorous control. In other words, a scientist and a creative engineer are called to operate along the patterns of increasingly better approximations to physical reality through general theories, their connections with experimental information, and suitable models for the implementation level.

The period between the end of the XIX century and the beginning of the XX century witnessed profound transformations in physics, with novel visions of nature adding to classical physics. The relativity theory and quantum mechanics were the most extant novelties; they changed our understanding of astrophysics and cosmology as well as that of the atomic and molecular levels of phenomena. The *field* concept expanded beyond the boundaries of its initial postulation, meant to provide correct interpretation of some experiments on magnetism, and extended to theoretical interpretations across many areas of science and engineering. Today's structural engineering comprises the concepts of *tension field*, *deformation field*, *displacement field*, *flexion field*, *shear field*, *torsion field*, and the like. In the case of seismic engineering, where traditional statics acquires the 'cyclic motion' notion leading to structural dynamics, one identifies a *velocity field*, an *acceleration field*, and a *displacement field,* which is no more referred to displacements that are reversible when the external action ceases, but to a vibrational motion depending on the force determining it. Many more examples could be recalled to emphasise the pervasive presence of the *field* concept in science and engineering. Structural engineering deals with forms of matter and energy that are easily observable and manageable, and their description does not conflict with the basic laws of classical mechanics. On the other hand, at the beginning of the XX century, it appeared clear that the field concept is not compatible with the laws of classical physics: whenever one uses classical mechanics for practical applications and simultaneously uses the field concept, one introduces an approximation. Then, the question arises whether the approximation is viable and legitimate. It is viable, without exceptions, in structural, hydraulic, mechanical and geotechnical engineering. The incompatibility is not negligible in the case of electronic engineering, which largely benefits from the incorporation of relativity-theory components. It is not negligible also for areas dealing with systems and phenomena that can only be described by quantum mechanics, like nanotechnologies.

The crisis in the decades close to the end of the XIX century and the beginning of the XX century was not the only crisis and palingenesis of physics, and of the human vision of nature in general: there had been others in the past, and it is reasonable to assume that there will be others in the future. A scientist or a creative engineer have the capability—stemming from their specific knowledge—to detect where an existing theory may be inadequate or misleading: when this happens, they have the technical and ethical duty to analyse the aspects for which that theory may lead to errors or malfunctionings in the objects that are produced on its basis. Researchers in physics or engineering ought to systematically examine the actual ability of the theory and models that they adopt to correctly explain what is customarily called 'physical reality' within implementation-oriented research, and to guarantee the quality of the 'constructed reality' (ensemble of items built by humans) that is based on them.

It is immediately clear that an approach purely based on observations is no more adequate for so complex a context like that of modern scientific, technical and technological research. Growing complexity has put an end to pure empiricism, whereas Galileo's scientific method—with its operational alternation of theoretical elaboration and experimental verification until total consistency is attained—remains valid. This method was the core of the first 'scientific revolution' in Europe after the Middle Ages, and has remained valid through subsequent scientific developments, from Newton's advances to Einstein's relativistic revolution. In order to familiarise all categories of scientists, engineers, and other professionals with the foundations of this method, it is necessary to revive the desire to understand nature and its reality, and the desire to find optimal ways to imitate it in the various human creations (objects, buildings, substances, etc.). Such revival needs also to become an integral component of education, as well as to be extended to the general public. A variety of examples suitable to highlight the importance of such approaches can be identified within current practices of science and engineering, aimed at designing environmentally friendly options. Chemistry research tries to imitate particularly efficient or benign natural processes, including the design of catalysts whose action mimics the mechanisms of natural catalysts [30]. Many engineers who gained important consideration in the history of engineering have formulated theories in which the art of design and construction benefits from the imitation of what can be observed in the flora, fauna and geology of our planet. In this way, engineering establishes close relationships between what we observe and understand in nature and possible design innovations. For instance, from the observation that trees do not fall during an earthquake and the inference that their root system, and how it expands in the soil, is a key factor enabling them to undergo the huge oscillations caused by an earthquake without being uprooted of otherwise destroyed, the structural engineer Giuseppe Ghinassi designed a new option for the foundations of buildings located in seismic regions; the buildings he had designed in this way in Guatemala survived the 1976 earthquake without collapsing.

Karl Popper was a keen interpreter of the crisis of physics in the early part of the XX century and attempted to embed into philosophy the requirements and the identity-uncertainties that had resulted from the fading of previous certainties. The seriousness of the uncertainties can be illustrated by the fact that, after the observation that classical physics could not provide mathematical models consistent with experimental findings and before the introduction of quantum mechanics, some physicists started wondering whether the very concept of *atom* might need to be dropped. The way in which Popper tackled the problem of the role of science in the immediate and long times is documented also through his correspondence with Albert Einstein, Niels Bohr, Erwin Schrödinger and other physicists of that period. Since engineering follows science through the incorporation and utilisation of any new information generated within science, it was influenced by the crisis and its inherent need for conceptual and operational re-foundation. Recent history highlights that engineering is all the more efficient and focused on its objectives the more it is founded in the science ground, i.e., the greater its affinity with science is. This is also a procedural affinity which minimises the risk—within its practical work—of duplications of already finalised heuristic searches.

As explained in [6],

> *Popper first of all promoted a new interpretation and a new characterisation of science, understanding that, in the light of the scientific revolutions of those years, it could no more be thought of as a system of certain assertions, of consolidated and unassailable conquests.*

Popper thus repudiated the idea that the objective of science is that of providing definitive answers. Similarly, engineering cannot and should not provide definitive solutions and absolute design and construction paradigms for the practical problems that it is called to solve. This is proved by the fact that the whole world of industrial production is continuously and rapidly evolving, so that what is vanguard today may become obsolete after few months or few years. Both science and engineering thus undertake a variety of roles, including the epistemological one. They both lose the authoritarianism character that had been envisaged for them by XIX century philosophy. The progress of science would entail continuous advancements, asymptotically coming closer to the nature of physical reality through new theories that increasingly better respond to an increasing number of theoretical and experimental verifications. Any new theory needs to comprise the previous ones, unless the latter contained fatal errors, i.e., any new theory expands the generality character of the previous one. Thus, Einstein's relativity theory comprises classical mechanics as a subset [6] and quantum mechanics coincides with classical mechanics for the modelling of the macroscopic world (as also codified by the correspondence principle [31]). Similarly, the progress of engineering entails a continuous, generally asymptotic advancement towards what could be considered a 'perfect status' that could be imagined through a series of parameters for each utilisable object. Considering any of the objects that we use in everyday life, we can envisage that improved versions would:

- Solve, as best as possible, the problems that have been identified while using the previous versions of that object;
- Entail the best quality and completeness for each of these solutions.

There will be a limit to each of these improvement characteristics, but there is no possibility of predicting it: that is why the improvement process can be viewed as asymptotic—continuous improvements of the object to approach the perfectly optimal solution.

In summary, both science and engineering tend towards an unknown 'perfect status', which progressive improvements try to approach asymptotically through a sequence of theories for the former and a sequence of enhanced versions of whichever engineering-designed system for the latter. The two routes are not uncoupled, because engineering needs science to attain and optimise its results—e.g., for the control of materials, for energy related aspects, for performance optimisation, and the like. An easily understandable example is offered by the evolution of cars. In its earliest stages, a car was an ensemble consisting of the engine, the framework, the body, some seats, and few rudimentary amenities. Step by step during more than one century, a considerable number of accessories have been added, meant for comfort (air conditioning, radio, music system, navigator, and the like), for safety (electronic controls, ABS breaking, power steering, obstacles' detector, running speed monitoring, and the like), for mechanical efficiency (more advanced engines, made from more reliable materials), and so on. This process can be viewed as a sort of evolution of the 'car' species, engendered and determined by human design and implementation; the process is not yet completed with the current status, as scientific discoveries on the one hand and manufacturers' competition on the other continue producing cars with one or another addition entailing some new improvement. Envisaged major improvements currently refer on the one hand to the possibility of a car that would be capable of reaching the desired address through electronically-determined motions, without needing the action of a human driver, and travelling in conditions of total safety and, on the other hand, to the minimization of environmental-pollution generation. When this becomes reality, there will still be new possibilities for further improvements—currently beyond what we would be able to imagine.

8.2.2 Impacts of the Introduction of Einstein's Relativity Theory and Reflections on the Ways in which Science Proceeds

The introduction of Einstein's relativity theory into modern physics had theoretical, technological and practical impacts. The nature and realisation of such impacts stimulate reflection on the ways in which science proceeds, and on the ways in which key innovations progressively influence both theoretical investigations and practical activities.

The steps involved in the elaboration of Einstein's relativity theory can be summarised as follows:

- Prior relevant knowledge: Michelson and Morley's experiments proving that the speed of light is constant in all directions (1879); Faraday and Maxwell's electromagnetic field theory (1873); Lorentz' attempt to reconcile Newton's physics and Maxwell's theory.
- The publication of Einstein's restricted relativity theory (1905)
- The publication of Einstein's general relativity theory (1915). The theoretical proof utilised the formalism of tensor analysis (absolute differential calculus).
- The first experimental verification in May 1919, when a group of British astronomers led by A. Eddington observed the deflection of light during a solar eclipse, from a suitably selected location in the African continent: the mass of the Sun deflected a photon beam coming from stars located 'behind' the Sun.

According to Popper, a scientific revolution starts with a 'deluded expectation' which may refer to theoretical or experimental issues. Existing theories prove inadequate to account for new observations. This discrepancy between expectations and evidence engenders doubts about pre-existing theories; the doubts, in turn, prompt a new chapter of scientific research, where a new theory is formulated, then critically analysed in order to verify its correctness and improve it. Popper synthesised the overall process with the three words «problems—theories—critique»; in a more articulated way, it can be summarised as «problem statement; its solution through logical-physical–mathematical proofs; experimental verification of the resulting new theory». Einstein considered that we do not know whether this route to knowledge would ever yield a concluded system (an answer fully corresponding to true reality), but we can hope to approximate it as close as possible. He thought that the purpose of science was that of attempting to find a way for which the chaotic variety of our sensible experience gets matched to a logically uniform thought. With regard to his own theory, Einstein thought that it was a better approximation to reality than classical mechanics, but did not think that it was the ultimate truth. Popper views Einstein as representing the end of a vision of science as pursuit of theories that are true and final, replaced by a vision of a critical science, ready to interpret the overcoming of its own consolidated theories as a great success, instead of a failure [2].

As already mentioned, Einstein's theory constituted an expansion, rather than an overcoming, of Newton's mechanics: it could explain all what was explained by Newton's mechanics and also many theoretical issues and experimental observations that the latter was unable to account for. Therefore, the shift from classical to relativistic physics constitutes a clear example of expansion of existing theories to embrace a broader scope. A new theory does not replace the previous ones, but expands the general vision to broader borders, while incorporating and integrating the previous ones, and allowing them legitimate utilization-ranges wherever they can offer satisfactory approximations (a utilization that is usually less costly—at design level—than the one based on the more sophisticated new approaches).

Einstein considered it important to identify the features responsible for the inadequacy of classical physics to continue serving as the general foundation for the entire physics. Classical physics had based its theoretical speculation on the "material point" concept, whereas real objects—either existing in nature or built by humans—can only rarely be treated as material points without too rough approximations being introduced in the solution procedures regarding their behaviours. Light is one of the material entities that surely cannot be treated as material points. Classical physics proved inadequate to account for phenomena concerning light, which were discovered in the second half of the XIX century, like the photoelectric effect (Einstein was actually awarded the Nobel prize for his explanation of the photoelectric effect, which introduced the idea that a light beam consists of a swarm of energy-packets, which were called *photons*, and that the energy of photons can only take certain values, i.e., is quantised). Classical physics continues having an important role in the applied sciences for all those systems and phenomena where the velocities of moving bodies are much lower than the speed of light; these comprise all the objects of the macroscopic world on our planet; in these cases, classical physics or relativistic physics lead to the same working equations and the same results.

The route for the transition from previous theories to new expanded theories can be outlined in more details in terms of three phases:

- **Phase I**. New experiments, or a new analysis of some theoretical and physical–mathematical features, prompt doubts about the validity of the existing theory in its entirety (validity of all its components);
- **Phase II**. The doubts prompt reflections leading to the formulation of a new theory, which incorporates the previous one and also expands on it, so as to be able to provide explanations for a higher number of questions;
- **Phase III.** The new theory spreads, is accepted by scientists, and enters some common domains; for instance, it gets incorporated into university education, it prompts changes and novelties in practical applications, and may affect the laws of a country, above all in terms of regulations on constructions and on the quality of material objects.

As far as the innovation brought by relativistic physics with respect to classical physics is concerned, the first two phases have already been outlined in previous paragraphs. Greater attention is here given to phase III, both because it is still ongoing and because it entails effects on the applied sciences, thus being directly linked to the theme of this chapter. Where expedient, references will be made also to the other great revolution expanding the boundaries of knowledge with respect to classical physics—quantum mechanics, which, although impacting less directly on engineering, has brought a radical epistemological change in the way in which we study the world of atoms and molecules, is currently at the bases of new achievements in chemistry as well as the new nanotechnologies area, with its massive application potentialities, and often plays important roles in other research areas.

The introduction of the relativity theory has been followed by a huge amount of physics research, still ongoing. The theory has also played substantial roles in the achievements attained by the "sciences of the Universe", astronomy, astrophysics and cosmology. Important discoveries like the presence of black holes (theoretically predicted by Einstein), the presence of background radiation going back to the cosmos birth, the identification and classification of exoplanets (planets revolving about other stars, and having characteristics not too different from those of Earth) have been made possible by powerful telescopes and by spaceships rotating around the Earth. Both of them are the products of high level electromechanical engineering. Their construction cannot make allowance for errors, implying the need for high level technology of the materials from which they are constructed and of the coordination of the pieces to be assembled into the overall product, and high level of the theoretical and computational tools aimed at minimising error risks so that they tend to zero. In this way, scientific research needs to rely on highly advanced engineering and a technology level aiming at approaching perfection. This also sheds new light on the relationships between science and engineering: engineering and technology are deductive derivatives of science (Sect. 8.1.1) and, in turn, provide essential supports to science investigative activities, resulting in interconnections constituting actual cross-disciplinarity, whose extent increases as the complexity of the research objects and objectives increases.

The relativity theory and quantum mechanics have set the conditions for practical exploitation of nuclear energy, both for civil uses, as a source of energy, and (regrettably) for war-related uses, in terms of atomic bombs. Both uses require huge amounts of equipment and human and material resources for their implementation. The creation and management of these resources pertain to engineering. This, in turn, has fostered the birth of a new branch of engineering, called nuclear engineering.

The relativity theory owed its birth to a focused and masterful utilization of mathematical tools. The use of advanced mathematics for the solution of the problems of interest has been continuously increasing in modern physics, and it can be considered one of its crucial epistemological foundations. The same use can determine important changes in engineering and technology. The case of a recent change in the approach to static problems in structural design is here analysed rather in detail for illustration purposes. Up to few decades ago, the hypothesis that reinforced concrete (a widely used building material) had linear elastic behaviour involving both the hardened mixture and the metallic framework was considered a good and legitimate approximation. A calculation procedure utilising formulas and tables and based on this hypothesis had been developed since when constructors had started tackling the problem of predicting the behaviour of materials. It engendered the verification method called *method of admissible stresses*, apt for the determination of the correct amount of metallic framework for a certain resistant section, expressed in section square centimetres per current quantity, and being a function of the stress field defined by the structure in question. The elastic theory of reinforced concrete dated back to the beginning of the XX century. At the very beginning of the century, the articles and books devoted to the building of such structures contained only empirical formulas of dubious validity, which however corresponded to the first efforts to provide working criteria to the design, thus reducing the resort to crude estimations and improvisation for aspects connected to the resistance and stability of structures meant to be used by human beings. The handbooks published between WWI and WWII show considerable improvement for the design and verification of these aspects, based on more expert use of physics and mathematics. The utilisation of tensor analysis (absolute differential calculus) in the generalised theory of tensions and deformations enabled the derivation of various useful formalisms. Tensor analysis was also the mathematics utilised in the development and proof of Einstein's relativity theory: Einstein had to use 4×4 tensors, because he had added time to the three space coordinates, whereas 3×3 tensors are sufficient for structural engineering. The shift from 3×3 to 4×4 tensors entails extreme complications in several aspects of the procedure. Regardless of the different complication levels for the two cases, it is important to note the attempt to adopt the highest attained levels of mathematics for the mathematical treatment of all physics questions. In the last decades of the XX century, research in specialised laboratories for tests on materials' properties found out that the hypothesis was not in agreement with the new experimental results: reinforced concrete is not elastic, and the method of admissible stresses yields results that are too roughly approximated in various classes of cases. About 30 years' research was needed (e.g., Leonhardt in Stuttgart, Migliacci in Milan, and various others) to ascertain that the physical reality of reinforced concrete is better approximated by ascribing an elastic–plastic-viscous behaviour to it, what entailed a different design and verification criterion, called "limit state method" (ultimate functioning state, safety state, damages, seismic conditions, etc.). Step by step, the new approach developed to completeness, both conceptually and for applications, and was incorporated into technical software. Since it involves more extensive and more articulated calculation procedures with respect to the method of admissible stresses, it had to necessarily be implemented through the use of software (computer-aided calculations), whereas the previous method could have also been implemented 'manually', through the use of formulas, tables and diagrams. In parallel, the new approach entered legal regulations about reinforced concrete, while the method of admissible stresses was progressively phased out up to being banned.

Considering a tangible example can also highlight the components of the major steps of this change. The Italian case is selected for illustration. The 1970–1996 years correspond to the gradual recognition of the new method by practising engineers, including training activities to attain better familiarization with it as the new state-of-the-art option. Regulations issued in 1996 allowed both methods, with preference for the new one. Regulations issued in 2008 made the new method compulsory for permanent buildings; the method of admissible stresses was tolerated only for provisional structures in the construction site, whereas it was practically forbidden for structures meant to have a long life. Regulations issued in 2018 confirmed that the new method was the only legally acceptable one, and improved regulatory details, filling the gaps that had appeared in the 2008 regulations, and that were detected through practice in the 2008–2018 years. It can be noted that the three periods, 1970–1996, 1996–2008 and 2008–2018, largely correspond to the three phases earlier identified for the transition from previous scientific theories to new ones. Similar development parallelisms can be recognised for many trends-changes, innovations, discoveries and inventions in science and technology areas. This may also prompt some reflections on the different ways in which important changes in science and technology influence the overall body of science following key revolutions such as the shift from the geocentric to the heliocentric vision in astronomy in the XVII century, or the introduction of the relativity theory and of quantum mechanics in physics in the XX.

The relativity theory has brought fundamental contributions to the development of worldwide real-time communications. This has enabled researchers in all areas to be in contact and exchange ideas easily and frequently, and many investigations have become shared activities across long distances. The way of doing science has progressively changed, engaging a continuously growing number of researchers across the world, and enjoying the benefits of real-time worldwide communication. Research in science and technology has increasingly become a shared endeavour, with scientists from different institutions and different countries working together on the same project. Understanding and monitoring the effects of these novel circumstances on the way in which science proceeds and progresses could be an interesting question for the philosophy of science.

8.2.3 *Current Epistemology and Scientific, Technical and Technological Progress*

8.2.3.1 The Mode of Proceeding of Science and the Nature of its Results

As already mentioned, epistemology is the area of philosophy more specifically concerned with the investigation of the nature of knowledge and of the ways in which it is pursued and attained. In ancient times and across the Middle Ages, science and humanities were integrated, and philosophical investigations extended to all the areas in the same way. The origin of what we call "modern science" can be traced back to the definition of the scientific method by Galileo Galilei (XVII century) and the seminal physics theories born in that period thanks to Copernicus, Galileo, Newton, Kepler and some other scientists. In the second half of the XVI and the first half of the XVII century, there was not yet a clearly defined boundary between science and philosophy, and the latter tried to provide the former with foundations legitimating the knowledge that it was trying to attain. Things changed in the XIX century, when [6]:

> *… positivism completely overturned the relationship between philosophy and science, taking for granted that the latter was a sure conquest of our civilisation and did not require any external philosophical assurance. Philosophers therefore were no more asked to find an absolute foundation to the discovered mathematical, physical and biological truths; their task was deemed to be the effort to try and expand the scope of scientific knowledge to include disciplines like psychology, sociology and the like.*

The latter thought may be viewed as an anticipation of the now-a-day tendency for which mathematical approaches are entering many other disciplines, besides their traditional ones like physics and chemistry.

In the first decades of the XX century, logical neopositivism (the Vienna circle) assumed that the areas of knowledge that can be expressed through strictly logical language are the ones that can be considered scientific (the language that these areas use is completely systematised on the basis of logic). The logic of the thought-flow and empirical verification became scientificity criteria to be satisfied simultaneously. The neopositivist position influenced the philosophy of science in the first part of the XX century, and also the style of scientific and technical literature, entailing strictly logical disourses.

Still in the first decades of the XX century, new developments entered many areas of science, including non-euclidean geometries, absolute differential calculus, quantum mechanics, the relativity theory, progresses in chemistry and biology, as well as innovations in the applied sciences and technologies, such as rational mechanics, construction science, and others. The uncertainty principle (pertaining to the roots of quantum mechanics) sets restrictions to our possibility of knowing specific pairs of quantities with unlimitedly high precision at the same time; this is a limitation that has theoretical and mathematical bases, and does not depend on the precision of our instruments; consequently, it cannot be overcome by any improvement of our technology. The restriction has impacts on investigation approaches. For each such pairs, we must choose which quantity we wish to estimate with as high precision as possible, knowing that the other one will remain largely undetermined. The position and the linear momentum of a particle constitute a pair of this type. Since we are usually more interested in the best possible estimations of the energy of a particle, and kinetic energy is related to the linear momentum, quantum mechanics approaches evaluate the probability of finding a particle in a certain space, but do not give an exact position for it. Furthermore, elementary particles are simultaneously particles and waves (can behave in both ways), making it impossible to rigidly classify them as one or the other. All this contributed to a perspectives' reversal, in which scientific knowledge lost its character of absolute certainty that positivism had ascribed to it. New debates opened for the epistemology of science, and are still evolving.

A major research area focuses on the identification of the place of science within the scope of human knowledge. Some reflections on applied sciences may contribute relevant insights. Technologies and derived techniques can introduce a product into the market only if it is consistent with the required science-determined characteristics for its components, structure, and materials of which it is made. This is true both for unrepeatable prototype systems, such as buildings, infrastructures, industrial plants and the like (items that are produced one at a time, with individual characteristics) and for mass-produced items such as common-use objects. This control of science over technologies is institutionalised through legally-binding implementation protocols, which are defined by committees of experts on the basis of the state-of-the-art knowledge-progress in the relevant sciences. As science advances, the protocols get updated. In this way, science, besides engendering engineering and technologies, defines the characteristics of their products with reference to optimal utilization by communities for which they are meant. Fluctuations in the philosophy of science are unavoidably transmitted to its applications, correspondingly influencing technologies and products.

As already recalled in previous paragraphs, one of the most critical concepts among those pertaining to epistemology is that of 'scientific truth'. The first doubts about the dogmaticity of its existence or possibility were expressed by the French philosopher and mathematician Henri Poincaré (1854–1912), [32–34]. According to his 'conventionalism' vision, if, at a certain historical moment, science comes to a discovery, or technology comes up with an invention useful for the community, the theoretical or practical innovation is accepted as true not only if it is verified through experience, but also if it is supported by the choices by scientists and, subsequently, by the market. Since, however, these choices and decisions can always be revised or corrected, any statement formulated by science or by technology must be always considered as hypothetical, or something whose validity is temporary. An example pertaining to structural engineering is offered by anti-seismic engineering. Every new earthquake (and in some regions, including Italy, they occur rather frequently) highlights dynamic interferences between the vibrating ground and the structures built on it, and also the materials of which they are made; some of these interferences are already known (have already been investigated), others may be totally new and need specific investigation and attention from the scientific, technical, technological, design and building points of view. As a result, each 'latest' earthquake may bring revisions, corrections, extensions and improvements in the anti-seismic regulations and consequent modifications in the design and implementation practices. This perspective on science and technology is more dynamic than the assumption of the possibility of an absolute 'scientific truth' and entails a higher degree of complexity.

Since the dispersion of the Vienna circle in the time immediately preceding WWII, and up to current times included, empirical validation has become a necessary but not sufficient condition for a discovery or an invention to be assigned the qualification of 'being true'. New questions and new thought threads appeared in scientific, technical and technological research, also in view of the new requirements of industrial production, and they entailed modifications and increased complexity in the epistemological scenario.

The validity of a theory can be verified through confirmations, like experimental ones, but also through the absence of phenomena that contradict it, i.e., through the absence of phenomena that prove that some aspects of it are 'false'. Einstein was looking for experimental confirmation of his relativity theory (like, e.g., Eddington's one), but was also watching out for facts that might not be in agreement with its predictions, because such facts, if existing, would have required a reformulation of the theory. Both Einstein and Popper thought that the task of a scientist is not that of transforming his/her ideas into truths, but that of considering those ideas as assumptions that are true only as long as they are not proved otherwise. The suitability of either of these two confirmation options, that can be called 'verification' (proving that something is true) and 'falsification' (proving that something is not true) depends on the nature of the theory and of the topic to which it relates. One of the two criteria may be more suitable than another for a given case. The two criteria are also not mutually excluding. For instance, given its extreme generality and thematic uniqueness, the relativity theory is suitable for confirmation through falsification, but experimental verification also proved highly important.

The issue of the confirmation of new theories or procedures pertains also to applied sciences and technologies. The verification criterion has been preferred, because they concern themes involving a high number of parameters. On the other hand, since they have practical nature and are often related to production, the appearance of a falsification (something that does not function as expected) unavoidably requires rethinking. The two criteria may also be integrated into a combined operational plan. For instance, the theory for the instability of the elastic equilibrium of thin metallic structures taken as reference in Italian technical regulations has changed more than once—every time an adopted theory, which had been confirmed through verification, was later disproved by falsification in significant classes of cases, or further analysis of its internal logical organisation highlighted the presence of too rough approximations. Similarly, various theories have subsequently been adopted for the diagnostic study and stabilization of landslides, with any subsequent theory corresponding to broader verification and absence of serious falsifications with respect to the previous one.

In summary, there is always a balance between certainty and uncertainty in the birth, apogee and fading away of scientific theories. A theory that has encountered some falsifications needs to be abandoned, or to be reserved to those classes of cases for which it has proved to be an acceptable approximation (like classical physics can be used for the macroscopic world on our planet), while it is replaced by a new, expanded theory for the other cases (like the relativity theory expands on classical physics and is utilised for the cases for which classical physics proved inadequate). In turn, the new theory remains valid until some new evidence contradicts it. Thus, the history of science entails a sort of recursion development, where each ‘loop’ involves the formulation of assumptions and a period of confirmations, until a ‘falsification’ comes up, requesting the formulation of modified or new assumptions. From an epistemological point of view, the key role of falsifications in the permanence of the validity of a theory (‘valid until a falsification appears’) prompts the inference that our knowledge cannot reach an ‘absolute truth’ about the world. Within the scope of science investigation, the modern scientific method does not pose major focus on the truth-falsity antinomy, but on the extent to which a given theory may reasonably approximate the truth. Science does not lead to absolute truths, but it provides increasingly enhanced approximations to physical reality, in turn enabling increasingly advanced practical applications.

In many areas of applied science and technologies, it is not easy to model physical reality because information from measurements may not be sufficient to build a univocal picture of a system and its behaviour, and it is not always easy to measure the values of the data needed in the inputs of computational procedures based on a given model. For instance, problems related to geotechnical systems are inherently uncertain, because they deal with soils and rocks, whose behaviours depend on a high number of parameters, most of which are difficult to measure or to compute. If one wishes to be strictly consistent with their inherent reality, geostatic and statistical analyses of tunnels, landslides, port works, mines, and the like, would require that the physical–mathematical procedures incorporate the uncertainties relative to initial data, intermediate values and design results. On the other hand, the sum of all the individual uncertainties is always greater than the uncertainty of the result, and carrying on uncertainties along the entire procedure would limit the range of options stemming from the selected model and which the designer or decision-maker needs to explore in order to select the most viable one. An entire branch of modern mathematics, the theory of fuzzy sets and systems [35], deals with these issues. In this way, uncertainty is not a pathology of the model, but a characteristic that can be measured in terms of minima and maxima, and can be managed up to attaining results that are acceptable from an engineering point of view.

8.2.3.2 The Dynamics of Scientific, Technical and Technological Progress

As outlined in the previous subsection, the history of science can be viewed as a sequence of assumptions, verifications and falsifications, both with regard to broad-scope theories and for theories focused on specific issues or associated with specific practical themes. On the other hand, the history of science identifies with its progress, in turn identifiable as coalescence of three major simultaneous components: the increase of empirical content in agreement with theories; the increase in the explanation and predictive power of new theories with respect to the previous ones; the increase in the rigour of the verification or falsification controls applied to the theory's assumptions. Popper categorises the progress of science in terms of 'knowledge growth' rather than the previous terms of 'growth of the knowledge of truth', and considers the former as "strong conception" and the latter as "weak conception".

The fading away of the idea of certainty for scientific knowledge entailed the fading away of classification criteria for scientific theories. Popper introduced the verisimilitude concept for a qualitative evaluation, where *verisimilitude* is meant to quantify the discrepancy between the truth content and the untruth content of a theory; if this discrepancy could be calculated, it would provide indications for the selection of the best theory on a given theme, for which more than one theory might exist. However, verisimilitude is a word that suggests a comparison between the content of a certain assumption and a scenario simultaneously attributable to the object of that same assumption and accessible to human senses, i.e. capable of being observed. In this way, empiric information is given a significant role in logical positivism.

Engineering and technology share with science the theorization-verification-falsification route and its recursion character; however, they are amenable to a simpler evaluation, although isomorph with the one just outlined for science. The progress of engineering—a strategic factor for now-a-day life of humans and for building their future life conditions—can currently (in the absence of known restrictions) be viewed as a limitless process for the following reasons:

(a) Engineering creates objects and tools advantageous for practical life, or which can minimise the hardest and more boring components of work. Humans will always want to pursue these objectives and, therefore, the human mind will continue imagining new items, which will then be realised.

(b) Engineering displays itself through shapes—the 3D shapes of the objects it creates. The variety of these shapes has no limits and, therefore, there is no limit to the possibility of imagining new configurations.
(c) Engineering is embedded in the selection of the materials to build the desired objects. The design of new materials, or of improved versions of existing materials, entails such a variety of possibilities that the ensemble of substances that can become available is practically limitless.

Therefore, the "theorization—verification—falsification—new design" iterations can recur without limits, and the imagination power of the human mind will continue being able to invent new options.

In the 1960s, Thomas Kuhn (1922–1996), [36], distinguished two categories of phases in the development of science through time: revolutionary phases, entailing paramount changes, and normal phases, during which scientists operate without radical modifications of the theories. Technology and applied sciences follow the same pattern, being influenced both by revolutionary changes and by the explorations and details-refinement following them. Thus, scientific progress and technical progress remain isomorphous, although with a time gap for the innovations from the former to prompt innovations in the latter. Furthermore, Kuhn envisages systematic alternations of long periods of 'normal science' between short periods of 'revolutionary science'; he also suggests that the variation potentials of the two phases are comparable, with the revolutionary one concentrated in a short period of time. For instance, the XVII century European scientific renaissance can be viewed as a revolutionary phase, and the XX century period when the relativity theory and quantum mechanics were introduced as the subsequent revolutionary phase. The period between the two revolutions was a phase of normal science, highly productive in many science areas (physics, chemistry, mathematics, biology, earth sciences, etc.) as well as in engineering and technology areas (for instance, baroque architecture marked the beginning of a new and more daring way of structure design in engineering). According to Kuhn's vision, scientific research in the periods of normal science proceeds within the foundations set by the latest revolutionary phase, which constitute a "paradigm", and continues as long as new discoveries and innovations enrich the paradigm but do not come into unresolvable conflict with it, i.e., as long as discrepancies that may arise can be resolved through re-adaptations of the paradigm, which remain compatible with its foundations. When this is no more possible, the ongoing paradigm encounters a crisis; a revolutionary phase sets in and ultimately engenders a new paradigm, i.e., it brings a "paradigm shift". The crisis of a paradigm is a period characterized by uncertainties and, simultaneously, by active search for new responses. A paradigm can be abandoned only after a new one has been developed and can replace it; the crisis alone is not sufficient motivation to abandon a paradigm, although it is a necessary one. Once science starts operating within the new paradigm, different mathematical tools are used, and new interpretations, new innovations and new applications are born. The old paradigm may remain within the new one as a subset, for applications to the cases for which it worked without problems.

A clear example of paradigm shift is offered by the crisis of classical physics in the last decades of the XIX century and the birth of quantum mechanics in the first part of the XX century. Scientists discovered phenomena that could not be explained or modelled by classical physics: the black body radiation, the photoelectric effect, the behaviour of the heat capacity of solids, the Compton effect, and the discrete nature of atomic spectra (sets of separate signals emitted by atoms and characteristic of each element). Rutherford's experiment proved that the mass and positive charge of the atom are concentrated in its nucleus and the electrons revolve around it, but this model was incompatible with classical physics, which would have predicted that the negatively charged electrons would fall on the positively charged nucleus. Altogether, it amounted to the inability of classical physics to account for phenomena concerning the small particles of which matter consists. Doubts became so serious that even the hypothesis that matter consists of atoms was doubted, as the experiment-based information about their structure was not in agreement with the existing paradigm. The development of quantum mechanics constituted a revolution and provided a new paradigm, capable of explaining what was known about atoms and molecules, and of prompting new advancements. In recent decades, the enormous development of computer power has enabled the possibility of performing increasingly more demanding quantum mechanical calculations, and this has offered powerful tools to chemistry research, enabling computer-aided design of new molecules with desired properties through the computational prediction of their properties before their actual experimental synthesis, so that the costly experimental work can focus mostly or solely on promising molecular structures in the design of new substances and materials, from drugs to treat diseases to catalysts for the chemical industry and to semiconductors for the electronic industry. Quantum mechanics has also engendered new application areas, like nanotechnologies. Other sciences use it for studies where the properties of atoms and molecules are somehow involved and need to be considered.

As already highlighted in the previous section, Einstein's relativity theory entailed a radical change with respect to classical physics, thus responding to the concept of revolution determining a paradigm shift. Paradigm shifts may concern general theories and have implications on large sectors of knowledge, like the relativity theory or quantum mechanics, and also concern specific issues relevant within a given branch of science or technology. For instance, the transition replacing the *admissible stresses* paradigm by the *limit state method* paradigm for the study of the behaviour of reinforced concrete can be viewed as a paradigm shift in structural engineering; the phases described in Sect. 8.2.2. can easily be matched to the typical phases of a paradigm shift.

In the 1970s years, the Hungarian philosopher Imre Lakatos tried to reconcile into a unified theory the largely diverging visions about the dynamic evolution of science (and, consequently, of applied sciences and technologies) formulated by Popper and by Kuhn [37, 38]. The question had remained open about the possibility of objective criteria to distinguish which one among two or more theories was the best and, therefore, was eligible as a paradigm. Lakatos was the first to introduce the "scientific research programmes" concept into the philosophy of science [37–39];

the concept responded to the new reality of scientific research as team work, rather than the work of isolated stakeholders. The "team work" and "research programme" concepts had actually entered the practice of theoretical and experimental research well before their recognition within epistemological exegesis. They had also entered industrial level technological investigation, and had modified both the history of discoveries and inventions and the relationships between individuals and communities. Lakatos recognises that the nature of relevant scientific theories entails a high level of complexity, which can have clarity and organicity and, therefore, the ability to lead to useful results, only within the research programme framework. He also attempts to outline a sort of institutional framework for research programmes; however, current research programmes worldwide initiate, grow and come to completion with a high variety of different patterns, tailored according to their themes and to contextual characteristics.

Lakatos distinguishes between progressive and regressive programmes, where theoretical advance develops faster than experimental observations in the former, whereas theory follows experimental observations at a later stage in the latter. Scientists engaged in pure research prefer the former, as they see experiments as a posteriori verifications of the validity of a theory. For instance, the relativity theory was completed as a self-consistent theory before experiments were conducted. Scientists engaged in engineering and technology prefer regressive programmes, as experimental facts have determining roles to calibrate the completeness, efficiency and efficacy of an invention or of an item meant for the market. For instance, when designing a new car model, meant to be an improvement of an existing prototype, above all in term of restyling and of upgraded mechanical configurations, facts like experiences communicated by customers and changes in customers' tastes or in the official regulations about private motorization constitute important information that needs to be known before initiating the design process; then, the creativity inherent in this process takes them into account and the final product responds to the expectations of the market as well as to regulation requirements (e.g., in terms of increased safety or reduced pollutants emissions).

8.2.3.3 Latest Developments of Scientific Progress and Current Philosophy of Science

Variety of Schools-of-Thought Since the Second Half of the XX Century

The philosophical views recalled in the previous sections, and various others that are not recalled here because of space reasons, highlight the presence of intensive and lively debates, throughout the XX century, about some questions that had arisen from the crisis of XIX century positivism and from the XX century paradigm shifts in physics. The main focus of the debate concerned the relationships between science and physical reality, and the ensuing issue of the validity of the approaches utilised by science and of the results it obtains.

None of the questions debated during the XX century has so far encountered an exhaustive and definitive answer. The questions investigated within the philosophy of science have not reached an organic and unitary picture, and different schools-of-thought propose different answers in relation to their specific views, resulting in a broadly diverse scenario. The very definition of science encounters many options. A definition that might be compatible with many philosophical views considers science as “a generator of models and a flexible intermediary for our understanding and description of the reality in which we are inserted” [40].

The following subsections provide a quick overview of selected present-day schools-of-thought that have investigated epistemological themes. Some of them have been developed only in recent decades, others have been present for many decades and are currently receiving continued or new attention. The overview is not exhaustive, but selects schools that can be considered representative of major trends.

Naturalism

According to naturalism (e.g., [41]), science stems from the direct suggestions that the reality around us insinuates itself into us by means of our senses and subsequent rational elaboration. Therefore, science would not need philosophical foundations or starting points, from which to carry out investigations according to its specific language. The role of science is preponderant and prior with respect to that of philosophy, to which science provides not only the questions on which to reflect, but also the tools to search for answers. The task of the philosophy of science would thus continue that of science for each of the themes that science investigates. On the other hand, their roles and objectives remain distinct and specific, even if they deal with the same themes.

According to the more radical streams of naturalism, philosophy should answer the questions posed by physical reality by utilising the results obtained by science, and also the presence and valence of the rational foundation of the knowledge-content developed by science. Thus, philosophy would have an essential regulatory role with regard to the discovery and achievements of scientific research.

As far as engineering and technology are concerned, this perspective may prove interesting in regulatory definitions as, by being based on a posteriori evaluations, it appears inherently capable of always selecting the best options among what science has produced in recent times with regard to a given topic.

Realism

Realism comprises a group of thought-streams agreeing on the assumption that there is a reality existing outside us and independent of us [42]. Scientific theories are “*attempts to give true descriptions of the mind-independent and possibly unobservable reality, where truth means correspondence between language and reality*” [43]. We need to distinguish the real world as it is inherently and the ideas that we develop about it through our investigations; on the other hand, the latter influence our subsequent perception of the former. Philosophy can interact with scientific investigation by posing question to which science will try to find answers and by evaluating the extent to which the answers are valid and rational. Scientific descriptions and interpretations of reality get closer to the truth as they reflect our observations more closely.

Critical realism adds four important components [44]:

> *Our access to the world is always relative to a chosen linguistic framework (conceptual pluralism); all human knowledge about reality is uncertain and corrigible (fallibilism); even the best theories in science may fail to be true, but nevertheless, successful theories typically are close to the truth (truthlikeness); a part, but only a part, of reality consists of human-made constructions.*

The basic assumptions of realism are compatible with the needs of engineering, which treats all its objects as parts of physical reality. However, engineering also needs to assume that the knowledge we acquire about them is sufficiently certain to be used for practical purposes such as design, construction, and other implementation-related ones.

Metaphysical Constructivism

Metaphysical constructivism comprises a group of thought-streams united by the belief that there may or may not be a physical reality outside of us, but it is not independent of us, thus opposing the core of realism conceptions [41]. The term *constructivism* relates to *construction*, as it views that humans are the ones who construct reality through their investigations and interpretations. Most constructivists consider that this construction activity at least partially determines the nature of the objects which, therefore, depends on humans.

Immanuel Kant (1724–1804) may be viewed as precursor of constructivism views, with his dualism between a *noumenic* world, which is the real world and exists independently of us, but which we are not able to know, and a *phenomenic* world, which is the world as we see it, which we can know, but which is also our construction, because it does not exist outside our observations, perceptions and interpretations. Our knowledge of the phenomenic world varies depending on who is observing it and when the observations are carried out.

Metaphysical constructivism is not suitable for the needs of epistemological formulations and supervision typical of technology, as it cannot provide a framework for it.

Instrumentalism

Instrumentalism assumes that scientific theories should be deemed of as useful instruments available for us to relate to our everyday experience, without having to be concerned whether they provide a correct picture of reality, about whose existence we are not certain. The main purpose of a scientific theory is that of enabling us to make correct predictions, regardless of the philosophical questions that might arise about the "true nature" of reality, because such questions have scarce practical relevance for the explanation of our experiences.

In the XX century, instrumentalism views were introduced by P. Duhem in relation to physics [45]:

> *A physical theory is not an explanation. It is a system of mathematical propositions which can be derived from a small number of principles that serve to precisely depict a coherent group of experimental laws in a both simple and complete way.*

Thus, instrumentalism focuses on the practical utility of scientific theories [46], mostly viewing them as instruments for human use [47].

Instrumentalism is pertinent for the scientific exegesis needs of engineering and technology—both for the problems in which the theoretical phase precedes the empirical one and for those in which the empirical phase precedes the theoretical one—thanks to the practicality of its operational conceptions. Technologies may frequently need to go back to the science to which they are related, when designing a new invention or a new product destined for the market, but they need practical and concrete answers, whose validity refers to the extent to which they target the given objectives and enable progress along the route that will lead to the final product. Instrumentalism epistemology is concrete, simple to formulate and to adapt to individual projects, thus matching the practicality requirements of engineering and technology.

Semantic View

Differently from most of the other epistemological views, the semantic school [48–53] does not privilege the roles of language or of verification, but considers that, if we want to correctly analyse the relationships between theories and physical reality and utilise it effectively to solve problems, we need to use adequate models. Models transcend languages or cannot be defined within them [52]. On the one hand, the scarce interest for the debate on language and for empirical verifications may be questioned, as they remain an integral component of modern scientific approaches. On the other hand, the utilization of models has become common practice in all types of research, theoretical and applicative, general or referred to particular cases.

The generalization of the use of models in research and in design pertains to the second half of the XX century. It is interesting to analyse a bit more in detail the role of models, both in the operational practice and in the associated epistemological exegesis. Mathematical models of theories in science, engineering and technology can be viewed as interpreters and intermediaries for the relationship between physical reality and the descriptions and explanation that we are able provide about them. A model is an abstract system that enables us to translate reality—whether physical or constructed—into something representing that reality and to which it is possible to apply scientific theories within science or engineering contexts. In specific cases, frequently used in engineering, models are material objects which mimic the objects to be constructed on a pre-selected smaller scale, reproducing all their details and enabling a sort of experimental and predictive study of their properties and behaviours. They span from models of buildings, or of water flowing within specific boundaries, to pilot plants for the chemical industry. The semantic of models, whether abstract or material, is more complex, articulated and conceptually deeper with respect to the semantic of an ensemble of logical propositions, and in most cases enables more complete and ready-to-use descriptions of physical reality or constructed items, thus also facilitating the pursuit of research and design objectives.

Important features to be considered for models are their adequacy, representativeness and verisimilitude with respect to a truth that is not completely known, and the possibility and correctness of interpretations based on the descriptive information that they provide. Approaches based on the semantic view follow exegetical and philological criteria congruous with, and suitable for, the items and system that are objects of interest in engineering and technology. There are investigations for a possible classification of models into categories, according to their purposes and functions [6]. The intermediary role of models has become so important as to prompt a tendency to consider science as a creator of models, rather than theories, thus changing the function and perception of scientific discoveries and of research results [6]. For this reason, they are devoted greater attention in the next section.

8.3 Models: Theory and Practice

8.3.1 Models and Some Traditional Epistemological Questions

The creation and use of models relate to the philosophy of science and to the epistemological significance of scientific hypotheses and theories. They can also offer reliable responses to some epistemological questions that have been objects of investigations by philosophers and by scientists for centuries, with the former aiming at a systematisation of epistemological theories, and the latter trying to find a philosophical framework for their quantitative and algorithmic deductions starting from the observation of physical reality. Examples are the questions that can be formulated in the following terms: "Do scientific theories actually represent the phenomena that occur in physical reality and that we perceive, or are they only a construction of the scientist's mind, an ensemble of assumptions and conventions"?, "Is experience the foundation of scientific theories, or does it only have the role of verifying those theories through the evidence provided by direct observation?" and "What is the role of hypotheses in the development of a theory, or in the choice among different theories?".

It is first of all necessary to define the two key concepts. A **model** is an abstract or material system which respectively enables us either to verify all the propositions of a theory or to highlight—through its behaviour on a reduced scale—the behaviour of the actual real-scale system, which may be already existing and exhibiting critical issues or is to be built and requires a priori verification of its potentialities and characteristics. A **theory** is an ensemble of propositions, expressed in a formal language and meant to define (predict) the general-character impacts, on the behaviour of a physical system, resulting from specific hypothesised factors or from experimental evidence; the operations to be developed on building a theory need to exclude any assumption that does not derive from available data or observation-based evidence.

Experiments do not have the purpose of demonstrating something, but only that of verifying something, to ascertain whether it is true or not. As Poincaré explains, if we were to assume that an empirical verification process could demonstrate a certain theory, then, for instance, if one wanted to confirm the celestial mechanics theory about planets' and stars' motions, one would need to observe the positions of all the celestial bodies in the universe at close and regularly spaced time intervals.

The relationship between physics and mathematics is at the basis of models. Poincaré stated that the purpose of mathematics is not that of revealing the ultimate structure of reality, but that of "coordinating the physical laws that we know through experience, but that we would not even be able to enunciate without the help of mathematics". This principle recalls Galileo's views, which founded the scientific method on continuous and necessary interchange between 'sensible experience' and 'certain demonstrations', i.e., on experimental observations mediated by mathematical reasoning and integrated with theoretical deduction. Both views establish a systematic relationship between the world that we perceive with our senses and the theories, which are expected to have direct links with tangible, observable and measurable reality.

Poincaré also observed that part of science, and above all most of mathematics, have a conventional character. A double character is inherent in applied mathematics: on the one hand, mathematical language has a formal and conventional character; on the other hand, when applied to any science, engineering or technology theme, it provides correct and reliable solutions for practical problems, including highly complex ones.

Concerning the intrinsic value of scientific work, Poincaré did not think that science consists solely of conventions and that, therefore, has to be taken solely as an ensemble of action rules. The all-permeating presence of mathematics in science and technology themes, including their end results, does entail the character of action rules, but also enables an operational control of the consistency of the model with the physical or constructed reality. Sensitive experience generally has two contact points with theory:

- The initial point, because it is the real world, made of matter and energy, which suggests a certain research programme to the scientist, who will conduct it following creative intuition and guided by mathematics.
- The final point, involving the experimental verification of the developed theory.

In any case, as Poincaré stated, science enables us to know only the relationships between things; we cannot know what is beyond these relationships.

8.3.2 The Outcomes of Modelling-Based Research

The outcomes of research, in terms of the results that can be obtained, can be summarised in a way that recalls and somewhat integrates the previously analysed conceptions:

- For science: it progresses by establishing laws about the behaviour of nature; these laws increasingly better approximate a way-of-being of nature that can be considered true but unknowable; the increasingly improved approximation is viewed by some scholars as greater probability of it being true, and by others as a way of asymptotically approaching truth.
- For engineering: it progresses by creating new objects useful for human life: buildings and infrastructures, machines, items used in everyday life, research equipment in the various disciplines, and many others. They attain increasingly enhanced functioning as their refinement progresses. Some of them are harmless for the environment, some others are not, and one of the major challenges in current research is that of minimising the harm to the environment.
- For technology: it progresses through continuous improvements of the routes with which the objects designed by engineering are realised, through synergies of human and instrumental resources and their management, until the final product is obtained. The improvements increasingly rely on automation and computer-based controls.

8.3.3 The Challenges of Models-Building

Mathematical modelling is currently the dominant form of models in science and engineering. In the initial stage of a modelling operation, the observation of the object or the phenomenon and the inferences from these observations lead to the formulation of a hypothesis or a description, which are expressed in mathematical terms. Thus, a mathematical model is an abstract outline of a system or phenomenon under investigation, leading to a set of equations capable of being solved through appropriate algorithms, and whose solutions provide a reliable description of the system or phenomenon of interest and/or enable predictions about its behaviour, or provide guidelines about the best way of building a structure. The results are then verified with independently-obtained information, which can be statistical data, or experimental information about the parameters that can be measured directly, or results obtained from a different type of model (using two models with different natures for the same system or phenomenon is a way of verifying the reliability of the obtained results, through the extent to which they are mutually consistent).

Quite often, the practical use of mathematical models entails the search for a reasonable compromise between results accuracy and computational efforts. Many models start with sets of equations that, because of the complexity of the object considered, cannot be solved exactly (some of the terms related to some characteristics of the object prevent an exact solution). Then, physical and/or mathematical approximations are introduced in the solution procedure and associated computational component, and the main task is that of finding a balance between the desire to obtain as accurate results as possible and the feasibility of the needed calculations. The increasing power of computers is increasingly enabling computational feasibility of increasingly complex algorithms, thus enabling the use of more refined approximations and the obtainment of better quality results.

In view of this, one might consider two major components in the utilisation of models in science, engineering and technology: mathematical modelling and numerical simulation. The former entails the observational and interpretative recognition of the physical reality to be investigated, or intervened upon for anthropic purposes, and the formulation of the starting equations; the latter interprets the model in terms of the specific needs or objectives of a specific problem, identifying suitable relationships or simplifications that can maintain adequate reasonability of the ensuing approximations, and eventually translating them into numbers to perform the computations leading to the final outcome. For instance, if some integrals cannot be solved exactly, suitable numerical approximations enable reliable solutions. Figure 8.1 illustrates this through a schematic representation of possible approximations for the computational study of molecules. In this way, physical–mathematical-numerical modelling constitutes an important branch of applied mathematics and, simultaneously, a scientific approach consistent with the way in which Galileo's method entails balanced synergies between theory and experiments. Building, testing, comparing, revising and applying models constitutes a considerable part of scientific research and of engineering design.

Building a model for a given physical-sciences problem requires careful attention to each and every analytical and logical step, to avoid the risk of errors affecting the correspondence between the model being built and the system or phenomenon that is the object of the modelling. In many cases, it may be expedient to initially prepare an 'idealised' model for a narrowed-scope problem, through which one may verify whether the selected solution route is functional or a different approach is needed, and also verify the typology of the solution algorithm and the order of magnitude of the expected results. This provides a basic model which can subsequently be enriched and expanded by deleting imposed restrictions, introducing suitable generalizations, increasing the number of variables and factors relevant for the given problem, outlining variants to the core theme according to the details of specific physical or constructed realities that one wishes to represent, refining the results and improving their approximation-closeness with respect to the actual unknown 'truth'. This approach corresponds to a cautious and safe criterion for the initialisation and development of research in engineering and technology fields, as well as for a number of pure-science investigations. The progressive and gradual removal

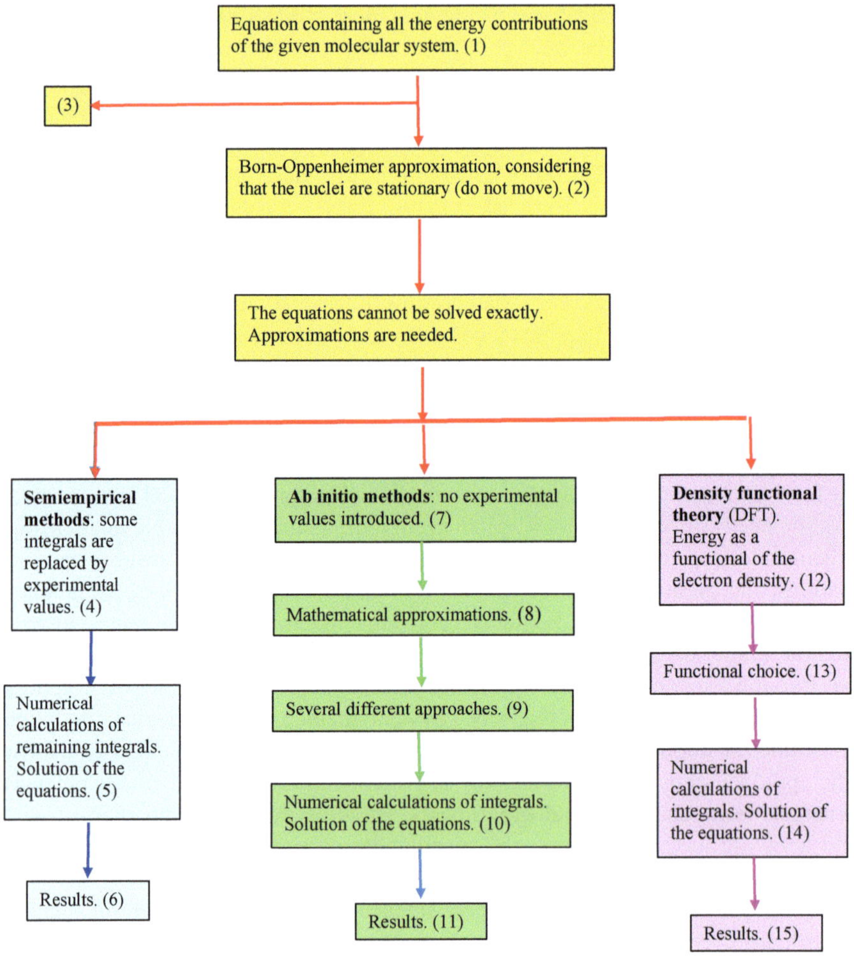
Equation containing all the energy contributions of the given molecular system. (1)
(3)
Born-Oppenheimer approximation, considering that the nuclei are stationary (do not move). (2)
The equations cannot be solved exactly. Approximations are needed.
Semiempirical methods: some integrals are replaced by experimental values. (4)
Numerical calculations of remaining integrals. Solution of the equations. (5)
Results. (6)
Ab initio methods: no experimental values introduced. (7)
Mathematical approximations. (8)
Several different approaches. (9)
Numerical calculations of integrals. Solution of the equations. (10)
Results. (11)
Density functional theory (DFT). Energy as a functional of the electron density. (12)
Functional choice. (13)
Numerical calculations of integrals. Solution of the equations. (14)
Results. (15)

◀**Fig. 8.1** Schematic illustration of the possibility of different models in the quantum mechanical computational study of molecules. The illustration highlights the meaning of mathematical modelling, focusing on the modelling aspects. A simplified explanation of the underpinning physical concepts is offered in [54]. The numbers in parentheses in the various boxes of the diagram refer to the explanations in the next sencences. (1). A molecule consists of positively charged nuclei and negatively charged electrons. The complete equation (Schrödinger equation) contains all the energy terms of the given molecule (kinetic energy of each nucleus and of each electron, Coulomb attraction between each electron and each of the nuclei, Coulomb repulsion between each pair of nuclei, and Coulomb repulsion between each pair of electrons). The equation is a second order differential equation. Solving this equation gives a set of functions describing the different states of the electrons in the molecule (molecular orbitals, MOs), and the energy values corresponding to each state. (2). The Born-Oppenheimer approximation is a commonly used approximation. It enables the removal of the terms for the kinetic energy of the nuclei and for the Coulomb repulsion between each pair of nuclei, thus simplifying the equation. (3). In some cases, the Born–Oppenheimer approximation is not used, and a different mathematical procedure is used to solve the equation. (4). Semiempirical methods are the most simplified methods. (5). The computation is fast, because of the simplifications introduced. (6). The accuracy of the results is not high. These methods are useful for quick comparisons, or for other particular purposes. (7). Ab initio methods do not introduce any experimental value, and proceed 'from first principles'. (8). Some mathematical approximations are introduced, like expanding the MOs in terms of simpler functions, whose type and number must be chosen for each calculation. Mathematical requirements for the MOs are also taken into account, e.g., the fact that the MO must change sign if the labellings of two electrons are exchanged. (9). There are several ab initio methods, differring by the conceptual approach, the computational time and the accuracy of the results. More accurate methods require more computational time. (10). The computational time depends on the accuracy and sophistication of each method. (11). The results are reliable. The more sophisticated the method, the more accurate the results. The more sophisticated ab initio methods are the ones capable of giving the best results among all the methods. (12). the Density Functional Theory utilises functionals (a functional is a function of a function). The energy is a function of the electron density, which is a function of the spatial coordinates. (13). A high number of functionals have been developed. The functional considered more suitable for the given molecule or research objectives is selected. (14). The calculations are faster than most ab initio method (only Hartree Fock calculations are faster than DFT for the same molecule). (15). The results are sufficiently reliable for most cases

of the initial idealisation and restrictions enables the researcher to follow the evolution of the model as the number of parameters involved increases, by identifying and keeping track of the transformations of its formulations as it gradually expands from the simplified basic case to more general cases incorporating more components of the entire updated theory. Figure 8.2 offers a schematic representation of this process. It has to be noted that the same framework applies to refinings that become important with time, as newly acquired knowledge prompts improvements, as is the case of the example outlined in the next paragraph. It also applies to cases where a main operative framework, like that illustrated in Fig. 8.1, needs adaptations for specific cases, expressed in terms of additions of specific features to the procedure; for instance, if a molecular system contains heavy metal atoms, it is advisable to add relativistic effects, which are added through a specific input instruction guiding the computational procedure to a route incorporating these effects.

An example from structural engineering, with clear correspondence with the approach illustrated in Fig. 8.2, is offered by the 'convergence-confinement' analysis method used for the design and testing of tunnels as underground areas [55].

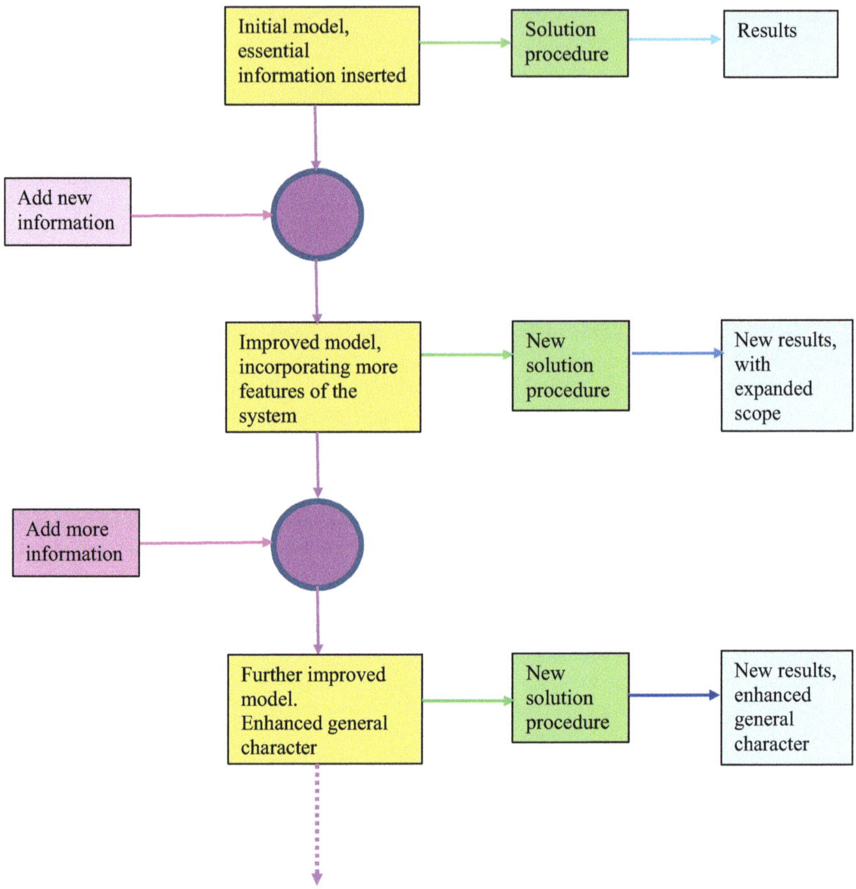

Fig. 8.2 Illustration of subsequent scope-expansion from a starting basic model, by adding new information. The new information may be related to the requirements of specific systems, or to the historical evolution of the knowledge about a given system, which needs to be incorporated in the model to obtain more accurate results. As indicated by the dashed arrow at the bottom, new expansions are always possible. Each new expansion increases the general-character of the model. It is also to be noted that different types of information additions may be entered for different types of problems. Thus, the same initial model may be the 'parent' of several refined models, responding to different necessities

The method was developed in the second half of the XX century and was inspired by the generalised theory concerning a vertical thin plate having a circular hole in whichever position, constrained on the two faces so as to be subjected to a planar deformation state (and, therefore, also subjected to normal tension along its thickness), and subjected to loads parallel to the average plane and to the conventional boundary planes, in a position identifiable as 'remote contour'. The restricted initial approach made the hypothesis that the constituting material was perfectly elastic, and the model developed through a system of differential equations easy to solve, yielding simple descriptive formulas for the tension and deformation fields, fully plausible from an intuitive perspective of the phenomenon. In the 1970s years, the hypothesis of elastic–plastic behaviour of the rock mass was introduced: plastic from the lining to the radial coordinate marking the elastic limit for the circumferential tension, and elastic for the major radial coordinates; this led to the convergence-confinement method, which became the reference method for underground works, widely used worldwide and yielding good correspondence between predictions and verifications. The breaking criterion marking the transition between the deep elastic zone and the plastic vertical zone towards the excavated cavity was later adapted to enable the consideration of various lithotypes (Mohr–Coulomb criterion for fractured rocks and loose grainy soils; Hoek–Brown criterion for competent-continuous rocks with comparatively high, but not very high, elastic module, or for clays; and others). Each breaking criterion led to a different system of differential equations, but the algorithmic framework held well for a high variety of formulas apt to represent plasticization as a physical phenomenon of overstressed natural materials. Between the 1980s and 1990 years, other factors were inserted into the same model: the mechanical effect, on the structure, by oversaturation of the lithotype with groundwater or deep artesian water; low coverage related asymmetry-inducing anomalies in the geomorphological environment of the slope; the effect, on the soil-structure interference, induced by the deviation of the cross section of the completed structure from the perfectly circular theoretical shape (deviation that is frequently unavoidable in the lower part, to enable access to the tunnel). The mathematical model maintained the same fundamental nature developed for the initial model, through all the—even substantial—complications added along the route from the narrowed-scope (restricted) initial model to a model expanding to increasing generalisation by including additional factors.

Models and simulations are becoming objects of epistemological investigation, for their increasingly expanding and irreplaceable roles in science and technology, and in many other areas in which mathematics can be applied. All mathematical models are based on idealisations of physical reality, thus inherently constituting simplifications of this reality. They need to match the details of the given system or phenomenon as closely as possible and, on the other hand, ensure that the model is suitable for computational handling up to the solution. A scientist or an engineer needs to distinguish the details that are essential, those that are relevant, those that can be neglected and those that would introduce unnecessary complications into the solution procedure (or make it unfeasible) without contributing relevant information with regard to the objectives of a given study. In this way, a scientist or an engineer needs to have sufficient epistemological foundations to be able to identify the pieces of knowledge that are more important for a given problem, both at model-building level and in terms of the desired results, which obviously depend on how a model is built. It is possible to build more than one model for each system or phenomenon, and they differ by the aspects that are ascribed greater importance; these, in turn, depend on the practical objectives that one wants to attain. In general, model-building depends on the expert's creativity which, in turn, largely depend on his/her theoretical knowledge and epistemological awareness.

Another epistemology-based skill is the ability to identify analogies between systems or phenomena, which may enable the use of important parts of the model already built for one of them to build the model for the other. Actually, analogy-classes can be identified, and the models for each item of a given class entail modifications with respect to a framework model suitable for the whole class. Several such classes are identified in engineering. For instance, the previously mentioned convergence-confinement model was initially intended for the design of continuous-type tunnels for roads, railways and pipes; some rearrangements in the algorithms enabled it to describe also the rock-mass/structure interference in the design of underground cavities with spherical symmetry, like artificial caverns meant to host power plants, or to be used for storage or other purposes. Analogous expansions are possible for the more advanced model calibrated on the hypotheses of elastic–plastic behaviour of the rock, with its varieties associated with different braking criteria of the rock types. Expansions are also possible for different shapes of the underground space, besides the tunnel and spherical ones.

8.3.4 Epistemology of Computer-Aided Modelling/simulation

As already mentioned, in many cases the equations expressing the initial mathematical description of a model or a phenomenon cannot be solved analytically, i.e., there are no mathematical procedures that enable a solution from the starting equations. Then, one has to introduce physics-related and mathematics-related simplifications, to attain equations that can be solved through approximation methods (e.g., Fig. 8.1). The associated calculations are performed using computers; the type of computers depends on the demands of the problem, and may range from personal computers to the powerful clusters of High Performance Computational Centres. Computational solutions of integrals and differential equations utilise numerical methods, implemented through algorithms that are translated into computer programs.

Numerical mathematics [56] has developed fast parallel to the development of computers' power. Numerical methods provide procedures to find approximate and consistent solutions to the equations that are relevant within a model. Numerical analysis provides a theoretical analysis of numerical methods, in particular studying the discrepancies (error) between computed numerical solution and the exact solution. In general, it can be said that numerical simulation is the application of a given mathematical model to a specific case; the latter is represented by data that must be compatible with the potentialities of the model, and must be prepared in such a way that they do not engender critical conditions within the algorithmic formulation pertaining to the model.

Numerical simulation enables the study of phenomena that cannot be studied through direct experiments (for ethical, safety or other practical reasons), thus having the role of a numeric experiment. It can also provide indications about the more suitable experiments to perform (for instance, which molecules are sufficiently promising for a certain purpose that it is reasonable to synthesise them and test their properties experimentally), or help in the critical analysis of experimental results.

The computational components of science and engineering are recent, as their development has been associated with the development of computers. They are already present in most research and design activities, and regulations for areas like engineering codify their use and the presence and type of computational results in the reports submitted before a given intervention (building, bridge, tunnel, etc.) is approved. In the case of non-standard structures, like large bridges or dams, the design stage entails both a numerical simulation and the construction of a reduced scale material model, on which static conditions are determined up to the breaking point; then, these experimental results are compared with the simulation ones, and their consistency is an indication of the validity of the design (together with the physical reasonability of the values, which is always an essential evaluation criterion). The combined use of experimental verification on reduced-scale material models and computational simulations is currently adopted in all branches of engineering: aeronautic engineering, for the design of new aeroplanes; space engineering, for the design of spaceships; mechanical engineering, for the design of cars, trains, machinery for the industry, and the like; naval engineering, for the design of ships; chemical engineering, for the design of production plants. Many other types of analyses also resort to numerical simulations: electricity generation and distribution in a given region; traffic control (road, trains, aviation, ships); water flow in natural and artificial water bodies; waves, streams and tides in seas and oceans; landslides or landslide risks; geohydrological analyses of filtering motion in water tables or artesian aquifers; and so on.

Given the historical novelty of computational components, epistemological considerations concerning them are also new. On the basis of the modes in which they are applied today, they are sometimes viewed as occupying an intermediate position between theory and experiment, practically bridging them, but they can also be viewed as one of the vertexes of a triangle, with the theoretical and experimental components constituting the other two vertexes [57], to leave space for the cases in which the interplay between theory and experiment is not necessarily mediated by computations.

Model-based simulations are often viewed as simulated experiments and their results are considered analogous to experimental results. This choice requires careful attention to ensure that simulations do not replace the experimental components beyond what can ensure full reliability of the overall design process. Simulations must obviously replace experiments in all the cases when experiments cannot be performed. They should also be performed parallel to material (laboratory) experiments as frequently as possible, ideally in all cases. The possibility of simulations enables considerable reduction of the material experiments, because they can 'guide' the experiments [58], in the sense that they can provide indications about which experiments can be relevant for verification and testing purposes, and how many of them. An experimental component remains however essential. The optimal selection of the number and types of material experiments to be performed depends on the design object and on the type of information needed from the results. The engineer's epistemological awareness—awareness of the type of knowledge involved in the project, in its simulation and in the material experiments—is fundamental for a balanced, informative and safe selection.

The all-permeating use of mathematical models and numerical simulations in civil engineering leads to the 'Semantic View' automatically becoming the dominant epistemological view for this area, because it theorises and privileges the use of models. Models are more complex, articulated and flexible than other forms of description, thus enabling more complete and realistic descriptions of the objects that they characterise. The model concept is common to the two domains: it can be analysed within the sciences in terms of its physical–mathematical nature and its correct and efficient use, and within epistemology in conceptual terms related to language, form and substance. Various questions remain open, such as a suitable classification of models into categories, with reference to the high variety of objectives for which they are functional.

The epistemology of mathematics is reflected in the epistemology of engineering as well as in the epistemology of science. A mathematical model that acts as implementation basis for various numerical simulations for various themes, establishes a language in which consistent and rigorous syntax and semantics are crucial. Syntactically considering both the instrument (mathematics) and the case in question (the physical–mathematical model of a given problem and theme) entails assigning order and organisation to connections, relationships, and dependence links, which relate language expression in both the instrument and the model. Once a unitary syntax has been established for both mathematics and the algorithmic structure of the model, the specific theme of the model becomes a domain of semantics. Hilbert underlined that the proofing process in mathematics is largely a syntactic process, because it is the study of the relationships between different propositions. The proofing chain—both in the mathematical basis and its application in the model—takes the typical features of a deductive chain expressed through symbols and operation-signs; the deduction process proceeds through various steps applying the rules of logic, and the typical automatism of logic entails a syntax which transfers from mathematics to the model and to the physical theme that it represents. For a syntax to be present, the mathematics need to be axiomatised. Now-a-day, all classical mathematics are axiomatised. The thread of a model contains physical–mathematical parts and logical parts, which together represent a problem or describe an object (system or phenomenon). Both mathematics and logic are disciplines that are not connected to specific physical realities. New mathematical theories are often purely 'abstract' when they are developed, but they may find applications at a later stage. Because of its characteristics, mathematics constitutes a basic language common to various sciences, both theoretical and applied; it enables them to make statements about reality—which is not a purely hypothetical-deductive system—and the statements are always expressed through a structured and rigorous language.

A classical epistemological question examines how it is possible that a model can say something meaningful although it utilises a language which does not have concrete meanings. Centuries-long experience confirms that mathematics enables pure and applied sciences to provide factual explanations which can be utilised directly in the concrete, material world, using its abstract language which has the advantages of being controlled and controllable, unambiguous and inherently unable to contain errors. Mathematics by itself cannot directly provide information on reality; but science can utilise it as a language and logical guidance to try and understand the world. Its total logical abstractness enables mathematics to be the most suitable language to investigate reality.

8.4 Discussion

The previous sections have considered that all the components of engineering (theoretical, design, technological) derive from science and, therefore, the epistemology of engineering can be built using the epistemology of science as starting or reference point. There is, however, an important difference. The theoretical branches of science study a 'physical reality' whose nature does not change with time, at least, within the period corresponding to human history, and does not include the impacts of human activities (e.g., environmental pollution); therefore, scientific knowledge attempts to approach 'truth', and comes increasingly closer to it in an asymptotic way with time (historical time) on the axis. Epistemological investigation referred to science follows the development of science knowledge, but with a delay, as it investigates issues that arise every time science makes a significant step forward. In the case of engineering (as for all the other disciplines associated with human activities), a new question arises. For clarity reasons, it will be illustrated with an example from structural engineering. All the theories focusing on the efforts to continuously enhance the knowledge of the loads acting on a structure, such as wind, snow, seism, the push from the soil, and others (and we need to recall that each of these terms entails a huge and complex accumulation of knowledge), follow an investigation route similar to that of the sciences. All the theories focusing on design techniques and the art-of-building follow an investigation route aimed at the construction of a material object meant to be useful to humans; in other words, the object will be built according to the wishes and needs of those who will use it, so, it is a constructed reality; the building techniques vary with time and, within the same civilization, improve with time, as scientific and technical knowledge progresses.

The epistemology of engineering is delayed with respect to the epistemology of science. It entails problems that have emerged since long, but have not been objects of deep reflexions both because of lack of interest for these questions by people devoted to practical activities, and because of lack of interest for practical and production-related themes by practicing philosophers. The relationships between science and philosophy have been recognised by many throughout centuries and up to modern time. For instance, Carlo Rovelli [59] recalls that philosophical thought has had extensive influence on western science, and that it "*opens possibilities, removes prejudices, reveals inconsistencies and logical leaps, suggests new methodological approaches and, in general, opens the mind to new possibilities*"; he also emphasises that methodology is not static and is the area where interplays between philosophy and science are most crucial; science has greatest need of philosophy when it goes through periods of fundamental changes challenging accepted basic concepts; he mentions quantum gravity (one of his research focuses) as an example "*where the space and time concepts are once more questioned and the old argumentations on space and time, from Aristotle to Kant and to David Lewis are resumed*". He also recalls that the benefits of the interplays are mutual, that philosophy is a source of inspiration and ideas for science, and that philosophy has been enriched by science throughout its history: if science ignores philosophy, it will "*wither through superficiality*"; if philosophy ignores the scientific knowledge of its time "*it is dull and sterile; it betrays its own deep root, its etymological meaning of «love of knowledge»*".

Different epistemological views may be more suitable for different sciences or different branches, and one might identify correspondences between the various disciplines and the philosophical schools-of-thought that are closer to their *modus operandi*. This is a correspondence that comes a posteriori: scientists usually develop the epistemology of their field in close connection with their work, whereas identifying and developing schools-of-thought pertains largely to philosophy. A previous paragraph recalled that the use of models largely relates engineering practice to the Semantic View. On the other hand, forensic engineering appears to be closer to Popper's philosophical approach. Forensic engineering entails the investigation of a physical problem with strong engineering characterisation (for instance, the collapse of a bridge or a building) utilising theoretical and experimental approaches that match legal requirements and provide all the details that are necessary for a court to reach a judiciary decision. Popper shared Einstein's approach of searching for crucial experiences that would confirm his theory, but also for experiments that might oppose it, in order to verify its validity, and that a single negative experiment would outweigh all the positive ones. This finds correspondence in the "proof" and "proof to the contrary" that are fundamental in law. A forensic engineer needs to start from facts, and interpretations need to stem from the facts in a strictly logical way, with the deductive component playing a major role.

Another huge area inherently linked to epistemology is the use of language in science and engineering, whether for scientific communication or for the development of scientific thought. The language needs to be rigorous, consistent with all the details of the concepts concerned, including the way or extent to which we know them. The examples analysed in Chapters 4 and 5 clearly highlight how the correct use of language is intimately linked to adequate epistemological awareness about the concerned theme.

Historically, epistemology has devoted significant consideration to the classification of sciences. Now-a-day, however, the number of disciplines characterised by individual autonomy and conceptual physiognomy is very high, and the growing tendency towards interdisciplinarity and multidisciplinarity establishes interface areas which also have their physiognomy. A first basic classification could distinguish between abstract-formal sciences and real-empirical sciences. The former traditionally comprise mathematics and logic, which often work synergistically and are applied in a variety of modes to a variety of contexts; a third science has added recently, informatics, where mathematics and logic are intertwined, and which is providing instruments for all the sciences and basically for most aspects of private, civic and work life. The real-empiric sciences comprise all the physical and natural sciences and all the applied sciences, thus including engineering. Separating physical and applied sciences would not be sufficiently motivated from an epistemological point of view. For instance, engineering is based on physics, from which it derives its theory, and to which it provides verifications and practically-oriented research questions. Building a mathematical/numerical model for an engineering problem or design is tantamount to building an applied physics model. Experiments in engineering cover a broad range of possibilities, all referring to the laws of physics. In the case of structural engineering, experiments refer to the laws of classical physics; some examples can be recalled for illustration purposes: (i) crushing tests of cubic concrete specimens to evaluate their mechanical resistance; (ii) tensile tests of linear steel specimens to evaluate their mechanical resistance; (iii) crushing and tensile tests of various building materials, such as wood, fiberglass, carbon fibre, special plastics, etc.; (iv) tests on real scale structural samples to investigate their response to seism-type vibrations and the kinematic and dynamic nature of these responses; (v) tests on small scale models of structures like a long cable-stayed suspension bridge or a tall antenna wedged to the ground, to evaluate their tendency to develop vibrational self-excitations, and be affected by them, under the action of a persistent wind; and many others.

To better emphasise the close links of engineering to the 'pure' sciences, and how the sciences are thoroughly embedded in engineering, it may also be useful to recall that it is possible to be a physicist without having extensive knowledge of engineering, but it is not possible to be a civil engineer without having extensive knowledge of physics; similarly, it is possible to be a chemist without having extensive knowledge of chemical engineering, but it is not possible to be a chemical engineer without having extensive knowledge of chemistry.

8.5 Conclusions

The relationships between engineering and humanities are various. For instance, Chapter 5 highlights how language pertains to the humanities but is fundamental in science and engineering and Chapter 9 presents a case in which cross-disciplinarity with literature and with history plays relevant role in a specific engineering project. An entire branch—forensic engineering—is inherently a cross-disciplinary area between engineering and law.

Epistemology is a fundamental cross-disciplinary area because of its roles in relation to the methods with which science proceeds and to the understanding of the nature of the attained knowledge and its validity scope. The previous sections have outlined an overview of the various approaches to the philosophy of science through its historical development, and how they reflect on the epistemology of engineering. While practical implications of epistemology have been implicitly incorporated into engineering design when needed, the awareness leading to focused investigations of its roles is rather recent and explorations are ongoing. The previous sections attempted to relate known epistemological approaches and schools-of-thought to engineering practice, first of all design, which is the component requiring theory-based investigations, complemented by simulations and experiments, to identify the best options to build the item that is the object of each individual project. The engineering-practice components utilised as examples have been recalled in their basic terms, without getting into more technical details, because the latter would require the use of highly technical language understandable only by engineers, whereas this chapter is meant to ensure readability by a broad range of readers. On the other hand, it is considered that works reporting the full bodies of projects which have utilised innovative design, including all their technical details and adding epistemological considerations and analyses wherever they have a role, would contribute to emphasise the value of epistemology in engineering practice.

Building awareness of the epistemology of engineering in the training of engineers is bound to have relevant roles in their professional practices. Although a large proportion of the design of standard structures now utilises standardised software for the needed calculations, the engineer ought to possess the knowledge that can enable sure evaluation of the reliability of the obtained results. In addition, the design of non-standard structures entails the design of the relevant mathematical models taking into account all the details of the complexity of such structures, and this requires the integration of various types of knowledge, which is productively guided by epistemological awareness.

References

1. D. P. Michelfelder, N. Doorn (Eds), *The Routledge Handbook of the Philosophy of Engineering* (Routledge, 2022).
2. K. D Beddoes, C. T Schimpf, Undisciplined Epistemology: Conceptual Heterogeneity in a Field in the Making, American Society for Engineering Education, (2015), Paper ID #11421. sociologyofengineering.org/wp-content/uploads/2021/03/Beddoes-ASEE-2015-lees.pdf
3. https://en.wikipedia.org/wiki/Philosophy_of_engineering, accessed 20–09–2022
4. C. Mitcham, The Importance of Philosophy to Engineering. Teorema **37**(3), 27–47 (1998)
5. S.H. Christensen, M. Meganck, B. Delahousse, *Philosophy in Engineering* (Academica, Aarhus, 2007)
6. E. Di Bernardino, R. Lucchetti, V. Schiaffonati (2011) *Temi filosofici dell'ingegneria e della scienza* 23, 23–24, 28, 44, 45 (Maggioli Editore, Milano, 2011)
7. E. Terrone, *Filosofia dell'ingegneria* (Il Mulino, Bologna, 2019)
8. V. Marchis, *Dall'arte...allo zero, Piccolo dizionario filosofico dell'ingegneria* (Mondadori, 2020).
9. A.C. Zambroni de Souza, M.J. Verkerk, P.F. Ribeiro, *Studies in applied philosophy, epistemology and rational eth: interdisciplinary and social nature of engineering practices: philosophy, examples and approaches* (Springer, Cham, 2022)
10. P. Vermaas, V. Stéphane (Eds.) *Advancements in the Philosophy of Design. Design Research Foundations* (Springer, 2018).

11. A. D. de Figueiredo, *Toward an Epistemology of Engineering (November 1, 2008). 2008 Workshop on Philosophy and Engineering* (The Royal Academy of Engineering, London, November 10–12, 2008). SSRN: https://ssrn.com/abstract=1314224
12. F. Doridot, Towards an 'engineered epistemology'? Interdisc. Sci. Rev. **33**(3), 254–262 (2008). https://doi.org/10.1179/174327908X366941
13. N. Cruz Zúñiga, E. Centeno Mora (2019) La construcción epistemológica en Ingeniería Civil: Visión de la Universidad de Costa Rica (The epistemological construction in Civil Engineering: the vision of the University of Costa Rica). *Revista Electrónica "Actualidades Investigativas en Educación"* 19(1):1–30.
14. V. Alcántara Alza, On The Epistemological Basis of Engineering Knowledge. J. Multidiscip. Eng. Sci. Technol. (JMEST) **4**(1), 6519–6522 (2017)
15. V. Kant, E. Kerr, Taking stock of engineering epistemology: multidisciplinarity perspectives. Philos. Technol. **32**, 685–726 (2019). https://doi.org/10.1007/s13347-018-0331-5
16. J.W. Ramírez Cano, W.M. Mora Penagos, Epistemological reflection of science and technology leading to the integration of science, technology and engineering. Int. J. Appl. Eng. Res. **14**(1), 212–219 (2019)
17. M. Boon, P.D. Center, In defense of engineering sciences. Techné: Res. Philos. Technol. **15**(1), 49–71 (2011). https://doi.org/10.5840/TECHNE20111515
18. R. Ratcliff, Applying epistemology to system engineering: an illustration. Procedia Comput. Sci. **16**, 393–402 (2013)
19. L. Koskela, E. Pikas, J. Niiranen, A. Ferrantelli, B. Dave, On epistemology of construction engineering and management, in K. Walsh, R. Sacks, I. Brilakis (eds.), *LC3 2017 Volume II—Proceedings of the 25th Annual Conference of the International Group for Lean Construction (IGLC)* 169–176 (Heraklion, Greece, 2107). https://doi.org/10.24928/2017/0336
20. R. A. Gonzalez, H. G. Sol, Validation and design science research in information systems. In: M. Mora, O. Gelman, A, Steenkamp, M. S. Raisinghani (Eds.), *Research Methodologies Innovations and Philosophies in Software Systems Engineering and Information Systems*, 403–426 (Springer, Hershey, 2012).
21. C. Eckert, R. Hillerbrand, Models in engineering design: generative and epistemic function of product models. In P. Vermaas, V. Stéphane (Eds.), *Advancements in the Philosophy of Design. Design Research* Foundations 219–242 (Springer, 2018). https://doi.org/10.1007/978-3-319-73302-9_11
22. A. Nordmann, Object lessons: towards an epistemology of technoscience. Sci. Stud. **10**, 1 (2012). https://doi.org/10.1590/S1678-31662012000500002
23. M. Boon, Measurements in the engineering sciences: an epistemology of producing knowledge of physical phenomena, in N. Mößner, A. Nordmann (Eds.) *Reasoning in Measurement*, Series "History and Philosophy of Technoscience" 203–219 (Routledge, London, 2017).
24. Yu, J. & Strobel, J., *Engineering-specific Epistemological/Epistemic/Ontological Belief*s
25. T. Cosgrove, J. O'Reilly, Theory, practice and interiority: an extended epistemology for engineering education. Europ. J. Engin. Educ. (2018). https://doi.org/10.1080/03043797.2018.1544226
26. J. Bernhard, A. Carstensen, J. Davidsen, T. Ryberg, Practical epistemic cognition in a design project-engineering students developing epistemic fluency. IEEE Trans. Educ. **62**(3), 216–225 (2019). https://doi.org/10.1109/TE.2019.2912348
27. A. Hashim El-Eqapy, D. Salim Resen Al Asadi, Epistemology and Creativity in Architecture. IOP Conf. Ser.: Mater. Sci. Eng. **928**, 234 (2020)
28. S. Frezza, D. Nordquest, R. Moodey, Knowledge-generation epistemology and the foundations of engineering. IEEE Front. Educ. Conf. (FIE) **2013**, 818–824 (2013). https://doi.org/10.1109/FIE.2013.6684940
29. K. Popper, *Logik der Forschung (The logic of the scientific discovery)* (Springer, Vienna, 1934)
30. Y. Lyu, P. Scrimin, Mimicking enzymes: the quest for powerful catalysts from simple molecules to nanozymes. ACS Catal. **11**, 11501–11509 (2021)
31. R.L. Liboff, The correspondence principle revisited. Phys. Today **37**(2), 50 (1984). https://doi.org/10.1063/1.2916084

32. H. Poincaré, *La science et l'hypothèse* (Flammarion, Paris, 1902)
33. H. Poincaré, *La valeur de la science* (Flammarion, Paris, 1905)
34. H. Poincaré, *Science and méthode* (Flammarion, Paris, 1908)
35. L.A. Zadeh, Fuzzy sets. Inf. Control. **8**(3), 338–353 (1965)
36. T.S. Kuhn, *The Structure of Scientific Revolutions* (University of Chicago Press, Chicago, 1962)
37. I. Lakatos, *The Methodology of Scientific Research Programmes* (Cambridge University Press, Cambridge, 1978)
38. I. Lakatos, *Mathematics, Science and Epistemology* (Cambridge University Press, Cambridge, 1978)
39. I. Lakatos, Falsification and the methodology of scientific research programmes, in *Criticism and the Growth of Knowledge*. ed. by I. Lakatos, A. Musgrave (Cambridge University Press, Cambridge, 1970)
40. C. Ranzoli, *Dizionario di scienze filosofiche* (Hoepli, Milano, 1963)
41. A. Goldman, in *Philosophical Perspectives*, vol. 13, ed. by J.E. Tomberlin (Blackwell, Cambridge, 1999)
42. I. Niiniluoto, Realism in epistemology, in I. Niiniluoto (Ed.) *Critical Scientific Realism* 79–108. (Oxford University Press, Oxford, 2002). https://doi.org/10.1093/0199251614.003.0004
43. I. Niiniluoto, *Abstract, Critical Scientific Realism* (Oxford University Press, Oxford, 2002)
44. D. Goswick, Constructivism in Metaphysics, *Internet Encyclopaedia of Philosophy*, https://iep.utm.edu/constructivism-in-metaphysics/
45. P. Duhem, *La Théorie Physique. Son Objet, sa Structure* (Chevalier & Riviére, Pari, 1906)
46. F. Roda, Systems engineering and modeling: some epistemological remarks, *Proceedings of the Posters Workshop at CSD&M* (2013), http://ceur-ws.org/Vol-1085/10-paper.pdf
47. https://www.newworldencyclopedia.org/entry/Pierre_Duhem
48. A. Tarski, Der Wahrheitsbegriff in den formalisierten Sprachen, *Studia*. Philosophica **I**, 53–198 (1935)
49. A. Tarski, Grundlegung der wissenschaftlichen Semantik, in Actes du Congrès international de philosophie scientifique, Paris 1935, fasc. 3: Semantique 1–14 (Herman, Paris, 1936).
50. A. Tarski, Über den Begriff der logischen Folgerung, in Actes du Congrès international de philosophie scientifique, Paris 1935, fasc. 7: Logique 1–11 (Herman, Paris, 1936).
51. A. Tarski, The Semantic Conception of Truth and the Foundations of Semantics. Philosophy and Phenomenological Research 4, 341–395; reprinted in Tarski 1 Collected Papers, v. 2, 665–699 (Birkhäuser, Basel, 1944).
52. J. Woleński, The Semantic Theory of Truth, *Internet Encyclopaedia of Philosophy*, https://iep.utm.edu/s-truth/
53. A. Cusmariu, Toward a semantic approach in epistemology. Logos Episteme **3**(4), 531–543 (2012). https://doi.org/10.5840/logos-episteme2012341
54. L. Mammino, Computational chemistry: studying the properties and behaviours of molecules, in L. Mammino (Ed.) *Green chemistry and Computational Chemistry: Shared Lessons in Sustainability* 1–39 (Elsevier, 2021)
55. J. Sulem, M. Panet, *Convergence-Confinement Method for Tunnel Design* (Springer, Berlin, 2022)
56. A. Quarteroni, R. Sacco, F. Saleri, Numerical mathematics. Springer (2007). https://doi.org/10.1007/b98885
57. L. Mammino, Computational chemistry and green chemistry: Familiarizing chemistry students with the modes and benefits of promising synergies. Sust. Chem. Pharm. **29**, 100743 (2022)
58. G. L. C. de Souza, Reducing undesirable products: computational chemistry guiding the experiments, in L. Mammino (Ed), *Green Chemistry and Computational Chemistry—shared lessons in sustainability* 245–262 (Elsevier, 2021)
59. C. Rovelli, *Helgoland* (Adelphi, Milano).

Chapter 9
Convergence of Different Disciplines in the Recovery and New Destination of an Ancient Iron Mine: A Case Study

Armando Mammino and Liliana Mammino

Abstract Multidisciplinarity may take many forms, depending on what it deals with. This chapter utilises a concrete case—the restoration of a dismissed mine to turn it into an underground museum of mining history—to illustrate how many disciplines can converge and contribute to define the characteristics and quality of a project. The case study highlights how a project that would at first glance appear to be purely technical benefits from the interfaces with history, literature, and also the legends and traditions of the location. The contributions of each individual discipline are first analysed individually, and then the synergies of each possible pair of disciplines are outlined. It is concluded that multidisciplinarity can be extremely fruitful to enhance the efficiency and quality of a project, and also its articulation and scope.

Keywords Convergence of many disciplines into a mining-museum project · Interdisciplinarity analyses for pairs of disciplines · Interdisciplinarity between engineering and humanities · Interdisciplinarity between engineering and other sciences · Technically-valuable information from history and legends

9.1 Introduction

Multidisciplinarity may take many forms, depending on the object or project of interest and on the disciplines whose contributions are relevant for it. This chapter considers a concrete case, in which disciplines from both sciences and humanities contributed information relevant to the design of an engineering project. It is particularly interesting for the scope of the present book because it is not frequent that

---

A. Mammino
Consultant Structural Engineer, Treviso, Italy

L. Mammino (✉)
Faculty of Science, Engineering and Agriculture, University of Venda, Thohoyandou, South Africa
e-mail: sasdestria1@gmail.com; mammino@univen.ac.za

L. Mammino (ed.), *Insights into the Relationships between Humanities and Sciences*,
https://doi.org/10.1007/978-3-031-97900-2_9

Photo 9.1 One of the main tunnels of the Fursill mine, supported by the original wooden ribs. Photo credit Arch, Antonio Pollazzon

areas like history, literature and folklore contribute technical information relevant for engineering design.

The project entailed the restoration of an old dismissed iron mine, to turn it into an underground museum of the history of mining activities in the region. The mine, *miniera del Fursill* (Fursill mine, Photos 9.1, 9.2, 9.3, 9.4, 9.5) is located in Colle Santa Lucia, a village in Cadore, in the Dolomites mountains, Belluno Province, Italy [1]. The project utilised well tested techniques and falls within a rather frequent scenario: after the activity in mines that are no more suitable for large scale production has been closed, the mines that appear suitable for it get transformed into tourists' attractions for a special category of tourists, who are interested in mines and their history.

A number of disciplines correlated and converged to ensure the best outcome, considering feature such as structural quality, optimal layout, the most suitable set-up, visitors' safety, the pleasantness of the visit, attractive advertising, and the like. The interdisciplinarity relationships that were relevant during the design and conduction of the project involved the following disciplines:

1. Architecture;
2. Structural engineering;
3. Geotechnical engineering;
4. Geology and geomechanics;
5. The chemistry of natural materials, of structural materials, and of the materials of which the technological set-up consists;
6. History and politics;
7. Literature;
8. Topography;
9. The science of safety in the work place under difficult conditions.

Photo 9.2 A tunnel in the Fursill mine, which is not supported by ribs because the rock is more resilient. However, the rubble from a local collapse is visible in the background. The rail on the bottom of the tunnel dates back to the activity of the Breda company in the first half of the XX century. Photo credit Arch. Antonio Pollazzon

Of these, the last one pertains only to the building stage and is not involved afterwards, while all the others play roles in the design stage and leave their marks in the morphology of the completed work. Their roles are analysed individually in the following section, and their mutual interplays are analysed in Sect. 9.3. They are considered equally important throughout. The outlines consider the needs of the project as references.

9.2 The Role of Each Discipline

9.2.1 The Role of Architecture

Architecture has primary importance because it takes care of the visibility components of the whole complex. It has to ensure the quality of contemplative fruition and emotional response by the visitors, so that, at the end of their visit, they can feel happy to have devoted time to visit something in the underground darkness rather than something in the open air.

Photo 9.3 The main tunnel of the Fursill mine, located between the external (above ground) area and the historical mine cultivation area. The tunnel part in the foreground is not protected by ribs because of being excavated in an iron-bearing rock with fairly good mechanical characteristics. The part in the background is reinforced with heavy and occluding structures because it crosses an important and active fault. Photo credit Arch. Antonio Pollazzon

9.2.2 The Role of Structural Engineering

Structural engineering is extremely important in this specific case because the rock is soft and crumbly, although very ancient, and, therefore, it tends to collapse, even extensively, if it is not supported by appropriate ribs. In order to prevent architectural disfigurements, the only rib prototype aesthetically valuable and compatible with the tunnels is a system of wooden arch frames, similar to those that were used on equipping new excavations to ensure their stability before miners would go in, when the mine was operational. The loads on the ribs are very high. In other words, we are speaking of highly stressed structures which, however, cannot be excessively massive because they should not block the visibility of the rock, i.e., of the element that the average visitor wishes to observe as the natural component of the place. This requires careful calculations of all the structural components.

Photo 9.4 A descendery connecting two different cultivation levels in the Fursill mine. The wooden structures date back to the activity of the Breda company in the first half of the XX century. Photo credit Arch. Antonio Pollazzon.

9.2.3 The Role of Geotechnical Engineering

Geotechnical engineering brings a functional contribution to the application of structural engineering, as it entails the expertise of the methods to calculate horizontal external actions (lithostatic push) and vertical external actions (direct load from above and counterthrust from below) acting on reinforcement structures (ribs, rock pitons, frameworks of deep wells, and the like).

9.2.4 The Role of Geology and Geomechanics

Geology studies lithotypes that are encountered in the various tunnels, and provides analytical indications on how they may interfere with the reinforcing works, and about the risk factors that may be embedded in their structures.

Photo 9.5 Fursill mine: a tunnel not protected by ribs, excavated in a rock with fairly good mechanical characteristics. Cluttered in the bottom by steep material fallen in several rounds and small quantities. Photo credit Arch. Antonio Pollazzon

Geomechanics was born as part of applied geology, but is increasingly asserting itself as an independent discipline. It studies the secondary structure of rocks, that is, the ensemble of faults, interlayers, fractures, disconnections, and the like, which—inherited from long past geological eras—may today generate critical equilibrium situations entailing real looming risks that need to be resolved before the mine-museum is opened to the public. In addition, geomechanics is extremely important to define the characterisation of rock masses, that is, their mechanical characteristics, which need to be inserted, as input data, into the algorithms that calculate the stress fields of the rock mass around the tunnels.

9.2.5 The Role of the Chemistry of Materials

The chemistry of natural materials is needed to verify the absence, or their resolvability in case of presence, of the following problems:

(a) Whether the rocks surrounding thc tunnels, and remaining visible to the visitors' eyes, are compositionally permeated or prone to release gases harmful to the human organism, such as hydrogen sulphide or products generated by the transformation of carboniferous or oil-rich masses, or other products, depending on the nature and age of the lithotypes enveloping the tunnels.
(b) Whether the waters permeating the rocks carry dissolved substances that may be aggressive for the utilised structural materials such as wood, iron, concrete, and others.

The chemistry of structural materials and of the constituent materials of architectural and technological equipment is important to ensure their needed durability degree, as it would be unthinkable that one would need to carry out additional renovations not long after the mine is opened to the public, interrupting its economic and cultural fruition. For instance, it is known that wood has very short duration in the underground world of the mine; since aesthetical and consequently architectonical reasons make it necessary to utilise wooden ribs to support the rock (which is heavy, although soft, because of being an iron-rich rock in a friable gangue), the wood needs either to be mineralised in an autoclave (that is, impregnated with inertisising organic materials), or to be 'acetylated' (that is, impregnated with acetyl and/or its derivatives, which can create a chemical environment resistant to aggression by moulds and insects), or to be treated with other chemical and/or physical processes that eradicate for several decades (at least 50 years) the aggression by xylophage animal and vegetal organisms.

9.2.6 The Role of History

For a disused mine—a mine that has been deleted from the list of the industrial activities of a country—history [1] becomes a cardinal component of its value, of public attention, and of its ontological right to continue existing with some beneficial function (it may be recalled that, for instance, other mines are converted into mushroom cultivation sites, because their characteristics make them suitable for it).

The first mention of the Fursill mine dates back to 1177, when the emperor Federico Barbarossa recognised it as a property of the Novacella convent, together with a vast strip of mountain territory. The latter comprised the "Puchperc" or "Wersil" farm, donated to the Novacella convent by the bishop Artmanno in 1145; the deed of ownership was confirmed by pope Alexander II with a bull dated 10 May 1177, that is, slightly more than 3 months before the analogous emperor's document. The discovery of the mine likely occurred in the time between 1145 and 1177, if not in the brief three-months period between the papal and imperial documents, as the mine is not mentioned in the former, but is explicitly mentioned in the latter. After this enticing birth act, the history of the place and the mine's ownership continued through a long and complex series of contrasts and international law issues, involving Central European potentates and North-East Italian potentates, as the metalliferous orographic body (Monte Pore) was located in the border area between the two potentates. Abundant technical and historical literature documents this troubled mining activity, also owing to the fact that what was mined in that area was a sort of natural steel, rich in carbon and manganese, and not just iron: carbon greatly enhanced strength, and manganese guaranteed duration, up to the point that the railings of noble palaces in Colle Santa Lucia do not show any hint of rust after 500 years, and do not need any painting because their natural colour is a beautiful opaque and uniform black (Photo 9.6). For these reasons, this natural steel was traded all over Europe for the manufacture of swords, ploughs, railings, fine tools, keys, locks, and the like.

The mines were closed in 1753; by that time, the Prince Bishop of Bressanone had been the owner for one and a half century, while the Andraz family, from the neighbouring castle at the foot of Mount Pore, had been the agent [1]. Historical sources inform that, in the alpine context in those years, wood was used as fuel for

Photo 9.6 Chizzali-Bonfadini house in Colle Santa Lucia. Built in 1612, the house was the administrative centre of the Fursill mine. Its railings were built using the manganese-rich iron from the mine. It currently hosts the Ladino Cultural Institute "Cesa de Jan" [7]. Photo credit Dr. Vittorio Fenti

Photo 9.7 Monte Pore at the beginning of the XX century. Courtesy of the Technical Office of the Colle Santa Lucia Municipality, with technical assistance by Arch. Antonio Pollazzon

iron smelting, and enormous amounts were needed, because of the low heat content of fresh vegetable matter and because of the large amount of iron-containing minerals that were excavated from the rocks in the last period of the mine's functioning. Mount Pore (Photo 9.7) and all the surrounding mountains had been cleared of forests to provide wood for the smelters. At a certain point, the production chain got stuck because of the exhaustion of the fuel sources. The surrounding woods had been cut for so large a radius that providing wood from yet farther-away areas would have been unaffordable. This was the reason for which the mine was closed, although its iron veins were not yet exhausted.

Sporadic, short and unsuccessful attempts to restart the mining activities were carried out between 1753 and 1938. The most noteworthy experiment in this regard was attempted by the Milano "Breda" firm between 1938 and 1943: all the old tunnels were probed and tested, high level excavation equipment and tools were introduced, and a thorough search was carried out for that natural steel that had been an important resource for the local economy and employment for centuries. The collapse of Fascism in 1943 swept away also the reborn mining activity, condemned by the subsequent political class as a useless autarchic expedient, incompatible with the modern world. Thus, politics decreed the final cancellation of the mine from the raw materials production activities; the analysis of this socio-economic phenomenon, which ended with the irreversible termination of mining, can be viewed in terms of tripolar interdisciplinarity among mining engineering, economy and politics.

9.2.7 The Role of Literature

Literature has strong interactions with the project of the touristic and cultural reutilization of the Fursill mine, because the mine was the scene of one of the most beautiful alpine legends collected directly from the narrations of the mountain inhabitants by the writer Karl Felix Wolff. In the first edition [2], the legend about the Fursill mine takes 82 pages, thus sitting on the undefined border between a long 'short story' and a short novel. The book was the last one of a series (by the same author) of books collecting Dolomites' legends. The story set in the Fursill mine is titled "*L'ultima Delibana*" (the last Delibana) and belongs to ancient (although likely less ancient than others) legends of the Livinalongo Valley miners. Literary philology has highlighted that it was namely this epic, in its original oral version narrated by the descendants of the protagonists, which inspired the tale of "Snow-white and the seven dwarfs" by Northern narrators. Wolff's masterful reworking is posterior, but is based on narrative material which was more ancient by far. The narration is centred on a woman devoted to the miners—maybe a poetic transfiguration of the harlot in mining communities, reworked in the myth and in an almost priestly apotheosis of ancient customs, with the gentleness and the ability for implicit purification that only great writers manage to orchestrate through the scenarios that can be detected in between lines.

Delibana is a young and beautiful girl, from a noble family, who, by her own choice, lives in the mine for seven years, together with the miners, and the passing of time is marked by the cyclic appearance of the "iron flowers", spread across a waterlogged meadow and emitting light in the middle of the night. If the Delibana were missing, the mines would run out and the people of the area would be compelled to go back to the agricultural and pastoral activities of their ancestors—an indication that the long historical period in which the mine was active had granted the local population a previously unknown wellbeing, supported by a centuries-long international market. The last Delibana lasted "nine times seven years" in the deep underground, and the people of the area forgot the magic rituals—weaving the sacral habit of the priestess of the underground, who for such a long time had remained without replacement. She was then the last Delibana, and the mine ran out, and the surrounding area went back to being a place of shepherds and farmers, as it had been in the early days of settled communities. At the South-East of the Bolzano-Cortina road, above Andraz, one can see a reddish and stony top emerging above the far away green pastures, like a sad warning monument: it is Mount Pore, the old iron-rich mountain.

The legend thus narrates what happened on those mountains a bit more than a century before the mine was closed in the middle of the XVIII century, establishing interesting parallels with history. Many references point to previous centuries, starting from the XII century, because the mine was not known earlier. It is therefore a substantially recent legend, with respect to many others collected by Wolff [3–6], which gather oral narrations going back to the communities formed in the various stone and metals ages. By its very existence, the legend influences the architectural design of the project for the new destination of the mine, because the visitor centre needs to host a "study centre" devoted to Karl Felix Wolff—unjustly relegated among the 'minor' writers, although he expressed a high literary level and a narrative strength of great intensity; this project offers an opportunity to re-evaluate his works, also in view of the fact that the public has already expressed its opinion by keeping reading his books for hundred years.

9.2.8 The Role of Topography

Topography is a discipline of high strategic value for the considered project, as it enables the reproduction on easily searchable materials (hard copies or soft copies) of the complex geometry of the mine-tunnels' systems in their initial state and in the state envisaged by the project design. It is well known that an accurate topographic survey of the existing situation is fundamental prior to the design of engineering interventions. This survey provided the three-dimensional Cartesian reference on which the entire architectural-structural-technological design was based.

9.3 Focusing on Key Interdisciplinarity Features

9.3.1 The Identification of Synergies Between Pairs of Disciplines

The previous section outlined the individual contributions of eight crucial disciplines to the project for the touristic valorisation of the ancient Fursill mine. Each of these disciplines interacted with all the others through what can be viewed as interdisciplinary synergies. It would not be easy to consider multiple interactions (interactions among three or more disciplines) simultaneously. Therefore, the best option to highlight synergies entails the consideration of how each discipline interacts with each of the others. Their identification follows a mathematical pattern (combinatorial analysis) that is illustrated in detail by the graph in Fig. 9.1. The synergies are indicated by arrows, starting from an element and connecting it to all the other elements. The arrows are convenient for illustration purposes, including keeping track of the connections on building the graph, but they do not indicate directionality, because the synergy between two areas does not entail directionality: it involves the two areas simultaneously and is the same whether we mention one or the other area first. Thus, for instance, the arrow from element 1 to element 2 is not accompanied by a separate arrow from element 2 to element 1 in the graph, because they would have the same meaning, and each synergy needs to be counted only once. Double arrows would actually be more consistent with this conceptual aspect; on the other hand, the choice of one-headed arrows enables systematic tracking of the synergies-pairs in terms of the sequence of the initial items, and this is expected to add clarity.

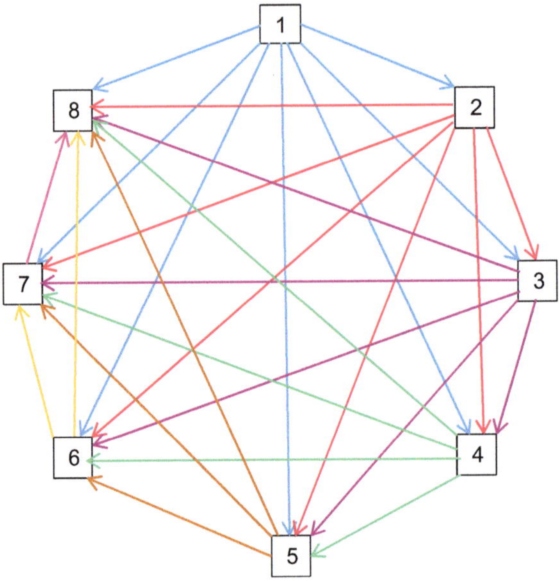

Fig. 9.1 Illustration of the identification of pairs of synergies between 8 elements. The synergies are indicated by arrows, starting from an element and connecting it to all the other elements. Each synergy is indicated only once, so that, for instance, since there is an arrow from element 1 to element 2, there will not be a separate arrow from element 2 to element 1. The graph has a general character, but the numbers may also be referred to the disciplines considered in this chapter, as they are listed in the introduction. The synergies are identified as follows: Consider element 1 and connect it with the other elements. Seven synergies are identified: {1,2}, {1,3}, {1,4}, {1,5}, {1,6}, {1,7}, {1,8}. Consider element 2 and connect it with the remaining elements. Six new synergies are identified: {2,3}, {2,4}, {2,5}, {2,6}, {2,7}, {2,8}. Consider element 3 and connect it with the remaining elements. Five new synergies are identified: {3,4}, {3,5}, {3,6}, {3,7},{3,8}. Consider element 4 and connect it with the remaining elements. Four new synergies are identified: {4,5}, {4,6}, {4,7}, {4,8}. Consider element 5 and connect it with the remaining elements. Three new synergies are identified: {5,6}, {5,7}, {5,8}. Consider element 6 and connect it with the remaining elements. Two new synergies are identified: {6,7}, {6,8}. Consider element 7 and connect it with the remaining element. One new synergy is identified: {7,8}

In total, 28 such synergies between pairs of disciplines are identified: {1,2}, {1,3}, {1,4}, {1,5}, {1,6}, {1,7}, {1,8}, {2,3}, {2,4}, {2,5}, {2,6}, {2,7}, {2,8}, {3,4}, {3,5}, {3,6}, {3,7},{3,8}, {4,5}, {4,6}, {4,7}, {4,8},{5,6}, {5,7}, {5,8}, {6,7}, {6,8}, {7,8}. They comprise synergies between different technical disciplines, between technical and humanities disciplines, and between different humanities disciplines. None of these 28 synergies was ‘empty’, i.e., each of them contributed to the project. The next subsections offer brief outlines for each of them, considering them in the same order in which they are identified in Fig. 9.1 and listed in the previous sentence; the identification in terms of pairs of numbers is also reported at the end of each subsection title for the sake of clarity. It is opted to consider all the synergies individually to better highlight the contribution that each of them brings to the project; this option corresponds to an analysis of details, whose sum builds the overall project. Because of obvious space reasons, only the basic conceptual characters can be outlined for each synergy.

9.3.2 Synergies Between Architecture and Structural Engineering ({1, 2})

In the visitors' centre, architecture and structural engineering interplay along the classical ways of the art of building; the details of the interplay are not going to be analysed here because, although the two disciplines influence each other, their mutual conditioning does not have specific (unusual) characters with respect to the average buildings with few storeys.

In the underground mine tunnels, architectural quality implies the maximisation of what can be observed by visitors, which, in turn, would entail the largest possible expansion of the areas in which the rocky surface is uncovered; on the other hand, ribs' statics would require considerable closeness of the ribs because of the high horizontal and vertical loads induced by the heavy covering and by the general weakness of the rock. Thus, the distance between consecutive ribs should increase because of architectural reasons and decrease because of structural reasons. The synergy of the two disciplines is necessary for an optimization analysis aimed at finding the best compromise between the two contrasting requirements.

The variables concerned have different natures. The static variables are of a physical–mathematical type, deterministic for some aspects and fuzzy for others, to the extent induced by the uncertainty in the estimation of the mechanical and rheological characteristics of the enveloping lithotypes. The architectural variables have inherently fuzzy character, as they correlate to the extent of the satisfaction perceived by the average visitor after browsing the details of the rock surface and the uncovered spaces that may indicate the presence of metals, or of baritone, or outcrops of rare minerals, or anomalies in the lithoid structure.

Another choice that depends on the interplay between architecture and structural engineering is the univocal choice of wood as ribs' material and the adoption of the arch frame as the static setting of the same ribs, because the latter need to conform to the geometrical styles and to the compositional exclusivities of the times when the mine was functioning.

9.3.3 Synergies Between Architecture and Geotechnical Engineering ({1, 3})

For this type of underground interventions, the interplay of architecture and geotechnical engineering entails the definition of an insurmountable threshold for the size of any new underground space that may be functional to the intended activity. In the case of the Fursill mine, some problems—later solved—were posed by the wish to maintain unaltered (except for some reinforcements meant not to remain visible) a rocky septum between two parallel tunnels eventually converging near the upper opening of a deep well, which was equipped with an overpassing small bridge as part of the intervention.

9.3.4 Synergies Between Architecture, Geology and Geomechanics ({1, 4})

Geology and geomechanics are the disciplines that enable the search, identification and valorisation of the characteristics of the rocky structure that are worthy to be seen by visitors. Then, architecture has the task of designing the altimetric routes that enable the viewing of the rock surface in the most extensive and satisfying manner.

9.3.5 Synergies Between Architecture and the Chemistry of Natural and Technological Materials ({1, 5})

The architectural choices concerning the materials for setting up the museum, i.e., those utilised in the accessible underground spaces, need to ensure that their chemical composition does not make them vulnerable to aggression by the fluid and solid substances present in the enveloping rock or soil masses. This entails chemical compatibility between the construction materials and the natural materials constituting the surroundings, in order to avoid that everything goes to ruin in a short time. Therefore, the architectural options cannot take shape without taking into account the chemistry of the two ensembles of materials.

9.3.6 Synergies Between Architecture and History ({1, 6})

The architectural design needs to take into account the fact that the visitor's centre ought to comprise a centre for the study of the history of the place, with the following major themes:

- the history of the role of the Andraz castle (which is located in the same geographical and orographic area) in the management of the mine, as well as in the political management of the overall territory;
- the history of the economic value of the mine, and how it influenced European history, also in relation to the fact that that area, during the late Middle Ages

and through the entire Renaissance and Baroque epochs, became the border belt between the Central European potentates and those of the Italian peninsula;

- the last attempt to reactivate the mine, between the end of the '30 s and the beginning of the '40's in the XX century, by the Breda company, with large investments and many good intentions. This historical information might be complemented by the controversial issue of whether those attempts should be viewed as part of the autarchic extremisms of Fascism and, therefore, part of that dictatorship, or whether valuing raw materials national resources had a meaning, and whether it might still have now, when new technologies might make it sustainable and profitable.

9.3.7 Synergies Between Architecture and Literature ({1, 7})

As already mentioned, the importance of the legend of the Delibana, for a complete picture of the perceptions and memories about the mine, suggests the opportunity to include a study-centre devoted to Karl Felix Wolf in the Visitors' Centre of the museum. From an architectural point of view, this requires the design of suitable spaces for scholars, the public, and the documentation, considering that the latter comprises the original one (already gathered before the opening of the museum) as well as the one that will be added as it gets produced by new investigations and initiatives.

9.3.8 Synergies Between Architecture and Topography ({1, 8})

Topography is the discipline that enables the numerical definition of the 'architectural metrics': the ensemble of all the measurements, transferred into design graphics, makes the meant-to-be-realised architectural system perfectly known and ready to be materially built by workers. This is even more important for underground architectures, because of the presence of often unknown pitfalls and limitations that are not present above ground.

9.3.9 Synergies Between Structural Engineering and Geotechnical Engineering ({2, 3})

Geotechnical engineering provides the experimental data and the theoretical deductions enabling the evaluation of the loads acting on the structures along the closed perimeter of the transversal sections of the tunnels. Once the loads have been analysed, the algorithms proceed according to the classical methods of structural analysis.

In addition, geotechnical engineering provides the criteria for the sizing of perforated anchors in the soil and the rock, that is, their length and resistant cross section, and also the algorithms to diagnose whether the preconceived distribution of the anchors reaches the conditions that ensure a mechanical behaviour corresponding to the 'soil nailing' style (ability to deform and to withstand forces like a bound medium, resistant to compression along the lithoid component and resistant to traction along the metal lines included through perforation and connected through sealings using either high-strength cement or epoxy or polyurethanic resins).

9.3.10 Synergies Between Structural Engineering, Geology and Geomechanics ({2, 4})

Only geology and geomechanics can provide the designer with a general picture of the problem and highlight the implications of the interactions between the rocky mass and the structures, which engender the ensemble of strengths and pressures to be inserted into the static calculation. The terms 'structures' here comprises both the structures which constitute the border between the rocky mass and the empty spaces of the mine tunnels and the structures that have been embedded into the rocky mass by perforation or sealing, starting from approximately radial positions with regard to the same mine tunnels.

9.3.11 Synergies Between Structural Engineering and the Chemistry of Natural Materials and Structural Materials ({2, 5})

The synergies between structural engineering, chemistry of natural materials, and chemistry of structural materials are fundamental for the durability of structures. As far as the load-bearing works inside the rocky mass (the anchored anchors) and all the constituent materials (steel, artificial resins, special mortars, and the like) are concerned, they are designed since the beginning and then built with all the requirements needed for them to last not less than 50 years (or even 100 or more) within rocky masses with whichever chemical composition.

The problem of duration arises more seriously for the ribs, which, as already explained, must be wooden due to architectural reasons. Wood is a material subject to rapid deterioration in a mine environment, both because of the high and diffuse humidity and because of the miscellaneous and aggressive chemical properties of the lithotypes that remain uncovered in the walls, in the vaults and in the basement of all the tunnels. In addition, natural wood rots within few years in a mine due to the aggression by vegetal organisms such as moulds and fungi, and animals such as specific insects, other arthropods and worms, which have evolved to be xylophages. Therefore, the wood to be utilised for the ribs must previously be mineralized in autoclaves or acetylated by impregnations with acetyl or its derivatives, or treated with other procedures and substances that have been patented and are technologically warranted.

The parameters that need to be taken into account in the selection of the most suitable material for a given lithotype context comprise the chemical, mineralogical and geohydrological parameters of the enveloping rocky mass. Since treated wood is highly expensive, and the overall volume of the ribs is large (both for the project considered here and in general), it becomes necessary to identify the cheapest option among the ones suitable for a given project. In the case of the Fursill mine, it was not necessary to establish a separate additional interdisciplinary route between structural engineering and economics, entailing complex estimations within a border area between the two disciplines, because the final selection of the material corresponding to the best quality/price ratio depended on a simple market survey among few alternative options. This fact indicates that it would not make sense to overuse multidisciplinarity and interdisciplinarity approaches: they are highly complex methodologies, often implying the opening of new cultural frontiers in terms of 'vanguard' as well as 'academia' and, therefore, it is wise to resort to them only when traditional one-discipline or two-disciplines-interface approaches are not adequate.

9.3.12 Synergies Between Structural Engineering and History ({2, 6})

The wooden ribs—to be designed and calculated in such a way as to create conditions of stable safety for visitors—need to resemble those that were actually used in the active mine (Photo 9.1), as much as possible. It is therefore important to study the historical documentation (descriptions in the texts of ancient accounts, ancient drawings, and various forms of collective memory) in order to design static schemes and structural forms as close as possible to those that were used in this specific mine. The fact that the new materials are more resistant and durable does not affect the value of the imitation, whose aim is that of bringing the mine to its original configuration, and 'reproducing the past within the present' through a scenographic reconstruction for the benefit of the visitors.

9.3.13 Synergies Between Structural Engineering and Literature ({2, 7})

The tale organised by Karl Felix Wolff on the basis of information from local inhabitants provides many details of how the mine was labyrinthine and intricate, how many outlets it had into the external ground, how many extraction levels had been identified and excavated across centuries, and how the mutual connections between these levels corresponded to wells which were up to many tens of meters deep. From Wolff's tale, Mount Pore appears thoroughly permeated by a maze of tunnels and externally pockmarked by outlets and landfills. Today—nearly a century after he outlined this picture—that singular volcanic and metamorphic orographic body, surrounded by soaring Dolomites mountains (the ones that can be seen from Giau Pass) is completely covered by woods. However, little has changed in its rocky interior since the times in which the mine was closed, and since the more recent times when Wolff visited it. The largely unexplored interior has certainly suffered the effects of macroscopic collapses, or of the slow collapses that, in old mines, result in the formation of thick and extruding mounds of debris. The plano-altimetric mapping of the tunnels that are currently known, and in the process of being ameliorated for tourism, appears to be just a small portion with respect to the brilliant ideography of that same maze provided by Wolff through narrative descriptions (not through measurements). On reading that native legend, one develops the idea that the maze extended through the entire orographic volume, crossing it for a height of many hundreds of meters. On the basis of what is now known of that mining system, careful comparisons confirm the truthfulness of the descriptions by Wolff, who used to be very well documented when he wrote about legends. The archaeologist Giuliano Palmieri (from Treviso) was able to attest, through findings-based evidence, the historical roots of many of his legends, the material traces of the events he narrated and the scenarios he described. There are therefore abundant knowledge elements to trust Wolff's writings.

A philological exegesis of the legend of the last Delibana can provide guidance in the search for other tunnels, other underground halls, other wells, and other analogous spaces, in turn taking the re-discovered ones as starting points. This would not only expand the underground space available for the tourist attraction, but will also provide the possibility to ascertain that currently unknown tunnels above or below those that will be opened to the public would not undermine the known and operational ones from a geostatic point of view, thus jeopardising their safety. Therefore, in this specific case, the analysis of the literature text becomes an instrument for structural engineering and geomechanics, i.e., for the art of design, along the lines of a completely new but potentially fruitful interdisciplinarity route. The stakes comprise not only an aid to the overall safety of the intervention, but also the prospect of being able to roam widely through the interior of Mount Pore, with its immense mining maze—the result of six centuries of cultivation and of the work of an entire miners' community spreading through the municipalities of Colle Santa Lucia, Livinalongo, the surroundings of the remnants of the Andraz castle (which has been stabilised without altering its outward appearance), and few other places next to the old iron-road. Wolff's documenting seriousness enables his work to go beyond its literature value, and to be valuable also for technical disciplines, and for a culturally and sociologically important initiative like the renovation of the Fursill mine.

9.3.14 Synergies Between Structural Engineering and Topography ({2, 8})

Topography is the discipline that enables the numerical definition of "structural metrics", i.e., the ensemble of measurements that enter the specific graphical representation of the structural design, accompanied by the elementary geometry figures/diagrams needed to define in detail the structural shape of the overall complex and of each of its parts. Many such figures are required, to ensure that workers can realize what is represented in them without needing any other source of information; altogether, the figures establish an a priori prefiguration of the work as it will be after completion. Such interdisciplinarity between engineering and topography is always present and is somehow institutionalised; it is even truer for structures to be set underground, as they may have to deal with often-unknown pitfalls and limits that do not exist above ground and that can seriously jeopardize the implementation, resistance and stability when the work is operational, if not adequately taken into account in advance. Additional information is presented in the next subsection.

9.3.15 Synergies Between Geotechnical Engineering, Geology and Geomechanics ({3, 4})

The synergy between geotechnical engineering, geology and geomechanics is institutional and recurrent in the infrastructure sector. Geotechnical engineering utilises the knowledge of the geology of a given site as information to be embedded into an engineering design; in other words, it takes the information that, within geology, has historically and traditionally been the object of qualitative and conceptual treatment (besides being descriptive on the basis of field observations) and translates it into quantitative information. Geomechanics is more frequently referred to rocks than to loose soils in technical jargon, as it investigates the mechanical behaviour of underground rocks; it also merges geology, soil mechanics and rock mechanics to build an overall picture. The boundary line between the two disciplines is not sharp, and there is increasing affinity in the physical–mathematical algorithms used as references at problem-solving level; the integration of the foundation concepts continues growing as the research on general principles and particular application topics progresses.

This {3, 4} synergy was active within the Fursill mine project, to the same extent and with the same theoretical and practical modalities with which it is increasingly getting embedded in the design of underground structures.

9.3.16 Synergies Between Geotechnical Engineering and the Chemistry of Natural Materials ({3, 5})

In the case of the Fursill mine, the synergy between geotechnical engineering and the chemistry of natural materials reflects the general terms with which it usually appears, rather than having specific characters. The chemistry of natural materials (soils, rocks, intermediate lithotypes, and the like) constitutes a key to understanding many of the mechanical, deformation and rheological behaviours that are the objects of study of geotechnics and geotechnical engineering, as it relates the behaviours of a material to its composition.

9.3.17 Synergies Between Geotechnical Engineering and History ({3, 6})

The history of the cultivation of the mine enables the investigation of the extent of the maize created within Mount Pore during the six centuries of its activity, on the basis of written documents and oral testimonies transmitted from one generation to the next. A process of philological research on the distribution, arrangement and specific dimensions of main tunnels, secondary tunnels, wells, caverns, and the like, enables—at least in terms of clues—the localization of the levels at which iron was extracted and of their mutual connections; from these, it is then possible to deduce valuable information on possible geotechnical and geomechanical criticalities engendered by such an extensive and all-pervading man-created drilling of the orographic body. In turn, the localization and anamnesis of those criticalities enable the project of the tourism-associated development of the mine to be managed in such a way as to ensure maximum safety for the visitors.

9.3.18 Synergies Between Geotechnical Engineering and Literature ({3, 7})

The information ensemble analysed in Sect. 9.3.17 can benefit also from the legend of the last Delibana—as reconstructed by Karl Felix Wolff on the basis of the fragmentary recollections of the inhabitants of Livinalongo—as it is comparatively rich in references to the intricate and vast underground structure of the mine. Wolff was always extremely rigorous in being faithful to the sources and in declaring to readers where the narrative theme coincided with history supported by written documentation, where it derived from oral recounting by the descendants of the protagonists, where the reference was the ancient pre-Christianity alpine mythology, and where he had personally taken care of filling unresolved gaps with his writer's imagination, fully respecting consistencies at the interfaces between what he had read or listened and what he had added to enable narration and plot continuity. Giuliano Palmieri's studies attest the correspondence between archaeological data and the narrative content of many of Wolff's tales. It follows that Wolff's writing can be considered a highly reliable source of interesting pieces of information and knowledge about the details of the underground structure.

9.3.19 Synergies Between Geotechnical Engineering and Topography ({3, 8})

The rock's structure, i.e., the set of faults, interlayers, fractures, inclusions, compositional anomalies, areas with particular mechanical disturbances, and so on, must be carefully mapped. The visual and material marks that substantiate that structure must be transcribed and represented in the plano-altimetric charts of the various cavities, large and small, so as to have position coordinates for each of them, and descriptions in terms of inherent nature and potential effects on the overall mechanics. This is a recurrent type of general-character interdisciplinarity, aimed above all at differentiating the components of the structural entity—the dimensioning of ribs and anchors as a function of their structural stress, i.e., of the forces acting on them as a consequence of the local physical state of the rocky mass and the ways in which they had been disturbed by the pre-existing cavities.

9.3.20 Synergies Between Geology + geomechanics and the Chemistry of Natural Materials ({4, 5})

The interdisciplinarity between geology and geomechanics and the chemistry of natural materials also constitutes a general case, not only a specific feature of the Fursill mine project. The chemistry of soils and rocks (also termed geochemistry—a term already consolidated by use) is a key to understand many of the deformation, rheological, degenerative and reactivity behaviours responding to anthropic interventions, which are objects of study by geology and geomechanics.

9.3.21 Synergies Between Geology + geomechanics and History ({4, 6})

As explained in Sect. 9.3.17, the history of the cultivation of the mine enables the localisation of the various types of empty spaces (tunnels, wells, caverns) and the recognition of their dimensions. This information enables useful inferences on the unavoidable geological and geomechanical impacts on the orographic body, as those empty spaces impoverish its texture, i.e., affect its original continuity and structural integrity. Such geological impact has surely occurred, and it is not easy to know whether, after two and a half centuries, the mountain has fully stabilised or not yet. It is therefore appropriate to analyse this aspect in detail, in order to detect and understand possible impacts on the geotechnical characterisation of the rocky mass. Since the sources of information on this aspect are fragmentary and uncertain, history can provide significant contributions.

9.3.22 Synergies Between Geology + geomechanics and Literature ({4, 7})

The interdisciplinarity between geology/geomechanics and literature is isomorph and conceptually analogous to that between geotechnical engineering and literature, already described in Sect. 9.3.18.

9.3.23 Synergies Between Geology + geomechanics and Topography ({4, 8})

The role of topography in the representation of the geological and geomechanical scenario of the rocky mass and its internal cavities is analogous to the role of topography in the representation of the geotechnical scenario of the same rocky mass and its internal cavities, described in Sect. 9.3.19. In general, geological, geomechanical and geotechnical characteristics are mapped together in a single topographic set of charts derived from direct survey.

9.3.24 Synergies Between the Chemistry of Natural Materials and History ({5, 6})

The history of the materials extracted from the mine during its long cultivation period and, above all, the sporadic descriptions in the available documents, enable the possibility to deduce at least part of the lost information on the raw material deriving both from the caverns and tunnels that have been rediscovered and made passable, and from the other caverns and tunnels that surely exist in the vast orographic body, but are not connected to the above-ground areas or to the known tunnels, due to numerous collapses and the cloggings resulting from piles of debris.

9.3.25 Synergies Between the Chemistry of Natural Materials and Literature ({5, 7})

The tale of the last Delibana contains information about the characteristics of the extracted iron (behaviour when heated, resistance, diffusion in the mass, etc.). Its consideration adds to the interdisciplinarity reflections already outlined in Sects. 9.3.18 and 9.3.22.

9.3.26 Synergies Between the Chemistry of Natural Materials and Topography ({5, 8})

The chemical composition of the rocks enveloping the internal cavities must be mapped in the same charts reporting the rock types, the data on the lithoid structure, the geological, geomechanical and geotechnical characterization of the lithotypes present in the walls of the underground routes that will be open to the public. The combination of their information is essential to predict the behaviour of the entire structure and its individual components.

9.3.27 Synergies Between History and Literature ({6, 7})

Among Karl Felix Wolff's vast literature production, the tale of the last Delibana is likely the one most extensively characterised by robust and detailed historical references, also in terms of consistency with written documents. We are actually speaking of one of the classical parallels between history and legends. The history side comprises archive documents, which are rather abundant both because six century of mining exploitation are bound to leave many traces on paper in the offices of the owners and dealers-managers and in the palaces of the powerful leaders, where opposing policies were defined for the possession of that important mining site. In addition, the Novacella Convent—the owner to whom Federico Barbarossa granted the concession for the cultivation of the mine—was extremely meticulous in keeping absolutely all the paperwork related to the history and places of the monastic possessions.

The legend side comprises the oral transmissions through succeeding generations. They are wide-ranging, as confirmed by the fact that Wolff managed to reconstruct a picture of considerable narrative breadth and completeness. As already mentioned, the narration mostly concerns the last century and a half in which the mine was operative, but it also contains frequent historical and mythological references, sometimes mixing with fairy tales and magic, dating back to more ancient times, although posterior to the initial grant by Federico Barbarossa.

In this underlying binomial between history and legend, the bifurcation in the interpretation of characters and objects becomes unavoidable. The historical persons, the count of Andraz (the noble agent on behalf of the Bishop-Prince of Bressanone) and his wife become the king and the queen in the popular imagination and consequently in the legend, following a tradition rather frequent in sociological scenarios of that type and stemming from the convergence of the direct experience in everyday reality and the fantasy legacy suggested by ancestral myths.

With reference to the last decades before the mine was closed, the tale hints to the presence, within the mine's maze, of iron-dwarfs and the iron-giant—the former cunning and little reliable, but not expressely malevolent or dangerous for humans, the latter perched in the lower levels of the mine, bloodthirsty and ferocious towards the miners, scary up to the point that it was considered advisable to avoid meeting with him. This narrative metaphor relates to the introducion—in the first decades of the XVIII century—of mechanization in important mining realities, which, in the eyes of the humblest miners, appeared as self-moving iron organisms, as they were driven either by hydric power or by the first introduction of steam power. Some of them were small and easy to handle by workers who had been specially trained. There was also a big one at the bottom of the mine, likely a water pump or a crusher, in which some workers might have been crushed by accident, thus originating the legend of the extreme ruthlessness of the iron-giant, and the need to avoid it as much as possible. The timing actually coincides. The last period in which the mine was cultivated was particularly fortunate in terms of the amount of extracted material, and the last Delibana—to whose presence the richness of the metal veins was ascribed—had remained with the miners through the previous half century. This anomalous productivity of the mine might have encouraged investments on the new machines produced in England (initially, the ones invented by Newcommen, later the more advanced ones invented by Stevenson and Watt) and utilized in European mines as pumps and crushers since the first decades of the XVIII century. The use of these machines, i.e., the ingress of the iron-dwarfs and iron-giant, was likely the swan's song of the Fursill mine, which had to close not because of being exhausted, but because of lack of available fuel within reasonable (economically affordable) distance, to smelt the extracted iron.

The entire text of the tale of the last Delibana could be analysed from the perspective of parallels between history and legends. The examples outlined in the previous paragraphs offer adequate illustrations of the potential and actual interests of such analysis.

9.3.28 Synergies Between History and Topography ({6, 8})

The history of the cultivation of the mine can provide useful information to distinguish the tunnels, wells and caverns that are so far known, have been topographically surveyed and are ready for inclusion in the tourism-oriented museum, and to have an idea of how many, and of what types, have not yet been rediscovered in the immense interior space of the mountain. The comparison of the alleged plano-altimetric topography of the entire mine, as broadly inferred from historical sources, and the much narrower one put together on the basis of the results of recent explorations and surveys, can provide valuable indications for a more plausible reconstruction of the topography of the part that is still unknown, above all for the areas bordering the explored ones.

9.3.29 Synergies Between Literature and Topography ({7, 8})

The same reflections outlined in Sect. 9.3.28. hold for this case.

9.4 Discussion and Conclusions

The previous sections have shown multidisciplinarity at work, in a project where different branches of engineering interact among themselves, with other sciences like geology and chemistry, with a discipline—architecture—that is by its nature a discipline where sciences and humanities converge, and with humanities-based disciplines such as history, literature and folklore. The outcome of the project is the creation of a museum that offers visitors and scholars a comprehensive picture of the history of mining, of the specific mine turned into a museum, and of the site where the mine was located, and offers scholars opportunities for further research into that history. Thus, the museum itself offers a synergy of information from different areas and is open to the collection and generation of new information.

Considering the synergies between all the possible pairs of the disciplines involved in the project, as done in Sect. 9.3, provides a first important outlook on how different disciplines can converge to build new knowledge from the combination of their contributions. In the real work, synergies are not limited to two disciplines at a time, as all contributions come together to build the actual project. On the other hand, the information in Sect. 9.3 can be used as a basis to identify and build up the outcomes of the synergies between three or more disciplines at a time, until the overall outcome is glimpsed.

In terms of the general theme of this book, the content of the previous sections shows a concrete case in which history and literature can provide information important for the design of technical approaches. It is concluded that multidisciplinarity can take many forms, which have to be identified and tailored for each individual context or project. The best results are obtained when all possible contributions from different disciplines are explored, and the new information generated by their synergies is taken into account. The explorations need not to be restricted to only one domain (either science-based disciplines or humanities-based disciplines), as cross-domain synergies can bring new valuable perspectives.

References

1. M. Baldin (Ed.), *Il castello di Andraz e le miniere del Fursil: un itinerario storico culturale nelle Dolomiti* (Marsilio, Venice, 1997)
2. K.F. Wolff, *Ultimi fiori delle Dolomiti* (Licinio Cappelli, Bologna, 1953)
3. K. F. Wolff, *I monti pallidi* (Bologna, Licinio Cappelli)
4. K. F. Wolff, *L'anima delle Dolomiti* (Bologna, Licinio Cappelli)
5. K. F. Wolff, *Il regno dei Fanes* (Bologna, Licinio Cappelli)
6. K. F. Wolff, *Rododendri bianchi delle Dolomiti* (Bologna, Licinio Cappelli)
7. http://www.galaltobellunese.com/wp-content/uploads/2013/09/ST_A4_73-74.pdf

Chapter 10
Art and Chemistry: Partners in Progress

Mary Virginia Orna

Abstract The discovery, use and development of various materials for artistic purposes is the subject of this chapter. Beginning with the Stone Age, ancient humans applied heat to plant, animal and mineral substances, effecting chemical transformations into products useful for artistic expression. Some of these products, like metals and glass, gave rise to flourishing artistic specialties like sculpture and ceramics. A key component of almost every work of art was color, a property that remained associated intimately with the material in which it inhered; how it was produced stretched the artist-chemist's versatility to the limit. Discovery of new artistic materials also went hand in hand with the discovery of the elements, particularly in the area of pigment production. At every step along the way, new discoveries by chemists placed new artistic materials into the hands of artists so quickly that the two disciplines actually co-evolved as partners in progress.

Keywords Art-chemistry partnership · Ceramics · Glass · Metals · Organic colors · Pigments · Polymers

> "…the Future of Art Is Only Limited By the Human Imagination and What the Artist Strives to Do Is Experiment With All of the Materials Available to Express a Personal Perception About the World." [1]

10.1 Introduction

If ever a union was made in heaven, one of the earliest would have to have been that between art and chemistry. There is nothing in art that does not have something to do with chemistry. All art objects are material substances, and as such, are subject to the laws and to the manipulations of chemistry. At the same time, chemistry, in

M. V. Orna (✉)
ChemSource, Inc, 309 Bradley Ave, Mount Vernon, NY, USA
e-mail: maryvirginiaorna@gmail.com

L. Mammino (ed.), *Insights into the Relationships between Humanities and Sciences*,
https://doi.org/10.1007/978-3-031-97900-2_10

some limited instances, can also be subject to the manipulations of the artist [2]. Once the reality of chemical change in ever more sophisticated modifications was introduced into the routine activities of *Homo sapiens*, the products of these changes found their way increasingly into the cultural expression we call art [1]. This chapter will document this progression.

10.2 Roots of Art

Art is a mirror of culture and of a culture's values. Using the materials available to them, artists can interpret their experience from a variety of perspectives: historical, socio-historical, symbolic, cultural, behaviorist, communal, environmental, functionalist, and structuralist. Thus, when examining a work of art, a chemist must keep these factors in mind and not concentrate solely on the material substance. Otherwise it would be possible to miss understanding the cultural context out of which the object came, and thus reduce the examination of art to an exercise in analytical chemistry rather than an appreciation of the cultural value of the object in question [3].

Art is material matter supported on the one hand by inspiration, on the other by motivation, both on the part of the creator of the art. But without the intelligent manipulation of that matter by way of chemistry, whether empirical or theoretical, there would be no art. Chemistry, in its turn, is supported by the three traditional branches of knowledge known as technology, medicine and alchemy. Visual representations of these relationships are shown in Fig. 10.1. Hovering unspoken over both disciplines is the indefinable aura of creativity—a *sine qua non* for pursuit of both.

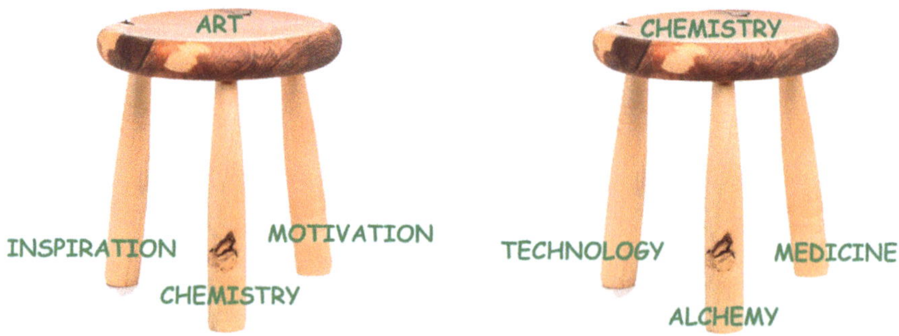

Fig. 10.1 Support Systems of Art and Chemistry. Graphic: Mary Virginia Orna

10.3 Roots of Chemistry

Often, chemistry is seen as a direct descendant of alchemy, but that is a very narrow view. Ancient technologies such as metallurgy, stone working, ceramics and glass manufacture fed into the development of chemistry independent of searches for the Philosopher's Stone. Medicine, initially based on the study of plant, animal and mineral compositions applied to human well-being, played a role in the progress of chemistry as well.

In addition, chemical science, from its very beginning, has interwoven relationships with many other disciplines such as philosophy, literature, poetry, the fine arts, physics, and even religion. The cultural journey of this discipline can be traced, in particular, by way of the works of engravers, designers and great artists who have illustrated, explained or interpreted the concepts, the theories and the experiments of alchemy, of chemistry and finally of modern science.

Every form of artistic expression has taken alchemy and chemistry as its subject matter, portraying their symbols and apparatus through painting, engraving and in the high reliefs that adorn many architectural structures. Works that represent the gradual evolution of alchemy right up to what, in the 1700s, becomes chemistry in the modern sense, that is, a rational and quantitative science, are boundless [4]. For example, alchemical and early chemical apparatus such as that displayed in Fig. 10.2 has been a rich font of inspiration for generations of artists.

10.4 From Smelting Furnace to Fine Art: Bronze Sculpture

When and where Stone Age craftsmen discovered metals is shrouded in mystery. Working earth and clay substances in a hearth undoubtedly would yield some intriguing new substances that were fluid and shiny and that hardened on cooling. These metals found in their native state were gold, silver, copper and, rarely, mercury. Gradually, metalworkers made the paradigm shift from working with native metals to extracting them from their ores. This discovery, called smelting, was an "intellectual and technical landmark in the history of the human race." [5] For example, at a mere 400 K (127 °C) and a limited supply of air, copper can be released from its ore, cuprite, Cu_2O, when heated in the presence of carbon according to the reaction sequence in Reaction Scheme 10.1. The scheme also shows the free energy (ΔG) of the reactions – a quantity that tells us about the tendency of the reaction to occur: the more negative ΔG, the more spontaneous the reaction.

As the temperature is raised beyond 400 K, the ΔG for both reactions A and B become more negative, but the ΔG for reaction A + B becomes decidedly more negative and therefore the reaction becomes thermodynamically more favorable. Smelting led to the availability of additional metals, chiefly lead, tin, iron and zinc, and the other metalworking arts, smithing, annealing, forging, alloying and casting, developed in parallel. Copper, alloyed with tin or lead, formed lower melting, more

Fig. 10.2 Alchemical Laboratory Apparatus. Alchemy Museum, Kutná Hora, Czech Republic. Photograph: Mary Virginia Orna

Reaction	Reaction Equation	ΔG (kJ/mol) 400 K	ΔG (kJ/mol) 800 K
A	$2C(s) + O_2(g) \rightarrow 2CO(g)$	-280	-350
B	$2Cu_2O(s) \rightarrow 4Cu(s) + O_2(g)$	+280	+250
A + B	$2C(s) + 2Cu_2O(s) \rightarrow 4Cu(s) + 2CO(g)$	0	-100

Reaction Scheme 10.1 Reaction of Copper Ore with Carbon at 400 K and 800 K

easily workable materials called bronzes. A typical bronze, a copper-tin alloy, was 10–12% tin with small percentages of other metals and nonmetals as impurities. The Bronze Age is estimated to have lasted from 3000 BCE to 500 CE, overlapping with our present Iron Age with its start date of about 1000 BCE.

While metallurgy was developing, the metallic arts were not far behind. By the thirteenth century BCE, the Bronze Age was at its technical and artistic height. Numerous highly sophisticated artifacts, both useful and decorative, entered the archaeological record [6]. And many of them continue to grace major museums

Fig. 10.3 Boxer at Rest, ca. 350 BCE. Museo nazionale romano, Palazzo Massimo alle Terme, Rome. Photographs by Mary Virginia Orna

around the world. One spectacular example is a life-sized bronze formed by the lost wax method at around 350 BCE (Fig. 10.3).

The "Boxer at Rest" depicts at once a figure both monumental and full of pathos. Possibly a slave, the expression on his face is a mixture of curiosity and resignation; a Roman entertainer of this sort never knew which contest would be his last. The very fine modeling, especially of his boxing "mitts" as seen in the right-hand detail, was probably done in a clay mold negative. Bronze has the curious property of expanding slightly while setting so that it fills in every tiny nook and cranny of the mold; then as it cools, it most cooperatively shrinks a little so that it is easy to extract the bronze from the mold [7].

Before the hearth gave way to smelting furnaces, it provided many an artist's material as long ago as 45 000 years, according to the archaeological record. Naturally occurring yellow ochre, FeO(OH), yielded up red ochre, Fe_2O_3, on heating. Both have been found in prehistoric cave paintings along with charcoal, C, formed from charring wood in a limited supply of air. The latter, synthesized almost exclusively from the tree *Pinus sylvestris* by the Ice Age painters of southern France (Fig. 10.4), remains to this day the most commonly used pigment [8], mainly because it is a waste byproduct from the production of coke [9].

Thermochemistry continued unabated in ancient Egypt, and with the development of the bellows, it was possible to heat furnaces to as high as 950 °C, allowing for the formation of the first known synthetic pigment, Egyptian Blue. We now know that the major ingredients in Egyptian Blue manufacture are copper (as the metal, alloyed

Fig. 10.4 The famous “Lion Panel” in the Grotte Chauvet, or Caverne du Pont d’Arc, southern France. Discovered in 1994, the charcoal-drawn images from this cave, dated to the Aurignacian Period (35,000–23,000 before present), are the best-preserved and the earliest known in the world

with other metals, or in oxidized form), silica in the form of sand (SiO_2), limestone ($CaCO_3$) and a flux, natron, which helps lower the melting point of the mix. The Egyptians had all four of these ingredients in good supply thanks to the copper mines in the Sinai, the fine sands on the shores of the Mediterranean, ubiquitous limestone, and the Wadi Natrun that supplied the eponymous natron, a mix of common sodium salts. Roman author and engineer, Vitruvius (d. 15 BCE), left us the recipe, but omitted a key ingredient, limestone, a common ploy to maintain secrecy in ancient recipes [10]. Even more mysterious, there is a total lack of written documentation on Egyptian Blue production, trade or usage among the ancient Egyptian papyri that have come down to us, although they cover every other conceivable subject [11]. The ancients knew how to keep a secret when it behooved them.

When the specified mixture is heated, as given above, to about 900 °C for a long period of time, preferably overnight, Egyptian Blue is formed with the established chemical formula of $CaCuSi_4O_{10}$, which corresponds to the relatively rare mineral, cuprorivaite, as shown in Reaction Scheme 10.2.

The formula was established by X-ray powder diffraction analysis [12, 13]. Delamare [14] provides the details for this remarkable work that experimentally linked ancient Egyptian Blue to modern cuprorivaite. Egyptian Blue remained the blue pigment of choice in the ancient world until the early medieval period. After that its use declined, possibly because of Vitruvius’s egregious omission although examples of its use have been documented to as late as the sixteenth century.

$$\underset{\text{Malachite}}{Cu_2CO_3(OH)_2(s)} + \underset{\text{Sand}}{8SiO_2(s)} + \underset{\text{Lime}}{2CaCO_3(s)} \xrightarrow[800\text{-}900\ ^\circ C]{\text{Flux}} \underset{\text{Egyptian Blue}}{2CaCuSi_4O_{10}(s)} + 3CO_2(g) + H_2O(g)$$

Reaction Scheme 10.2 Formation of Egyptian Blue

While metallurgy provided artists with free metals to work with, a bit more sophisticated chemistry provided them with pigments too. A side benefit from smelting that burgeoned into an indispensable technology and an art form in its own right arose out of simple sand and soda—we call it glass.

10.5 Glass: An Art Form that Changed the World

How glass manufacture came about is the stuff of legend, but may have more humble beginnings. Nature visits us from time to time with glass of her own making in the form of obsidian and pumice from volcanoes, fulgurites from lightning strikes and tektites from outer space. These curiosities may have intrigued ancient metal workers, but it is far more likely that the slag coming off from copper smelting, filled with sparkling glasslike material that could be fashioned into jewelry, captured more attention [15]. The archaeological record, at any rate, tells us that by 2500 BCE, glass objects and beads were abundant in Syria and Mesopotamia, and by 1500 BCE, it was a major industry in Egypt. This makes glass-working, apart from metallurgy and ceramics, the oldest technology known to us. This is a not so surprising fact given that glass is made up of four of the six most abundant elements in Earth's crust: silicon, oxygen, sodium and calcium.

It is also not surprising that Egypt was an early major glass-working center: fine Mediterranean sand, consisting mainly of silica (SiO_2), coupled with availability of natron, a mix of sodium salts from the nearby evaporitic lakes of the Wadi Natrun, combined conveniently in a civilization likely to take advantage of every opportunity afforded it by Nature. When silica is heated to beyond its melting point of 1710 °C, it becomes a workable fluid that can be shaped at will into various forms. However, such a high temperature was impossible to achieve in ancient furnaces and hearths, necessitating addition of a fluxing agent, like natron, that would lower the melting point to a reasonable degree, about 1000 °C (for a charcoal fire) and perhaps lower to about 650 °C (for an ordinary campfire). The glass formed by silica and soda is noncrystalline, but amorphous, with all the properties of a fluid but it is too viscous to flow. Since sodium ions, which are intimately mixed with the silica, are soluble in water, calcium ions, usually in the form of lime, CaO, are necessary to render the glass insoluble. Ancient glaziers were ignorant of this necessity but added lime unwittingly in the form of an impurity; intentional use of lime as a stabilizer probably did not come about until the Roman glassmaking period (fourth century BCE—ninth century CE) [16].

Color in glass is a paradox. It was rare for ancient glass to be colorless because of the ubiquitous presence of iron as iron(II), the 4th most abundant element. Decoloration was finally achieved in about 1400 BCE in Thebes by the following ingenious three-step process (Scheme 10.3.)

The naturally occurring mineral, stibnite, Sb_2S_3, is oxidized in a furnace to Sb_2O_3, which is further oxidized to antimony tetroxide, Sb_2O_4. Sb_2O_4 is a powerful oxidizing agent that can oxidize the blue-greenish FeO present in the glass mix to Fe_2O_3 which, although yellow, has much less molar absorbance than FeO and at low concentrations appears colorless in glass.

There were numerous compounds available to the ancient glazier to produce color (Table 10.1.).

$$2Sb_2S_3(s) + 9O_2(g) \rightarrow 2Sb_2O_3(s) + 6\ SO_2(g)$$

$$2Sb_2O_3(s) + O_2(g) \rightarrow 2Sb_2O_4(s)$$

$$Sb_2O_4(s) + 2FeO\ (s) \rightarrow Sb_2O_3(s) + Fe_2O_3(s)$$

Reaction Scheme 10.3 Decoloration of Ancient Glass

Table 10.1 Typical Colorants for Glass

Color	Metal Additive	Coloring Oxide
Yellow	Lead/Antimony, or Iron	$Pb_2Sb_2O_3$; Fe_2O_3
Orange	Chromium	CrO_3
Red	Copper and/or Lead	Cu_2O; Pb_3O_4
Purple	Manganese	Mn_2O_3
Blue	Cobalt or Copper	CoO; CuO
Blue-green	Iron	FeO
Green	Chromium	Cr_2O_3

Fig. 10.5 Left: Greek glass perfume bottle, fifth century BCE. Metropolitan Museum of Art. Right: Decanters containing aventurine glass threads, 2nd half of the 19th C. century. Glass museum, Murano

Glassmaking in Rome held sway for over 1300 years, but following the fall of the Roman Empire, Venice took over and perfected the art to the highest degree. The word "Venetian" or "Murano" applied to glass is the world benchmark for artistic glass (when genuine). Here are some examples (Fig. 10.5) from Rome and Venice [17].

In the medieval period, the open architecture of the Gothic cathedrals simply begged to be filled with art, and the stained glass workers gladly obliged. Their creations brought about a need for conservation of this precious heritage, which did not always "work" as we can see in the left-hand image in Fig. 10.6. As time went on, modern technology allowed for the creation of outdoor monumental art as well. (Fig. 10.6 right). Glass as a useful art form made life livable with the nonporous containers it afforded. Used as a glaze on ceramic materials, it would vitrify in the kiln, rendering whatever it covered nonporous as well. It is beyond the scope of this chapter to enlarge upon other ways that glass has changed our world, from eyeglasses, telescopes, solar cells, to all-glass modern buildings. However, it would be remiss not to mention that without the development of inert, shock and heat resistant, durable laboratory glassware, modern chemistry could not have developed to the extent that it did [18].

In the next section, we will see how the isolation of mineral acids led to a famed specialty pigment with a glorious history.

Fig. 10.6 Left: A "restored" stained glass window in Cartmel Abbey, Cumbria. The pieces don't fit because they were found outside on the ground after a major metal failure on the part of the lead placeholders. Right: A monumental work by the glass sculptor, Dale Chihuly. New York Botanical Garden. Both photographs: Mary Virginia Orna

10.6 Art and Chemistry Hand in Hand. From Ancient Assyria to the Nobel Prize: Purple of Cassius

Ancient physicians, hidden under the cloak of alchemy, perfected the art of fermentation and distillation leading to the isolation of alcohol [19, 20]. It was a short step from there to vinegar, probably the first acidic substance ever isolated. In warmer climes, citric acid undoubtedly followed.

Over the centuries, proto-chemists learned that heating various minerals could produce different "airs," what we now call gases. If they mixed these airs with water, liquids similar to vinegar, but far more powerful, were produced. These acidic substances, derived from minerals, were the mineral acids [21]. Shown in Reaction Scheme 10.4 is a hypothetical sequence that likely led to the trial and error production of the major mineral acids, H_2SO_4, HCl and HNO_3.

Knowledge of how to manipulate acids, and by the same token, bases, led to the synthesis of simple and complex salts, some of which became staples of the artist's

Reaction Scheme 10.4 Hypothetical Origins of the Major Mineral Acids

$$FeSO_4 \cdot 7H_2O\,(s) \rightarrow FeSO_4(s) + 7H_2O(g)$$

$$2FeSO_4(s) \rightarrow Fe_2O_3(s) + SO_2(g) + SO_3(g)$$

$$SO_2(g) + H_2O(l) \rightarrow H_2SO_3(aq)$$

$$SO_3(g) + H_2O(l) \rightarrow H_2SO_4(aq)$$

$$H_2SO_4(aq) + 2NaCl(s) \rightarrow Na_2SO_4(aq) + 2HCl(aq)$$

$$H_2SO_4(aq) + 2NaNO_3(s) \rightarrow Na_2SO_4(aq) + 2HNO_3(aq)$$

palette for many centuries. Among them are lead white, $(PbCO_3)_2{\cdot}Pb(OH)_2$, and blue-green verdigris, $Cu(C_2H_3O_2)_2{\cdot}2Cu(OH)_2$.

The latter two mineral acids in Reaction Scheme 10.4, nitric and hydrochloric, when combined, constitute *aqua regia*, Royal water, a mixture capable of dissolving even recalcitrant gold to form a colloidal gold pigment sometimes referred to as Purple of Cassius. The earliest attribution of this substance is to the ancient Assyrians (2500–600 BCE), which presumes that they also knew how to make *aqua regia* [22] although no analytical information exists to back up this claim [23]. In fact, a 2009 study found that the ruby red glass of Nimrud, Assyria, was actually due to dendritic copper and Cu_2O [24]. So we have to look elsewhere for *aqua regia*, most likely in the work of the alchemist, False Geber (ca. 1300 CE), *De inventione veritatis* (On the discovery of truth) [25].

Once the existence of *aqua regia* is established, we can look further for its use in dissolving gold to form Purple of Cassius. As early as 1597, Andreas Libavius (1555–1616) was aware of the beautiful red color that could be obtained from gold dissolved in *aqua regia*, and in 1660, Johann Glauber (1604–1670) discovered an effective and economic way of manufacturing the red color. Although Glauber himself never availed himself of his find, another distinguished chemist of the time, Johann Kunckel (1630–1703) produced large quantities of it over a period of about ten years (1679–1689) at the Potsdam Glass Factory. Many samples survive to this day. See Fig. 10.7. However, Andreas Cassius the Elder (1605–1673) of Leyden, by publishing the first workable recipe for its production in 1685, seems to get all the credit [26].

The nature of Purple of Cassius continued to mystify chemists for centuries. In 1797, Jeremias Benjamin Richter (1762–1807), famous for propounding the law of definite proportions, dismissed it as nothing but finely divided gold mixed with tin calx [27]. In 1806, Joseph Louis Proust (1754–1826), who discovered the law of constant composition, thought so too [28]. In 1857, Michael Faraday (1791–1867) hazarded the guess that Purple of Cassius was essentially finely divided gold associated with more or less of oxide of tin [29]. Not until the early part of the twentieth century did the Viennese chemist, Richard Zsigmondy (1865–1925) show conclusively that it consisted of very finely divided gold with colloidal stannic acid and that a mixture of colloidal substances was capable of behaving like a chemical compound [30].

Zsigmondy tells us that he "was dealing with gold orpiment glass and with certain ceramic dyes which were based in principle on extremely fine gold divisions. It seemed to me particularly remarkable that the effect of very closely allied chemical compounds could be fundamentally different in regard to the preparation and appearance of these dyes. This observation seemed to me all the stranger as compounds with completely opposite properties frequently produce similar effects on the dyes of extremely finely divided (i.e. colloidal) gold. These results could not be brought into harmony with the previous experience of chemistry. Consequently, I sought to discover the reason for these phenomena in a way other than the purely chemical one. The first special object of such investigations was the *purple of Cassius*." [31]

Some outstanding chemists, like J. J. Berzelius (1779–1848), regarded this mysterious substance as a true chemical compound. Others felt that it was a mixture of

Fig. 10.7 Gold ruby glass. Schatzkammer Wittelsbacher, München Residenz. The piece on the right is attributed to Johann Kunckel

finely divided gold and stannic acid. Zsigmondy's goal was to find out which of these views was correct. Some scientists regarded it as a true solution; others said that it was a suspension. It did not behave like a suspension by eventually settling out; osmotic pressure determinations were inconclusive regarding its nature as a true solution. Zsigmondy realized that he was dealing with a substance somewhere in the never-never land between solutions and suspensions called the colloidal state. To investigate, he had to devise new instrumentation like an ultramicroscope so that he could measure and count individual particles. Using this instrument, he found that by counting the number of colloidal gold particles in a small volume of a dilute solution of known concentration, he could calculate the average size and mass of individual particles. This direct observation of solute particles marked a turning point in chemistry, eventually convincing the most intransigent of chemists, including Nobel laureate Wilhelm Ostwald (1853–1932), of the reality of molecules and the fundamental discontinuity of matter. This methodology made it possible to study proteins and other biological macromolecules, thus launching the new discipline of molecular biology. For this classic work, Zsigmondy received the Nobel Prize in Chemistry in 1925 [32].

While not every art and chemistry related investigation eventually led to a Nobel Prize, there are enough examples to show that their marriage produced many more outstanding and notable offspring. What follows are several additional "trajectories" that illustrate the partnership between chemistry and art.

10.7 Clever Greek and Etruscan Ceramicists

Attic black-figured and red-figured vases and other forms of pottery are a staple in every museum that displays Greek and Roman artifacts [33]. The rich black satiny surface of astonishing durability is its most outstanding feature [34]. However, how the ancient Greek potters, and by extension, the Roman and Etruscan ceramicists who mastered the same technique, achieved this art form has been the subject of research right into the mid-twentieth century. Truly one of the lost arts of the ancients, it has been an ongoing technical mystery for a very long time [35].

Early investigators, most notably the French mineralogist and Director of the famous porcelain manufactory at Sèvres, Alexandre Brongniart (1770–1847), posited two different pigments, red iron oxide and black manganese oxide, with not much more evidence than his own eyes. "[T]he coloring principles of the black luster, [consist] of oxide of iron and oxide of manganese; these are the essential elements."[36] Unfortunately, this gratuitous error gained a firm place in the literature, but was challenged about fifty years later by Émile Durand-Gréville (1838–1914) who claimed: "If we heat out of contact with air - for example in a crucible full of charcoal - a shard of Greek vase covered with black paint, this paint retains its color; but if the fragment were strongly heated in the presence of air, its black color would overoxidize and. would turn red in the form of iron sesquioxide, which is colcothar, hematite." [37] Demonstrating that the Greeks achieved iron with two different oxidation states on the same vessel was left to the work of Charles Binns and A. D. Frazer, ceramicists at Alfred University who re-discovered the three-stage oxidation–reduction-oxidation scheme of the ancient Greeks [38].

But the key to *how* to realize the red and black on the same piece of pottery still eluded investigators. In 1943, the turning point arrived when the ceramic chemist, Theodore Schumann, re-discovered the process [39]. Schumann realized that the technical problem of Attic pottery was twofold: the secret of the colors and their relationship to the black glossy surface that was initially called a black "glaze." But the Greek "glaze" was different from ordinary glaze, a vitreous coating that fused with the clay body upon firing in a kiln: it did not fuse in spite of the thinness of its application. Schumann, by experiment, found that the "glaze," which should really be called a "gloss," consisted of very finely divided clay particles held in a jelly-like suspension characteristic of a colloid [40]. Part of the ancient potters' relatively sophisticated chemical technology was controlled additions of potassia (K_2O), iron oxide and tannins to obtain desired gloss performance [41]. The tannins, derivatives of gallic acid, $C_6H_2(OH)_3COOH$, would tend to hydrogen bond or esterify with the surface OH groups on the clay, thus holding the platelets of clay minerals in suspension, allowing them to be physically separated from the rest of the clay body [42]. Further, analysis showed that there was no element found in the Greek black "gloss" that was not also found in the iron-rich clay body. Joseph V. Noble (1920–2007), a self-trained ceramic archaeologist, described how he reproduced the classic black "gloss" during his tenure on the administrative team of the Metropolitan Museum of Art [43]. It was a very painstaking process that took several weeks. The end result was a dark-brown to black liquid with the consistency of heavy cream; this concoction is also often called a "slip" as well as "gloss." It is necessary for the clay particles in the slip to be very fine with an average size of about a few tenths of a micrometer. Later studies showed that the black gloss composition was quite different from that of the clay body, containing only a third of Al_2O_3, twice as much Fe_3O_4 and three times as much K_2O. It is the K_2O that confers the "sintering" property, i.e., little or no molten liquid at the firing temperature, on the gloss, allowing it to be the agent that protects the black clay body underneath it from re-oxidizing back to red in the third firing stage. The ceramic piece itself is made of an iron-rich clay body in the normal way, usually by throwing on a potter's wheel. It is then allowed to dry to the consistency of leather and at this time the slip, or gloss, is brushed onto an outlined surface to cover the area that is destined to remain black (black-figure technique). Once painted, the piece is fired in a three-stage firing process. Stage 1 takes place in an oxygen-rich atmosphere in a wood- or charcoal-burning kiln to about 850 °C, turning the entire vessel the color of the red clay, i.e., of Fe_2O_3. For Stage 2, the temperature is raised to about 925 °C, and while being held at this temperature, the kiln is converted to a reducing atmosphere by shutting off the outside oxygen supply and introducing green wood or wet sawdust in order to generate carbon monoxide, CO. CO will suck the oxygen out of its entire surroundings, thus reducing the Fe_2O_3 to black ferrous oxide, FeO, and if there is moisture in the furnace, formation of black Fe_3O_4, magnetite, will form as well according to Reaction Scheme 10.5.

At this stage, the entire vessel is black. In Stage 3, the furnace is allowed to begin cooling down slowly while a small opening is made, permitting oxygen to re-enter. The oxygen reacts with both the FeO and Fe_3O_4, turning them both into red Fe_2O_3 again *except* the black glazed area which has become such a dense sintered (vitrified) mass that it "protects" the black surface underneath from re-oxidizing by not permitting contact with the newly introduced oxygen. Once cooled, the gloss area remains irreversibly black. To reverse the whole process and make red-figured pottery, the background to the figures would have to be protected with the gloss. While this technique was purely empirical for the ancient ceramicists, there was most certainly a

$$Fe_2O_3(s) + CO(g) \rightarrow 2FeO(s) + CO_2(g)$$

$$3Fe_2O_3(s) + CO(g) \rightarrow 2Fe_3O_4(s) + CO_2(g)$$

Reaction Scheme 10.5 Stage 2 (reduction stage) of the firing process for attic black ceramics

Fig. 10.8 Left: Red-Figure Amphora with Musical Scene, Greek, 460–450 BCE; Walters Art Gallery. Right: Black-Figure Pottery, Etruscan, fifth century BCE; Museo Villa Giulia

tradition that was handed down over the centuries [44]. Recent research has revealed the necessity for the "right" type of clay to form the desired properties of the gloss. Illite, from among the three major minerals associated with clay, kaolinite, illite, and montmorillonite, is the best choice to produce the submicroscopic particles that form the best slip (gloss).The geographic distribution of black-figure pottery manufactories reflects this geological reality: illitic deposits abound in Attica, Corinth and Etruria, major centers of production in the ancient world [45]. Figure 10.8 displays some examples of this ancient technique.

The black-figure is really quite striking and was a very desirable color for ancient pottery. The Etruscans were quite skillful at producing a renowned all-black type of ceramic called "bucchero." They made it by carrying out only the first two stages of the firing process described above. The dense blackness of Fe_3O_4 in all of these pottery types is due to the fact that it contains iron in two different oxidation states, Fe(II) and Fe(III). Consequently, charge transfer takes place between the two valence states with such a high probability that almost the entire visible spectrum is absorbed by multiple electron transfers—almost eight per formula unit [46, 47].

This technology is one of the prime examples of how modern chemical analysis can elucidate and even replicate a long-lost art form, one that was systematically developed over the course of centuries. In the next section, we will see some examples of precisely the opposite approach: accidental discovery.

10.8 Some Serendipitous Colors

10.8.1 Accident # 1: Drebbel's Scarlet

Dutchman Cornelis Drebbel (1572–1633), a contemporary and likely acquaintance of William Shakespeare (1564–1616), Ben Jonson (1572–1637) and Peter Paul Rubens (1577–1640), erupted into the court of England's James I (1566–1625) in 1605. His claim to fame was a series of ingenious inventions, all of which now lie largely forgotten, overshadowed by his one accidental discovery, Drebbel's Scarlet. Though it was a new color that took the European world by storm, it was the product of some sloppy alchemy that seemed to pervade the practice if contemporaneous paintings are accurate. (Fig. 10.9).

Fig. 10.9 Thomas Wijck (1616–1677), Interior with an Alchemist. Wellcome Collection. An accident waiting to happen

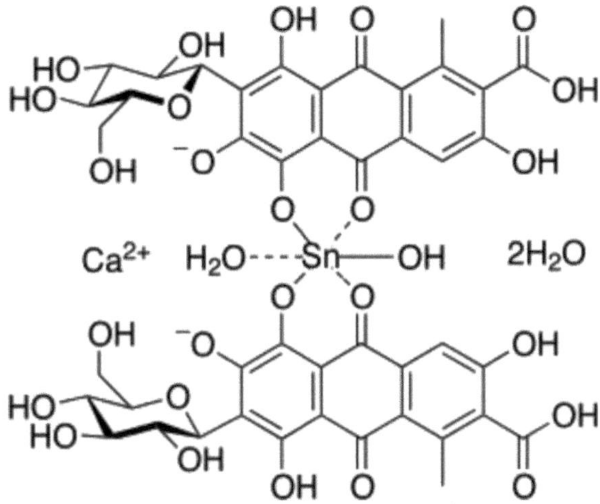

Fig. 10.10 To form a chelate, which means a claw-shaped complex with the tin, the C=O and the O–H in each of the two molecules of carminic acid approach the tin ion (Sn^{2+}) like a pincer movement and form bonds. The new compound is brighter and redder than carminic acid alone

The oft-cited story has it that Drebbel placed some cochineal red dye extracted from a New World scale insect on his windowsill intending to use it to make a thermometer. But somehow, a falling tube of *aqua regia* (a mix of nitric and hydrochloric acids) spilled all over the tin window frame and then dripped into the dye. To Drebbel's surprise (and delight), the dye intensified into a remarkable fiery scarlet. Further inquiry revealed that it was the presence of the dissolved tin that bestowed the color [48]. Armed with such knowledge, Drebbel consciously added tin to his subsequent formulations of cochineal, thus initiating the age of scientific dyeing [49]. Chemists now know that anthraquinoid dyes such as cochineal will form chelate complexes with many metals such as tin and copper, thus changing their physical and chemical properties. Figure 10.10 is a structural representation of the complex formed by two molecules of carminic acid with tin, presumably the formula for Drebbel's Scarlet.

Drebbel's Scarlet went on to worldwide fame as the dye that colored the British Redcoats. However, the recipe that had developed in rural England had very stringent requirements: in addition to the expected cochineal and tin liquor, stale urine from a beer drinker was reputed to give the best scarlet color [50]. In addition to this stale urine, the recipe called for wine dregs after fermentation and cutbear, a form of lichen found on sea rocks. Not exactly a convenient process for the contemporary tie-dyeing workshop [51].

10.8.2 Accident # 2: Prussian Blue

Every pigment under the sun has a certain unique "personality," but until Prussian Blue came on the scene, it was always possible to classify each as either organic or inorganic. For a long time, this new and unique pigment straddled the line as a one-off reality. The problem began with its unusual mode of synthesis. According to the given legend, one day somewhere between 1704 and 1707, a bumbling Berlin color maker, Johann Jacob Diesbach, was trying to make his signature red color, Florence Lake, but ran out of an ingredient. His recipe called for alum, cochineal and green vitriol (ferrous sulfate, $FeSO_4$) mixed in an alkaline medium. Out of alkali, his laboratory partner, alchemist Johann Konrad Dippel (1673–1734) supplied him with some waste potash, which turned out to be tainted with animal matter. This made all the difference: when Diesbach added the potash, the mix developed a gorgeous deep blue color. This was a turning point for both men: they realized that they had a fortune bubbling away in the pot. Diesbach went into business and for almost 20 years maintained the secrecy of the recipe. Meanwhile, Dippel skipped out to the Netherlands and marketed what he called "Prussian Blue" there [52].

The actual chemistry behind the discovery took centuries to resolve; the niggling problem was the unknown role of the animal material (flesh or blood) in the synthesis. French chemist Joseph Louis Gay-Lussac (1778–1850) found that the pigment released hydrogen cyanide, HCN, upon heating, but the true nature of the compound remained undetermined until 1970. The accepted formula is now $Fe_4[Fe(CN)_6]_3 \cdot xH_2O$, named iron(III) hexacyanoferrate(II). The pigment is a mixed-valence compound, containing iron in two valence states, Fe(II) and Fe(III) which accounts for its intense blue color via the charge transfer mechanism, as discussed in Sect. 10.7 (above) for magnetite. And it is clearly inorganic—the requisite cyanide, originally coming from animal matter, can be supplied by many mineral sources.

Prussian Blue, which also eventually went under other names such as Berlin Blue, Milori Blue, and iron blue, was an instant success. Artists and colorists immediately added it to their palettes. By the mid-eighteenth century, it had become the blue pigment of choice for oil painting. It was present in the paint box of Jean-Honoré Fragonard (1732–1806) in two different formulations; other noted artists such as Thomas Gainsborough (1727–1788), Vincent Van Gogh (1853–1890) and the noted Japanese artist Katsushika Hokusai (1760–1849). (Fig. 10.11) It also found its way into household items and laundry blueing. Because it is also photosensitive, it is the blue pigment in blueprints that changed the way architecture was done [53, 54].

Because of its open crystal lattice, Prussian Blue can harbor large singly-positive ions like toxic thallium and cesium, making it an excellent sequestering agent for this type of poison. Pharmacies market it for this purpose. It has justly been dubbed "a last gift from alchemy," which it truly is.

At the turn of the twentieth century it was the most commonly used blue pigment; fifty thousand metric tons were in demand annually. It was displaced as the world's workhorse pigment only recently by phthalocyanine blue, which is the topic of our next accidental discovery [55].

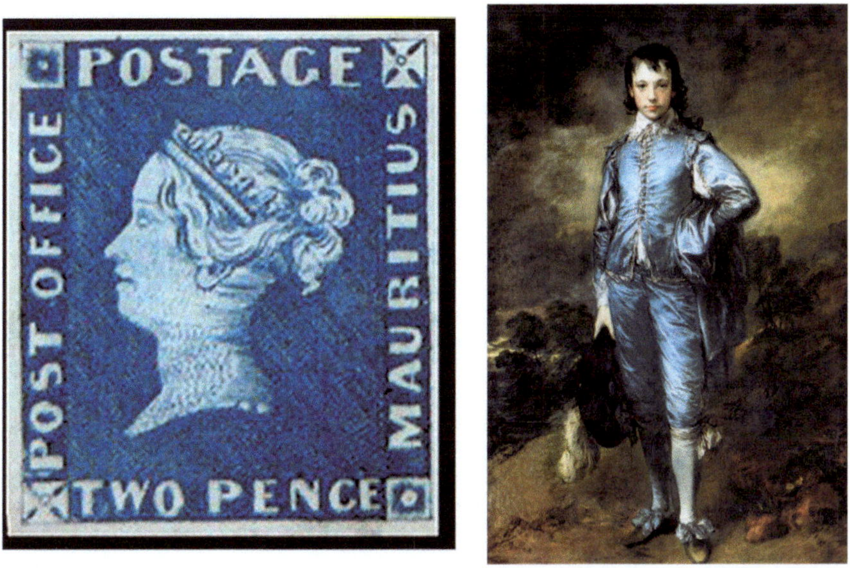

Fig. 10.11 Left. Mauritius two pence stamp printed in Prussian Blue, 1847; it is one of the rarest postage stamps in the world. Right. Thomas Gainsborough, The Blue Boy (1870), oil on canvas; Henry E. Huntington Art Gallery, San Marino, CA. Gainsborough deliberately highlighted the color as Prussian Blue in response to a criticism of Sir Joshua Reynolds (1723–1792)

10.8.3 Accident # 3. Phthalocyanine Blue

Although “sightings” of what came to be called phthalocyanines happened as early as 1907 when chemists were heating *o*-dinitriles (compounds with two –C≡N groups on adjacent carbon atoms of a benzene ring) and similar compounds in the presence of metals, the happy accident that spawned phthalocyanine blue did not come about until 1935. Eight years earlier, a group of chemists working at the British firm, Imperial Chemical Industries (ICI), found a blue-green substance in a reactor that should have contained only colorless material. The anomaly was due to a chip in the reactor’s lining that exposed the chemicals to the underlying iron. The group seized upon this gift from Dame Fortune and were soon making a bluish colorant—but they noticed that the color deepened into an intense blue if they substituted copper for the iron, producing copper phthalocyanine. This was the object of their 1935 patent.

On further study, the chemists found that copper phthalocyanine was very easy to make: just heat phthalonitrile in the presence of elemental copper to 200 °C and eureka! (Fig. 10.12, left) As astounding as its discovery was, the determination of its structural formula was even more astounding. It turned out to be a compound called a macrocycle, one of the strongest and most important types of colored compounds Mother Nature ever devised. Copper phthalocyanine turned out to be in the good company of its close relatives, heme and chlorophyll, two energy engines without which life would probably not exist. (Fig. 10.12, center and right respectively).

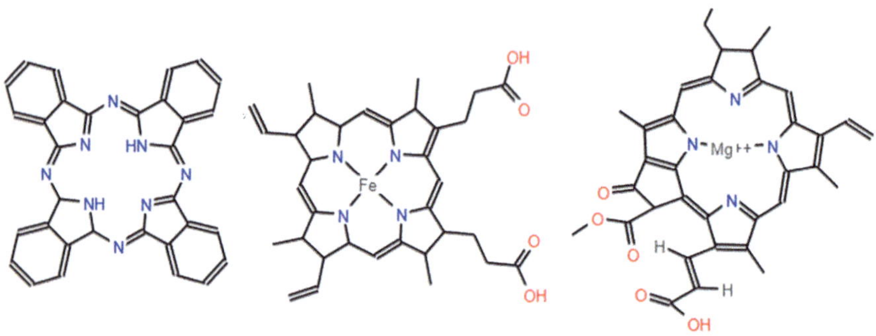

Fig. 10.12 Three macrocyclic compounds. Phthalocyanine is the parent compound for some natural macrocyclics like heme in blood and chlorophyll in green plants

Copper phthalocyanine has almost completely dominated the blues market wherever a pigment is needed—in printing, paints, inks, plastics, cements, and wall and floor coverings [56]. It is also the primary blue pigment used in artists' materials as well [57].

From this accidental discovery, a whole series of colored pigments could be made. With copper in the center of the macrocycle, the pigment is blue. Substituting nickel or cobalt for the copper yields a turquoise and violet pigment respectively. Figure 10.21 (A) shows the basic structure of phthalocyanine blue (Pigment Blue 15); a derivative on the right, (B), with ten substituted chlorines, gives a green color and is marketed under the name of phthalocyanine green (Pigment Green 36).

10.8.4 Accident # 4. YInMn Blue

A serendipitous discovery at the University of Oregon gave rise to the twenty-first century's first blue pigment. While working with the white oxide $YInO_3$ and the black oxide $YMnO_3$, a research team headed by Mas Subramanian wondered if a mixed oxide would exhibit properties that might be useful in the electronics industry. They started by adding tiny bits of Mn^{3+}, a process called doping, to $YInO_3$, making oxides with the formula $YIn_{1-x}Mn_xO_3$, and much to their surprise, they created an intensely blue compound even when x was as low as 0.05 mol, and the intense blue color persisted over much of the solid solution range. The team concluded that the blue color was due to crystal field splitting of the manganese in a trigonal bipyramidal coordination with short apical Mn—O bonds [58]. The authors were delighted to have discovered and contributed to the development of inexpensive, earth-abundant, environmentally benign blue pigments based on manganese [59].

Discovered in 2009, it took almost a decade to bring it to market as a paint, but its claim to fame is that it is the first new blue pigment to grace the artists' palette in 200 years. Its brilliant intense blue color has inspired Crayola™ crayons to debut a new shade to its collection. Appropriately enough, they are calling it "bluetiful." [60]

Though Subramanian's new pigment was not suitable for the electronics industry, it was the very inception of this industry at the end of the eighteenth century that led to a movement that changed the way that pigments were used and art was viewed. Read on.

10.9 How an Italian Spawned Impressionism: The Voltaic Pile Leads to *Plein Air* Art

In 1799, at the height of the disruption of the Napoleonic wars across all of Italy, Alessandro Volta (1745–1827) was hard at work on his research on galvanism at the University of Pavia. Two years later, he was invited to demonstrate his "pila" to the Emperor Napoléon at the Académie des Sciences at Paris. Consisting of 88 pairs of alternating copper and zinc discs separated interstitially with brine-soaked cardboard, his pile's wire leads, when properly connected, gave off great sparks, melted a steel wire, shot off a pistol, and decomposed water by electrolysis. Napoléon was duly impressed, but Volta's greatest triumph was the overthrow of Luigi Galvani's (1737–1798) hypothetical "animal electricity." [61]

Within a year of this event, Humphry Davy (1778–1819), working at the Royal Institution in London, availed himself of the "pile," the first ever continuous and dependable source of electricity. (Fig. 10.13).

Building more and more powerful batteries, Davy failed in his aim of isolating the components of alkali metal compounds from their aqueous solutions until he abandoned solutions altogether and simply electrolyzed the damp solid compound instead. Success came when, electrolyzing solid KOH, he produced mercury-like globules of potassium that immediately burst into flame. Isolation of sodium from NaOH quickly followed [62]. Davy's 1807 Bakerian Lecture, that encompasses the discovery and isolation of potassium, sodium, barium, calcium, strontium and magnesium, published in 1808, marks the beginning of "fused salt" electrolysis [63]. This not-so-humble beginning eventually led to the isolation on an industrial scale of lead, tin, and hard-to-get aluminum.

What does the isolation of these metals have to do with fine art apart from the likes of sculptors in fabricated aluminum like Roy Lichtenstein (1923–1997)? In the eighteenth century, it was common to find pigment vendors advertising "bladder colors" for which practicing artists needed no explanation. Bladders, strong, flexible and watertight, are the anatomical containers par excellence for liquids; pig bladders, a byproduct of many a slaughterhouse, had some very useful applications. Filled with beans or peas, they made great rattles for infants. Inflated, they were as good as any polyethylene ball sold in today's toy stores. Uninflated, they were the ideal material

Fig. 10.13 Left: Model of Alessandro Volta's "Pila." Photograph by Luigi Chiesa; Right: Sir Humphry Davy. Detail of an oil painting after Sir Thomas Lawrence; in the National Portrait Gallery, London

to contain artists' pigments. Vendors would cut pieces of bladder, place their ready-made pigments ground in oil media on top, gather the corners and secure the whole with a clasp or twine. Back in the studio, artists would access the colors by piercing the bladder with a tack, squeezing out the desired amount of paint, and replacing the tack firmly. Some pig's bladders have been shown to keep the paints fresh for decades. Studio work was one thing, but try to imagine an artist carting a collection of pigs' bladders along with all the other gear needed to do a day's work outdoors. Oddly enough, it was an American portrait painter, John G. Rand (1801–1873) who solved this problem in 1841. His invention—a collapsible tin tube (later replaced by aluminum) with a screw-cap closure—was both groundbreaking and historic [64]. The artist could now dispense with bulky, awkward packaging that employed both hands for manipulation. From then on, painting was a breeze—with a squeeze. But it never would have happened unless there were a ready and abundant supply of the appropriate metals from which to manufacture the tubes.

This development was so important that Jean Renoir (1894–1979) quotes his father, Pierre-Auguste (1841–1919) as saying: "Without paints in tubes, there would have been no Cézanne, no Monet, no Sisley or Pissarro, nothing of what the journalists were later to call Impressionism." [65]

Electrochemical deposition of metals is not the only contribution of electrochemistry to the arts. Electrosynthesis of pigments is a viable and efficient pathway and is employed routinely on an industrial scale, particularly to synthesize workhorse pigments like the phthalocyanines [66]. Since electrosynthesis enables several steps to be combined into a one-pot reaction in many cases, it is a "green" method for obtaining purer starting materials for pigment synthesis, resulting in pigments with a higher color purity [67].

An exciting new electrochemical method for producing environmentally friendly photonic pigments over the entire color range of the visible spectrum has been developed by a team working at Tianjin University in China. The process involves controlling the polymer chain stretching of the constituent block copolymers via an oxidation reaction. Photonic pigments produce color by constructive interference of light waves; they are useful colorants that could replace traditional dyes and pigments in some applications [68]. A 2021 special exhibit at the American Museum of Natural History, New York City, called "The Nature of Color," invited visitors to "paint" without the mess just by moving their hands in a floor-to-ceiling color play interactive, certainly a fine public display of photonic colors!

While electrochemistry was an ideal method for isolating new elements, element discovery was proceeding at a rapid rate using various other techniques. During the eighteenth and nineteenth centuries, fractional crystallization was the preferred technique followed, after the invention of the spectroscope in 1851, by spectroscopic detection and confirmation. Discovery of several new elements during this period gave rise to a flowering of new inorganic artists' pigments as described in the next section.

10.10 From Arsenic to Titanium: A Panorama of New Elements and Their Pigments

10.10.1 Arsenic

Volcanic ore and hydrothermal deposits were a great source of artists' pigments from ancient times, and one of the most used, but most overlooked in modern times, is the ruby red pigment, realgar. Its name derives from the Arabic, Rahj al ghār, i.e., "powder of the mines." Curiously, its molecular formula is As_4S_4, causing some chemists to scratch their heads, knowing that arsenic is trivalent or pentavalent in its compounds. However, because arsenic is a metalloid, it can form covalent bonds with itself, and in realgar, four arsenic atoms are covalently bonded to each other with an overall charge of 8+, so it is reasonable for the four sulfur atoms, each with a charge of 2−, to give us electrically neutral As_4S_4. This makes much more sense if we look at its structural formula in Fig. 10.14.

Historically, realgar has been used as a depilatory, a cosmetic, a medicine and a pigment; in none of those roles has it ever been considered benign. It was a highly valued ingredient of paints, pigments and dyes.

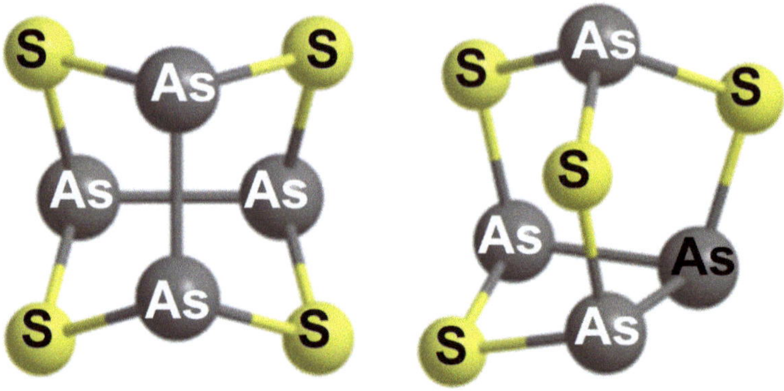

Fig. 10.14 The two forms of As_4S_4 Realgar, left, is bright red; Pararealgar, right, its polymorph, is yellow-orange. Figure courtesy of Liliana Mammino

Although arsenic compounds like realgar have played a role in human culture from prehistoric times, the element itself was only "discovered" in 1250 by the German Dominican friar and alchemist, Albertus Magnus (ca. 1206–1280).

Realgar's first cousin, As_2S_3, called orpiment ("golden pigment") by the ancients, was also a staple of the prehistoric palette. But once identified, modern chemists began to "tinker" with the properties of arsenic and came up with two coveted brilliant green pigments, Scheele's green in 1775 and Emerald green in 1814. Both were highly toxic compounds of arsenic. Carl Wilhelm Scheele (1742–1786), a German-Swedish pharmacist and chemist, synthesized his eponymous green in a two-step process by slowly adding arsenious oxide to an aqueous solution of sodium carbonate to form sodium arsenite, which was then added to a solution of copper sulfate. The resulting precipitate was $CuHAsO_3$; Scheele did not publish the procedure until three years later. While he acknowledged the pigment's toxicity, he publicized its existence because he wanted priority for its discovery [69]. Unfortunately, it was soon used for coloring wallpaper, artists' pigments, dyeing textiles and (horrors!) as food coloring and in children's toys. It was a favorite green of Edouard Manet (1832–1883) and J. M. W. Turner (1775–1851). Mercifully, it gradually fell out of use because it tended to blacken over time due to its copper content, but not soon enough to avert a highly publicized case of arsenic poisoning. Over a period of about a year in the mid-1950s, U.S. Ambassador to Italy, Clare Boothe Luce (1903–1987), suffered from a mysterious illness. It began when the apartment above hers installed a washing machine. When, eventually, she was diagnosed with arsenic poisoning, she surmised publicly that it was probably due to vibrations from the laundry overhead shaking small flakes of the green arsenic-containing ceiling paint onto her bed and into her food [70].

Emerald green (sometimes called Schweinfurt green or Paris green) made its commercial appearance in 1814. It is copper aceto-arsenite, $Cu(C_2H_3O_2)_2{\cdot}3Cu(AsO_2)_2$, prepared in a variety of ways with starting materials of copper, acetic acid, sodium carbonate and arsenic trioxide. It was the brainchild of Wilhelm Sattler (1784–1859) in an attempt to improve upon Scheele's green. Although it was not much used in easel painting, it has been identified as the green pigment used for making an imitation patina for repairing ancient Chinese bronzes [71]. The toxic history of both pigments sent their use into decline.

10.10.2 Cobalt

Although Swedish chemist Georg Brandt (1694–1768) gets the credit for discovering cobalt in 1732–1735, glassy cobalt oxide, CoO, has been in use in ceramics since the fifteenth century CE [72]. However, the ancient Egyptians got the jump on this and many other resources long before they were "discovered." From at least 1480 BCE and continuously until about 1300 BCE, they were using cobalt blue in glass artifacts and cobalt blue pigment, $CoO{\cdot}Al_2O_3$, for painting on faience [73, 74].

It took a while for chemists to get their act together to exploit this promising new "discovery," but by 1802, thanks to Louis Jacques Thénard (1777–1857), it was back on the artists' palette as "Thénard's Blue,", to be followed before 1805 by cerulean blue, $CoO{\cdot}SnO_2$, or Höpfner's blue, in Germany [75]. By 1835, a new green cobalt pigment, Rinmann's green, was on the market. It was discovered by Swedish chemist Sven Rinman (1720–1792) in 1780 as essentially cobalt aluminate in which zinc oxide, ZnO, has replaced some of the aluminum oxide and formed a solid solution with what remained. It was largely unknown until about the mid-nineteenth century because that was when zinc oxide became available in industrial quantities. It is sometimes referred to as zinc green. Colorist A. H. Church refers to it as "one of the too-rare pigments which is at once chemically and artistically perfect." [76]

The next cobalt pigment, cobalt yellow or aureolin, appeared in 1848. A complex material, potassium cobaltinitrite, $CoK_3(NO_2)_6{\cdot}H_2O$, it was discovered by Nikolaus Wolfgang Fischer (1782–1850) in Breslau, Germany (presently Wrocław, Poland) and introduced as an artists' pigment in 1860 [77]. Because of its great expense, it has not found much use as an artists' pigment beyond occasional watercolor use. Winslow Homer (1836–1910) worked with it often in his earliest opaque paintings of the 1870s such as his 1878 work, *The Green Hill*, where he applied it as a glaze to create the color of sunshine on a grassy bank (Fig. 10.15). Curiously, it disappeared from his works in 1881 only to be rediscovered in his later works in 1890. Homer favored it because of its pleasant hue and versatile handling properties [78].

Lest you thought we were finished with the cobalt pigments, you have not seen the entry in *The Pigment Compendium* labeled "Cobalt Phosphates Group" and those that follow up to "Cobalt Violet." [79] These pigments yield colors largely in the blue-violet range, but some are pink. They are double salts of cobalt including cobalt aluminum phosphate, cobalt ammonium phosphate, cobalt magnesium phosphate and cobalt zinc phosphate. Including the cobalt compounds under "Cobalt Violet," we can add cobalt magnesium arsenate, anhydrous cobalt phosphate, cobalt phosphate octahydrate, ammonium cobalt phosphate hydrate and cobalt lithium phosphate.

Fig. 10.15 Winslow Homer. The Green Hill, 1878. Watercolor, gouache, and graphite on gray-green paper faded to brown. National Gallery of Art, Washington, DC

Cobalt violet only begins to appear in pigment terminology in 1890, but pigments answering to this name were identified in paintings as early as the mid-nineteenth century.

This rich array of new additions to the artists' palette is only in mid-flight. We have yet to hear from titanium, chromium and cadmium, elements discovered in 1791, 1797 and 1817 respectively.

10.10.3 Titanium

About 1920, a new white inorganic pigment, titanium white (titanium dioxide, TiO_2, or any white pigment containing a compound of titanium) became available as an artists' pigment. Titanium had to wait almost 130 years to come into its own as a contributor to the artists' palette. Why this is so was due to the many recalcitrant impurities present in titanium's principal ores, rutile and anatase. The element itself was discovered by the Cornish pastor, Rev. William Gregor (1761–1817) in 1791 and named in 1795 by Martin Heinrich Klaproth (1743–1817), who recognized it in its oxide. But it was not until the chemical technology of the early twentieth century devised a viable process to resolve the problem of titanium ore's impurities that commercial exploitation could take place [80].

White is not considered a "glamourous" pigment like sexy iodine scarlet that can excite like no other red ever could. The reason for titanium white's popularity is that the traditional darling of all white pigments, lead white, fell out of favor because of its toxicity. Titanium white is a welcome nontoxic, opaque and stable substitute; it is now the most widely used pigment of all time except for carbon black. It can be found in paper goods, inks, enamels, textiles, plastics, rubber, glass, ceramics, leather, and, of course, in artists' pigments. Among the first artists to incorporate it into their works were Jean Arp (1886–1966) by 1924 and Georgia O'Keeffe (1887–1986) by 1928 [81].

10.10.4 Chromium

In 1789, two revolutions were proceeding along on parallel tracks in France. The chemical revolution ignited by the groundbreaking proposals of Antoine-Laurent Lavoisier (1743–1794) intersected with the 14 July storming of the infamous Paris Bastille that set off the French Revolution. Meanwhile, Nicolas-Louis Vauquelin (1763–1829), a young man from a tiny Norman village, was pursuing his study of chemistry, physics and philosophy at the Jardin du Roi. Both Vauquelin and Lavoisier got caught up in the political upheaval and, at the height of the Reign of Terror in late 1793, Vauquelin was forced to flee Paris and in the spring of 1794, Lavoisier lost his head. Once a semblance of normalcy returned, Vauquelin obtained a professorship at Paris's School of Mines. As part of his research, he tackled the analysis of a scarce, but beautiful red crystalline mineral once called Siberian red lead, but presently crocoite. No one had yet succeeded in determining its elemental composition, and with good reason: it contained a hitherto unknown element that Vauquelin uncovered for the first time in 1797. This mysterious new substance yielded beautiful red, yellow and green precipitates, which prompted him to call it chromium, based on the Greek term for color, *chroma* (χρώμα).[82] It was only a matter of time before a dazzling bouquet of chromium colors began to embellish the artists' palette. Chrome yellow ($PbCrO_4$), chrome red ($PbCrO_4{\cdot}Pb(OH)_2$), chrome orange ($PbCrO_4{\cdot}PbO$), chromium oxide green, opaque (Cr_2O_3), viridian, Guignet's Green or hydrated chromium oxide green, transparent ($Cr_2O_3{\cdot}2H_2O$), zinc yellow ($K_2O{\cdot}4ZnCrO_4{\cdot}3H_2O$), basic zinc yellow ($ZnCrO_4{\cdot}4Zn(OH)_2$), molybdate orange ($PbCrO_4{\cdot}PbMoO_4{\cdot}PbSO_4$), lemon yellow or strontium yellow (SrCrO4), light lemon yellow or barium yellow ($BaCrO_4$) made their appearances one by one [83, 84]. Vauquelin hardly had time to follow up on his discovery before chrome yellow began to appear in paintings. By about 1830, chrome yellow was a fixture in watercolors and in oils. Due to the fact that its bright yellow hue on a black background is one the easiest color combinations to see, it has since become one of the major commercial pigments for signs, vehicles and highway markings.

Fig. 10.16 Left: Statue of Nicolas-Louis Vauquelin. Sculptor: Pierre Hébert (1804–1859). Faculté de Pharmacie, Rue de l'Observatoire, Paris. Vauquelin is holding a phial labeled "chrome" in his left hand. Photo: Mary Virginia Orna, Right: Pierre-Auguste Renoir, *The Seine at Asnières* (1875), The National Gallery London

During his initial studies of chromium compounds, Vauquelin (Fig. 10.16, left) prepared chrome orange. Possibly because of its sensitivity to sulfur compounds, it was rarely used commercially or by artists during the nineteenth century. One notable exception was Pierre-Auguste Renoir who dazzled the eye with his rendition of an orange skiff against a bright cobalt blue background (Fig. 10.16, right).

The availability of artists' pigments virtually doubled between 1800 and 1850 largely due to the increased number of chromium-based pigments that came on the market in the early nineteenth century.

10.10.5 Cadmium

The Harz mountain range in Germany shelters a little town, Göttingen, that is very special to the art historian. It is the birthplace of the beautiful red, orange and yellow pigments derived from cadmium, the element discovered there by chemist Friedrich Stromeyer (1776–1835) in 1817. Because of the element's close association with zinc ores smelted industrially as "cadmia fornacum," Stromeyer named his find "cadmium." [85] Stromeyer suggested cadmium's use as a pigment very soon after thereafter [86], but the metal was so scarce at the time that it could not be commercially exploited until several decades later [87]. Though cadmium's chief pigment, CdS, is a bright yellow solid, its color can vary depending on its particle size. Substituting some of the sulfur atoms with its close relative selenium can also change the color depending on the degree of substitution as the selenide. Colors of these cadmium sulfo-selenides can range from orange (100 S/12 Se), to orange-red (100 S/15 Se) to dark red (100 S/30 Se) to dark maroon (100 S/40 Se) as the atom ratio of selenium increases [88]. Other formulations of cadmium-based pigments with other compounds are marketed as artists' pigments as well. Among these are the lithopones (a mix with barium sulfate, $BaSO_4$) and cadmium/mercury pigments, solid solutions with variable Cd/Hg ratios [89]. Cadmium pigments must be meticulously purified by the color vendor to avoid discoloration. The culprit is any free sulfur present that can react with atmospheric moisture to form H_2SO_3 that, in turn, reacts with sulfide to form the colorless sulfate, causing fading [90].

Cadmium's career as a pigment began when cadmium-based pigments were early-on favored by artists due to their warm colors, good hiding power, and lightfastness. Melandri was the first artist on record to use cadmium sulfide as a pigment in oil in 1829 [91]. It made its first public appearance at the Great 1851 Exhibition in London [92] and by the turn of the century was one of the most important artists' pigments in use. Among its eager devotees were Vincent van Gogh (1853–1890) in his Self-Portrait of 1886–1887, Henri Matisse (1869–1954) in his Bathers by a River of 1916–1919 and numerous works by Claude Monet (1840–1926) spanning 35 years of his long career from 1873 to at least 1908 (Fig. 10.17).

Though the availability of these pigments was a great boon to creativity at the time, the modern museum curator is much more aware of an ongoing problem: the fugitive nature of many of the new offerings. Winslow Homer was famously cavalier about the known impermanence of some of his colors [93]. In the case of J. M. W. Turner, a little tale from *The Turner Society News* highlights this problem [94]. "Mr.

Fig. 10.17 Claude Monet. Grainstacks, 1891. A painting dripping with cadmium reds, yellows and oranges

Winsor (of the color manufacturing company, Winsor and Newton) during his life was pretty intimate with the great artist, Turner, who used to get his colours from us and noticing from time to time the very fugitive colours that Turner bought from us, plucked up courage one day to remonstrate with him for so doing. Turner's answer, in spite of the friendship between the two men, was somewhat uncompromising; he is alleged to have said, 'Your business Winsor is to make colours for Artists, mine is to use them.' I am afraid that poor Mr. Winsor had nothing to say in reply." Nor do we. To many an artist, color is an end in itself, and to worry about its permanence is someone else's job.

10.10.6 The Legacy of New Metals: The twenty first Century's High Performance Pigments.

Traditional pigments are defined as insoluble particulate colorants that are stable during processing and during their useful lifetime. This definition applies most especially to a workhorse group technically called Complex Inorganic Colored Pigments (CICPs). What makes members of this group "high performance" are qualities that are highly desirable in today's economy: sustainability in terms of manufacture, environmental friendliness, economy of scale, increased functionality (e.g., corrosion inhibition, optical variability, heat-reflecting characteristics) and better technical performance (e.g., ease of application, higher tinting strength). Chemically, as a class, they can be called "mixed metal oxides," namely oxides containing more than one metallic element in their formulas. Some examples are bismuth vanadate, ($BiVO_3$), a vivid yellow substitute for toxic yellow pigments [95], zinc ferrite brown spinel ($ZnFe_2O_4$) as a magnetic catalyst [96] and cobalt(II) aluminate/aluminum oxide nanoparticles ($CoAl_2O_4$ /Al_2O_3) as a ceramic blue pigment useful for non-clogging properties in ink-jet printer nozzles [97]. Figure 10.18 displays some of these high performance pigments along with some traditional ones [98].

Since the color range of these pigments is broad, the aesthetics are vivid, and the utility is high, these pigments will continue to be in great demand. Though artistic use is limited, industrial and architectural use will burgeon with their use as protective coatings, preventing the damaging effects of ultraviolet light, or averting heat buildup by reflecting infrared light [99]. It is also quite possible that some really ancient pigments, like ultramarine blue, could be revitalized by improved manufacturing technology, and that they too could join the ranks of high performance pigments with their enhanced physical properties [100].

Just beyond the 19th mid-century mark, two events were to change artists' pigment usage forever—the slow migration from inorganic to organic colors began.

Fig. 10.18 An array of modern synthetic pigments in use by many artists because they are stable, uniform and widely available. Top row, from left: cobalt silicate blue, cobalt violet phosphate, zirconium vanadium blue zircon, cobalt titanate green spinel, titanium vanadium antimony gray rutile. Bottom row, from left: nickel antimony titanium yellow rutile, zinc ferrite brown spinel, cadmium orange (CdS, CdSe), cadmium light red (CdS, CdSe), cadmium dark red (CdS, CdSe). Photo by J.J. Jacobsen and R.J. Wildman. Reprinted with permission, American Chemical Society, copyright 2001

10.11 Perkin's Discovery and Kekulé's Dream: The Inexorable March to a Different Drummer

August Wilhelm Hofmann (1818–1892), Professor at London's Royal College of Chemistry, was arguably the world expert on organic synthesis and an excellent research director. On a fateful day in March, 1856, he sent one of his precocious students, William Henry Perkin (1838–1907), on a hunt for a desperately needed commodity. Quinine, the only known effective medicine for the killer disease, malaria, could only be found naturally in the bark of the cinchona tree. Thinking to synthesize it in the laboratory, he advised Perkin to start with allyltoluidine, $C_{10}H_{13}N$—the formula needed only an oxygen to square up with quinine, so a powerful oxidizing agent like dichromate was just the ticket. When that idea failed, Perkin decided to depart from the playbook and use aniline, $C_6H_5NH_2$, (Fig. 10.19) as a starting material. Aniline was the "child" of coal tar, a byproduct of the manufacture of coke from coal, a necessary process of England's growing industrial revolution. When only a black gunk developed from this next try, Perkin almost just threw it down the drain, but first he decided to clean his vessel with alcohol. Lo! As the alcohol gobbled up the gunk a gorgeous purple solution appeared—and Perkin recognized a good thing when he saw it. He called his color mauve, quit school to develop it, and retired in his thirties as a very rich man. From this historic event, the synthetic dye industry grew to envelop the fashion world and eventually the whole world for from these colors came not only dyes, but also pharmaceuticals and high explosives.

In 1865, while Perkin was stirring his dye pot, another chemist, Friedrich August Kekulé von Stradonitz (1829–1896), was pondering atomic binding capacity especially as applied to the puzzling molecular formula of benzene, C_6H_6. Dozing off, he had a dream that the carbon atoms in benzene whirled about, formed a circle,

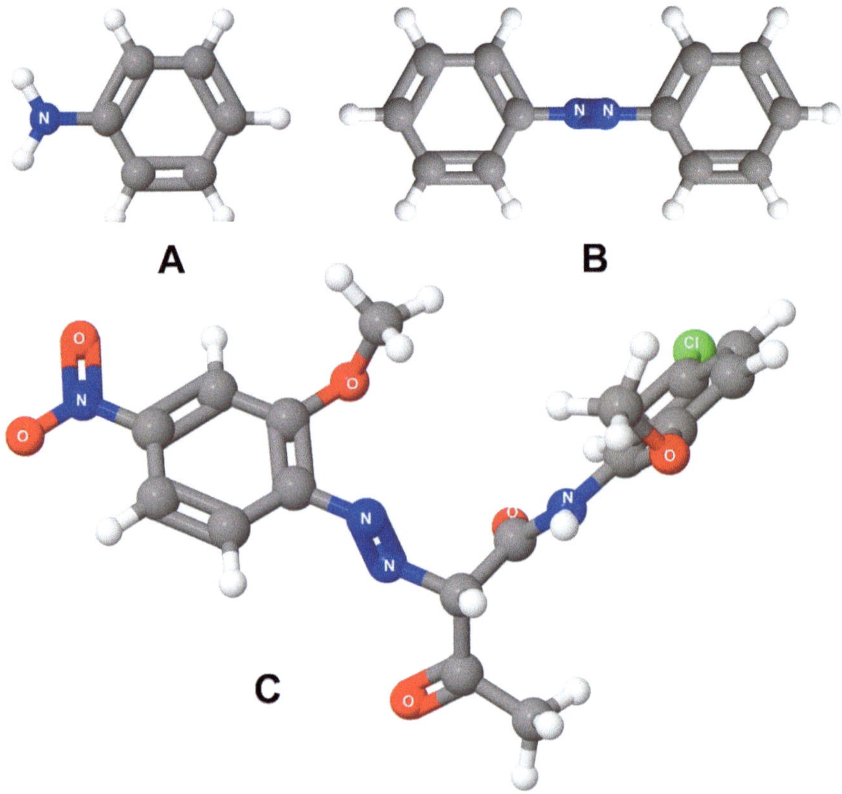

Fig. 10.19 A. Aniline, Perkin's starting material; B. Azobenzene, starting material for many modern dyes and pigments; C. An azopigment, Pigment Yellow 111 (aka Hansa Brilliant Yellow). Large gray atoms are carbon; small white atoms are hydrogen; all other atoms as labeled. (Note the N = N double bond in B and C—characteristic of azo compounds)

and then like a snake, bit their own tail. This flash of insight led to the structure of benzene that we ascribe to even today; it also led to the birth of structural organic chemistry [101]. Kekulé's structural ideas, published almost simultaneously by Archibald Scott Couper (1831–1892), a young Scot working in Charles-Adolphe Wurtz's (1817–1884) Paris laboratory, drew the same conclusions:

- that carbon atoms were tetravalent and
- that they had the ability to bond with one another.

As this idea began to take hold, chemists started to think and speak about the structure of molecules, realizing that how the atoms related to one another in a molecule was a key to understanding their identity and their properties. Eventually, this approach would give rise to theory-based synthesis of colorants rather than the dominant hit-or-miss method of earlier years [102]. These new ideas bore so much fruit that by the end of the century, over 650 dyestuffs had been created. Thousands

Fig. 10.20 Display of Bright Pigments for Sale at a Market in Pisaq, Peru. Photograph: Mary Virginia Orna

more would appear as a result of this systematic approach, and many of these new compounds would land as pigments on the artist's palette (Fig. 10.20)—for weal or for woe. The march away from inorganic pigments was almost complete [103].

Aniline is the building block for most of the synthetic organic pigments (SOPs). Some major groups are the azo pigments, containing –N=N–, the azo functional group; others are the dioxazines, DPP (diketopyrrolo-pyrrole), indanthrene [104], the triarylmethanes, the quinacridones, the phthalocyanines, the perylenes, the pyrroles [105] and the arylide (Hansa) yellow pigments [106]. One particular azo pigment, Lithol Red, became famous as the choice pigment of Mark Rothko in his Seagram and Harvard murals of 1950s and 1960s. Although it is a very bright pigment, its major drawback is its poor lightfastness, presenting a huge problem to the art conservationists trying to deal with Rothko's work [107]. Figure 10.21 shows some structures of SOPs commonly used as artists' pigments.

10.12 Sticking with It: Synthetic Polymers in Artists' Media

Making a pigment adhere to a wall or other surface has always been a "sticking point" for artists. An ancient technique called "buon fresco" wherein lime reacts with atmospheric CO_2 to form an insoluble carbonate that embeds the pigment in the wall was a major technique for millennia. Other materials found in cave paintings and wall paintings dating back 45 000 years have been various waxes, fats, urine, dung, plant

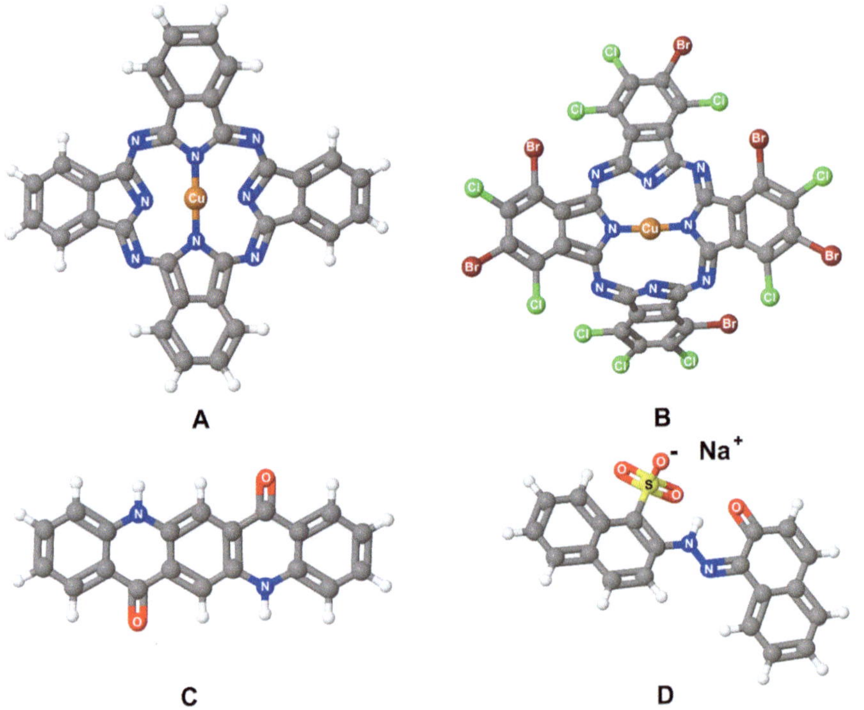

Fig. 10.21 Some Common SOPs. A. Pigment Blue 15 (Phthalocyanine Blue); B. Pigment Green 36 (Phthalocyanine Green); C. Quinacridone Violet (Pigment Violet 19); D. Lithol or Toluidine Red (Pigment Red 3). Large gray atoms are carbon; small white atoms are hydrogen; all other atoms as labeled

resins and even earwax! As easel painting developed with wood panel and canvas surfaces, the preferred binding media were egg tempera and oil (usually linseed and walnut). Their use marked the first consistent use of materials that would slowly polymerize to form a protective film on the surface of the painting, a process taking days, weeks, months and even many years to complete. The fresh egg yolk used in tempera is about 47% water, 33% lipids and 17% proteins by mass. As the water evaporates, denaturing the protein in the process, initial drying and hardening take place to form the paint film. The proteinaceous components are the major film-formers, but both the lipids and the pigments are involved in the cross-linking that leads eventually to formation of the polymeric network. The drying oils form their polymeric network exclusively by the chemical process of cross-linking among polyunsaturated fatty acids that are components of the triglycerides that comprise the oils [108].

Obviously, the use of egg tempera and oil was empirical and serendipitous. That they stood the test of time over many centuries is witness to the expertise of the practicing artists who perfected them. They only gave way in the twentieth century to polymeric materials after the groundbreaking work of Hermann Staudinger (1881–1965) and Wallace Carothers (1896–1937). These two chemists tackled the polymer problem from opposite ends of thc spectrum: Staudinger began by focusing his experimental work on naturally-occurring polymers, whereas Carothers began with smaller molecules to build up entirely synthetic long chains. Together they laid the groundwork for one of the most important developments of the twentieth century, both economically and theoretically. They probably would have shared the Nobel Prize if it were not for Carothers' untimely death by suicide [109, 110].

Building on the experimental and theoretical work of these two giants, and some who went before and those who came after, polymeric media such as acrylics and alkyds opened up new vistas for artists, both professional and amateur. Chemically, the acrylic polymer is polymethylmethacrylate. Especially popular are the acrylic emulsions developed by the Rohm and Haas Company in 1953. They are water-based, quick drying, flexible, impervious to cracking, mix well with other media, and are inexpensive. With an almost limitless spectrum of vivid colors embedded into an acrylic binder, many artists looked no further for their perfect medium.

Acrylic formation is straightforward once the variables are worked out. Reaction Scheme 10.6 shows how.

However, emulsion polymerization is more complicated. It involves adding an initiator to a suspension of monomer droplets in a water–oil based mix that disperses the droplets into tiny micelles that then copolymerize. The advantage is that as the molecular weight of the polymer increases, the viscosity of the emulsion remains virtually unchanged, whereas in a simple water-borne copolymerization scheme, the viscosity increases exponentially with the molecular weight, a situation untenable for an artists' pigment binder.

Alkyd formation involves reaction of an anhydride such as phthalic anhydride with a monoglyceride ester formed from glycerol and a fatty acid according to Scheme 10.7

These two binders have changed the way art is done. Professionals like Josef Albers (1888–1976), David Hockney (b. 1937), Loris Cecchini (b. 1969), José Gutiérrez (1900–1968) and Gunther Gerzso (1915–2001) have taken advantage of the many qualities of these and other polymeric materials to give free rein to their

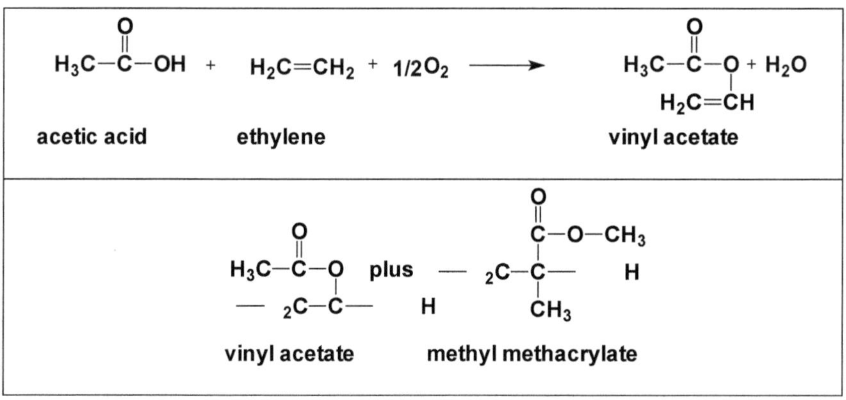

Reaction Scheme 10.6 Copolymerization Reactions to Form an Acrylic Polymer

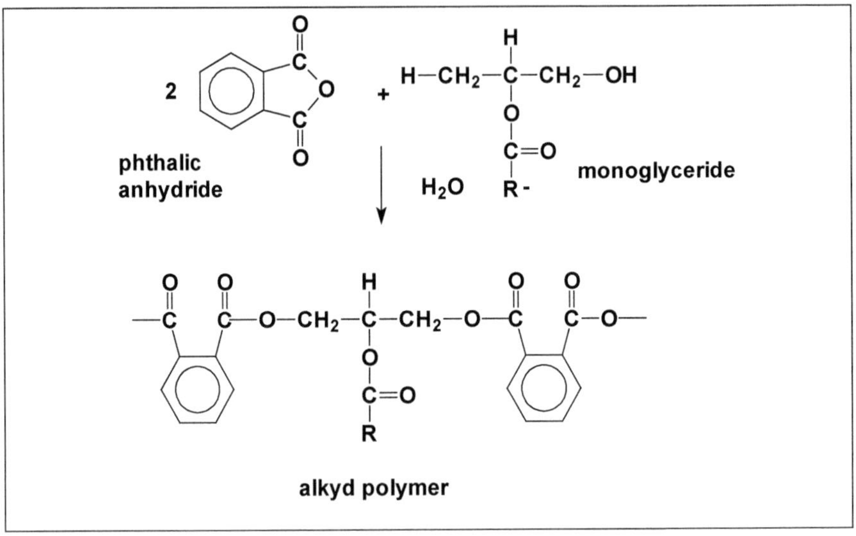

Reaction Scheme 10.7 Formation of an Alkyd Polymer from Phthalic Anhydride and a Generic Monoglyceride

creativity. An iconic user of alkyd resins was Jackson Pollock (1912–1956), a medium that allowed for his famous drip paintings [111]. Inveterate users of the phthalocyanines were René Magritte (1898–1967), Pablo Picasso (1881–1973), Wassily Kandinsky (1866–1944), Roy Lichtenstein (1923–1997) and Piet Mondrian (1872–1944), as well as Pollock [112]. The alkyds and acrylics have also opened the door to thousands of amateurs for whom oil painting would be otherwise most daunting.

10.13 Conclusion

An initial perusal of the Bible makes it obvious that God creates in a cosmic craft, bringing forth the sun, the moon and the stars. This glorious array was captured in an almost visceral, and godlike, way by Vincent van Gogh (1853–1890) in the painting that has come to epitomize all of modern art, *Starry Night* (Fig. 10.22).

It is hoped that this chapter has shown, in an equally obvious way, that art could not have progressed in any material approach without the efforts of the ancillary discipline of chemistry. With the trend, at present, to photonic colors generated "on the spot" by lasers and electronics, physics may begin to play such a role into the future. This insight into the relationship between the arts and the sciences has been all too brief and has left many areas where art and chemistry intersect untouched. Volumes exist and are yet to be written that will document more fully the evolutionary nature of this partnership.

Fig. 10.22 Vincent van Gogh: Starry Night, oil on canvas, 1889. Museum of Modern Art, New York, NY. Public domain

10.14 Postscript

Due to space limitations, some of the details regarding color technology and nomenclature have been left to the reader to pursue. Among them are color order systems, laid out superbly in the volume, *Color Ordered* [113] and the *Colour Index* system [114]. How and why we see color in the first place is beautifully addressed in an essay by Robert Finlay [115]. Interaction of light with matter, the nature of the pigments and dyes that produce color, and their cultural history are treated in this author's latest book [116]. Enjoy.

References

1. A.P. Abbott, From test tube to Turner—The role of the chemist in art. Sci. Prog. London **96**(4), 398–416 (2013)
2. M.V. Orna, Chemistry and art: Ancient textiles and medieval manuscripts examined through chemistry. Educ. Quím. **22**(3), 191–197 (2011)
3. P. Hill, D. Simon, Chemistry, art, and cultural diversity, in *Paper presented at the 21st Biennial Conference on Chemical Education*, p. 3 (University of North Texas, Denton, Texas, 2010).

4. L. Campanella, F. Cardone, G. Olivetti, *Chimica e storia dell'arte* (UniversItalia, Roma, 2012), p.5
5. J.B. Lambert, *Traces of the Past* (Addison Wesley, Reading, 1997), p.173
6. M.V. Orna, P.J. Smith, Metals of archaeological interest: new directions, in *Archaeological Chemistry: A Multidisciplinary Analysis of the Past*. ed. by M.V. Orna, S.C. Rasmussen (Newcastle upon Tyne, Cambridge Scholars Publishers, 2020), pp.14–51
7. G. Savage, *A Concise History of Bronzes* (Frederick A. Praeger, New York, 1968), p.17
8. I. Théry-Parisot, S. Thiébault, Le pin (Pinus sylvestris): préférence d'un taxon ou contrainte de l'environnement? Étude des charbons de bois de la grotte Chauvet. Bull Soc Préhistor Française. **102**(1), 69–75 (2007)
9. Abbott, p. 400
10. Marcus Vitruvius Pollio, De Architectura VII, 11, 1
11. J. Riederer, Egyptian blue, in *Artists' pigments: a handbook of their history and characteristics*, ed. by E.W. FitzHugh. National Gallery of Art, vol. 3 (Washington, D.C., 1997), pp.23–45
12. A. Pabst, Structure of Some Tetragonal Sheet Silicates. Acta Crystallogr. **12**, 733–739 (1959)
13. F. Mazzi, A. Pabst, Reexamination of Cuprorivaite. Am. Mineral. **47**, 409–410 (1962)
14. F. Delamare, *Blue Pigments: 5000 years of art and industry* (Archetype Publications, London, 2013), pp.285–288
15. Lambert, p. 105–109
16. R.H. Brill, Ancient glass. Sci. Am. **109**(5), 120–130 (1963)
17. S.C. Rasmussen, *How Glass Changed the World* (Springer, Heidelberg, 2012)
18. S.C. Rasmussen, Development of chemical glassware: Evaluating historical narratives via chemical archaeological data, in *Archaeological Chemistry: A Multidisciplinary Analysis of the Past*. ed. by M.V. Orna, S.C. Rasmussen (Cambridge Scholars Publishers, Newcastle upon Tyne, 2020), pp.146–175
19. P.E. McGovern, Alcoholic beverages as the universal medicine before synthetics, in *Chemistry's Role in Food Production and Sustainability: Past and Present*. ed. by M.V. Orna, G. Eggleston, A.F. Bopp (American Chemical Society, Washington, DC, 2019), pp.111–128
20. A.J. Liebman, History of distillation. J. Chem. Educ. **33**, 166–173 (1956)
21. C. Cobb, H. Goldwhite, *Creations of Fire, Chemistry's Lively History from Alchemy to the Atomic Age* (Plenum Press, New York, 1995), p.87
22. C. Cobb, M.L. Fetterolf, H. Goldwhite, *The Chemistry of Alchemy: From Dragon's Blood to Donkey Dung, How Chemistry was Forged* (Prometheus Books, Amherst, 2014), p.16
23. R.J. Forbes, *Studies in Ancient Technology*, vol. 3 (EJ Brill Press, Leiden, 1955), p.218
24. D.J. Barber, I.C. Freestone, K.M. Moulding, Ancient copper red glasses: investigation and analysis by microbeam techniques, in *From Mine to Microscope: Advances in the Study of Ancient Technology*. ed. by A.J. Shortland, I.C. Freestone, T.H. Rehren (Oxbow Books, Oxford, 2009), pp.115–128
25. V. Karpenko, J.A. Norris, Vitriol in the history of chemistry. Chemické listy. **96**(12), 997–1005 (2002)
26. A. Cassius, De extremo illo et perfectissimo naturae opificio ac principe terrae norum sidere auro (Georg Wolff, Hamburg, 1685).
27. J.B. Richter, Über die neuern Gegenstände der Chymie. 11: Vorzüglich über die Glucine, Agust-Erde und einige besondres Eigenschaften des Goldes, (1802); p. 91. https://www.bavarikon.de/object/bav:BSB-MDZ-00000BSB10707959?cq=Thon&p=1&lang=de. Accessed May 2021.
28. J.L. Proust, Faits pour server à l'histoire de l'or. J. Phys. **62**, 131–147 (1806)
29. M. Faraday, Experimental Relations of Gold (and Other Metals) to Light. Phil. Trans. R. Soc. London **147**, 145–181 (1857)
30. L. Hunt, Gold-based glass and enamel colours. Endeavour New Series **5**(2), 61–67 (1981)
31. R. Zsigmondy, Properties of Colloids. Nobel Lecture, (December 11, 1926). https://www.nobelprize.org/prizes/chemistry/1925/zsigmondy/lecture/. Accessed April 2021.

32. G. Fleck, Richard Zsigmondy (1865–1929), in *Nobel Laureates in Chemistry 1901–1992*. ed. by L.K. James (American Chemical Society, Washington, DC, 1993), pp.151–157
33. The pottery was also a staple of the Athenian economy—one of its major exports and the one on which they heavily relied to obtain grain from the Ukrainian breadbasket. It was so renowned that a Scythian prince would gladly exchange a boatload of grain for one such vase. Each one was an original creation by a master craftsman.
34. G.M.A. Richter, *The craft of Athenian pottery* (Yale University Press, New Haven, 1923)
35. W.C. Alexander, A pottery of the middle Roman imperial period in Augst, Forschungen in Augst 2, p. 10 (Pro Augusta Raurica, Basel, 1975).
36. A. Brongniart, *Traité des Art ceramiques ou des Poteries*, vol. 1 (Béchet Jeune, Paris, 1844), p.552
37. E. Durand-Gréville, La Couleur du Décor des Vases grecs, Revue Archéologique, Troisième Serie (1891, 18, July-December), 99–118, at 101
38. C.F. Binns, A.D. Frazer, The Genesis of the Greek Black Glaze. Am. J. Archaeol. **33**(1), 1–9 (1929)
39. T. Schumann, Oberflächenverzierung in der antiken Töpferkunst—Terra sigillata und greischische Schwarzrotmalerei. Ber. Dtsch. Keram. Ges. **23**, 408–426 (1942)
40. C.P.T. Naudé, The glaze technique of the Attic vase. Acta Classica **2**, 106–116 (1959)
41. W.D. Kingery, Attic Pottery Gloss Technology. Archeomaterials **5**, 47–54 (1991)
42. U. Hofmann, The chemical basis of ancient Greek vase painting. Angew. Chem. Int. Ed. Engl. **1**(7), 341–414 (1962)
43. J.V. Noble, The Technique of Attic Vase Painting. Am. J. Archaeol. **64**(4), 307–318 (1960)
44. J. Paul Getty Museum, Making Greek Vases, https://www.khanacademy.org/humanities/special-topics-art-history/creating-conserving/ceramics-glass/v/making-greek-vases. Accessed May 2021.
45. M. Bimson, The technique of Greek black and terra sigillata red. Antiquar. J. **36**, 200–204 (1956)
46. K. Nassau, The origins of colors in minerals. Am. Mineral. **63**, 219–229 (1978)
47. A.M. Bruck, C.A. Cama, C.N. Gannett, A.C. Marschilok, E.S. Takeuchi, K.J. Takeuchi, Nanocrystalline iron oxide based electroactive materials in lithium ion batteries: the critical role of crystallite size, morphology, and electrode heterostructure on battery relevant electrochemistry. Inorg. Chem. Front. **3**, 26–40 (2016)
48. G. Tierie, *Cornelis Drebbel, 1572–1633* (H. J. Paris, Amsterdam, 1932), pp.76–77
49. D. Cardon, *Le Monde des Teintures Naturelles* (Belin, Paris, 2003), pp.51–52
50. W. Partridge, *Practical Treatise on Dying (sic)* (H. Wallis, New York, 1823), p.47
51. E.H. Edwards, Military Illustrated: Past and Present **81**, 15–16 (1995)
52. A. Kraft, On the discovery and history of Prussian blue. Bull. Hist. Chem. **33**(2), 61–67 (2008)
53. J.W.F. Herschel, On the action of the rays of the solar spectrum on vegetable colours, and on some new photographic processes. Philos. Trans. R. Soc. Lond. **132**(1842), 181–215 (2021). https://doi.org/10.1098/rstl.1842.0013
54. M. Ware, Prussian Blue: Artists' pigment and chemists' sponge. J. Chem. Educ. 85(8), 612–613, 615–621 (2008)
55. Delamare, p. 147–161
56. Delamare, p. 261–266
57. S.Q. Lomax, Phthalocyanine and quinacridone pigments: Their history, properties and use. Stud. Conserv. 50, Issue supplement 1– Rev. Conserv. 6, 19–29 (2005)
58. A.E. Smith, H. Mizoguchi, K. Delaney, N.A. Spaldin, A.W. Sleight, M.A. Subramanian, Mn^{3+} in Trigonal Bipyramidal Coordination: A New Blue Chromophore. J. Am. Chem. Soc. **131**(47), 17084–17086 (2009)
59. D. Reinen, T.C. Brunold, H.O. Guedel, N.D. Yordanov, Die Natur des Farbzentrums im Manganblau. Z. anorg. Aallg. Chem. **624**(3), 438–442 (1998)
60. E.N. Brown, It's Not Every Day We Get a New Blue. The New York Times, February 8 (2021)
61. U.F. Molteni, Alessandro Volta: Como e il Lario nel bicentenario dell'invenzione della pila, 1799–1999. Università Terza Età "A. Volta," Como, p. 41 (1999).

62. J.T. Stock, Electrochemistry in retrospect: an overview, in *Electrochemistry Past and Present*. ed. by J.T. Stock, M.V. Orna (American Chemical Society, Washington, 1989), pp.1–17
63. H. Davy, The decomposition of the fixed alkalies and alkaline earths. Phil Trans. **98**(1–44), 333–370 (1808)
64. J. Ayres, *The Artist's Craft: A History of Tools, Techniques, and Materials* (Phaidon Press, Phaidon Press, 1985), p.111
65. J. Renoir, *My Father* (Collins, London, 1962), p.73
66. B.I. Kharisov, L.M. Blanco, L.M. Torres-Martinez, A. García-Luna, Electrosynthesis of phthalocyanines: influence of solvent. Ind. Eng. Chem. Res. **38**(8), 2880–2887 (1999)
67. H.J. Schäfer, Contributions of organic electrosynthesis to green chemistry. C. R. Chim. **14**(7–8), 745–765 (2011)
68. Y.-L. Li, X. Chen, K.-G. Geng, Y. Dong, B. Wang, A. Ma, L. Pan, G.-Q. Ma, D.-P. Song, Y.-S. Li, Oxidation control of bottlebrush molecular conformation for producing libraries of photonic structures. Angew. Chem. Int. Ed. Engl. **1**, 1 (2021). https://doi.org/10.1002/anie.202011702
69. R. Harley, *Artists' Pigments c. 1600–1835* (Butterworths, London, 1970), pp.75–76
70. New York Times (Sunday, July 22, 1956) p. 22.
71. R.J. Gettens, G.L. Stout, *Painting Materials: A Short Encyclopedia* (Dover Publications, New York, 1966), p.113
72. J. Riederer, Die Smalte. Dtsch. Farben-Z. **22**, 387–389 (1968)
73. Y. Abe, R. Harimoto, T. Kikigawa, K. Yazawa, A. Nishisaka, N. Kawai et al., Transition in the use of cobalt-blue colorant in the New Kingdom of Egypt. J. Archaeol. Sci. **39**, 1793–1808 (2012)
74. A.P. Laurie, *The Materials of the Painter's Craft* (T.N. Foulis, London, 1910), p.19
75. N. Eastaugh, V. Walsh, T. Chaplin, R. Siddall, *The Pigment Compendium* (Elsevier, Oxford, 2004), p.90
76. A.H. Church, *The Chemistry of Paints and Painting* (Seely Service & Co., London, 1901), p.196
77. A.P. Laurie, *New light on old masters* (The Sheldon Press, London, 1935), p.44
78. B. Berrie, F. Casadio, K. Dahm, Y. Strumfels, M. Tedeschi, J. Walsh, A vibrant surface: Investigating color, texture and transparency in Winslow Homer's watercolors, in *Science and Art: The Painted Surface*. ed. by A. Sgamellotti, B.G. Brunetti, C. Miliani (Royal Society of Chemistry, Cambridge, 2014), pp.404–428
79. Eastaugh, pp. 115–117
80. N. Heaton, The production of titanium oxide and its use as a paint material. J. R. Soc. Arts **70**(31), 552–565 (1922)
81. M. Laver, Titanium dioxide whites, in *Artists' Pigments: A Handbook of Their History and Characteristics*, ed. by E.W. FitzHugh. National Gallery of Art, vol. 3 (Washington, DC, 1997), pp.295–355
82. N.L. Vauquelin, Analysis of the red lead of Siberia with experiments on the new metal it contains. Nicholson's J. **2**, 387–393 (1798)
83. R.J. Lancashire, *Chromium Chemistry*, http://wwwchem.uwimona.edu.jm/courses/chromium.html. Accessed November 2020.
84. H. Kühn, M. Curran, Chrome yellow and other chromate pigments, in *Artists Pigments: A Handbook of Their History and Characteristics*, ed. by R.L. Feller. National Gallery of Art, vol. 1 (Washington, DC, 1986), pp.187–217
85. I. Fiedler, M.A. Bayard, Cadmium yellows, oranges and reds, in *Artists' Pigments: A Handbook of Their History and Characteristics*, ed. by R.L. Feller. National Gallery of Art, vol. 1 (Washington, DC, 1986), pp.65–108
86. F. Stromeyer, New details respecting cadmium. Ann. Philos. **14**, 269–274 (1819)
87. G.H. Bachhoffner, *Chemistry as applied to the fine arts* (J. Carpenter, London, 1837), p.117
88. P.J. Curtis, R.B. Wright, Cadmium pigments. JOCCA **37**, 26–43 (1954)
89. E.L. Moore, Cadmium/mercury sulfides, in *Pigment handbook*, vol. 1, ed. by T.C. Patton (Interscience, New York, 1973), pp.395–399

90. J.N. Friend, *An introduction to the chemistry of paints* (Longmans, Green, London, 1910), pp.38–39
91. A. Eibner, *Malmaterialienkunde als Grundlage der Maltechnik* (Julius Springer, Berlin, 1909), p.128
92. A.P. Laurie, *The pigments and mediums of the old masters* (Macmillan, London, 1914), p.16
93. B. Berrie, A vibrant surface, p. 406.
94. P. Bower, The fugitive Mr. Turner. The Turner Society News **53**, 5 (1989)
95. T. Masui, T. Honda, Wendusu, Imanaka I, Novel and environmentally friendly (Bi, Ca, Zn)$_V$O4 yellow pigments. Dyes Pigm. **99**(3), 636–641 (2013)
96. N.M. Mahmoodi, Zinc ferrite nanoparticle as a magnetic catalyst: synthesis and dye degradation. Mater. Res. Bull. **48**(10), 4255–4260 (2013)
97. M. Román-Aguirre, L. De la Torre-Sáenz, P. Pizá-Ruiz, M. Bocanegra-Bernal, Synthesis of $CoAl_2O_4/Al_2O_3$ nanoparticles for ceramic blue pigments. Ceram. Int. **43**(17), 15254–15257 (2017)
98. M.V. Orna, Chemistry, color and art. J. Chem. Educ. **78**(10), 1305–1311 (2001)
99. M.C. Comstock, Complex inorganic colored pigments. J. Surf. Coat. Aust. (2016)
100. G. Buxbaum, Introduction to inorganic high performance pigments, in *High Performance Pigments*, 2nd edn., ed. by E.B. Faulkner, R.J. Schwartz (Wiley, New York, 2009), pp.3–6
101. F.R. Japp, Kekulé memorial lecture. J. Chem. Soc. **78**, 97–138 (1898)
102. A.J. Ihde, *The development of modern chemistry* (Dover Books, New York, 1984), p.218
103. Z. Hao, A. Iqbal, Some aspects of organic pigments. Chem. Soc. Rev. **26**, 203–213 (1997)
104. A. Dal Fovo, M. Oujja, M. Sanz, A. Martínez-Hernández, M. Vega Cañamares, M. Castillejo, R. Fontana, Multiananalytical non-invasive characterization of phthalocyanine acrylic paints through spectroscopic and non-linear optical techniques. Spectrochim. Acta Part A **208**, 262–270 (2019)
105. Abbott, p. 413
106. S. Lake, Lomax SQ Arylide (Hansa) yellow pigments, in *Artists' Pigments: A Handbook of Their History and Characteristics*, ed. by B.H. Berrie. National Gallery of Art, vol. 4 (Washington, DC, 2007), pp.179–222
107. G. Poldi, C. Anselmi, A. Daveri, M. Vagnini, Josef Albers' use of 20th century pigments: a noninvasive analytical approach, in *Science and Art: The Contemporary Painted Surface*. ed. by A. Sgamellotti, B.G. Brunetti, C. Miliani (Royal Society of Chemistry, Cambridge, 2020), pp.67–94
108. D.A. Peggie, The chemistry and chemical investigation of the transition from egg tempera painting to oil in Italy in the 15th century, in *Science and Art: the Painted Surface*. ed. by A. Sgamellotti, B.G. Brunetti, C. Miliani (Royal Society of Chemistry, Cambridge, 2014), pp.209–229
109. American Chemical Society, The foundation of polymer science by Hermann Staudinger. An international historic chemical landmark, (19 April 1999). https://www.acs.org/content/acs/en/education/whatischemistry/landmarks/staudingerpolymerscience.html. Accessed February 2021.
110. M. Hermes, *Enough for one lifetime: Wallace Carothers, inventor of nylon* (Chemical Heritage Foundation, Philadelphia, 1996)
111. A. Sgamellotti, B.G. Brunetti, C. Miliani, (eds.) *Science and Art: The Contemporary Painted Surface*, p. 1–17, 67–94, 316–337, 374–389, 404–430 (Royal Society of Chemistry, Cambridge, 2020).
112. Dal Fovo
113. R.G. Kuehni, A. Schwarz, *Color Ordered: A Survey of Color Order Systems from Antiquity to the Present* (Oxford University Press, New York, 2008)
114. Pigments are identified by both their common names and their Color Index Constitution Number when available. The Color Index (CI) is a list of identification numbers and names given to individual pigments by the Society of Dyers and Colourists in the UK and the Association of Textile Chemists and Colorists in the United States. A useful database is http://www.artiscreation.com/Color_index_names.html#.YCVdNvlOmM9. Accessed February 2021.

115. R. Finlay, Weaving the rainbow: visions of color in world history. J. World Hist. **18**(4), 383–431 (2007)
116. M.V. Orna, *March of the Pigments* (Royal Society of Chemistry, Cambridge, 2022).

Chapter 11
Architecture in the Atomic Microcosm: The Riveting Case of Metal Organic Frameworks

David Dell'Angelo, Hari Ramesh, Zeyu Zhang, and Farnaz A. Shakib

Abstract Similarly to real-world buildings made of suitable materials, with construction order and me-chanical strength that must meet the demands of its functionality, a molecule is characterized by its constituent atoms, by the succession and spatial order of chemical bonds and by the strength of the bonds themselves. With time, chemists have learned how to synthesize new molecules not existing in nature, thus inventing new geometries and new functionalities. In this view, chemistry is the equivalent of architecture in the atomic microcosm. The aim of this chapter is to highlight the connections between Art & Science, in the belief that openness to esthetic sensibility and genuine curiosity towards disciplines beyond theory may also be beneficial for doing high quality scientific research. Metal organic frameworks epitomize these connections, as they are known for their properties and intriguing applications but also for beautifully designed tessellation composed of repeating patterns of molecular building units.

D. Dell'Angelo (✉) · H. Ramesh · Z. Zhang · F. A. Shakib
Department of Chemistry and Environmental Science, New Jersey Institute of Technology, Newark, NJ, USA
e-mail: david.dellangelo@univ-artois.fr

D. Dell'Angelo
Univ. Artois, CNRS, UMR 8181, Unité de Catalyse et de Chimie du Solide (UCCS), Univ. Lille, Centrale Lille, Lens, France

L. Mammino (ed.), *Insights into the Relationships between Humanities and Sciences*,
https://doi.org/10.1007/978-3-031-97900-2_11

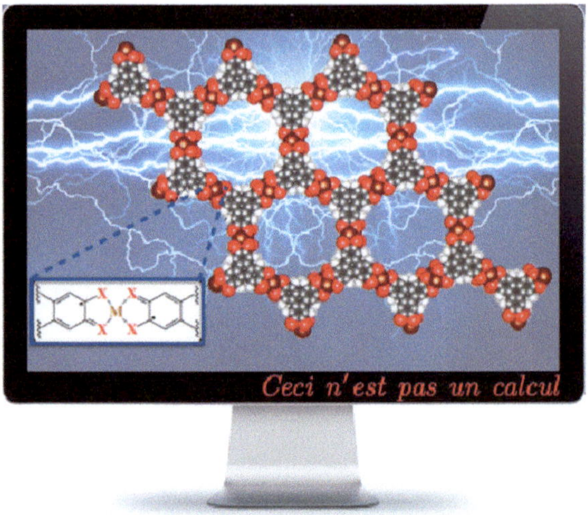

11.1 Introduction

"If it is far from anything I do, I must be interested."

—Kenneth Arrow.

In addition to their technological significance, metal–organic frameworks (MOFs) [1–3] are distinguished by exquisite balance inherent in their symmetric elegance (see Fig. 11.1), which renders them visually appealing to observers. Indeed, a meticulously ordered composition conveys a feeling that is both exquisite and serene. Further, symmetrical images deliver a sense of stability which can generate a sense of presence. From Byzantine to Baroque paintings, the sense of presence evoked by symmetry has clearly played a role in religious figures. The cross in Christianity, the Star of David or the Gobosei in Japanese culture are symmetrical images. And tribal masks and sculptures are often symmetrical, suggesting both human and supernatural presences. But can symmetry and balance in some way offer the experience of *beauty* ? The latter defined by Aristotle's components: order, symmetry and definiteness [4]. Nonetheless, if its comprehension is conditioned by the cultural environment in which we are immersed, is beauty an objective quality for a successful artwork ? Shall we have art without beauty ? Albeit, the prior question is: what do we mean by *Art* ? Which like most words has multiple meanings. If we are talking about *fine* art, in its narrowest sense the term refers to visual arts of painting, drawing, and sculpture. More broadly, it also applies to other major art forms, such as poetry, drama, music, or dance. The visual fine arts have traditionally been contrasted with *applied* arts, the latter referring to the design of attractive objects for practical use and physical function. Symmetry and balance play a larger role in applied arts but also in *Science*. For instance, as represented in Fig. 11.2, in 2D materials periodicity can be exploited to model the energy transfer between consecutive building units (Fig. 11.2b) and

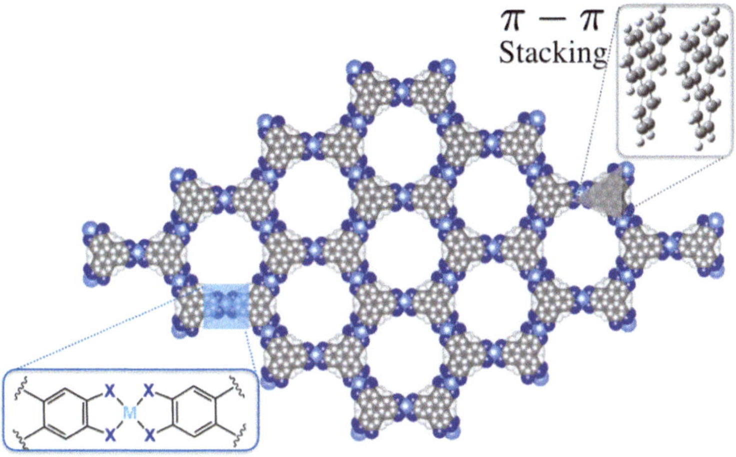

Fig. 11.1 Representative layered architecture of a conductive π—stacked metal organic framework (MOF), highlighting the "secondary building unit (SBU) and π–π stacking of the layers via organic linkers [5]

eventually study the optical properties in these materials (Fig. 11.2c). Nowadays, a large part of computational chemists and engineers dedicate their research activity to the study of the structure–property relationship in such materials [6], in other words how molecular design and molecular structure may affect the properties of the device.

Science in the broadest term relies on evidence of data and falsifiability of hypotheses rather than on subjective feelings commonly experienced by artists. Thus, where may we start to search for the commonalities between Art(s) & Science ? Likely, taking a stroll in the park. Everyday life experience shows that an insight in our work and creative ideas may often come outside the office rather than sitting behind a desk. The chemist A. Kekul´e for instance fell asleep in his chair in front of the fire place and woke up with the revolutionary idea of the circular structure of a benzene molecule. Since research suggests that a random group of people may often be better at solving complex problems than a selection of experts [8], this chapter was conceived after interesting, informal discussions with people from other backgrounds.

Sections—discuss the only apparent dichotomy between Art & Science. Section emphasizes the connection between visual and applied art, epitomized by Escher's interest in symmetry and mathematics of tessellation. The following sections gradually become more technical. Section introduces the weak out-of-plane interactions to generate layered solids along the vertical stacking of 2D materials. Section focuses on the unique area of Moir´e superlattices, which has revealed a diverse range of novel quantum phenomena and phases, often related to the specific way the layers are twisted with respect to each other (section). Section discusses the intense interest

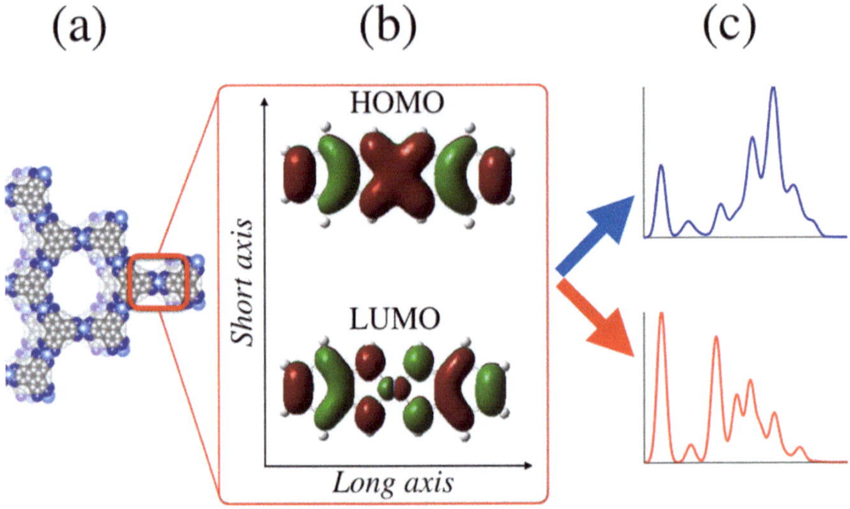

Fig. 11.2 Representation of the structure–property relationship in 2D material. **a** By exploiting *in*—plane balance, symmetry and periodicity, the out-of-plane interactions between consecutive layers in 2D material can be evaluated by calculating **b** the energy transfer between Highest Occupied Molecular Orbitals (HOMOs) and Lowest Unoccupied Molecular Orbitals (LUMOs). **c** The engineering of optical properties, analyzed by the optical spectra, can be used to develop materials for specific purposs. Details may be found in Reference [7]

in exploring the science and engineering of 2D MOFs structures. Finally, section concludes.

11.2 On the Dichotomy Between Art and Science

According to today's traditional thinking, Arts and Science (A&S) are separate with opposite values: imagination and freedom of thought for the former and rationality and rigorousness for the latter. Yet, the line that in the modern era tries to separate A&S is a superficial one. Art can stimulate intuitive and creative approaches to problems [9, 10], and A&S connections may trigger the invention of new techniques and foster scientific advances through the development of new aesthetic sensibilities [11, 12].

11.2.1 Creativity is a Recent Idea

A&S have in common the creative act, which requires the same attention to details, vision, contemplation, inventiveness, other than careful distinction between fact and illusion in what they observe [14]. However, only during the Renaissance *creativity* began to be considered a characteristic of human beings. The first valid attempt to analyze the creative process dates back only to 1926, when G. Wallas proposed the model [13] illustrated in Fig. 11.3. The core of this model is the primacy of observation and interpretation, as only critical judgment permits to disclose the only apparent inconsistency between selected observations, in the attempt to understand and interpret the world around us. Thus, artists and scientists both proceed through a decoding process and a subsequent recomposition of the reality they observe. At first, they both edit complex situations into a readable set of forms or ideas. As shown in Fig. 11.4, this process of simplification can be divided into four stages. The elements are then arranged in a way that furthers the artistic or scientific intentions. When all parts are integrated into a whole in which the relationships between the elements are mutually beneficial, balance and harmony are achieved. As mentioned in section and will be detailed in section, in an artwork the sense of balance is a source of pleasure for the viewer. Colors that stand in an orderly relationship and provide a sort of dynamic balance to a composition are perceived as harmonious. In science, mathematical relationships play a role in visual harmony where variations of shapes and angles exhibit mathematical consistency. Also elements that differ in scale by ratios that are consistent tend to confer harmony. In the tessellation process for instance (see section), grouping of shapes that have similar characteristics tend to form harmonius compositions.

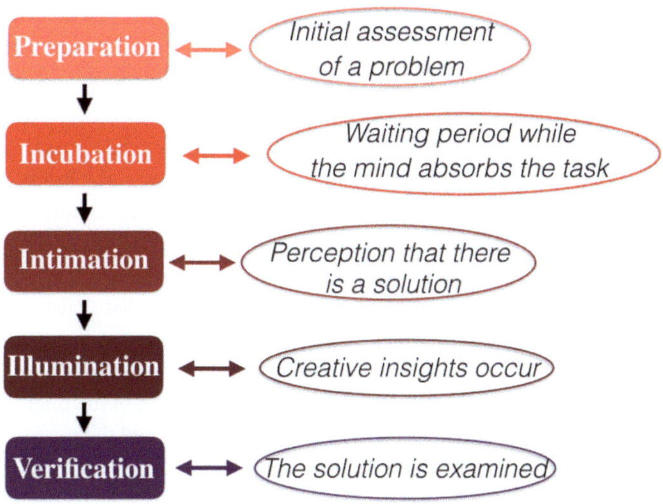

Fig. 11.3 The first valid model for the analysis of the creative process [13]. We can concede that it is perfectly suited to both Art & Science

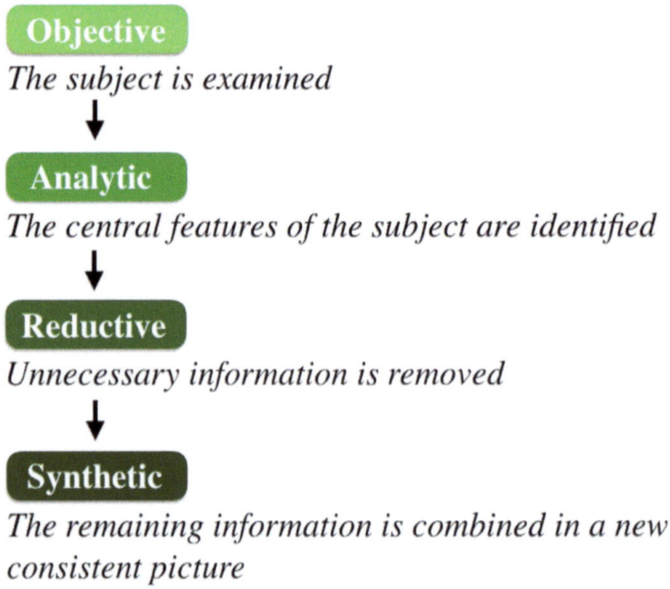

Fig. 11.4 The four stages concerning the process of simplification, which can be applied to an artwork or scientific project

11.2.2 The Term Scientist

Historically, A&S share their origins in the representation and interpretation of nature and have coexisted often indistinguishable from each other until the nineteenth century [15]. Their methodologies only diverged over time, consistently with earlier studies on the functioning of the brain. Accordingly, right brain is used for more intuitive (i.e., artistic) thinking while the left brain is used for more rational (i.e., scientific) thinking [16]. The separation between the two disciplines was acknowledged on June 24, 1833, in one of the first meetings of the British Association for the Advancement of Science. At that time there was the obsolete custom of calling anyone interested in the new sciences of nature by the name of *natural* or *experimental* philosopher. During the debate following the inaugural lecture, W. Whewell, a philosopher but also a passionate lover of sciences and aware of the ongoing process of specialization and separation of scientific disciplines, coined the term *scientist* for analogy with the term *artist*.

11.2.3 Difference Means Richeness

Despite the historical divergence about the method, scientists use art to help them vi-sualize abstract aspects of the environment. Artists in turn experiment with the materials generated by science and have their own ways to explicate ideas about the reality [17]. Eventually, the real divergence in A&S concerns how the message is delivered. Art commu-nicates to others the emotions the artist feels. Artists are skilled to generate an emotional response in their viewer. Art has also an important role in communicating scientific infor-mation, as they offer alternative ways to communicate science and have a role in helping people understand complex scientific information [18, 19]. By contrast, to satisfy our curios-ity about natural phenomena, science emphasizes objectivity and reproducibility, leaving out the personal, emotional side. Scientists often struggle to convey their research to the general public, and often find that their academic background does not encourage direct public engagement. On the other hand, artistic productivity needs science, otherwise it can become shallow or lacking in technique and precision [20]. A&S are then intertwined, and the interplay between the two disciplines as well as the continual exchange of views with which scientists and artists look at the world may have many advantages for scientists working on contemporary issues. For instance, it may address the growing awareness towards the great environmental challenges of the twenty-first century [21]. To this aim, it is highly desirable to go beyond individual disciplinary perspectives, or in other words, to transcend traditional A&S boundaries to generate accessible knowledge [22–24].

11.3 Actual Confrontation in A&S

The risk today is still to conceive A&S as separate, siloed disciplines. The confrontation between the two disciplines has a long history that begins with the first industrial revolution. At that time, the vision of a society governed by the new experimental philosophies was not shared by the humanists [25]. A dispute that still continues uninterruptedly.

11.3.1 The Two Cultures

Natural philosophers (section) such as F. Bacon exalted the potential of the new inductive natural science [26]. In the utopian novel *New Atlantis* (1626), he depicts a society based on.

the cognitive and realizing abilities of the *new science*. The plot is compelling from the very beginning: on the island of Bensalem (a portmanteau of Bethlehem and Jerusalem) an elite of new experimental philosophers who lead the community's most important institution, the House of Solomon, conduct scientific experiments to understand and dominate nature. The applications are then aimed at the progress and well-being of society.

The conflict between the emerging science and traditional culture deepened in the following centuries. Advocates of the new science emphasized the importance of the latter in a liberal education [27]. Critics believed that the new science had a limited perception on nature and emphasized the vital role of poetry in reciting the human experience [28, 29]. More recently, during a famous lecture in 1959 [15], C. P. Snow, chemist and physicist but also a novelist and writer of essays on various cultural topics, used for the first time the expression *the two cultures*. He pointed out that the intellectual life in Western society was increasingly being split into two polar groups. Snow's belief was that the lack of communication in A&S would have led to the failure to solve the great problems of humanity. At the same time, the spread of scientific culture—thanks to its ability to influence developments of the economy- would have had a significant impact on the political and social behavior of the population with higher education, directing them to a more mature understanding of the problems of underdeveloped countries.

11.3.2 Reconciliation

Part of the humanism has developed a sort of mistrust towards science, which has been linked to the idea of an irremediable separation between the two cultures. In turn, this has led people to look at scientific achievements with suspicion, for fear that they may orient the development of society towards a form of Orwellian power of technicians, capable of conditioning the freedom and humanity. This old-fashioned separation between scientific and humanistic culture still represents an important limitation to the educational development in Western countries. Nevertheless, we can look forward to the reawakening in recent years of the collaborative spirit and growing interest in A&S, with emphasis placed on observation and careful documentation [30, 31].

11.4 The Unitary Perspective

The purpose of a scientific experiment is to question about *Nature*, although the result is conditioned by how the question is asked: what tools, under what conditions, and so on. In turn, the answer is often a source of further questions. And even when everything seems clear, we know that emerging models and theories can always be called into question by new interpretations and more sophisticated experiments. One of the effects of this process consists of a growing specialization and diversification in the field of culture and knowledge. Thus, while the Renaissance was a period of philosophical, artistic, and scientific rebirth exemplified by the works of Leonardo Da Vinci, which were equal parts art and science, today we assist to an extreme specialization in all fields of knowledge. The reason for the latter is two fold: (a) when one falls in love with a hypothesis, it turns out challenging to take into consideration observations that fail to fall in harmony [32]. (b) Testing and refining hypothesis tends to draw scientific enquiry into ever smaller circles of interest, at the risk of creating islands of insight in a largely unexplored sea of ignorance [10]. It is therefore not surprising to find physicists who do not know what a gene is, and biologists who guess that string theory has something to do with violins [33]. In this scenario, the question is whether to prefer an approach aimed at identifying the differences between the different cultural expressions and then proceeding to search for possible interactions or rather an approach that highlights a common denominator that automatically highlights the possible differences and convergences. The first approach has led in the past, and perhaps still today, inevitably to a conflict in A&S. Since the aim is not a classification of the various types of A&S interactions but rather to demonstrate the inexhaustible reservoir of common experiences, retracing the path of interactions in A&S can be particularly significant in recovering a correct perception of science as a cognitive process.

11.5 Everything is Illuminated

Speaking of a unified perspective in A&S, light is the epitome. Light is used in art to convey a lasting impression of emotion, thought, or to emphasize the focal point of the piece. It is so well integrated in common and contemporary art that it has its own branch of art dedicated to it: luminism, an art form in which light is the primary mode of communication.

In science, light is heavily studied through the electromagnetic spectrum, and the scientist's interpretation of light depends on their formal background. The physicist is interested in the physical properties of light (see example in Fig. 11.5), the chemist in the synthesis of new dyes, and the artist in its aesthetic value. Nevertheless, light is a widely studied phenomenon across all of these professions as it constitutes the basis through which the world is experienced for many living organisms.

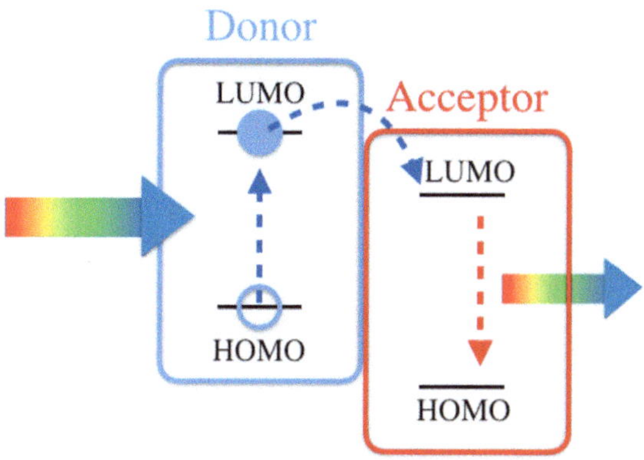

Fig. 11.5 Depiction of one of the central aspect in photoconductivity. After irradiation, the electron in the ground state (HOMO) of a chromophore may undergo a vertical transition to first excited state (LUMO). Under the influence of an external field, the electron may diffuse from the donor to the acceptor and decay to the ground state, generating in turn light

11.5.1 Shapes of Light, Play of Light

The study of light has been founded upon centuries of thorough and thoughtful investiga-tion into the essence of light itself. Pythagoras proposed that sight is caused by visual rays emanating from the eye, Ptolemy tabulated the pairs of angles of incidence and transmission for combinations of several media [34]. In Christianity, light shining on golden halos has been used to signify a contrast between the divine and the earthly [35]. The usage of gold in the creation of early christian manuscripts creates a stunning visual effect when the page is turned, helping capture a sense of awe and devotion from the reader. During the Middle Ages, light has functioned as a metaphor for the divine, and is evidenced in the stained glasses that still exist today. Leonardo Da Vinci was one of the first people to scientifically.

study the properties of light during the Renaissance where he studied the effect that distance has on how light and shadow is perceived in landscapes, employing his findings to create perspective and space. Newton speculated that transparent objects exert attractive forces on the particles, with the consequence that the speed of light in a transparent medium is always greater than the speed of light in a vacuum [36].

11.5.2 The Art of Lighting

In art, light is an essential tool in communicating a sense of shape and space serving to help build painting depth and perspective, and it endows objects with form and thereby purpose. The eye is naturally guided by light onto the canvas and more particularly the object it touches. In impressionist works, the features of light are fully explored and careful attention is paid to the behavior of light. The painter reproduces what he sees guided from his point of view and not by mere imitation of an objective reality. And the spectator after him, sees a show mediated by his own cultural conditioning. Claude Monet, used light to explore reflections in the water, showing shapes to the viewer through their reflected image [37]. In doing so, a surreal effect is created where things only become clear when they are studied through their relationship to the light. Monet is able to soften the distinction between what is real and what is imagined, using pleasing colors to lull the viewer into his world. Indeed, one of the beauties of light, lies in its ability to build up symbolism.

By "shining a light" on a piece of art or on certain elements, a story is constructed. Our eyes tend to assign varying senses of weight to elements within the composition (Fig. 11.6). When the elements are dynamic—i.e., they feel as they are in movement—and they seem to be balanced across the painting, the viewer experiences the pleasure which stems from this balance. In Renoir's painting *Le d´ejeuner des canotiers*, Fig. 11.7, the analysis may be further developed on the basis of the geometric construction of the scene, the perspective game and the correspondence of the gazes. The scene offers a multiplicity of reference centers which are not mutually exclusive but alternate during the observation. Because of the correlations and cross-references in the representation, the characters lose importance in their static nature. Likewise in a molecule, what counts is the relationships between characters and reciprocal interactions.

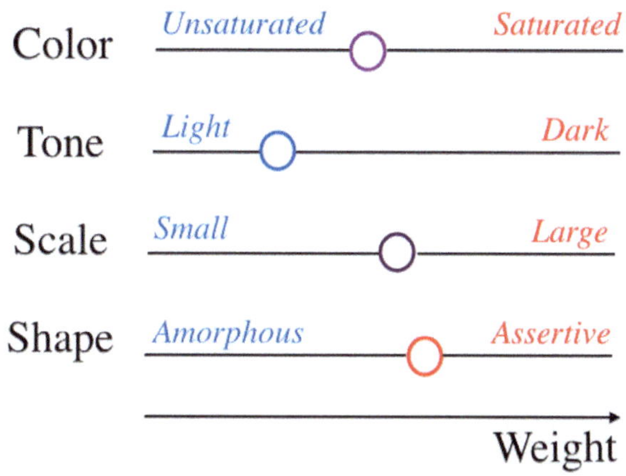

Fig. 11.6 Schematic representation of the research balance in a composition. The assignment of weight to a particular element can be affected by color, tone, scale (i.e., larger elements are read as heavier) and strength of shape. Further, the sense of weight assigned to an element may be balanced by its relationship with other elements within the painting

Fig. 11.7 *Le d´ejeuner des canotiers*, Pierre-Auguste Renoir (1881). The brushstroke conveys the feeling of vitality of a warm day at the *La Fournaise* restaurant, nearby Chatou (France). The painting reminds a snapshot of provincial life that crystallizes the actors of the scene in a precise position. However, what counts is the weight given to each element of the scene as well as their balance and reciprocal interaction

11.5.3 Light on Color

Despite the true nature of light not yet being fully understood, it has been employed for as long as art has existed to help create color, determine form, and create shading. But is colour a purely visual phenomenon—i.e., the result of placing paint on a canvas surface or it simply related to temperature, extending from the infrared to the ultraviolet and beyond [17]? Certainly, in architecture color is often the place where science meets arts to generate compositions of exquisite beauty. Let us consider for instance the glass windows that adorn many European cathedrals (see Fig. 11.8). The richness of the colors is obtained by a technique that is still widely used today, which consists of introducing metal ions from the family of transition metals, such as copper, titanium, vanadium, into the glass matrix, which is essentially made of silica. The addition of these ingredients lowers the melting point of the matrix, obtaining its coloring. Yet, the latter depends also on the degree of oxidation of the metal. To explain the phenomenon, when the metallic ions receive light, they absorb part of the energy transported by the photons, causing an excitation of the electrons in their external orbitals. These electrons then evolve from their ground energy state to higher energy levels. Depending on the gap energy between levels, the ion absorbs a certain light frequency and then emits the complementary color. Thus, the variety of the colors obtained by this technique depends on the available differences between energy levels. Typically, copper makes blue glasses, titanium makes violet, vanadium green, and so on. Glassmakers have a variety of technologies for introducing transition metals to produce the colors they want. Nowadays the collaboration between chemists and physicists has made it possible to understand and control the structure of glasses at the nanoscale, transforming the glass industry into a high-tech industry whose applications are used in transport, electronics, security, to name a few.

11.5.4 Light and Technology

Nowadays, light forms the basis of many communication technologies given its innate ability to transmit spatial and temporal information. Technological applications based on the various properties of light gave rise to lasers, holography, and fiber-optic telecommunications systems, all of which are essential components to the modern world and everyday life. For instance fiber-optic telecommunications have forever revolutionized the way in which data is transmitted globally, connecting continents through an intricate system of underwater cables that has been so well integrated into society, society itself has become dependent on it. With so much dependency on the properties of light to function there is naturally heavy interest in exploring its capabilities to the fullest extent. Ultimately, light is a mechanism through which all things pertaining to life are perceived by the human eye. Despite the complex mathematical models and the advanced physics used to describe its properties, light

Fig. 11.8 The stained glass windows of Chartres Cathedral (France), XIIIth century. An impressive mastery of glass coloring

has always been a simple concept at heart, and has always guided the human eye into the next step and into the burgeoning new horizon.

11.6 Applied Art

Suppose we want to design a two-dimensional object that is not only attractive from an aesthetic point of view but that is also of practical use. The first step is *to tile* a representation of the object, according to the intention on the part of the artist or scientist to achieve specific results.

11.6.1 *The Art of Tiling*

In the tessellation or tiling process, closed polygons or shapes are regularly arranged to completely cover a plane, without overlaps and with no gaps. As represented in Fig. 11.9, the result consists of uniform (or irregular) two-dimensional (2D) patterns. In particular, tiling deals not not only with the constituent tiles, but also with the edges and vertices where these tiles meet. Collectively the tiles, vertices and edges of a tiling are called the elements of that tiling [38]. Tiling patterns may be observed in our everyday life, from the (macro) modern architecture to the (micro) beautifully chemical designed 2D frameworks composed of repeating patterns of molecular building units. Tessellation has been used [40] to tile regular polygons for decoration in art since antiquity, likely reaching the apotheosis in the sumptuous traditions of the intricate tiling in Islamic culture [41] (Fig. 11.10a). An-cient Greek mathematicians such as Pappus of Alexandria (fourth century AD) were aware of the regular or Platonic tilings, where all tiles are identical regular polygons [38]. Tiling is not only a mathematic theme but it may find expression in crystalline materials (Fig. 11.10b). Regular tessellation is found in natural and synthetic crystals, such as metal–organic frame-works (MOFs) [42, 43]. Other tiling expressions comprise semi-regular tessellations, which have been reported in polymer assemblies [44, 45] or in interfacial self-assembly [46], liquid crystalline networks [47], self-assembled DNA structures [48] or nanoparticle and colloidal supperlattices [49]. Recently, the isolation of atomically thin crystal layers has given rise to the emergence of the so-called two-dimensional (2D) materials. The latter have shown a wide range of electronic and magnetic properties that can be affected by the dimensional-ity. Most of these materials derive from layered inorganic materials, formed by covalently bound layers interconnected by weak van der Waals interactions, allowing the isolation of monolayers by exfoliation (see section).

11.6.2 *Escher in Translation*

The Dutch artist M. C. Escher frequently used tessellation in his art and has often been a source of inspiration for those interested in geometry, including mathematicians, scientists and crystallographers. In the artist's "Day and Night" (1938, Fig. 11.11), the viewer's at-tention is first of all captured by the birds flying in opposite directions. Then, the striking symmetry obtained the sharp juxtaposition of abstraction is disclosed. Escher took the ad-vantage of mathematics to analyse objects and figures, composing regular divisions on the surface without any space between them by creating abstractions in one or more patterns, creating contrast effects with lines and white–black areas. Between his most striking works are the drawings in which he uses figures bringing us to the contradiction. Escher's con-tradiction primarily appears in works that deal with issues related to time and space. This often necessitated a system that dealt with the confrontation of logic and language, which Escher would later

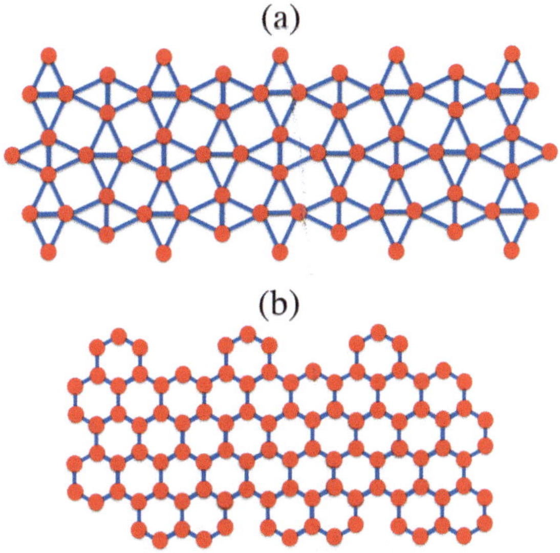

Fig. 11.9 Schematic models of **a** snub square and **b** hexagonal lattices. In two dimensional metal organic frameworks, metal atoms and molecular linkers may be represented by red dots and blue frames, respectively

blend synonymously to create an order in which these two contradictions can be integrated respectively. Escher's contradiction primarily appears in works that deal with issues related to time and space. This often necessitated a system that dealt with the confrontation of logic and language, which Escher would later blend synonymously to create an order in which these two contradictions can be integrated respectively. For instance, in "House of Stairs" (1951), Escher's impossible reality comes to life with his employment of the penrose stairs: a two-dimensional depiction of a staircase, ninety degrees at each corner, that forms a closed loop such that a person on those stairs could never get higher or lower. The motif of infinity is commonly found in Escher's work and represents the confluence of his fascination with spatial reality. Escher's work would later go on to encompass additional challenging mathematical concepts including platonics and hyperbolic geometry that would later gauge him a reputation for being one of the few artists with the ability to represent in-finity on a two-dimensional plane [50]. Today several strands of research are inspired directly by Escher. Between them, the classification of regular tilings and counterchange symme-try or antisymmetry, color symmetry in crystallography, topological change in symmetric patterns, orderliness not induced by symmetry groups, and many more [51].

11.7 Beyond the Monolayer

So far we have mainly discussed *in-plane* artworks and tiled objects with potential application in science. Let us assume now to vertically stack the monolayers represented in Fig. 11.9. The new structure's properties are not just a physical superposition

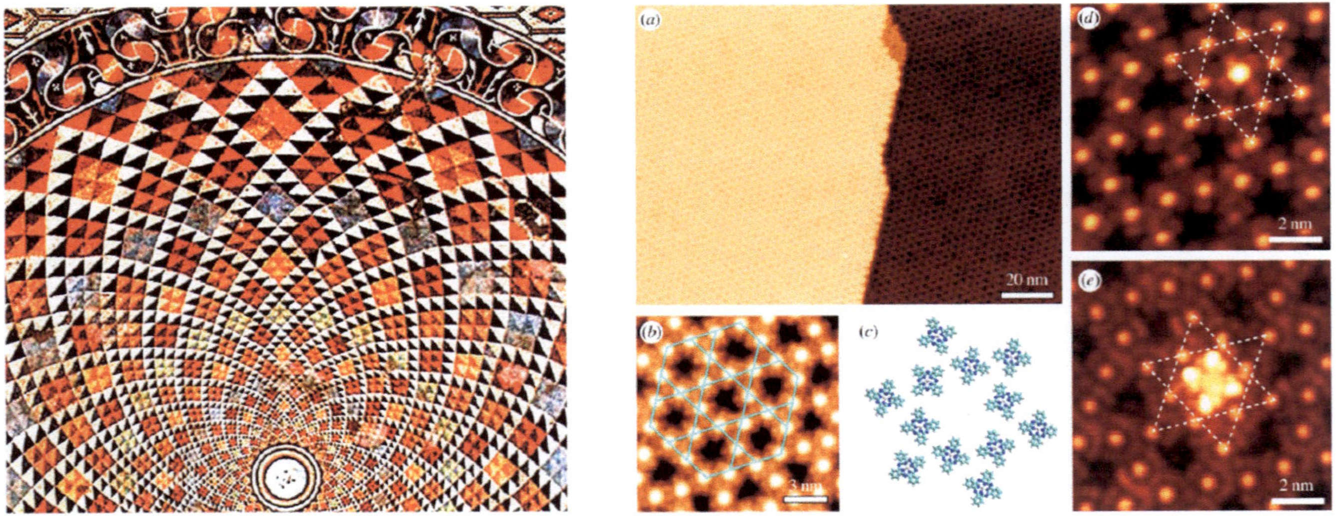

Fig. 11.10 (left) Example of Islamic tile patterns and (right) scanning tunnelling microscopy (STM) image of the lattice of iron phthalocyanine (FePc) molecule taken from Ref. [39]. We can infer that tessellation may be observed in our everyday life, from the (macro) architecture to the (micro) chemical designed 2D framework

Fig. 11.11 M. C. Esher, “Day and Night”, 1938

of those of their individual layers, but are a complex outcome of the interlayer coupling, lattice recon-struction and twist angle, which can manifest intriguing new phenomena.

11.7.1 Supramolecular Chemistry

Supramolecular Chemistry (SC) is the chemistry beyond the covalent bond [52]. Since the properties of a material depend both on the nature of its constituents and the interactions between them, SC has a strong impact on material science as it is able to manipulate the non-covalent forces that hold the constituents together, creating a wide range of ever more sophisticated molecules [53, 54]. As represented in Fig. 11.12, the constituents are linked through reversible weak connections. Hence, supramolecular materials are by nature dynamic materials [56] and in specific conditions they undergo assembly or deassembly processes. Concepts and findings in SC have been rapidly employed in different fields, such as pharmaceutical sciences, catalysis, chemical and biological sensors, or molecular electronics. In particular, two-dimensional (2D) SC and crystal engineering of solids have also evolved significantly over the years [57], providing a wealth of information on assembling molecules at solid interfaces. Although in 2D the number of possible arrangements is restricted, the additional interfacial interactions increase the complexity of the *self-assembly* process, the latter based on molecular *recognition.* Depending on temperature, molecules recognize other molecules or ions and associate with them via non-covalent interactions, such as hydrogen bonds, van der Waals forces, dipole–dipole and π–π stacking interactions (see Fig. 11.13). The forces between assembling units are often weaker than the covalent bond but collectively allow formation of supramolecular nanostructures. While this at first sight may appear as a disadvantage, the understanding of the

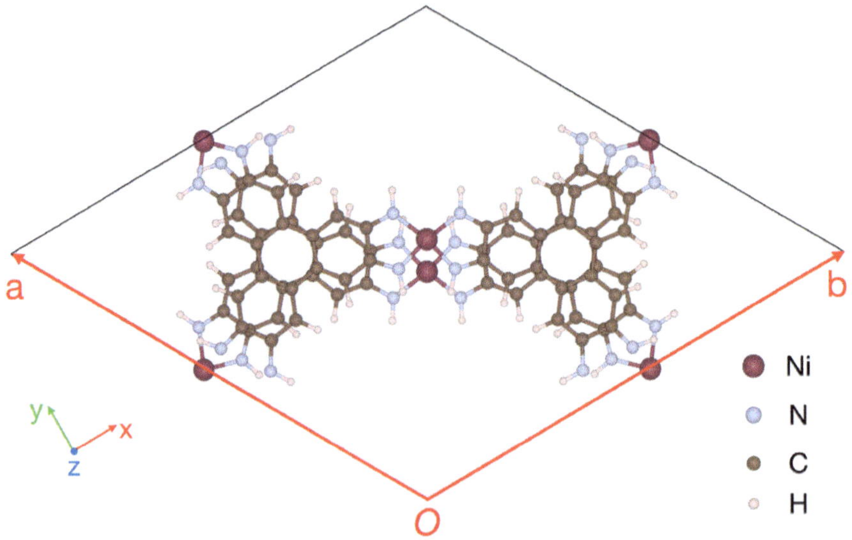

Fig. 11.12 Example of overlap stacking in the archetypal layered MOF Ni3(HITP)2 (HITP = 2,3,6,7,10,11–hexaiminotriphenylene) (see Section). *a* and *b* correspond to the cell vectors and the two layers are held together by π–π intractions. Further details may be found in Refer-ence [55]

interplay between these interactions can provide valuable informations for the fabrication of complex interfacial architectures with novel properties and function [58]. However, the big question in crystal engineering is how to design crystal structures by using the molecule as a building block. Further, in an unknown crystal structure it is still hard to solve the dichotomy between chemical interactions and geometrical close-packing. While all structures are nearly close-packed, great importance is provided by small deviations from close-packing, because these small deviations lead to the formation of crystal structures that can be engineered in a systematic manner. In line with an evolutive chemistry, the next step in complexity consists of the design of chemical "learning" systems, systems that are not just instructed but can be trained [59]. In this way, SC may open new perspectives in materials science toward an area of supramolecular *smart* materials whose features depend on molecular information.

11.8 The Moire Superlattice

The word *moir´e* (French etymology) refers to two layers of watered silk pressed together (ref). When monolayer materials are vertically stacked, new properties only emerge if a sufficiently strong interlayer coupling is formed, which facilitates the hopping of electrons and redistribution of charge between the layers. When the interlayer coupling is made strong, the independence of each layer is reduced, and

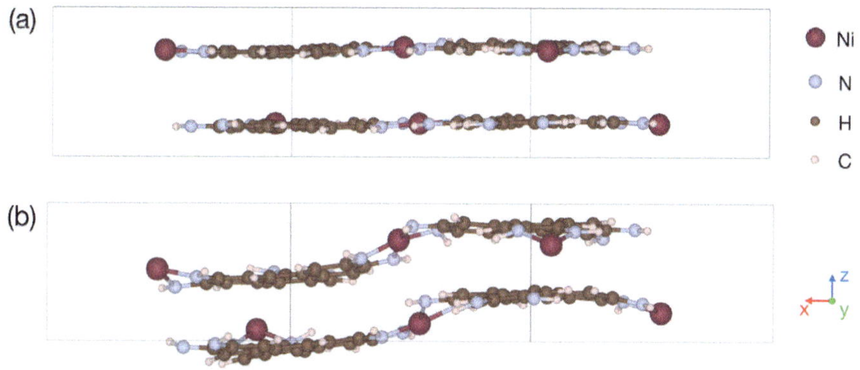

Fig. 11.13 **a** The optimized structure of Ni3(HITP)2 2D MOF at 0 K. **b** The equilibrated structure of Ni3(HITP)2 2D MOF at 293 K. The planar layers allow extended *in—plane* orbital conjugation while thermally–induced structural motions lead to deformation of the layers from planar geometry, changing the charge transport pathway and the electrical conductivity of the system

a superlattice forms that induces new properties. Thus, the interlayer coupling can facilitate the emergence of a new long-range periodicity, unique band structure, and dramatically new properties. When a small twist angle or lattice mismatch occurs between the layers of two-dimensional (2D) materials, the layers slide across each other, generating an interference pattern called the moir´e superlat-tice (Fig. 11.14). The latter is characterized by reduced electronic bandwidth may arise, providing a periodic modulation of the stacking [60]. In this situation, electronic interactions have a dominant role, leading to various correlated and topological phases [61–63], resulting in the appearance of many fascinating electrical, optical, and magnetic properties, such as gate-tunable unconventional superconductivity [64], photonic dispersion engineering, and ferromagnetism [65, 66]. Unexpected distortions in band-structure and topology lead to long-range correlations and charge-ordering [67]. The way in which the atoms of the two layers interact creates intricate geometric patterns, and these patterns have a decisive impact on the material properties. Interesting patterns may emerge—the well-known moir´e effect -By turning one of the layer of a small angle, so that there are places where meshes roughly match, and other places where they do not. The twist angle (i.e., the relative angle between layers) is crucial in regulating the correlated performances of moir´e structures. It determines the new periodicity (i.e., moir´e pattern) formed between the individual crystalline layers, while the interlayer coupling determines the magnitude of electron-particle interactions [68], hybridization, charge redistribution [69], and lattice reconstruction. Particularly, the regula-tion of the interlayer twist angle has the ability to control the length of the moir´e superlattice and the gap between its valence and conduction bands [61, 70]. If it is true that it was already known that electronic properties of a 2D material can be changed by joining two layers to-gether, the fact that the mechanical oscillations in the material can also be controlled by this opens to device design opportunities. Lattice vibrations of the atoms—called phonons -are significantly influenced by the

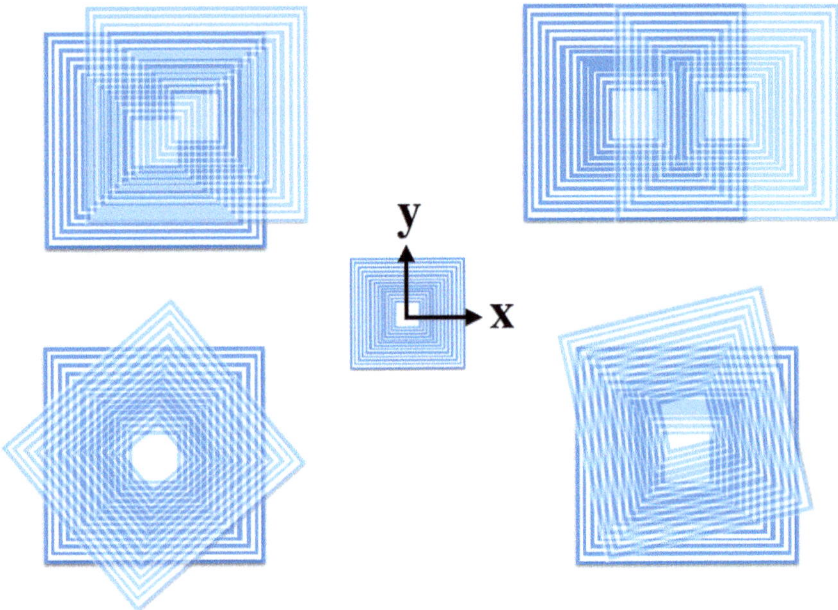

Fig. 11.14 Examples of Moir´e patterns obtained by *in*-plane sliding (top) and rotation (bottom)

angle at which the two material layers are stacked. Thus, whenever a tiny rotations of such a layer occurs, a change of the material properties can be engineered.

11.9 The Magic World of Twistronics

As introduced in the previous section, the moir´e pattern produced by the interaction between stacking layers induces the formation of a new periodic superlattice with unique band structure. Now, let us assume to twist one layer with respect to the consecutive one along the stacking direction: how does the twist angle affect the band structure and then the properties in 2D materials? When the interaction between layers is sufficiently strong, the moir´e pattern is influenced by atomic reconstruction, by which the atoms shift position due to an interplay between inter- and intralayer bonding. More specifically, depending on the specific 2D structure and the twist angle, the interlayer interaction induces a force that leads a repositioning of the atoms. The result is a new equilibrium configuration with a balance between the inter- and intralayer bonding. Further, the moir´e pattern evolves continuously with twist angle, suggesting the superlattice—and corresponding band structure—evolve as well. Twistronics is the field of how the relative angle (or twist) between the layers of a vertically stacked 2D structure affect its properties [71]. This effect is unique to 2D materials and in certain 2D structures, the effect is so transformative that the corresponding twist angles are described as “magic” [61, 62]. For instance, the twist angle has been shown to manipulate a wide-range of properties, including moir´e excitons [72, 73], Hall effect [74] strong electron–electron correlations [75], exciton photoluminescence [69], twist-dependent color [76], and spatially-dependent vertical conductivity [77, 78]. In particular, differences in twist angle as small as 0.1∘can induce astonishing transformative changes such as superconductivity [62, 64]. To better understand how the twist changes the physics, a model [79] was introduced to account for the way carbon atoms in the twisted sheets of graphene align more exactly in some places than in others. According to this model, as the angle varies, it alters the complex patterns of the regions where all the carbon atoms are fully aligned, and those where only half the atoms are aligned, thus modulating the tunneling between layers. As a result, in some regions the tunnelling is prohibited while in other electrons flow without resistance. The emerging large body of literature suggests that twistronics is as important as other major variables—such as doping or crystal structure—and holds significant promise for next-generation technologies.

11.10 Metal Organic Frameworks

Metal organic frameworks (MOFs) are crystalline coordination polymers in which metal nodes are connected by organic linkers, see Fig. 11.15. The coordination between the latter permit to design multifunctional MOFs with adjustable porosity and tunable functionality within a single material. In general, the construction principle of these frameworks is to predict the topological evolution by monitoring geometry and structural periodicity. As a result of their permanent porosity, tunability and multifunctionalization, MOF materials represent nowadays a hot topic in synthetic chemistry and materials science. However, precise control over the composition, structure and properties of MOF materials is still in its infancy. In this section the two main characteristics related to these compounds, namely porosity, photoconductivity and flexibility are outlined. The reader interested in each of these topics is referred to excellent articles and reviews.

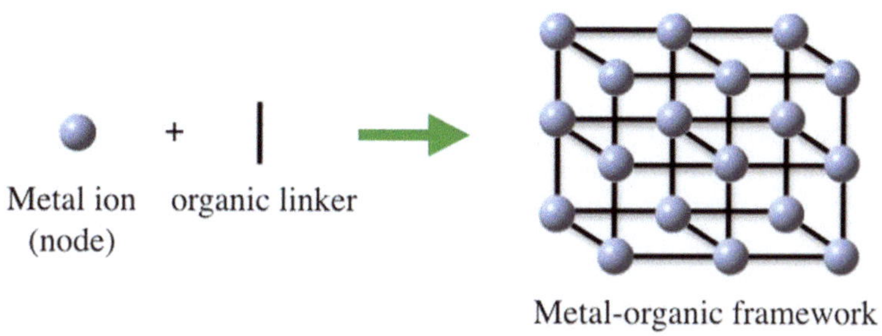

Fig. 11.15 Schematic representation of a metal organic framework (MOF)

11.10.1 Porosity

Environmental issues require new materials that can perform better, faster and are en-vironmentally friendly [21]. Porous materials [80–83]—i.e., solid matter permeable to fluids due to the presence of pores—are the most successful catalysts in the chemical process industry. Depending on the framework building blocks, they have been generally classified in two categories: purely inorganic [84] and purely organic or carbon-based materials [85]. MOFs exhibit a multitude of similarities with the well known zeolites [86], like crystalline structure and well-defined pore system, surface functionality and chemical reactivity. On the other hand, they differ substantially in terms of chemical nature, pore size and avail-able pore volume, thermal and chemical stability. Further, with respect to zeolites which have been used in the chemical process industry for over fifty years, MOFs more recently have begun to bridge the inorganic/organic gap. From materials-design perspective, MOFs can be achieved by controlling pore and skeleton design. MOFs can vary their size—from microporous to mesoporous ones [87–90]—by progressively lengthening the organic linker. Thus, due to their framework built of metallic atoms connected by organic linkers, MOFs exhibit much higher flexibility than inorganic materials and their surface chemistry can be modulated by in situ or post-synthesis methods and thus adapted to different applications. As a result, wide applications of MOFs have emerged in different fields, such as in catalysis, guest adsorption (molecular recognition), gas storage and water treatment, among others. Multivariate MOFs with increased complexity and functions can be obtained by integrat-ing multiple ligands with identical linking chemistry [91]. Variation in size and shape of the linker within the material implies synergetic coordination of the different linkers with the metal [92]. In this way, functional groups may be arranged within the pores of MOFs resulting in new properties that are not simply linear sums of those of the single compo-nents [93]. The reader further interested in the incorporation of two or more organic linkers into framework materials is referred to topic reviews [94, 95]. Theoretical works aimed to examine physical and chemical properties of MOFs by electronic structure calculations can be found in Refs. [90, 96–98].

11.10.2 Photoconductivity

Besides the received increasing research attention due to their unique conductive properties [2, 3], two-dimensional (2D) metal–organic frameworks (MOFs) have been recently investigated as potential photocatalytic materials due to their tuneable electronic and opti-cal behavior [99–101]. In 2D MOFs atoms are arranged into layers held together by in-plane covalent or coordinative bonds. These atomic layers are then stacked in the out-of-plane direction with rather weak van der Waals (vdW) interactions creating parallel 1D chan-nels. This architecture allows bulk exfoliation and nanosheet manufacturing, down to the single layer limit [102, 103]. Compared to their 3D bulk counterpart, 2D MOFs present ultra-thin thickness [102], large surface area [1] and efficient charge separation [104] which enhance the transfer of the captured energy to external acceptors, boosting the photocatalytic per-formance of the device [105]. Indeed, these systems may exhibit non-zero bandgaps [106], as proved by their intrinsic temperature dependence [107]. This semiconductive behav-ior was recently confirmed by our calculated electronic properties [55] for the archety-pal layered MOF Ni3(HITP)2 (HITP = 2,3,6,7,10,11–hexaiminotripheny-lene) [106, 108, 109]. Through band structure calculations, we have shown that finite temperature induces a com-petition between direct and indirect band gap openings in 2D layered MOFs [55]. This means that generation and recombination rates of the photo-induced electron–hole (e–h) pairs and then the photoconductive performance of the 2D MOF device would rely not only on the wavelength dependent absorption rate of the material but also on the balance between mo-mentum and energy conservation at the Fermi level (see Fig. 11.16). For instance, it is known that the photo-emission efficiency of light-emitting diode (LED) devices relies on the e–h recombination through a direct band gap [110]. On the other hand, in indirect band gap re-combination is mainly controlled by phonon-assisted nonradiative processes and the intrinsic luminescence is negligibly small. In this case, techniques such as optical near-field [111] in which the excitation is realized without phonon assistance may open new opportunities for manufacture of light-harvesting 2D MOF materials based on indirect band gap. Yet, the understanding of structure–property relationships in these systems is essentially in its early stages, suggesting that with new knowledge significant improvements on photoconductive properties may be achieved. As shown in Fig. 11.17, questions that can be addressed include determination of what kind of crystal packing along the stacking direction is most favorable for charge transport and whether the van der Waals (vdW) interactions can be enhanced by tuning the crystal packing of the 2D framework. In this respect, we have shown [7] that in 2D MOFs the intermolecular overlap of the electronic wave functions and thus the transfer integral (t) depend delicately on the intermolecular distance and orientation and it scales with the reorganization energy (λ) (see representation in Fig. 11.18). Our studies focused on how the control on the quantum confinement along the π— stacking of Ni3(HITP)2 to-gether with the engineering of t and λ parameters may extend the set of design principles for improving the photoconductive properties in 2D MOFs (see Fig. 11.19).

11.10.3 Flexibility

As shown recently [113, 114], π—stacked layered 2D MOFs are flexible materials with a wide range of dynamical motions especially in the presence of an external stimuli such as temperature, pressure, electric and/or magnetic field. Even the optimized Ni3(HITP)2 at 0 K shows layers that are slipped in the *ab* plane (~1.44 °A) and a small twist of ligands around the central Ni site (~177∘). In order to study the flexibility in these compounds, we carried out [55] ab initio molecular dynamics

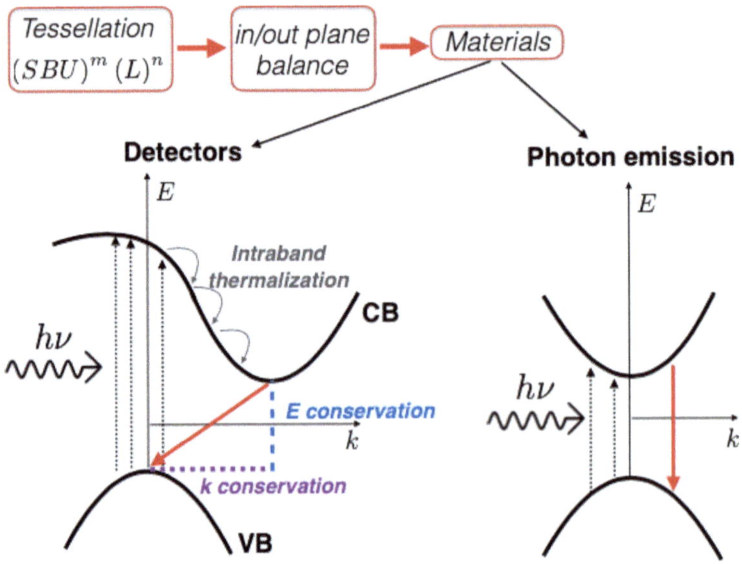

Fig. 11.16 Provided a certain degree of tessellation (Section), materials can be designed for photodector or photodiode purposes by controlling the momentum and energy conservation at the Fermi level

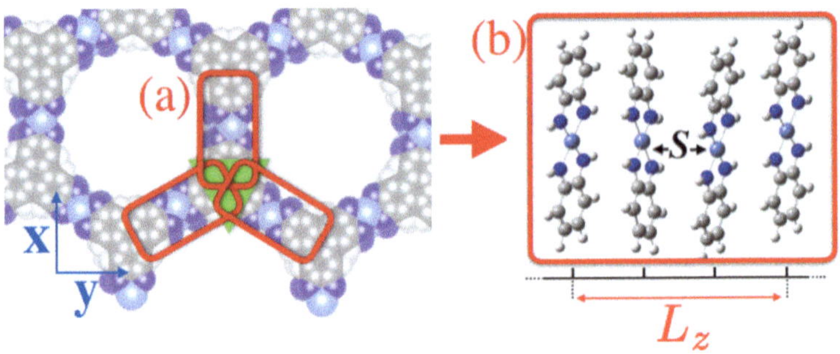

Fig. 11.17 Representation of the π-stacking model employed to fine-tune the vdW interactions in Ni3(HITP)2 [7]. **a** Since each unit contains two benzene rings, the triphenylene linker region highlighted by the green triangle is optically invariant to relative *in*—plane slides between layers. As represented in Fig. 1, in layered Ni3(HITP)2 the atoms are arranged into planes held together by *in*—plane covalent bonds. **b** The atomic layers are stacked in the out-of-plane direction with weak vdW interactions. Lz and S define, respectively, the the non–covalent π—stacking length and inter-layer spacing between two consecutive units

(AIMD) simulations at 293 K temperature and 1 atm pressure on Ni3(HITP)2 in the NPT ensemble (constant pressure/constant temperature), allowing the simulation box to vary, for a total production time of 8 ps.

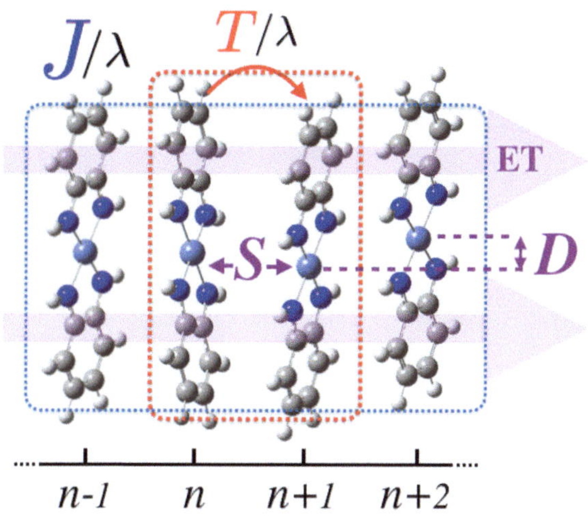

Fig. 11.18 Details on the model shown in Fig. 17. It consists of a linear array of N out-of-plane secondary building units (SBUs) [112] with periodic boundary conditions [7] along the π—stacking. The model used to study the energy transfer process along the π–π stacking direction in layered MOFs. Blue and red dashed frames represent the long and short range interactions determined by the J and T parameters, respectively. In turn, these quantities scale with the reorganization energy λ and depend on the interlayer spacing (S) and displacement (D) between consecutive units

The layers in the equilibrated system showed a stepped geometry unlike the planar ge-ometry of the 0 K optimized structure, with a more vivid range of dynamical motions. Figure 11.20 demonstrates schematic representation of these motions as well as their statistical analysis in Cartesian coordinates during the etire 8 ps AIMD simulations. As evident in Fig. 11.20a the in-plane slipping motion of the layers, represented by the position of the Ni atoms in each layer, is a continuous movement and can occur up to an extra 1 °A displace-ment in the x direction and up to 1.5 °A along the y direction. Another motion revealed in the equilibrated system is the breathing of the layers in the stacking direction as depicted in Fig. 11.20b. Changes in the inter–layer distance has a great impact on both the in–plane and out–of–plane electrical conductivity and hence one can envision the importance of including temperature and pressure in the simulations in order to paint a more realistic picture of CT pathways and mechanisms in these highly flexible materials. Finally, Fig. 11.20c depicts the out–of–plane motions of the N atoms around the central Ni centers, indicative of a stepped geometry for the layers as four amino groups are divided to two groups of two in the z di-rection. Moreover, the never–ending motion of the N atoms during the AIMD simulation as shown in Fig. 11.20c reflects a continuous SBU deformation at 293 K, resulting in a constant change of coordination environment from square planer to nearly tetrahedral.

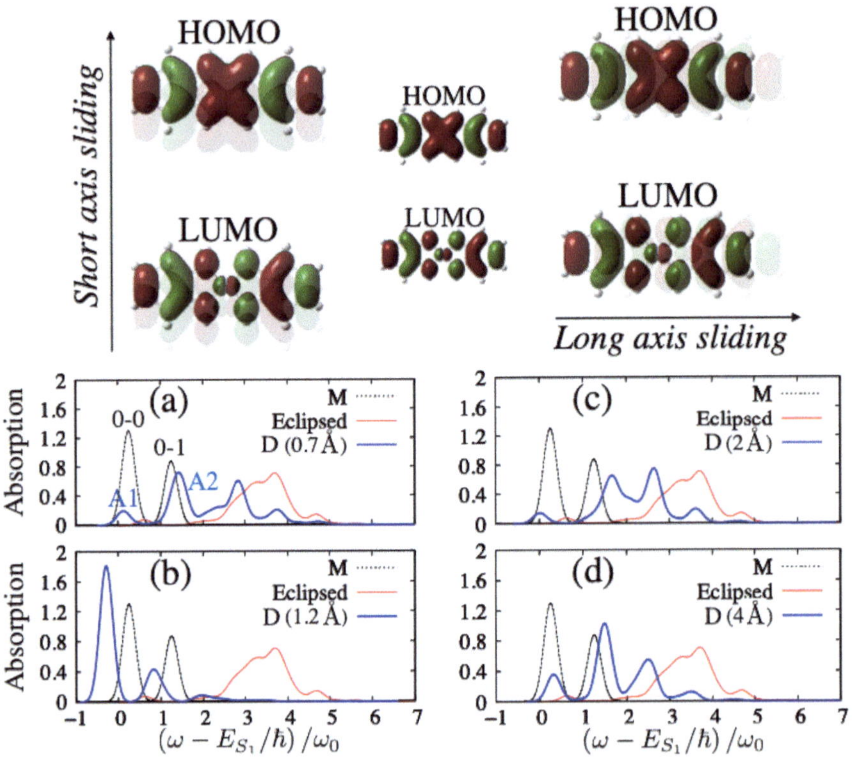

Fig. 11.19 Short (a-b) and long (c-d) axis sliding effect on the absorption spectra. In each panel, spectra for the monomer M (black, dashed) and fully eclipsed (red) are displayed. This study is detailed in Ref. [7]

11.10.4 Concluding Remarks

"Art is a province in which one finds all the problems of life reflected."

—Josef Albers.

In all cultures, artworks are separated by the world by formal boundaries. In Western paintings, the picture frame formalizes this separation, tribal artworks are confined to the structures on which they are painted. Yet, in the late twentieth century a few artists have started to blur the boundaries of the artwork. The reader for instance has possibly experienced a walk through the space occupied by installation art. Of course, one may argue that even this experience is confined by the gallery space. But what matters is the attitude. Recent skepticism towards the efficacy of vaccines or emerging theories on flat Earth should remind us that we live in a post-truth society, in which scientists must defend the objective interpretation of data but also learn from artists how to communicate facts and results. The good news is that the close methodological analogy between the artist and the scientist is as integral today as it

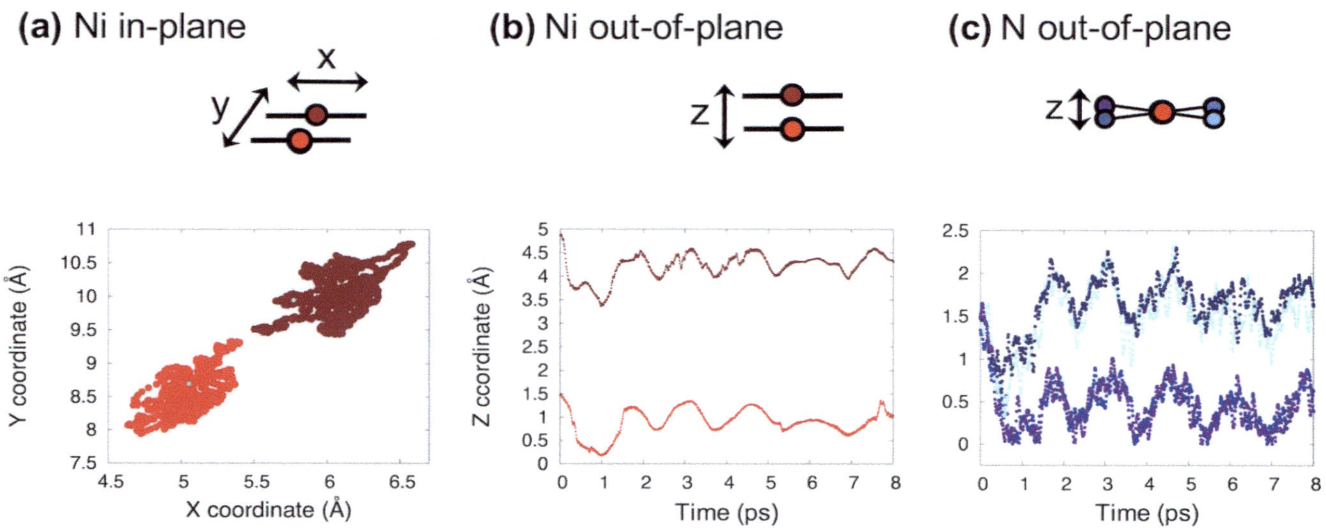

Fig. 11.20 In-plane vs. out-of-plane motions of the secondary building units (SBUs) in Ni3(HITP)2 during an entire 8 ps AIMD simulation. **a** Motion along the *x* and *y* axes for two Ni atoms belonging to the first (red) and second (brown) layer. Initial positions are shown as green points. **b** Out-of-plane motion of the Ni atoms in the first (bottom) and second layer (top). **c** Out-of-plane motion of N atoms around the central Ni metal sites. This study is detailed in Ref. [55]

was in the Renaissance. The hope is that the tenacity and perseverance showed by artists as M. C. Escher to learn and integrate new mathematical and philosophical ideas in their artworks may drive the imagination of future scientists.

Some of the greatest challenges of modern society are related to energy consumption and waste. Art can inspire the design of future technologies based on nanoscale materials which hold great potential for improved energy conservation, conversion, or harvesting. Among these, the diverse variety of atomically-thin 2D materials provide unique opportunities for creating new nanostructures through vertical and horizontal stacking with virtually limitless possibilities.

References

1. G. Chakraborty, I.-H. Park, R. Medishetty, J. Vittal, Chem. Rev. **121**, 3751 (2021)
2. L.S. Xie, G. Skorupskii, M. Dincˇa, Chem. Rev. **120**, 8536 (2020)
3. M. Ko, L. Mendecki, K.A. Mirica, Chem. Commun. **54**, 7873 (2018)
4. Aristotle, *Metaphysics* (387–322 BCE).
5. M.G. Campbell, S.F. Liu, T.M. Swager, M. Dinca, J. Am. Chem. Soc. **137**, 13780 (2015)
6. X. Li, L. Tao, Z. Chen, H. Fang, X. Li, X. Wang, J.-B. Xu, H. Zhu, App. Phys. Rev. **4**, 021306 (2017)
7. D. Dell'Angelo, M.R. Momeni, S. Pearson, F.A. Shakib, J. Chem. Phys. **156**, 044109 (2022)
8. J. Lorenz, H. Rauhut, F. Schweitzer, D. Helbing, Proc. Nat. Ac. Sci. **108**, 9020 (2011)
9. B. Uzzi, S. Mukherjee, M. Stringer, B. Jones, Science **342**, 468 (2013)
10. M. Scheffer, J. Bascompte, T. Bjordam, S. Carpenter, L. Clarke, C. Folke, P. Marquet, N. Mazzeo, M. Meerhoff, O. Sala, F. Westley, Ecol. Soc. **20**, 3 (2015)
11. R. Root-Bernstein, Nature **407**, 134 (2000)
12. J. Evans, P. Jones, Env. Plan. A: Economy and Space **40**, 1416 (2008)
13. G. Wallas, *The Art of Thought* (1926).
14. M. Kemp, Nature **434**, 308 (2005)
15. C. Snow, *The Two Cultures and the Scientific Revolution* (Cambridge University Press, Cambridge, 1959)
16. J. Jaynes, *The Origin of Consciousness in the Break-Down of the Bicameral Mind* (Houghton Mifflin Company, Boston, 1976)
17. K. McMullen, Nature **434**, 310 (2005)
18. D. Curtis, N. Reid, G. Ballard, Ecol. Soc. **17**, 3 (2012)
19. J. Vervoort, D. Keuskamp, K. Kok, R. van Lammeren, T. Stolk, T. Veldkamp, J. Rekveld, R. Schelfhout, B. Teklenburg, A. Borges, S. Janaskova, W. Wits, N. Assmann, E. Dezfouli, K. Cunningham, B. Nordeman, H. Rowlands, Ecol. Soc. **19**, 10 (2014)
20. F. Westley, C. Folke, Ecol. Soc. **23**, 31 (2018)
21. D. Dell'Angelo, in Green Chemistry and Computational Chemistry, edited by L. Mammino (Elsevier, 2022) pp. 299–332.
22. M. Stember, Soc. Sci. J. **28**, 1 (1991)
23. B.C.K. Choi, A.W.P. Pak, Cli. Inv. Med. **29**, 351 (2006)
24. D. Stokols, S. Misra, R. Moser, K. Hall, B. Taylor, Am. J. Prev. Med. **35**, S96 (2008)
25. J.C. Cartwright, Scienceand literature in the seventeeth-century england, in *Literature and Science. Social Impact and Interaction*, edited by J. Cartwright and B. Baker (ABC-CLIO Inc., Santa Barbara, 2005).
26. P. Rossi, *Francesco Bacone. Dalla magia alla scienza* (Mulino, Milano, 2004)
27. T. H. Huxley, *Science and Education* (1894).
28. J.A.V. Chapple, *Science and Literature in the Nineteenth Century* (Mac Millan, London, 1986)

29. M. Arnold, Literature and science, in *The Norton Anthology of English Literature* (Norton, New York).
30. D.P. Dixon, H. Hawkins, E.R. Straughan, Prog. Phys. Geo.: Earth and Environment **37**, 227 (2013)
31. A. Wulf, *The Invention of Nature: Alexander von Humboldt's New World* (Knopf, 2015).
32. T.C. Chamberlin, J. Geol. **5**, 837 (1897)
33. E.O. Wilson, *Consilience : The Unity of Knowledge* (Vintage Books, New York, 1998)
34. S. Wilk, Opt. Photonics News **15**, 14 (2004)
35. P. Basket, A history of light in art, https://onlineartlessons.com/tutorial/ (2019), Accessed on 2022 March 15.
36. Britannica, *Light* (2021). Accessed on 2022 March 15.
37. K. Muir and K. Sutherland, *Art Institute of Chicago* (2021).
38. D. Chavey, Comput. Math. Appl. **17**, 147 (1989)
39. G. Li, L. Huang, W. Xu, Y. Que, Y. Zhang, L. Jc, S. Du, Y. Liu, H.-J. Gao, Phil. Trans. R. Soc. A **372**, 20130015 (2014)
40. B. Grunbaum, G.C. Shephard, *Tilings and Patterns* (Freeman, New York, 1987)
41. R. Minder, *The New York Times* (2013).
42. M. O'Keefe, B.G. Hyde, J.S. Anderson, Philos. Tran. R. Soc. A **295**, 553 (1980)
43. B.-Q. Song, X.-L. Wang, Y.-T. Zhang, X.-S. Wu, H.-S. Liu, K.-Z. Shao, Z.-M. Su, Chem. Commun. **51**, 9515 (2015)
44. H. Iatrou, N. Hadjichristidis, Macromolecules **25**, 4649 (1992)
45. A. Takano, W. Kawashima, A. Noro, Y. Isono, N. Tanaka, T. Dotera, Y. Matsushita, J. Pol. Sci. Pol. Phys. **43**, 2427 (2005)
46. J. Mao, H. Zhang, Y. Jiang, Y. Pan, M. Gao, W. Xiao, H.-J. Gao, J. Am. Chem. Soc. **131**, 14136 (2009)
47. B. Chen, X. Zeng, U. Baumeister, G. Ungar, C. Tschierske, Science **307**, 96 (2005)
48. F. Zhang, Y. Liu, H. Yan, J. Am. Chem. Soc. **135**, 7458 (2013)
49. J.A. Millan, D. Ortiz, G. van Anders, S.C. Glotzer, ACS Nano **8**, 2918 (2014)
50. AMS, *Mathematics and Art* (2004). Accessed on 2022 March 15.
51. D. Schattschneider, *The Mathematical Side of M. C. Escher* (AMS, 2010), pp. 706–717.
52. K.S. Mali, N. Pearce, S. De Feyter, N.R. Champness, Chem. Soc. Rev. **46**, 2520 (2017)
53. J.L. Atwood, J.-M. Lehn, *Comprehensive Supramolecular Chemistry* (Pergamon, Oxford, 1996)
54. J.-M. Lehn, *Supramolecular Chemistry: Concepts and perspectives* (VCH, New York, 1995)
55. Z. Zhang, D. Dell'Angelo, M.R. Momeni, Y. Shi, F.A. Shakib, A.C.S. App, Mat. Int. **13**, 25270 (2021)
56. J.-M. Lehn, Proc. Nat. Ac. Sci. **99**, 4763 (2002)
57. G.R. Desiraju, J. Am. Chem. Soc. **135**, 9952 (2013)
58. J. Elemans, S. Lei, S. DeFeyter, Ang. Chem. Int. Ed. **48**, 7298 (2009)
59. W. Hohmann, M. Kraus, F.W. Schneider, J. Phys. Chem. A **102**, 3103 (1998)
60. J.M. Park, Y. Cao, K. Watanabe, T. Taniguchi, P. Jarillo-Herrero, Nature **590**, 249 (2021)
61. Y. Cao, V. Fatemi, S. Fang, K. Watanabe, T. Taniguchi, E. Kaxiras, P. Jarillo-Herrero, Nature **556**, 43 (2018)
62. Y. Cao, V. Fatemi, A. Demir, S. Fang, S.L. Tomarken, J.Y. Luo, J.D. Sanchez-Yamagishi, K. Watanabe, T. Taniguchi, E. Kaxiras, R.C. Ashoori, P. Jarillo-Herrero, Nature **556**, 80 (2018)
63. G.W. Burg, J. Zhu, T. Taniguchi, K. Watanabe, A.H. MacDonald, E. Tutuc, Phys. Rev. Lett. **123**, 197702 (2019)
64. M. Yankowitz, S. Chen, H. Polshyn, Y. Zhang, K. Watanabe, T. Taniguchi, D. Graf, A.F. Young, C.R. Dean, Science **363**, 1059 (2019)
65. G. Ni, H. Wang, B.-Y. Jiang, L. Chen, Y. Du, Z. Sun, M. Goldflam, A. Frenzel, X. Xie, M. Fogler, D. Basov, Nat. Comm. **10**, 1 (2019)
66. F. He, Y. Zhou, Z. Ye, S.-H. Cho, J. Jeong, X. Meng, Y. Wang, ACS Nano **15**, 5944 (2021)
67. Z. Hennighausen, S. Kar, Elec. Stru. **3**, 014004 (2021)
68. C. Lane, Phys. Rev. B **101**, 235138 (2020)

69. Z. Hennighausen, C. Lane, I.G. Buda, V.K. Mathur, A. Bansil, S. Kar, Nanoscale **11**, 15929 (2019)
70. M. Liu, L. Wang, G. Yu, Adv. Sci. **9**, 2103170 (2022)
71. S. Carr, D. Massatt, S. Fang, P. Cazeaux, M. Luskin, E. Kaxiras, Phys. Rev. B **95**, 075420 (2017)
72. K. Tran, G. Moody, F. Wu, X. Lu, J. Choi, K. Kim, A. Rai, D.A. Sanchez, J. Quan, A. Singh, J. Embley, A. Zepeda, M. Campbell, T. Autry, T. Taniguchi, K. Watanabe, N. Lu, S.K. Banerjee, K.L. Silverman, S. Kim, E. Tutuc, L. Yang, A.H. MacDonald, X. Li, Nature **567**, 71 (2019)
73. H. Yu, G.-B. Liu, J. Tang, X. Xu, W. Yao, Sci. Adv. **3**, e1701696 (2017)
74. M. Serlin, C.L. Tschirhart, H. Polshyn, Y. Zhang, J. Zhu, K. Watanabe, T. Taniguchi, L. Ba-lents, A.F. Young, Science **367**, 900 (2020)
75. E. Codecido, Q. Wang, R. Koester, S. Che, H. Tian, R. Lv, S. Tran, K. Watanabe, T. Taniguchi, F. Zhang, M. Bockrath, C.N. Lau, Sci. Adv. **5**, eaaw9770 (2019)
76. T. Ohta, J.T. Robinson, P.J. Feibelman, A. Bostwick, E. Rotenberg, T.E. Beechem, Phys. Rev. Lett. **109**, 186807 (2012)
77. M.R. Rosenberger, H.-J. Chuang, M. Phillips, V.P. Oleshko, K.M. McCreary, S.V. Sivaram, C.S. Hellberg, B.T. Jonker, ACS Nano **14**, 4550 (2020)
78. D.J. Trainer, A.V. Putilov, B. Wang, C. Lane, T. Saari, T.-R. Chang, H.-T. Jeng, H. Lin, X. Xi, J. Nieminen, A. Bansil, M. Iavarone, J. Phys. Chem. Sol. **128**, 325 (2019)
79. G. Tarnopolsky, A.J. Kruchkov, A. Vishwanath, Phys. Rev. Lett. **122**, 106405 (2019)
80. Q. Fang, J. Sculley, H.-C. Zhou, and G. Zhu, in *Comprehensive Nanoscience and Technology*, edited by D. L. Andrews, G. D. Scholes, and G. P. Wiederrecht (Academic Press, Amsterdam, 2011) pp. 1–20.
81. C.M.A. Parlett, K. Wilson, A. Lee, Chem. Soc. Rev. **42**, 3876 (2013)
82. A. F. Sierra-Salazar, A. Ayral, T. Chave, V. Hulea, S. I. Nikitenko, S. Abate, S. Perathoner, and P. Lacroix-Desmazes, in *Horizons in Sustainable Industrial Chemistry and Catalysis, Studies in Surface Science and Catalysis*, Vol. 178, edited by S. Albonetti, S. Perathoner, and E. A. Quadrelli (Elsevier, 2019) pp. 377–397.
83. S.L. Suib, J. Prech, J. Cejka, Y. Kuwahara, K. Mori, H. Yamashita, Mat. Today **32**, 244 (2020)
84. X. Wang, X. Bu, and P. Feng, Porous inorganic materials, in *Encyclopedia of Inorganic Chemistry* (American Cancer Society, 2006).
85. B. Li, H. Xiong, Y. Xiao, Int. J. Electrochem. Sci. **15**, 1363 (2020)
86. Y. Li, L. Li, J. Yu, Chem **3**, 928 (2017)
87. H. Furukawa, K.E. Cordova, M. O'Keeffe, O.M. Yaghi, Science **341**, 1230444 (2013)
88. M. Yoon, R. Srirambalaji, K. Kim, Chem. Rev. **112**, 1196 (2012)
89. S. Loera-Serna and E. Ortiz, *Advanced Catalytic Materials—Photocatalysis and Other Current Trends* (Intech, 2016) Chap. 5.
90. L. Zhu, X.-Q. Liu, H.-L. Jiang, L.-B. Sun, Chem. Rev. **117**, 8129 (2017)
91. H. Deng, C.J. Doonan, H. Furukawa, R.B. Ferreira, J. Towne, C.B. Knobler, B. Wang, O.M. Yaghi, Science **327**, 846 (2010)
92. K. Koh, A.G. Wong-Foy, A.J. Matzger, J. Am. Chem. Soc. **131**, 4184 (2009)
93. K. Hirai, S. Furukawa, M. Kondo, H. Uehara, O. Sakata, S. Kitagawa, Angew. Chemie **123**, 8207 (2011)
94. S. Henke, A. Schneemann, A. Wutscher, R.A. Fischer, J. Am. Chem. Soc. **134**, 9464 (2012)
95. D.N. Bunck, W.R. Dichtel, Chem.—A Euro. J. **19**, 818 (2013)
96. S.O. Odoh, C.J. Cramer, D.G. Truhlar, L. Gagliardi, Chem. Rev. **115**, 6051 (2015)
97. M.R. Momeni, C.J. Cramer, A.C.S. App, Mat. & Interf. **10**, 18435 (2018)
98. M.R. Momeni, C.J. Cramer, J. Phys. Chem. C **123**, 15157 (2019)
99. X. Wang, X. Zhang, W. Zhou, L. Liu, J. Ye, D. Wang, Nano En. **62**, 250 (2019)
100. Z. Wang, M. He, H. Jiang, H. He, J. Qi, and J. Ma, *Chem. Eng. J.* In press, 133870 (2021).
101. Z.W. Jiang, T.T. Zhao, S.J. Zhen, C.M. Li, Y.F. Li, C.Z. Huang, J. Mater. Chem. A **9**, 9301 (2021)
102. W.-M. Liao, J.-H. Zhang, S.-Y. Yin, H. Lin, X. Zhang, J. Wang, H.-P. Wang, K. Wu, Z. Wang, Y.-N. Fan, M. Pan, C.-Y. Su, Nat. Comm. **9**, 2401 (2018)

103. W. Pang, B. Shao, X.-Q. Tan, C. Tang, Z. Zhang, J. Huang, Nanoscale **12**, 3623 (2020)
104. Y. Liang, R. Shang, J. Lu, L. Liu, J. Hu, W. Cui, A.C.S. App, Mat. & Int. **10**, 8758 (2018)
105. M. Zhao, Y. Huang, Y. Peng, Z. Huang, Q. Ma, H. Zhang, Chem. Soc. Rev. **47**, 6267 (2018)
106. D. Sheberla, L. Sun, M.A. Blood-Forsythe, S. Er, C.R. Wade, C.K. Brozek, A. Aspuru-Guzik, M. Dincˇa, J. Am. Chem. Soc. **136**, 8859 (2014)
107. T. Chen, J.-H. Dou, L. Yang, C. Sun, N.J. Libretto, G. Skorupskii, J.T. Miller, M. Dincˇa, J. Am. Chem. Soc. **142**, 12367 (2020)
108. R.W. Day, D.K. Bediako, M. Rezaee, L.R. Parent, G. Skorupskii, M.Q. Arguilla, C.H. Hendon, I. Stassen, N.C. Gianneschi, P. Kim, M. Dinca, A.C.S. Cen, ACS Cen Sci. **5**, 1959 (2019)
109. K.N. Le, C.H. Hendon, Phys. Chem. Chem. Phys. **21**, 25773 (2019)
110. R. Geiger, T. Zabel, H. Sigg, Front. Mat. **2**, 52 (2015)
111. J. Jung, M.L. Trolle, K. Pedersen, T.G. Pedersen, Phys. Rev. B **84**, 165447 (2011)
112. M.J. Kalmutzki, N. Hanikel, O.M. Yaghi, Sci. Adv. **4**, eaat9180 (2018)
113. Y. Shi, M.R. Momeni, Y.-J. Chen, F.A. Shakib, Z. Zhang, Chem. Mater. **32**, 9664 (2020)
114. M.R. Momeni, Z. Zhang, F.A. Shakib, Chem. Commun. **57**, 315 (2021)

Chapter 12
Spotlights on Convergences of Science and Literature in the Italian Tradition

Armando Mammino and Liliana Mammino

Abstract The dichotomy between sciences and humanities has an official date of birth in Italy—the Gentile's reform of the educational system in 1923. Before that, it was unknown. Poets and writers could be interested in various sciences, and scientists were interested in philosophy and literature and pursued literary quality in their writings. This perception of the fundamental unitarity of human intellectual activities pertained to a long tradition, since Dante Alighieri's poem *Divina Commedia*, which set the foundations of the Italian language. The Renaissance was a period of maximum unitarity of all the areas, and most scholars were expert in several areas of humanities and sciences. This chapter offers selected highlights on this unitarity. It analyses the presence of the science of those times in *Divina Commedia*, and also the presence of intuitions that became part of science many centuries later. It proceeds with a quick excursus through the subsequent centuries, and then devotes specific attention to a XIX century poem ('on a fossil shell', by Giacomo Zanella) depicting the geological evolution of the planet, as a telling example of the interest of a literature person in the scientific development of his time. The section before the last quickly summarises how the dichotomy was artificially introduced by the Gentile's reform, the negative impacts on students' and public perceptions, and the fact that the most brilliant writers and scientists ignored it in their works. The conclusions highlight the importance of the synergies between humanities and sciences for greater completeness of their approaches, including the ability to foster creativity.

Keywords Gentile's reform and the birth of the dichotomy between sciences and humanities in Italy · Geological and historical evolution in Zanella's poem 'on a fossil shell' · Literature values of science and technical writings · Science in Dante's poem *Divine Comedy* · Science in Italian literature works · Scholars with expertise in both domains

---

A. Mammino
Treviso, Italy

L. Mammino (✉)
Faculty of Science, Engineering and Agriculture, University of Venda, Thohoyandou, South Africa
e-mail: sasdestria1@gmail.com; mammino@univen.ac.za

L. Mammino (ed.), *Insights into the Relationships between Humanities and Sciences*, https://doi.org/10.1007/978-3-031-97900-2_12

12.1 Introduction

Relationships between literature and science go back to ancient times. Science has always been an integral component of human knowledge, and as such it gets embedded in literature just as any other form of knowledge does. In earlier stages, poetry was used to express and teach science, just as it was used to express and transmit information about the history of a group (epic poems) and other types of knowledge that were relevant for a community. An example of a poem explicitly devoted to teaching a scientific model of those times is *De Rerum Natura* (on the Nature of Things), by the Roman poet Titus Lucretius Caro (towards the middle of the I century); an example of a poem teaching agricultural works is Ἔργα καὶ Ἡμέραι (Works and Days), by the Greek poet Hesiod, written in the VIII century b.c.

This chapter focuses more specifically on the presence of science in poems or novels that are officially (by general consensus in academia) considered as pertaining to literature. A comprehensive review of interactions between literature texts and science information would require years of research and thousands of pages (or many books) to report all the results. If one expands the search to those works that are often viewed as borderline between literature and other categories, such as science fiction or thriller and mystery stories, the amount of results would grow exponentially, even if one limits the search to those with generally acknowledged quality.

Science and literature have never been really separated, even after the dichotomy between humanities and sciences has kept growing. Many people who were professionally devoted to writing have also had keen interest in the sciences. The Italian poet Giacomo Leopardi was interested in astronomy and wrote a book on its history. The horror-stories writer Howard Phillips Lovecraft was an amateur astronomer and predicted the existence of Pluto on the basis of his observations of the sky. Some scientists became also famous writers. For instance, Isaac Asimov—one of the most famous science fiction writers—was a chemist; the Italian writer Primo Levi was also a chemist.

As already mentioned, considering worldwide literature works would require many volumes. The current chapter focuses on the presence of science in Italian literature, because it is the one with which the authors have greater familiarity. It also selects works that can be viewed as particularly representative. Greater space is devoted to two poetry works: Dante Alighieri's *Divina Commedia* (divine comedy), which was the first major work written in Italian, and Giacomo Zanella's poem 'on a fossil shell', which symbolizes the interest for science among poets and writers in the XIX century. The selected examples have the roles of spotlights showing that synergies between humanities and science are part of a long-standing tradition that would be worthy rediscovering. Poems containing or communicating science information are selected as particularly representative of interdisciplinarity because of the nature of poetic language.

In the following text, the considered poems are not translated into English, but only paraphrased, i.e., the excerpts of interest are expressed or summarised through normal prose. This is necessary because of the differences between the two languages.

The greater sentence-construction flexibility in Italian, and the frequent presence of several words differing by nuances, which are translated by the same English term, implies that it is possible to translate the content, but not all the features and nuances that make the composition to be perceived as poetry in the original text.

For the sake of conciseness and to maintain impersonal form throughout, the two authors of this chapter will be respectively denoted as A.M. and L.M. (pronouns he/him/his and she/her/hers respectively), whenever reference to one or the other of them is expedient.

12.2 Science in Dante' *Divina Commedia*

Dante Alighieri (1265–1321) offers an outstanding example of how humanistic culture and scientific culture can co-exist and harmonise in the same human mind. He was a poet, a writer and a humanist. He systematized Italian language up to the point of making it into a communication tool with communication efficacy comparable to that of Latin (which was then the language of all academic activities throughout Europe). He also bequeathed interested and sincere attention to the sciences of his times. Their state-of-the-art had regressed with respect to the great intuitions of the ancient Greek and Hellenistic period, and much of the available information had come through the translations of ancient works by Arab scholars and through their new contributions, among which algebra and alchemy (both terms retaining their Arab root words). Dante's times pertain to the late Middle Ages, prior to the awakening of the interest for the works of ancient Greece and Rome (the *Humanism*, which preluded and engendered the *Renaissance*); he can be viewed as one of those scholars who contributed to set the earliest bases for the epochal transitions marked by Humanism and the Renaissance.

Dante's main work, *Divina Commedia*, depicts an imaginary journey through the three kingdoms of the Catholic after-world, where Dante himself is the traveller. The journey starts from *Inferno* (hell), moves to *Purgatorio* (purgatory) and finally to *Paradiso* (paradise); Dante is guided by the ancient Latin poet Virgil through the first two kingdoms and by Beatrice (the woman that he idyllically loved) through *Paradise*. The *Paradise* book is also the part of the poem that includes more references to science. An overview of science in *Divina Commedia* is offered in [1, 2]. It is interesting to see how a poet incorporated the science of his time into his poetry, turning it into artistic content and expression.

The leitmotiv of the Divine Comedy is knowledge, intended with the same broad meaning of the ancient Greek word *sophia* (the root-word of the second part of the term *philosophy*), which combines the *knowledge* and *wisdom* meanings. Dante considers it so important that one of the worst punishments for the damned souls in hell is ignorance—being forever deprived of knowledge, including the knowledge of God. In (*Inferno*, XXVI: 119–120), Ulysses powerfully affirms the importance of knowledge by stating that human beings are not meant to live like animals, but to pursue virtue and knowledge.

Dante was keenly interested in sciences. In his work *Convivio* (I, I, 1), he states that science is the ultimate perfection of our soul, in which our ultimate happiness stays. He was particularly interested in astronomy and was familiar with the astronomical knowledge of his time, which derived from the views of the ancient Greeks, Plato, Eudoxus of Cnidos, Pythagoras and his school, Hipparchus, Eratosthenes, Aristarchus and Ptolemy. Eratosthenes had determined the Earth's radius with surprising accuracy for the instrumentation available in his times. Ptolemy systematised the astronomical knowledge of his time in a mathematical treatise that later became more known with its Arabic title *Almagest*. Dante was acquainted with the Latin translation of Almagest; he might also have been acquainted with some of the Arab star maps, which reported also the constellations of the Southern hemisphere. In *Convivio* (II, XIII, 30), he writes that astronomy is the noblest of the sciences, because its subject is the motion of the sky; and is also noblest for its flawless certainty, because it derives from highly perfect and regulated principles (these principles refer to the mathematical bases of astronomy). The way in which he writes about astronomical aspects indicates that his knowledge comes not only from books, but from direct observations; as Attilio Momigliano [3] comments, "Purgatory and Paradise are great indicators of the hours that Dante must have spent observing the sky". The next paragraphs highlight the main astronomical features of his poem.

In the last canto of *Inferno* (XXXIV), Dante expresses his surprise as, after having left him in the depth of the hell abyss, in the centre of the Earth, and moved to the other side through a side path, he now sees Lucifer upside down; he is also surprised that so many hours have passed for a journey that had seemed so short. Virgil explains that they have gone beyond the Earth's centre, which is the centre to which everything tends and also the reference for their prospective; therefore, now they are seeing Lucifer from the opposite point of view, and that is why they see him as being upside down; furthermore, they are now going towards the antipodes (Southern hemisphere), and time is measured differently at the antipodes. Virgil's explanation entails the idea that the Earth has a spherical shape—idea inherited from Eratosthenes and other ancient Greek scholars. In Dante's time, nobody knew what could be there in the antipodes region; in his poem, Dante locates purgatory in that region, exactly at the antipodes of Jerusalem. When Virgil introduces Dante to purgatory, he points out the different positions of the celestial bodies in the Southern hemisphere with respect to the Northern one. This indicates that Dante was familiar with the apparent motions of the stars according to the time of the day and the periods of the year, and with the constellations of both the Northern and the Southern hemispheres. The first 'canto' of Purgatory also mentions four bright stars—a description that might refer to the Southern Cross, or to the two brightest stars of the Southern Cross plus the neighbouring α and β *Centauri*.

Dante mentions that the light of the Moon makes the stars appear sparser. This is a phenomenon that can be identified with the naked eyes if one observes the sky regularly, so as to be able to note the changes through a Moon's cycle; it can be inferred that Dante did this personally, and might also have discussed it with other scholars of his time. The phenomenon is due to the different distances of the celestial bodies from the Earth (though this perception was likely not available in Dante's times) [4].

The study reported in [5] (where Sperello di Serego Alighieri is an astrophysicist and a direct descendant of Dante) confirms the correctness of the observation and adds that, even nowadays, when astronomers wish to observe objects with weaker light through telescopes, they avoid the days around full moon, and prefer the days around new moon; the desire to get observation-time around new moon may get so important that the telescope usage for those days is preferably scheduled for the best scientific proposal.

In *Paradiso* (II: 73–105), Beatrice explains the reasons for the moon spots by suggesting an analogy that can be summarised in terms of an experimental procedure [5]: take three mirrors; put two of them at the same distance from you, and the third one in the middle between them and further away from you; then light a lamp behind you; you will see that the lamp shines in the same way through all the three mirrors, but it appears smaller in the mirror that is farther away. This observation anticipates the theory of the independence of light intensity of the distance [5], according to which the light emitted by a light source per solid-angle unit does not depend on its distance; the reason being that the surface brightness is the ratio between the brightness of the emitting body and the solid angle; since both quantities are inversely proportional to the square of the distance, their ratio does not change, unless there is a light-absorbing item between the emitting body and the observer.

Other statements concerning astronomy can be summarised briefly. Dante considers the Milky Way as an ensemble of so many stars that we perceive it as something continuous (*Paradiso*, XVI: 97–108)—an inference that, in the absence of telescopes (which was invented by Galileo centuries later), could only be achieved by means of pure reasoning. He also states that tides are related to the moon (*Paradiso*, XVI: 82–83)—a hypothesis that had been present since ancient times, based on observations, but whose physical foundation was possible only after Newton introduced the theory of universal gravitation, which enabled the explanation that the Moon exerts gravitational attraction on fluid masses on the Earth.

Geography was already considered an important science in Dante's times, also in view of explorations of areas in Europe or close to it that were not yet well known. The geographic references in *Divina Commedia* are accurate as long as they concern places in Italy, or some other European countries, and the regions around the Mediterranean Sea. The way he mentions the relative positions of towns on the Mediterranean Sea with good approximation suggests that he must have been familiar with the nautical charts of his time—the only maps that could provide reliable information in those times, thanks to the expertise of seamen, who built them as guides that were also functional to survival when at sea. When geographic references concern areas supposedly beyond Hercules' Columns (Gibraltar Strait), or in the Far East Asia, they are mostly based on mythologies.

Dante views geometry as a science that is fundamental for the description of the universe, and also as a science that can offer useful analogies to tackle the difficulties in understanding some theological mysteries [6]. For instance, he compares the impossibility for the human mind to comprehend the mystery of Incarnation to the impossibility of "measuring the circle" (*Paradiso*, XXXIII: 133–135), which might be referred to the immeasurability of the length of the circle and the length of its radius (their ratio is not a rational number), or to the consequent impossibility of finding a square whose area is exactly equal to the area of a given circle.

When building the structure of Paradise, Dante tried to combine the Ptolemaic model, for which the Earth is the centre of the Universe, and the Cristian model, for which God is the centre of the universe. This attempt resulted in a vision that somehow goes beyond Euclidean geometry, envisaging a curved universe [6]. It is clear that this can be viewed as intuition, not as formal knowledge—a poet's intuition that gets intertwined within his verses. It can also be viewed as an additional confirmation that a poet's mind can conceive ideas that go beyond the knowledge of his times and may find formal development and confirmation centuries later. The Russian mathematician Pavel Alexandrovič Florenskij (1882–1937) was among the first to identify the presence of non-Euclidean geometry in Dante's works, and presented this view in 1921, on the 600th anniversary of Dante's death [7]. He recalls that other mathematicians had recently identified the same presence (Georg Bruce Halsted in 1905, Heinrich Martin Weber also in 1905, and Max Simon in 1912), analyses the details of the geometry of Paradise and concludes that the space outlined by Dante pertains to elliptical geometry; then relates it to relativity. Later on, similar analyses were carried out by the Swiss mathematician Andreas Speiser (1885–1970) [8] and the mathematical physicist Herman Weyl [9]. The American mathematician Mark Peterson noted a similarity between Dante's universe structure and a sphere in a 4-dimension space [10].

The overall description can be outlined as follows [5]. When Dante (*Paradiso*, XXVIII: 16–36) arrives in the presence of a strongly intense light coming from a dimensionless point (a point consistent with Euclidean geometry, and a geometric abstraction), a flaming diffuse halo revolves around it with a velocity that is greater than the high velocity of the First Mobile (the outer circle). It consists of nine circles, whose radiuses decrease as the rotation speed increases. The analysis of this part shows that a rational interpretation is possible only by resorting to the hypersphere concept, which was not part of the geometry's body of knowledge of those times. The First Mobile is the maximum circle, which surrounds all the planetary spheres, which, in turn, enclose the Earth—centred on Lucifer—and the Empyrean, where the angels' choirs progressively close through nine circles (as many as the skies) around a point that coincides with God. There are thus two antithetical focuses, enclosed by the First Mobile, from where it is possible to see both the focuses and the structures associated with them. The unifying feature is the rotation velocity, which increases uniformly from the Moon to the circle of the Seraphs; this is a sort of coordinate measuring an underlying property (the different reception of divine love) and a sort of cosmic law that Dante had anticipated earlier (*Paradiso* I: 1–3): "the glory of him who moves everything penetrates the universe and shines more in one area and less in another". According to [5], it is possible to read the structural organization of Paradise as a manner to fix the scale of the coordinate (velocity = divine virtue density), which scans the distances from Lucifer on one side and from God on the other. The resulting picture can be rationally interpreted by assuming that Dante had some intuition about the properties of the hypersphere and translated them into a model that he projected on the three dimensions that can be represented and narrated [5]. If it is possible to speak of 'model', it is a purely mathematical model, not a physical one. All the same, having conceived it is proof of great knowledge and intuition; the language of poetry contributed to the expression of the discourse with a power superior to what could have been attained through prose. A hypersphere is the 4-dimension geometric place whose points have the same distance from a point O, called its centre; considering fewer dimensions, the equivalent is the spherical surface in a 3-dimension space and the circle in 2-dimensions. When a hypersphere meets our 3-dimensional space, it defines a continuous sequence of spherical surfaces, from which a finite number of discrete spheres can be selected; their ordered sequence starts from a point, increases to a maximum, and decreases returning to the same point, which is therefore a sphere with radius equal to zero [5]. This happens because the hypersphere must sacrifice a coordinate. In practice, we could represent it through two hemispheres with equal radiuses and mutually tangent, with each of them incorporating circles with decreasing radiuses, until the last one is reduced to a point. This representation is strikingly similar to Dante's description of Paradise, where the coordinate labelling the individual circles is the increasing God's light. The circles are 'quantised', i.e., discrete: nine circles on each side, including the First Mobile (the outer circle), which is common to the two hemispheres and from which Dante can see in both directions in an equal way, wherever he is located. The two central points are symmetrically occupied by Lucifer and by God, and each of them is simultaneously 'focus' and 'border of the world'. From these descriptions, it appears that Dante's intuition perceived geometrical entities that entered the body of mathematics, with clear definitions, only five centuries later. The way in which Dante used them in his cosmology can be viewed as a kind of anticipation of Einstein's use of four coordinates—three space coordinates in both cases, while the fourth coordinate is time in the relativity theory and divine virtue in Dante [5]. In Dante's narration, after leaving the centre of the Earth, he crossed increasingly larger celestial spheres until reaching the First Mobile, from where he could see the Empyrean with smaller and smaller circles, up to the point where God is encountered. This vision recalls Einstein's general relativity's view of a curved and unlimited, but finite universe [11–14]. The fact that a great poem contains anticipations of views that will enter physics only after several centuries stresses the unitarity of human intellectual activities, as imagination ability can prompt ideas leading to similar interpretations—although in different forms—for questions with similar bases, such as the need to somehow taking four coordinates into simultaneous account.

Other science-related components can be recalled quickly. On various occasions (*Purgatorio*, XXIX: 73–78 and XXV: 91–93; *Paradiso*, XII: 10–18), Dante expresses concepts of geometrical optics and light theory with reference to the rainbow. He clearly knows that the rainbow colours are due to different reflections when the light of the sun goes through droplets of rain water. He notes that, when the rainbow appears, the sun is at the back of the observer and the rain in front, i.e., the sunlight is projected on the clouds that generate the rain. He also notes that the rainbow is nearly always double: the two arcs are parallel (concentric) and their colour-sequences are reversed.

Two other features can be mentioned, namely:

- Dante's interest in time as a variable, caught between a latent recognition of the need to overcome previous or current (for his times) definitions and the absence (then) of the type of knowledge that, centuries later, enabled modern definitions. The verses in (*Paradiso*, XXVII: 118–120) foreshadow a fractal character of creation, with the image of the upside-down tree having its roots—the roots of time—in the First Mobile and its branches spreading through the lower circles, where time exists.
- His awareness that measurements are conventional, i.e., they are defined in terms of comparisons with a pre-selected reference.

The multidisciplinarity of Dante Alighieri's culture has been outstanding in Italian literature and beyond. His broad interests for science made his mind an exceptional cross-road between humanities and sciences, above all for his time. Obviously, the science on which he focused remained linked to ancient models, even if, as far as mathematics is concerned, there are some signals and anticipations interfacing more with the vanguard than with academia.

Altogether, the previous paragraphs highlight how science, as well as all the other components of human intellectual activities, can become poetry content, and their synergies contribute to the poetic value of the work. This multidisciplinary integration benefits the poem, adding to its comprehensive outlook and, at the same time, to the ensemble of poetic imagery that builds it.

In general, poetry, with the greater potentialities and strength of its expression instruments, creates metaphors and images that can guide the reader to the comprehension of concepts, no matter how complex, along shortcuts that are not available to prose. Similarly, poetry can facilitate human psych to correctly grasp, through intuition, truth and discoveries while the theoretical and experimental cognitive tools that can (and will) lead to those results through pure reason are not yet developed. This appears evident in Dante's works. And this evidence suggests another important inference: a configurable synergy between poetry and sciences—the latter as an evocative component of the former, the former as an invigorating resource for the latter, with its expression-power opening unchartered possibilities to thought. It is not surprising that many scientists loved poetry and many poets loved science.

12.3 Some Quick Excerpts Through the Subsequent Centuries

The Humanism period and the subsequent Renaissance witnessed investigations in all the areas of science. Most scholars had expertise in multiple fields. Leonardo da Vinci is the most typical example of broad-scope multidisciplinarity: he was a painter, an architect, an engineer, studied the human body, animals and plants, and investigated the question of how knowledge is achieved. His writings have both scientific and literature value.

A particularly relevant example of synergy between science and literature is offered by Galileo Galilei's *Dialogo sui due massimi sistemi del mondo* (Dialogue Concerning the Two Chief World Systems, 1632). Galileo opted for a colloquial structure (dialogue) for the presentation of his new physics theories, thus composing a work that is simultaneously a scientific treatise and a prose whose high artistic level has been unanimously recognised for more than three centuries by literature scholars. How this book marked a new course of physics and, in particular, of astronomy, is well known. The lines from the Foscolo's poem "I sepolcri" (the sepulchres, 1807) devoted to Galileo's tomb summarise this role in a powerful way: "and the ark… of the one who saw a plurality of worlds rotating under the ethereal sky, and the motionless sun radiating them, whereby he first cleared the ways of the firmament to the Englishman who later spread his wings so broadly across it" (where the Englishman is Newton); these verses constitute an example of poetry exalting a crucial turning point in the history of science, engendered by the convergence of observation and deduction.

The XVIII century comprises anticipations of many science and technology innovations which will then ripen through the XIX century. The XIX century witnessed the highest expression quality in Italian literature, from novel writing (e.g., Alessandro Manzoni, Ippolito Nievo) and poetry (e.g., Giacomo Leopardi, Ugo Foscolo, Giosué Carducci) to treatises, including those on science and technology themes. Many publications (often broad-scope ones) are devoted to mathematics, physics, engineering, and other sciences. Their language is elegant, understandable and didactic, written with the purpose of guiding the reader towards comprehension without risks of boredom or over-seriousness. In other words, the excellence in poetry and novel writing are adopted as wording, logic and expression prototypes for all the other writings, both in the humanities (storiography, philosophy, law, etc.) and in the sciences. This trend continued into the first half of the XX century. An example is offered by the three-volumes book on rational mechanics by Levi Civita and Amaldi [15], which was innovative and deep-reaching in terms of its scientific content, and excellent from a literature point of view; the latter aspect was jointly verified through rigorous text-analysis by A.M. (structural engineer), Giuseppe Ghinassi (structural engineer, university professor and expert in the language and approaches of treatises focusing on themes specific to structural engineering), and Emidio Diletti (literature specialist). In Levi Civita and Amaldi's writings, the correctness of theoretical arguments always goes together with a clear prose making even complex concepts immediately evident, rhythmical and harmonious in the choice and combination of words, and rigorously logical in the deductive development of the discourse. They constitute examples of scientists aware of the resources of the art of composing a correct and effective prose.

In the XVIII century, poetry is often used by scholars of all areas. Several poets and writers were sensitive to the charm and the cultural contents of the sciences and included them in their works, combining good scientific knowledge with the communication and convincing abilities traditional of their area [16]. For instance, Mascheroni's poem *Invito a Lesbia Cidonia* (1793, [16]) is a guide through the marvels of the Museum of Natural History of the University of Pavia.

Poets' interest in science continues through the XIX century. Aleardo Aleardi's poem *Il monte Circello* (1863) contains also verses recalling that the Apennines mountains formed from a sea. Giacomo Leopardi (1798–1837) wrote a history of astronomy [17], which shows his ability to deeply understand the scientific knowledge of past times and his time, and to outline an organic historical picture of its development, up to the latest discoveries. When, nearly two centuries later, the Italian astrophysicist Margherita Hack published a history of astronomy [18], she wrote about the history after the time where the poet had stopped, and maintained Leopardi's part for the earlier times, clearly thinking that it was organised and expressed in the best possible way (Leopardi—although long dead—was thus co-author of [18], as the author of its first part).

12.4 Giacomo Zanella and the Geological History of the Earth

Giacomo Zanella's (1820–1888) multidisciplinarity spans across poetry, science and theology. A catholic priest, he lived in a historical period that was particularly fruitful for the progress of science in Europe. The works by Charles Lyell (1797–1875), Alfred Russell Wallace (1823–1913), other naturalists and, most of all, Charles Darwin (1809–1882), changed the vision of Earth and of life on Earth, introducing theories based on huge amounts of observations. Darwin's theories quickly spread through Italy, thanks to Michele Lessona, Filippo de Filippi, Paolo Mantegazza, Giovanni Canestrini and other scholars. One of the outcomes was a growth in the interest for palaeontology—the study of fossils embedded within soils and rocks—not only among academics, but also among self-taught persons passionate about recovering, collecting and studying fossils, as fossils started to being viewed as the footprints of evolution. On the other hand, this overall renovation of geology, zoology and botany knowledge engendered frictions between some religious authorities and the stakeholders of science vanguard. Giacomo Zanella attempted to smooth the frictions.

Zanella lived in the Vicenza region, an area in Northern Italy spreading from the Po valley to the western side of Mount Grappa and up to the eastern layers of the Alpone valley. The area is rich in fossils; it also comprises the "Pesciara" in Bolca Veronese—a worldwide known deposit of perfectly preserved fossil plants, fish and other marine creatures.

The ancient Greeks had realised that fossils are likely the remnants of animals and plants of past epochs. After the Middle Ages, Leonardo da Vinci was the first to study fossils systematically, and is therefore considered the father of palaeontology [19]; he understood that fossils were remnants of animals that lived in the areas where they were found, and hypothesised that there is some fluidity in the earth, and rocks might have replaced the sea, or vice versa, in the same place [19, 20]. Scholars' interest in fossils continued through the following centuries. In 1517, Girolamo Fracastoro (a physician and poet, also interested in mathematics, geography and astronomy) acknowledged them as remnants of sea animals and hypothesised that the mountains where they are found had been formed from previous seas. Giovanni Battista Brocchi's 1814 treatise [21] presents detailed stratigraphic analysis, compares fossils and introduces the idea of the possibility of species extinction; his work became part of the debate on species in European academia of those times, and may have influenced some of Darwin's reflections [22]. By the end of the XVIII century, fossils from Bolca were present in many European museums. Jean-François Séguier devoted one of his unfinished manuscripts to them, Pétrifications du Véronois ("Petrifactions from the Verona Province") [23]. Scholars used to give each other fossils as gifts. The German poet Wolfgang Goethe had his own collection, which comprised also fossils from Bolca [24].

The bed and shores of the Chiavon stream, north of Fara Vicentina, in Lonedo (Lugo di Vicenza, Photo 12.1), constitute another rich fossil site. The first villa designed by the Venetian architect Andrea Palladio (1508–1580) is located in this area (*Villa Godi Valmarana*, now pertaining to the Malinverni family; Photo 12.2). In Zanella's times, a fossils' museum was being built in the villa, under the care of Andrea Piovene, Achille di Zigno and others, maybe including also Giovan Battista Brocchi. These fossils comprised the bone remnants of tropical fish and other sea animals, such as clams and tortoises, and the charred, but isomorph to the originals, remnants of plants typical of tropical areas, mostly palms; a whole 9-m-long palm had been reassembled on the floor of one of the rooms in the villa (Photo 12.3). These fossils—which had remained hidden in the rocks for nearly 60 million years—were 10–15 million years older than those from Bolca. Zanella (himself from a noble family) was a frequent guest in this villa, where the owners had made a study-room available to him, and where he wrote several of his works. He used to interact with the palaeontologists who were organising the museum, and understood the deep meaning of those objects that were embedded in the rocks. The reality of fossils prompted him to accept the ideas of evolution, biological as well as geological [25], with the only exception of human beings, whom, as a priest, he perceived as different from the rest of nature, because of their spiritual component [26, 27].

His poem *Sopra una conchiglia fossile* (On a fossil shell, 1864, [28]) was inspired by a fossil shell that he had himself collected from the area surrounding the villa. The adjectives with which he describes it (*marmoreal, twisted*) suggest that it might have been a *Melangena*, a shell type still frequently found in the deposits of volcanic ashes below the stone material in the quarries of the area. The poem is an example of interdisciplinarity between literature (poetry) and earth sciences and, in some respects, also between religion and historical geology. The latter aspect, by combining a vision of geological history with a vision of God-guided objectives for mankind's efforts, attempts to bridge the gap, or even the contraposition, between science and religion, which was rather frequent in the XIX century. The poem was received with great appreciation by poets and writers, including Giosué Carducci and Alessandro Manzoni [29].

After briefly recalling a vision of the shell's wandering together with other sea creatures in a primordial sea, the poem emphasises the brevity of human history with

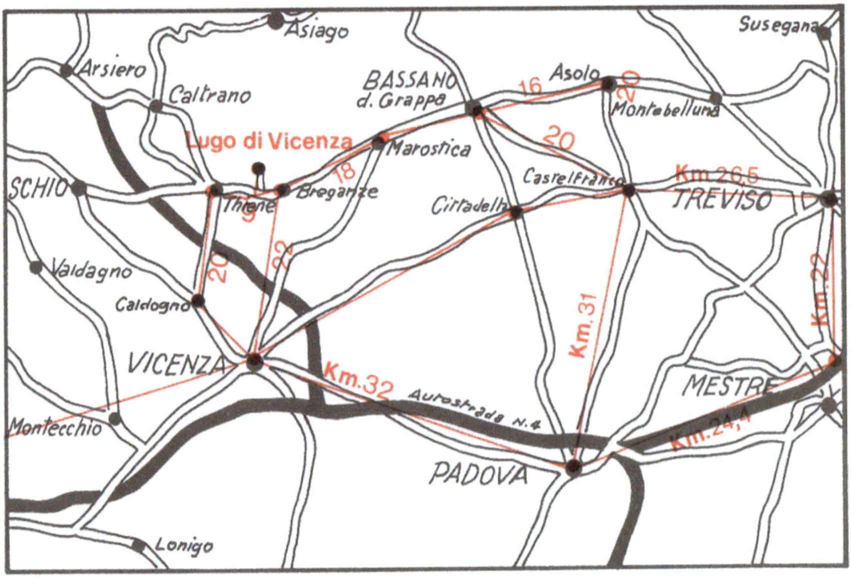

Photo 12.1 Map of the area in the Veneto region where the Lugo di Vicenza area is located

Photo 12.2 Villa Godi Valmarana, now Villa Malinverni, built by Palladio between 1538 and 1542, UNESCO heritage site

Photo 12.3 The 9 m long fossil palm reconstructed on the floor of one of the halls of the museum in Villa Malinverni. *Source* A. Mammino, *Museo dei fossili di Lonedo*, Foto Edizioni Pasqualotto

respect to geological history by mentioning some milestones of the development of human civilization (the ancient Asian empires, the Trojan war, the Roman empire) and repeating that these all happened "yesterday", i.e., their distance in time from us is incomparably shorter than the distance from the time when the shell was alive and roamed the sea. That time also preceded the time when Italy emerged from the sea; these verses show the knowledge that Italy emerged comparatively late (it emerged towards the end of the Tertiary era, together with the rising of the Alps and the Apennines). The poem then depicts quick images of the world in those times: the flashes from hundreds burning volcanos reflecting on the planes—a scenario that was true for broad eras; and powerful quakes tearing apart the dams of unknown seas. The latter scenario is a potent intuition by the poet: towards the end of the XX century, the analysis of the sediments in the sea bottoms corresponding to the deposits of the geological period subsequent to the glaciation era proved that the massive melting of continental glaciers and the rising of the levels of seas and oceans implied that the higher level of the surface of the latter with respect to the former caused the breaking of isthmuses, like the one between the Atlantic Ocean and the Mediterranean sea, or between the Mediterranean sea and the internal seas of Western Asia; such sudden huge redistributions of water and earth masses were unavoidably accompanied by strong earthquakes. After depicting these scenarios, the poem continues with what can be considered a purely scientific approach, by bringing the proofs for these hypotheses: the palm trees buried in the bottom of the lakes (some of which had been recovered and reconstructed in the museum within Villa Godi Valmarana), the petrified remnants of 'dragons' (dinosaurs), or their footprints engraved in the rocks.

The second part of the poem focuses on humankind—the last beings to arrive on Earth, destined to toil with huge efforts to pursue knowledge, until reaching the final, and so far mysterious, goal set for it by God. This part is fundamentally optimistic, largely in connection with the religious faith of the poet; the optimistic perspective of continuous progress and enhancement of humankind might have been further encouraged by the optimism inherent in the positivistic views of the last decades of the XIX century.

The poem "on a fossil shell" is by large the most famous by Zanella; but other poems also express his positive views on science and his conviction that science and religion are not in conflict. The poem *Natura e scienza* (Nature and science, 1868, [30]) celebrates the efforts of science to unveil the secrets of nature, until at the end it catches a glimpse of God creating new worlds. *Il taglio dell'istmo di Suez* (the cutting of Suez isthmus, [30]) celebrates an outstanding achievement of human technology.

Besides using poetry to express science, Zanella uses it also to express key issues of the economy of his time. In the face of the budding, but clearly meant for fast-growing, industry in his region, he perceives a contraposition between nature (including the agricultural landscape of the region) and industry, and the poems' collection that he titled *Astichello* (from the name of a river in the area) reflects hopes in a compromise that would balance the requirements of both. Similarly, he used poetry to express hopes in the possibility of harmonization between the requirements of the bourgeoning industry and the importance of humane conditions for workers (*Possagno*, 1854) —an issue that was very crucial around the mid of the XIX century [25], and that, in the face of excessively poor conditions for the working class in countries where industry had already grown massively, had prompted the birth of the strongest theories against the ways through which capitalism was building itself (Marx and Engels' *Manifest of the Communist Party* was published in 1848). Zanella hoped in the possibility of harmonising all requirements without losing humane perspectives and the humanistic bases, and interacted with all classes (labours, agriculture and industry workers, craftsmen, entrepreneurs) to try and identify routes for harmonisation [31, 32]. By using poetry to discuss the problems of the budding industry, Zanella also transmitted the hope that poetry and literature will not fade as the new economic systems grows. Luigi Luzzatti (1841–1927), an economist who had been a student to Zanella, commented on this fading in a letter to Angelo Messedaglia, in the sixties of the XIX century: "Even in this country, with the exception of some outstandingly intelligent persons, the sound principles of economy are totally ignored ... And the faith in the supreme harmonies of science is missing, the care for beauty is languishing, as if poetry should die smothered under the pile of the new economic laws …" [33].

Zanella's writings also contain numerous parts devoted to the history of various sciences (astronomy, physics, geometrical optics, electricity, hydraulics, zoology and botany, mathematical sciences, and others), sometimes embedded within formally-literature treatises. Altogether, these offer one of the best profiles of the history of science written in the XIX century and, simultaneously, an outstanding example of the unitarity of human intellectual activities, where science and literature, the art of writing and the knowledge of physical reality, converge into a uniting thread.

12.5 The Official 'Birth' of the Dichotomy Between Humanities and Sciences in Italy

As highlighted also by the previous sections, a net separation between the sciences and technology on the one side (including technical disciplines based on them, like engineering) and the humanities on the other had not pertained to the centuries-long Italian cultural tradition. The dichotomy was introduced in a specific year, 1923, with the school reform brought by Giovanni Gentile ('Gentile's reform'), and entailed a rigid, practically impenetrable barrier between the two domains. The reform reflected the strong antiscientific and anti-positivist reaction that developed within the Italian academic culture at the beginning of the XX century and found its theoretical blueprint in Benedetto Croce's idealistic philosophy. Croce's philosophy privileged the humanities and undervalued the sciences, considering their information content to be solely referred to particular cases, and the sciences themselves to be incapable of attaining an overall organic picture with universal character. He thought that the latter could be attained only from an idealistic and spiritual perspective, which is capable of understanding the universality of reality in its historicity, whereas the sciences can measure and classify reality, but they do not know it deeply, and knowing it is not part of their task. In Croce's view, science replaces concepts by schemes, the universal by mere classification of "small heaps of news", thus getting based on pseud-concepts, producing abstractions, and eventually building an empiric ensemble that is not thought, but built through formulas, through regularities' ideas which have no match in real life, and can be effective at most as practical but fictitious tools. In [34], he writes that "Natural and mathematical sciences have gladly surrendered the privilege of truth to philosophy and resignedly, or even smiling, confess that their concepts are concepts for convenience and practical utility, but have nothing to do with the search for truth". Croce's anti-scientific integralism largely stems from his misunderstood vision of English empiricism, which acted like an opaque filter preventing him from seeing both the potentialities of science and technology and the theoretical and applicative intrinsic value of a philosophy of science, both in its general terms and through the various specialised areas.

Having been born during fascism, this dichotomy was transferred practically unchanged to the decades following WWII, on the basis of a state-continuity criterion which tended to maintain foundation-type structures and approaches (like the school system) through political changes, and continued up to comparatively recent times. Although antithetical to the Italian cultural heritage, the dichotomy influenced pedagogical approaches and engendered rigidity in the way of viewing schools and professions both within instruction and in the public perceptions. One of the implications was an incomplete presentation, within instruction, of the real cultural heritage. For instance, there was no mention of Dante's interest in science, or the presence of science in Divine Comedy. Zanella's poem on a fossil shell was presented as part of the literature of the XIX century, but in-class explanations did not mention its relationships with the study of fossils, or with all the interest for palaeontology and evolution in that period, or with the actual geological history to which it refers. Leopardi was presented as an idyllic, or melancholic, or disillusioned poet; no mention was made of his interests and actual expertise in astronomy. The vision that was transmitted was a compartmentalised vision applied also to individual writers or poets—a vision unable to recognise how knowledge components of different types and from different areas were integrated in those poets' and writers' perception of the world and were, together, embedded in their productions. In this way, it also neglected the intellectual unitarity of scholars, whose research and production activities were not compartmentalised, but constituted one ensemble; thus, it failed to present a holistic picture of individual scholars as human beings, failing the human-side reality that they assumed to privilege.

Of course, particularly gifted persons soon overcame the limitations of this type of instruction. For instance, Primo Levi was a chemist, and also the author of several books, including a powerful denunciation of the reality of Nazi concentration camps, where he had been sent as an Italian Jewish (*Se questo e' un uomo*, If this is a man, 1947) and a series of short stories narrating selected chemist's experiences (*Il sistema periodico*, The periodic table, 1975). But too many intellectuals remained within one or the other set of 'boxes' exclusively pertaining to one domain. It is not easy to guess how much has been lost, in terms of creativity and the production that stems from creativity, because of the rigid compartmentalisation acquired through that type of formal instruction.

12.6 Conclusions

The inferences from the content of the previous sections are quite evident. Synergies between poetry and science yield results that are precious both for the messages conveyed by poetry and for the way in which expression through poetry may stimulate imagination in the sciences. Similar benefits may derive from synergies between literature in general and the sciences. Imaginations and intuitions have important roles in the outcomes of both areas. The care needed to pursue quality writing when writing about sciences favours the types of reflections that enhance understanding and creativity.

Science fiction (which formally pertains to literature) offers interesting sets of examples highlighting the enhancement of intuition fostered by the engagement in quality writing. It is a type of novel-writing that projects current science into future possible developments. By doing so, it has anticipated a variety of discoveries or inventions that have become reality at a later stage. When the projection comprised also the impacts on human societies, science fiction has anticipated developments and depicted scenarios that are increasingly turning into reality (many examples could be cited; Isaac Asimov, Ray Bradbury and Frederick Pohl are among the most extant).

As far as the educational level is concerned, highlighting the synergies of the past and their benefits, and encouraging students to engage in analogous synergies, is apt to promote creativity that will be valuable in whichever future career a student chooses.

References

1. V. Pappalardo, *La divina commedia tra fisica, matematica, astronomia* (self-published, 2012), https://www.liceoseverisalerno.edu.it/wp-content/uploads/2021/03/La-Divina-Commedia-tra-fisica-matematica-astronomia.pdf
2. V. Pappalardo, *Una nuova visione della cosmologia dantesca: La struttura quadridimensionale del Paradiso* (Kindle Edition, 2018)
3. Dante Alighieri, *La Divina Commedia. Inferno Purgatorio Paradiso.* Commento di Attilio Momigliano, Commento al Paradiso (Sansoni, Firenze, 1966), p. 670
4. N. Zingarelli, *Dante; III volume in Storia Letteraria d'Italia* (Vallardi, Milan, 1903)
5. S. di Serego Alighieri, M. Capaccioli, *Il sole, la luna e l'altre stelle* (Edizioni La Repubblica e GEDI Gruppo Editoriale, Torino, 2021)
6. G. De Cecco, *Dante e l'ipersfera* (Viaggio nella Scienza, Ithaca, XVII, 2021).
7. P. A. Florenskij, *Il simbolo e la forma, scritti di filosofia della forma*, a cura di N. Valentini e A. Gorelov, (Bollati Boringhieri, Torino, 2007). The Russian original, titled *Mnimosti v geometrii. Rascirenye oblasti dvukhmernykh obrasov geometrii*, was published in 1922.
8. A. Speiser, *Klassische Stücke der Mathematik, Orell Füssli* (Zurich, 1925), 58–59
9. H. Weyl, *Mind and Nature, Selected Writings on Philosophy, Mathematics, and Physics, Chapter 4: The Open World: Three Lectures on the Metaphysical Implications of Science* (Princeton University Press, Princeton and Oxford, 1932), 34–82
10. M.A. Peterson, Dante and the 3-sphere. Am. J. Phys. **47**, 1031–1035 (1979)
11. W. Egginton, On Dante, Hyperspheres, and curvature of the Medioeval Cosmos. J. Hist. Ideas **60**(2), 195–216 (1999)
12. S. Scoppolini, *Dante e l'Empireo riemanniano,* https://www.academia.edu/37618216/Dante_e_lEmpireo_riemanniano
13. R.F. Ggisu, Dante e Einstein - inserto del libro "La luce della fisica", (2019), ISBN 9–791220–054317—affinità della cosmologia dantesca con quella einsteiniana a partire dalle considerazioni di Pavel Florenskij.
14. A. Oppo, Se la misura di un corpo è un numero immaginario: Florenskij e il concetto di spazio in Dante. Theologica & Historica. Annali della Pontificia Facoltà Teologica della Sardegna **24**, 171–184 (2015)

15. T. Levi Civita, U. Amaldi, *Lezioni di meccanica razionale* (Lessons in Rational Mechanics) (Zanichelli, Bologna, 1923)
16. L. Mascheroni, *L' invito a Lesbia Cidonia* / Lorenzo Mascheroni; a cura di Marco Bernuzzi. (Bergamo, Liceo scientifico statale Lorenzo Mascheroni, 2004) Monografia—Testo a stampa [IT\ICCU\MIL\0641171]
17. G. Leopardi, *Storia dell'astronomia dalla sua origine fino all'anno MDCCCXI.*
18. M. Hack, G. Leopardi, *Storia dell'astronomia: dalle origini ai nostri giorni* (Edizioni Dell'altana, Roma, 2011)
19. https://www.leonardodavinci-italy.it/paleontologia/leonardo-da-vinci-paleontologia-e-sui-fossili
20. Leonardo da Vinci, *Codex Leicester*.
21. G.B. Brocchi, *Conchiologia fossile subappennina con osservazioni geologiche sugli Apennini e sul suolo adiacente* (Dalla Stamperia Reale, Milano, 1814)
22. S. Aranginu, *Giovanni Battista Brocchi e la questione sullo statuto delle specie nella "Conchiologia fossile Subappennina* (thesis at the University of Cagliari, 2017)
23. J. Gaudant, Histoire des sciences / History of Sciences Jean-François Séguier (1703–1784), premier historiographe de la paléontologie. C. R. Palevol **4**, 295–310 (2005)
24. G. P. Marchi, Una lettera di Goethe a Gian Giacomo Dionisi sui fossili di Bolca. Belfagor. **59**(3), 275–300 (2004). https://www.jstor.org/stable/26150132.
25. G. Guarda, *Giacomo Zanella contemporaneo di Darwin e Marx*, Comune di Chiampo, Assessorato alla Cultura, Centenario Zanelliano 1888–1988, Quaderno N° 6
26. G. Zanella, Il darwinismo, inediti a cura di I. F. Baldo (Editrice Vicentina, Vicenza, 2019)
27. L. Perazzolo, *Scienza, religione e fede nella poesia di Giacomo Zanella*, Comune di Chiampo, Assessorato alla Cultura, Centenario Zanelliano 1888–1988, Quaderno N° 3.
28. A. Gianni, M. Balestreri, A. Pasquali, *Antologia della letteratura italiana; III: dall'Ottocento alla prima meta' del Novecento* (G. D'Anna, Messina-Firenze, 1964), 184–188
29. A. Belloni, G. Bertoni, A. Galletti, G. Mazzoni, G. Natali, V. Rossi, N. Sapegno, G. Toffanin, A. Viscardi, N. Zingarelli, *Storia letteraria d'Italia; l'Ottocento (parte seconda)* (Milano, Vallardi 1949), 1381–1384
30. https://it.wikisource.org/wiki/Versi_di_Giacomo_Zanella
31. F. Bandini (ed.), *Atti del convegno di studi su "Giacomo Zanella e il suo tempo, nel 1° centenario della morte*", tenuto in Vicenza nei giorni 22–24 settembre 1988 (Accademia Olimpica, Vicenza, 1994)
32. A. Bedin, L'educazione sociale in Giacomo Zanella, Comune di Chiampo, Assessorato alla Cultura, Centenario Zanelliano 1888–1988, Quaderno N° 5.
33. L. Luzzatti, *Memorie autobiografiche e carteggi* (Zanichelli, Bologna, 1931), pp.207–214
34. B. Croce, *Indagini su Hegel e schiarimenti filosofici* (Laterza, 1952)

Chapter 13
The History of Science Within History Studies and Its Relevance for History Teaching

Francesca Bearzatto

Abstract Humanity has made new acquisitions in all the fields since ancient times. These acquisitions often influenced each other. In particular, developments in scientific knowledge and acquisitions have influenced the way of developing historical knowledge. Historians have applied scientific method and technics to its own research methodology and have broadened the concept of historical source. Historical sources and the way of analysing them have acquired methods from sciences. Historiography is prompted to complexity and interdisciplinarity. Science, on the other hand, has developed some fields of investigation integrated with history, for example ethnobiology. The history of science has also changed its characteristics, intersecting historical methodologies that are constantly developing. Currently, the history of science includes scientific discoveries and context in which they were developed; historical context is considered in an integral perspective, as political, social, and economic. The way in which science and history influence each other currently requires the ability to dialogue between the two disciplines. Therefore, it is essential to share a language that can be understood by both areas. There is a growing need to develop an interdisciplinary language, beside the technical one. History teaching reflects this need to improve interdisciplinary language. Historical subject requires students to have at least basic scientific knowledge for a non-superficial understanding of history. The history of science is not just a set of knowledge closed in specific focuses, but it integrates history. Education is important in order to pave the way to an interdisciplinary language that allows an integrated and in-depth understanding of disciplines.

Keywords Relationships between history and science · Historiography and science · Historical sources · History of science · Sciences in historical context · History teaching · Interdisciplinary language

---

F. Bearzatto (✉)
Liceo Giosuè Carducci - Dante Alighieri, Trieste, Italy
e-mail: francesca.bearzatto@libero.it

Istituto regionale per la storia della Resistenza e dell'Età contemporanea nel Friuli Venezia Giulia, Trieste, Italy

L. Mammino (ed.), *Insights into the Relationships between Humanities and Sciences*,
https://doi.org/10.1007/978-3-031-97900-2_13

13.1 Introduction

Both history and science aim at observing and understanding the reality: in the case of the history, the reality of the humankind of yesterday and of today; in the case of the science, the natural reality in relationship also with the humankind. The fact that both the disciplines have a common task, that is that to know the reality in which the humanity lives, has stimulated some scholars, of different epochs, to reflect weather a relationship has developed between the study of the science and the study of the history. The relationship between the science, considered as development of the scientific knowledge, and the historical studies, still takes great significance to the light of the roles that the scientific acquisitions have for the development of human societies. Scientific acquisitions and the consequent technical applications are undoubtedly important components for the historical reconstruction of the past operated by historians.

This chapter outlines some reflections on the relationships between the history of science and history studies, and how these relationships influence the didactics of history today. These relationships can be considered from several points of view that will be outlined in the current text.

First of all, the history of science is the history of how the way to generate scientific knowledge has changed with time [1]. The development of scientific knowledge can be studied both from an inside point of view, in its intrinsic developments, and in its relationship with other knowledge areas, including the historical one. Scientific knowledge has developed through centuries: from antiquity, humanity has made new acquisitions in all the fields. As their knowledge related needs changed, new methodological tools were adopted to respond to those changes. These tools were not developed in closed compartments, but with mutual influences across disciplines. In particular, a dialogue has often been opened between the research methods of historical discipline and the methods of scientific research. This dialogue is of interest in the current chapter for several reasons.

The first reason relates to the fact that reflections on the characteristics and the mutual differences of two research method—scientific and historical—has changed the approach to history investigation. The observation of new advances in the scientific method has prompted historians to reflect on their own investigation methods, and this resulting in new methodological tools. Attention to scientific research has allowed historians to develop new methodological tools, and also to acquire new analysis tools, derived from mathematics and statistics, and with new technologies to examine any type of material pertaining to the past. In this way, historians have widened the scope of historical research, by escribing increasing importance to historical source—documents and materials that serve as foundation of historical research. Historical sources have increasingly including scientific information and data, and can also be analysed through scientific-based tools, mathematics, statistics, computers. Some forms of historical research are currently moving toward interdisciplinary perspectives, having remarkable scientific component.

The second reason for which the dialogue between historical studies and sciences is important is the fact that the just mentioned changes in historical research method, also triggered by comparison with the sciences, has changed the way in which research about the history of science is conducted. The history of science also comprises the history of the methods employed by scientists to pursue the knowledge of the natural world. Consequently, the history of science becomes a meaningful aspect of the history of cultures and attitudes. The study of evolution of scientific thinking contributes to form the cultural-historical picture of the different epochs: the history of science not only inform about the scientific knowledge achieved in a specific period, but also provides information how different social groups in a culture perceive and reutilize scientific progress. Scientific and technological discoveries often have immediate consequences on the human life: they influence habits, daily life, the way to work, the way to make wars, that is they influence material life. In turns, changes in knowledge and in material life also influence the beliefs systems.

Conversely, the history of science is often adopting new interdisciplinary methods, by viewing scientific innovation in a broader context, such as a framework of social, economic, cultural, political factors [2]. This approach also entails the study of the conditions that have allowed the development of scientific innovation, and their large-scale consequences.

The third reason for which the dialogue between historical studies and sciences is important is associated with the features of communication between the two areas. Dialogue entails the use of the language that makes communication accessible between the two areas. Two opposite trends are presently dominant in the world of research. A field is tightly specialised, focusing on a particular aspect of a discipline and using a specialised language, which is hardly understandable to different areas of knowledge. And then, a different trend strives to interdisciplinary work: it starts from the formulation of questions that are interesting for more than one discipline, and searches for answers that integrate those provided by various disciplines; therefore, it requires a possibility of communication among different fields of knowledge.

Which relationships exist between science and history today? Public school can offer a suitable ground to verify the current developments of these relationships. Didactics has the task to communicate contemporary knowledge to younger generation. The practice of teaching not only gathers the most recent knowledge of the different disciplines, but it also gathers aspects related to how different disciplines relate to each other. The teaching of the history of science, within the teaching of history, is affected by the dichotomy between the sectorial and the interdisciplinary approaches. The dichotomy is also evident by compering teaching guidelines and better teaching practices. In the formal, many guidelines or teaching supports, the history of science is frequently proposed in the form of a series of thematic insights, within the history subject, to comply with isolated focus boxes. In good teaching practices, basic scientific culture is not treated as an external element separated from the tread of the historical information, but as a necessary component for a not superficial understanding of history by students. Conversely, the ability to gather the bases of historical research is necessary for a non-superficial understanding of the history of science.

To begin the discussion of the relationships between the history of science and history studies, and of the implications for the teaching of history, the next session recalls the definitions of history and science.

13.2 Defining 'History' and 'Science'

As prerequisite to the discussion that will follow it is important to define the two key-terms "history" and "science".

13.2.1 What is History?

History is the ordered explanation of past human facts, based on a critical analysis of source materials. The explanation is correlated with the interpretation of the causes of those same events; this interpretation constitutes historiography. It aims at ascertaining connections between events and recognizing development threads [3–5].

The object of historical studies has changed through the ages: for hundreds of years, historians have made different choices for the selection of human events on which to focus. For example, until the late eighteenth century, the evidence from popular culture did not fall into the objects studied by historians.

Historians have also been building different approaches. For instance, recently, the 2000 Congress in Oslo (Norway) marks the birth of new historiography methodology: the global history that analyses local, regional, and supra-regional histories within global or transnational networks. For example, within this approach, the part of history that earlier was mainly focused on Europe moves to be focused on Eurasia, considering Europe as its extreme western appendix.

Historians also have been forming various general visions, deeply conditioned by social models, political events, and different cultural backgrounds. For instance, the main approaches of the European historiography can be considered for illustration. The Ancient Greek historians aimed at separating myth from reality. During the Middle Ages, christian historians tried to paint the history of the whole humankind, through times and empires, as a continuous struggle between good and evil. The nineteenth century historicism, and its main figure G.W.F. Hegel, divided past Eurasian history into four worlds, Eastern, Greek, Roman, Germanic, and considered history as an expression of the spirit of time and the incarnation of the principle necessary for the progress of humankind [6].

Since mid-twentieth century, historians had begun adding a number of new research frameworks: studying from a gender prospective; studying different social categories, such as people hospitalized in psychiatric wards; considering different sources, such as oral ones; paying attention to other kinds of events, beside diplomatic or military matters, for instance including daily life habits. This new way of approaching history is continuously opening historical studies to interdisciplinary approaches in a way that is increasingly complex. On the other hand, the opening to interdisciplinary studies is making the boundaries of historical studies increasingly less sharp. For instance, the high specialization required from historians of science means that many of them currently have a scientific background, beside the historical one.

The fact that the approach to history studies has undergone significant changes over centuries prompts the question as to whether history may have a unifying system across these changes. Considering the meaning of the term "history" can give significant indications. "History" comes from the Greek word ἱστορία, that means research, investigation, knowledge, and derives from the Indo-European root *vid* [*vi* + *d*], "to distinguish [*vi*] in the light [*d*]". The related verbal form *ístōr* means "he knows by seeing"; the verbal form (second singular person) *oĩstha* means has a series of consecutive meanings: "you have seen", then "you know", then "you tell" [7]. The Greek word ἱστορία formally derives from the noun ἵστωρ, "knowing, expert", intern related to the verb ἱστορέω, "to be witness or expert, to give testimony, to recount, to get testimony, to find out, to search" [8]. The action aspect of the word roots of the Greek verb comprises both present and past: "(now) I tell, because I have seen and known". Yesterday liaises with today, the action of telling bridges times. "History", ἱστορία, in its meaning as "research", is intended as a way of understanding the reality of the past. The Greek historian Herodotus (fifth century BC) was the first to use the expression ἱστορίης ἀπόδεξις "exposition of the research". In addiction, "history" owns semantic duality: it indicates either the "events" (in Latin, *res gestae*) or the "narration of events" (in Latin, *historia rerum gestarum*) [9].

Research, data gathering, critical analysis of records, interweaving past human events with more recent human events within long-term history, statement of the facts, are the fundamental aspects of history. These aspects are relevant both in history research and in teaching history.

13.2.2 What is Science?

The term "science" comes to us from the Latin *scientia*, which means "knowledge". *Scientia* derives from the verb "sciō", "to know". According to M. de Vaan, "to know" is developed from "to distinguish, make a decision", and belongs to a root meaning "to cut off, incise" [10]. According to Rendich the verb "sciō" is related to the Indoeuropean root chā, chyoti, in Latin ch = sc and y = i [11] meaning "to cut", "to know" and "to decide", and consequently, as inference, "to distinguish", that means a kind of knowledge that implies clarity, a distinction between different kind of entities [7].

The root of the term "science" is therefore closely related to the meaning of "knowledge". The way of accessing scientific knowledge, like the approach to history,

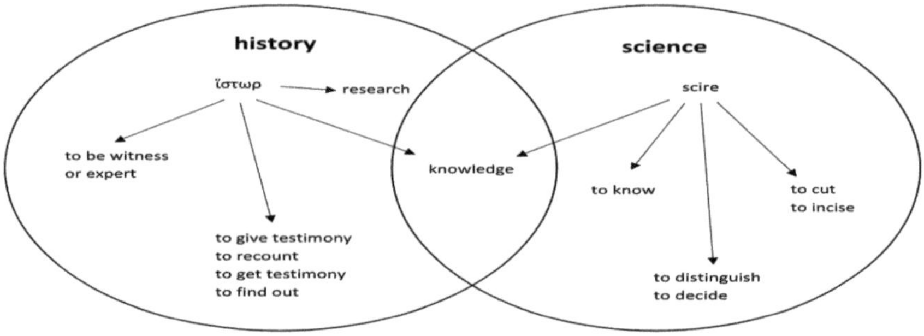

Fig. 13.1 The root meanings of the two key-terms "history" and "science". Both have an intersection in the meaning "knowledge"

has undergone significant changes over centuries. For instance, in ancient civilizations geometry and arithmetic were oriented towards the solution of practical problems, like those of builders and land surveyors, and towards the observation of the sky. For Greek scientists and philosophers, the logical arguments were combined, and sometimes opposed, with the empirical observations. Since the scientific revolution of the seventeenth century, science has been conceived as alternative knowledge to traditional knowledge and doctrines (like those related to the Aristotelian-Ptolemaic model for astronomy), as a synthesis of experience and reason.

At present, science is defined as a system of knowledge that is concerned with the physical world and its phenomena, that entails unbiased observations and systematic experimentation, and that uses formalized languages [12, 13] (Fig. 13.1).

13.3 The Relationships Between History Research and Science

Science, as a system of knowledge that is concerned with the physical world and its phenomena, is relevant for historians. The term "history" and the term "science" are connected by the meaning "knowledge". This connection has prompted historians to look with increasingly interest to scientific method through the ages.

13.3.1 The Relationships Between Human Beings and the Environment as Focus Areas for Both Historians and Scientists

Science, as a system of knowledge of the physical world, relates human beings and environment. This aspect of scientific knowledge is relevant for historians. From the history point of view, the interdependence of the humankind with nature ranges from the way of thinking about nature, to the way of interacting with it. These aspects change over time and space, following the development of human history. For instance, before the availability of refrigerators food storage systems were closely linked to the resources of the territory. In Europe the communication of food preservation techniques across generations strengthened family ties. An example is offered by the correspondence dating back to World War II, 1941, between Antonio Brun Pascutta, resident in Fossalta di Piave (province of Venice, Italy) and his niece, Elena Tramontina, resident in a mountain valley in north-east Italy, Val Colvera (province of Pordenone). In a letter to Elena, the uncle describes the method of preserving mushrooms by drying, and he advises her to stock them up, both for the family and for sale. This was important because the mushrooms were one of the main edible resources of the territory where the niece lives. The knowledge of how to preserve them was precious in the context of food crisis that hit Italy in those years, and it helps strengthen mutual assistance bonds between the two branches of the family [14]. Following the dissemination of refrigeration systems, the transfer of information on conservation methods became less important for the survival of individual; this information, therefore, also lost the cohesive role that it had in traditional societies.

The relationship between humans and the environment, the development of scientific knowledge system; the manipulation of the environment with technology, are all aspects of the history of science that have many implications for historians.

13.3.2 The 'Knowledge' Meaning as a Connection Between the History and Science Meanings

As previously mentioned, the term "history" and the term "science" are connected by the meaning "knowledge". But, looking at the roots of two words, the way in which we know in a historical approach, and the way in which we know in a scientific approach, are different (Sects. 13.2.1 and 13.2.2). "History" means "to know by seeing", from the Greek word ἱστορία; "science" derives from the Latin "sciō", "to know", in the sense of "to distinguish, make a decision". This connection in the roots of the two terms has prompted historians and scholars to reflect on the relationship between history and science. The awareness that both history and science refer to the observation and understanding of reality contributed to stimulate historians to compare the two ways of understanding reality. Since the fifteenth century, the methodological reflections of historians and scientist have challenged each other. After the growth of the scientific revolution in modern age, historians have looked with increasingly interest to the scientific method: comparison between the methods of historical research and the methods of scientific research has undoubtedly enriched the historiographical debate.

The consideration on the differences and similarities between the two ways of knowing reality are interwoven with epistemology, the philosophical study of the nature, origin, and limits of human knowledge. The philosophical questions about knowledge (such as whether knowledge is innate, present in the mind from birth, or acquired through experience; whether human can reach certain knowledge; whether the ultimate source of human knowledge is the faculty of reason, or experience) [15] are intertwined with both the methodology of science and the methodology of history.

13.3.3 Historiography and Science in the Age of Humanism

The connections between historical and scientific knowledge were identified since the dawn of the scientific revolution of Modern age, during the age of Humanism. Humanism is an historical period, between the 13th and the fifteenth century, characterized by a flourishing of studies on classical languages and literatures, i.e. ancient Greek and Latin. Classical culture was considered as a tool of spiritual elevation for humans, envisaging that poetry, rhetoric, philosophy, history provide guidance to citizens, scholars, rulers, politicians, in order to develop a new civilization. The efforts of humanist scholars aim at the recovery, study, and publication of classical texts. Humanism marks a transition to a different way of considering classical works. Medieval scholars interpreted classical authors as if they belonged to their era, as if they shared medieval culture. Humanist scholars interpreted classical authors by retrieving the classical culture, and contextualizing sources within their original cultural frame. Humanist-historians begin thinking of past philosophical and conceptual systems no longer as the oracles of God, but as the products of specific cultures [16]. Recovering as many ancient sources as possible and philologically respecting ancient texts thus become the basis of humanist-historians' studies.

The scientists of the fifteenth century are Humanism' children. For example, Leonardo da Vinci (1452–1519) adopts a humanist method for his scientific studies, which involve the return to genuine reality. In his opinion, the scientist needs to return to physical reality that is not satisfied with the opinion of past authoritative scholars, but to be guided by experience: «Knowledge is the child of experience». Experience is an open door to look at the reasons of things, the work of God. Leonardo had the same approach as humanists who read and interpreted ancient texts with humble respect: he wanted to read the work of God, the book of physical world with respect [16].

Galileo Galilei (1564–1642) inherits this attitude for the study of physical reality. Galileo, considered as the father of the modern scientific method, looks at experiments as a key-tool for discovering the facts of nature. Alongside this, he believes that the book of nature is written in the language of mathematics. He expresses this view in "The Assayer" (1623) with the following words, were philosophy means knowledge: «in philosophizing, it is not necessary to rely on the opinions of authoritative scholars. Philosophy is written in the great book that is continually open in front us (I say, the universe), but we cannot understand it if we do not learn its language, and the characters in which it is written. It is written in the language of mathematics, and its characters are triangles, circles, and other geometric figures. [17]

13.3.4 The Descartes's Influence on Historiography

In the seventeenth century, scholars and historians look closely at the cognitive method that scientists are shaping. Descartes' thought is particularly important for the elaboration of historiographical method.

René Descartes (1596–1650), mathematician, scientist, and philosopher, promotes methodical doubt as a tool to dismiss apparent knowledge derived from passively recognised authority. Descartes aims at building new epistemic foundations, both for the senses-based knowledge of reality and for the innate ideas of the mind, such as mathematical truths. Even in his early works, as "Rules for the Direction of the Mind", Descartes maintains that the human mind is capable of grasping the truth, if it is used correctly. For this purpose, it is necessary to reject all knowledge that is only probable, and accept only the knowledge that is perfectly owned, of which it is impossible to doubt [18]. Koyré comments: «Doubt is the touchstone of truth: the acid that dissolves errors. We must therefore make it as strong as possible, and doubt everything, and do it so wherever possible. Only after doing this we create the certainty of keeping only the pure gold of truth» [19].

In the seventeenth century, French libertine philosophers begin to profess Cartesian methodical doubt. Their critiques soon target rationalism, to which they oppose a way of thinking influenced by scepticism and by Montaigne's thought. The result is to suspend judgment and take no part in controversies regarding the possibility of certain types of knowledge. In the field of historiography, the suspension of judgment produces a critical view of the truths proposed by traditional historiography. A number of libertine philosophers develop a historical relativism that prompts them to compare the Western world with other civilizations, both contemporary and ancient [20]. P. Bayle' historical work, "Historical and Critical Dictionary" (1697) [21], originates from methodical scepticism. The historical-critical method adopted in the "Dictionary" is the application of doubt to various subjects—history, morality, religion and philosophy. The methodological doubt is complemented by an attitude on absolute tolerance, defined by Bayle in the fiery field of religious struggles of his time. Bayle based his historical-critical method on three fundamental aspects: (1) the detailed verification of historical data (the precise determination of facts); (2) the dialectical comparison of opposing theses (the definition of the topic); (3) the analysis of various philosophical and theological systems with major focus on error rather than truth. According to Bayle, an history fact is verifiable in principle, and, under certain circumstances, it can even be argued that history truths has a certainty comparable with the most evident truths of sciences.

Cartesianism influences the debate about history also in other ways. Philosophers following René Descartes distinguished sure knowledge with has rational bases and probabilistic knowledge, among which they include history. As Paul Hazard writes:

«In their inmost consciousness, men had come to look on history as a very broken reed. The very notion of historicity was tending to disappear. If, now, men turned their backs on the past, it was because they thought it something evanescent, Protean, something impossible to grasp and retain, something inherently and inveterately deceptive» [22] (Fig. 13.2).

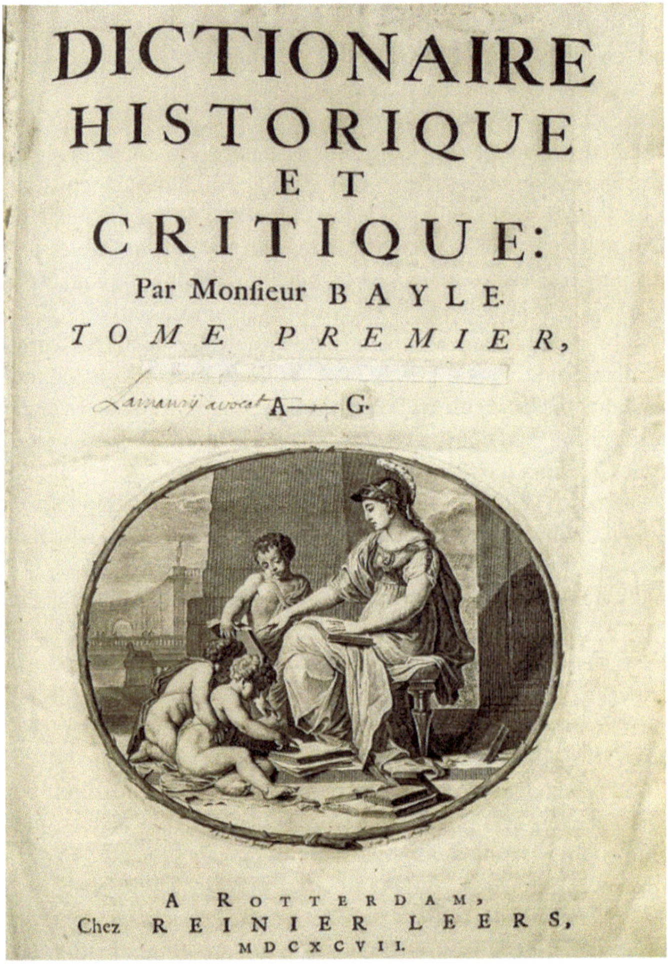

DICTIONAIRE
HISTORIQUE
ET
CRITIQUE:
Par Monſieur BAYLE.
TOME PREMIER,
A—G.
A ROTTERDAM,
Chez REINIER LEERS,
MDCXCVII.

Fig. 13.2 The frontispiece for a copy of the Pierre Bayle's book "Dictionaire historique et critique par Monsieur Bayle, Tome Premier A-G", published in Rotterdam, in 1697. The drawing on the frontispiece shows the Greek goddess Athena—goddess of knowledge—teaching three children in a public square. *Source* https://commons.wikimedia.org/wiki/File:Pierre_Bayle_-_Dictionaire_historique_et_critique,_Tome_Premier_A-G,_Reinier_Leers,_Rotterdam_1697_-_frontispice.jpg

13.3.5 Historiography and Newton' Method

After Descartes, doubt becomes a methodical base for a new analysis of reality. The sceptical point of view and the reflections of philosophers like P. Gassendi or A. Arnauld has consequences also on the methodology of historical studies. On the one hand, history is regarded as a form of probable knowledge, but not certain (unlike scientific knowledge), and therefore unreliable. On the other hand, thinking on the differences between the two forms of knowledge leads scholars to believe that historical truth is different from mathematical truth, but not less possible. This view on history methodology is related to the definition of Newton's method. The physicist, mathematician and philosopher Isaac Newton (1642–1727) elaborates a definition of scientific method, fundamental to the evolution of scientific revolution in the seventeenth century. This method consists in a synthesis between experimental investigation, considered as the basis of research, and mathematical reasoning, considered as an instrument, a language, that allows deriving from phenomena the laws that regulate them. These aspects of Newton's method become important for the way in which historical studies were considered.

Between the 17th and the eighteenth century, on one side, in every field of knowledge the aim is an objective certainty, a clear and distinct knowledge: all the branches of human research seek immutable laws, through strictly scientific procedures. In the humanities area, including historiography, the methodical and systematic study of sources is applied, similarly to the physical sciences method. The studies on diplomatic and archival documents, epigraphic sources, palaeography, numismatics, and the like, provide basic tools for humanities studies [23]. Methodical and sceptical doubt prompted the search for critical tools to investigate historical reality; these tools are further developed in the eighteenth century. Historians elaborate a logical system of norms and techniques, which will allow historiography to acquire new epistemological status [24]. This will allow history research to approach new Enlightenment categories of factual evidence, empiricism, and probability.

On the other hand, some scholars feel the need to develop a comparative method, on an experimental basis, also for humanities studies. The Montesquieu's "Esprit des Lois" tries to organize the diversity of uses, customs and ideas of different civilizations, classifying them into a limited numbers of categories. This philosopher wants to extend Newton's experimental method for examining the laws that govern the material world to the study of human society. Even if it is not a history book, "The Spirit of Low" introduces the application of comparative views, the historical contextualization of laws, attention about environmental background, and categorization into typologies; these operations will become typical to European historians in the eighteenth and nineteenth century [25].

13.3.6 Historiography and Positivism

The purpose of applying scientific models to historical study becomes more evident during Positivism, alongside with the changes in scientific thinking. In the nineteenth century, scientists aim at establishing conceptual foundations and empirical knowledge for rational scientific understanding, which they view as accurate, dependable, and universal [26]. Positivist scholars escribe to empirical sciences the function of providing the content and methods of knowledge; and to philosophy the task of collecting, systematizing and generalizing natural laws discovered by sciences. M.J. Nye observes: «All varieties of position insist that *positive knowledge* should be the determinant of what practices and claims are accepted in science. [...] What is positive knowledge? It is knowledge of what can be really known, where "what can be really known" is what happens in the real word» [26].

Positivist philosophers consider society as a part of reality, which can, therefore, be investigated and the resulting information can be organised in a way similar to the other sciences. They believe that sociology and the other human sciences might have the same predictiveness as exact sciences, if considered with the same attitude as astronomers, physicists, or physiologists use in their studies [27].

When society is studied as a part of reality, like an experimental subject, it is initially considered as a mechanism, and progressively as an organ.

The mechanistic view is derived from the Newtonian tradition, applied to humanities: society is meant as a mechanism, in which every aspect is the product of deterministic causality. Positive philosophy and historiography ascribe primacy to experimental research; and the cult of facts and the methods of exact sciences, extended to human affairs, slowly change the historical approach. The new historiographical method that is born has two countenances: scientific and philological. The scientific expression, attentive to environmental, social and racial factors, aims at composing the overall picture and understanding the events in their multiple connections. It increasingly incorporates also sources that derive from scientific and statistical data. H. Taine, E. Renan, T. Buckle and many others, applied criteria derived from economics, sociology, and the theories about collective psychology. Philological countenance enhances the gathering, critical analysis and publishing of sources and repertoires. The emblem of this season is T. Mommsen (1817–1903), who publishes a comprehensive collection of Latin inscriptions preserved on stone, iron, and other enduring materials, arranged them according to the principles of philological methodology; this "Corpus Inscriptionum Latinarum", CIL, was published from 1861 and still today is being updated.

In the mid-nineteenth century, Darwin (1809–1882) opens another scientific revolution: «the social scientists, taking their cue from biology, began to think of society as an organism. But the real importance of the Darwinian revolution was that Darwin, completing what Lyell had already begun in geology, brought history into science» [28].

13.3.7 Historiography and Scientific Method in the Twentieth Century

The historian J. G. Droysen (1838–1908)—a colleague of Mommsen—represents a link between the views of the history in the 18th and in the nineteenth century. Droysen expresses the same conviction of many of his contemporaries through these words: «The science of History is the result of empirical perception, experience and investigation» [29]. In his view, developed in "Historik", both natural sciences and history are based on observation; they proceed empirically, but their aims are different, as well as their methods. «The natural sciences try "to find in the restless stream of changes that which is permanent, namely, laws". History is concerned with changing forms and with their individual aspects; whereas natural scientist looks for analogies, historian ought to look for anomalies» [30]. Droysen sees a danger for the development of historiography in the positivist outlook of natural sciences. The inductive search for laws, carried out by natural sciences at the German universities, may affect historical studies, driving them towards establishing historical laws. History, therefore, must preserve its autonomy by acquiring its own method.

Droysen, acute observer, anticipates the controversy with positivism, which becomes more intensive in the twentieth century. F. W. Nietzsche (1844–1900), for example, looks at modern time as a time in which the usefulness of history is endangered by the claim that "history is a science". Modern scientific history, wanting to be objective and exhaustive, instead becomes erudition. A modern historian believes himself to be objective, «but it is only a superstition to say that the picture given to such a man by the object really shows the truth of things. [...] But this would be a myth, and a bad one at that. One forgets that this moment is actually the powerful and spontaneous moment of creation in the artist, of "composition" in its highest form, of which the result will be an artistically, but not a historically true picture. To think objectively, in this sense, of history is the work of dramatist: to think one thing with another, and weave the elements into a single whole, with the presumption that the unity of plan must be put into the objects if it is not already there» [31].

In the twentieth century, not all scholars and historians agree with the positions expressed by Nietzsche. He writes while Historicism is spreading in Europe. Historicism states the historicity of all thinking, and proclaims the superiority of historical knowledge over all the other disciplines: only history would be capable of contemplating individual aspects and values that constitute the deepest essence of life and spiritual reality in its dynamism.

A number of different theories on the relationships between history and science develop during the first half of 1900s. Some of them assume that laws should be identified for both sciences and human affairs. Some others think that the application of the inductive method borrowed from sciences to historical studies could result in a misrepresentation of history. Still others maintain that history could provide us with a system of knowledge more reliable than what the sciences could do.

While this debate grows, science undergoes a new and profound revolution. The Heisenberg's uncertainty principle leads to the development of a new physics and changes models. After W. K. Heisenberg (1901–1976), Newton's physics constitutes the physics that explains the macroscopic world. A physics completely different from the Newtonian one, quantum mechanics, is developed to study atoms and molecules. The uncertainty principle limits what humans can know. Such a radical change in physics' perspectives influences philosophy, beginning with the theory of human knowledge. While Positivism promote the idea that knowledge has no limits, the uncertainty principle sets some limits the human knowledge. Several areas of knowledge are influenced by these views. Absolute systems of knowledge become the object of spreading critics. Historiography gets involved through the critical approach of the Annales School. Lucien Febvre and Marc Bloch, two French historians, established this history school in the second decade of the twentieth century. The Annales school promoted a new form of history, focused on the lives of ordinary people. The humans' life is broadly contextualised through the studies about «climate, demography, agriculture, commerce, technology, transportation, and communication, as well as social groups and mentalities» [32].

Bloch links the new needs of historiography with the acquisitions in the scientific field of his time. He expresses it within these words: «We are much better prepared to admit that a scholarly discipline may pretend to the dignity of a science without insisting upon Euclidian demonstrations or immutable laws of repetition. We find it far easier to regard certainty and universality as questions of degree. We no longer feel obliged to impose upon every subject of knowledge a uniform intellectual pattern, borrowed from natural science, since, even there, that pattern has ceased to be entirely applicable» [33]. The historiography that results from the Annals' approach is a «broadened and deepened history».

A part of contemporary historiography tends to broaden contextualization. Some historical surveys include a series of information and methodologies from many disciplines. Anthropology, economics, statistics, the various branches of ethnology, psychiatry, and other disciplinary fields contribute in providing a rich set of information for historical research. The notion of context thus widens to include the social, cultural, family, political, economic, military, environmental contexts, and so forth. The various types of data are integrated in formulating historical scenarios. Such integration leads both to a combination of information from different research fields and, in some cases, to a combination of research methodologies. These topics will be dealt within the paragraph 3.

Whatever the future evolution, the comparison between the scientific and historical methods of research was, and is, a source of enrichment to the method of historical studies.

13.4 The Relationships Between Historical Source Material and Science

13.4.1 Interdisciplinary Outlooks—The Root of Historiography

The previous sections dealt with the question of how historians' views on the method of scientific research have enriched the historiographical method. Historians' thinking about the methodology used in scientific research, across the ages, has been contributing to a parallel and careful consideration about the theory of research, and about the method of processing and using source materials: collecting, classifying, analysing, and relating the different types of sources. From the end of 1800s, the increasing interest of historians towards sciences has allowed historians to equip themselves with new tools of analysis, taken from mathematics and statistics, and new technologies to analyse the realities of the past. This extensive ensemble of analysis tools has increased the contextualisation of the subjects of historical studies. The study of contexts has broadened, producing a deeper and more comprehensive awareness of the connections between humankind and its environment.

Actually, the issue of the connections between humankind and environment initiated together with historiography itself, but remained latent for centuries. Western historiography is born in Greece, in the Ionia region, in the fifth century BC. Herodotus, author of the first great narrative history produced in the Western ancient world, *Historìe*, about the Greek-Persian Wars, conceives historiography as observation and complex investigation. His research is focused on the relationships between humans and environment, and is woven with geographic, anthropologic, economic and sociologic information. For instance, Herodotus' interest in human geography and ethnology can be illustrated by the excerpt of *Historìe*, where he describes Nasamones, a Berber-speaking population who lived along the Gulf of Sidra: «Next Avestward of these Auschisae is the populous country of the Nasamones who in summer leave their flocks by the sea and go up to the land called Augila to gather dates from the palm-trees which grow there in great abundance, and all bear fruit. They hunt locusts, which when taken they dry in the sun and after grinding sprinkle them into milk and so drink it» [34]. In his descriptions, Herodotus does not use what we call scientific data, as entomological or botanical information. He uses mostly oral sources, distinguishing what he sees as a witness, what he reads in texts of logographers, and what he conjectures. His book attests a straightforward interest for connections between population and ecosystem.

Herodotus has traced the path to interdisciplinarity in history. This attitude, however, will remain a marginal approach for centuries: prevailing historiographical patterns focus on general narration about political, military and institutional events, until the twentieth century [25]. Herodotus start been re-appreciated in the middle of the sixteenth century, and is considered as major historian in contemporary historiography. As Momigliano writes: «The varying fortunes of Herodotus in the eighteenth and nineteenth centuries are a symptom of the tension existing between the supporters of political history and the supporters of the history of civilisation. [...] There is no need to add that today Herodotus is perhaps more generally appreciated, certainly more generally loved, than Thucydides. The need for a comprehensive, extra-political history is admitted by almost everyone» [35].

Alongside the development of the interdisciplinary approach, scientific information and tools become gradually more important for historians, and the application of scientific data in history has given several outcomes. We might distinguish two issue: historical data processing using mathematical-statistical models or computer technology and the incorporation of scientific data in historical research.

13.4.2 Historical Data Processing Using Mathematical-Statistical Models or Computer Technology

The combination of statistical models with historical research develops along the first trial of interdisciplinary history, after Herodotus. The first laboratory of "cross-discipline" history is considered the School of Göttingen, Germany, in the University of Göttingen, founded in 1737 by George II of England, in his capacity as Elector of Hanover. The University of Göttingen will become to the fore as one of the main centres of dissemination of Enlightenment ideas. The characteristic of the School of Göttingen is an education model based on history, law, statistics and economy, to train public officers, giving them all the expertises in their profession. Similar models were adopted in the nineteenth century in major continental universities in several European countries, including Berlin and Paris, to train public officers in the era of foundation of national States. From this time on, statistical methods for data analysis became an integral part of historical research [36]. K. G. Lamprecht, for instance, was a German historian who studied at the universities of Göttingen, Leipzig, and Munich between 1874 and 1879; and was the author of "Deutsches Wirtschaftsleben im Mittelalter, 3 vol.", (1885–86; "German Economic Life in the Middle Ages"). Lamprecht applied statistics to historiography, and investigated the parallels between social structures—such as demographic, juridical, and economic structures—and the manifestations of the intellectual culture—such as art, literature, music, and philosophy [37].

During the nineteenth century, positivist scholars think that history has a structure similar to physics, with its own laws, to be discovered and understood. Positivist historians look for relationship between data, such as cause-and-effect, by analysing, classifying, and using statistical laws in these operations. Historians develop particular interest in the application of statistics to economic phenomena, studying them with a historical perspective. The economic history purposes to link history and economics. This approach spreads in Europe. The first university chair of Economic History is established in United Kingdom in the late nineteenth century (the first in the English-speaking world was at Harvard, held by W. J. Ashley from 1892 to 1901 [38]).

In the mid-twentieth century, The American school of New Economic History goes further along this path. Also known as cliometrics, this approach makes systematic use of quantitative sources to study the history of work, slavery, agriculture, transport. One of the great exponents of this school was D.C. North, who was awarded the Nobel Prize for Economics in 1993 together with R.W. Fogel. North developed a growth model of American economy from 1790 to 1860, and a general model of institutional changes in Western economic history. Studies focusing on quantitative data increased between 1960 and 1970s. The studies of crime and disease statistics, tax rolls, demography, attracted increasing attention as can be seen also from reports of the conferences of those years on quantitative data [39]. The Annales School elaborates and promotes long-term social history, and applies cliometric research methods to social historiography.

Since the early 1980s, European historians have opened a debate about quantitative methods. Stone considers the focus on diagrams and tables, which is peculiar to quantitative history, as a loss of the narrative role of history [40]. Other doubts arise about the interpretative evaluation of results: not even all archival documents, for instance, are adequate for quantitative examination, because they may contain heterogeneous data [41]; furthermore, serial documentation homogenizes human reality, to the detriment of specificity [42]. The application of statistical methods to historical research provides models extremely significant to understand economic trends or social development. However, aggregate numbers do not account for other information, such as response strategies given by individuals or small communities to specific problems, or to important historical changes. For example, statistics on food shortage sufferings in Italian rural communities during World War II do not tell what strategies individuals developed to overcome these difficulties. Survival strategies are linked to political and military context and may be the motivation behind different political behaviours.

Mathematical and statistical data, however, can integrate historical sources in a determinant way. An example could be Lewi Stone's study about railway transportation records to Belzec, Sobibor, and Treblinka death camps, during World War II, in the context of Nazi "Final Solution to the Jewish question" in Europe. Lewi Stone is a mathematical biologist interested in employing mathematical models to the study biological systems. He utilized a dataset originating from railway transportation records for his study "Quantifying the Holocaust", and «identifies an extreme phase of hyperintense killing, when > 1.47 million Jews—more than 25% of the Jews killed in all 6 years of World War II—were murdered by the Nazis in an intense, 100-day (~3-month) surge» [43]. The Holocaust studies are complicated by the destruction of information by the Nazis during the war, for fear of future incrimination. Stone' study, therefore, utilized the data extrapolated from railway records as a substitute of missing documentation about Operation Reinhard, the code name for German plan to murder Jewish communities living in German-occupied Poland, between march 1943 and november 1943.

The interest in quantitative methods is further spreading in historiography since the mid-1980s, with the mass diffusion of personal computers and database management software. Systematic and publicly available databases, which can be exploited for a wide range of research, are increasing in historian communities. The web allows the new practices of source dissemination, like the digitized archives or the collection of digital object, including papers, books, recorded oral sources, interactive and thematic atlases. The spreading of digital information systems has further modified the historians' approach to data. Index source materials across internet, query forms, search engines to examine records, are tools by which historical researchers pursue widespread studies and compare data with a new smartness. An example is the application of computational linguistic to historical studies: the concordance detection systems allows observing the variation of key-words meaning in texts, which are relevant to historians [44]. Recent oral history studies submit interviews to the quantitative approach of computational linguistics software. Historians exploit data analysed by software for studying the linguistic content of oral memories and developing complex historical and cultural contexts [45].

The application of digital technologies to history research is still developing, but it issues some problems. The volatility of digital documentation and fast changes in technology, computer software and other assets, raise a delicate point over the integrity, authenticity, access and identifiability of computerized records over the time. Added to this, the challenges in determining the scientific quality of data available on the web, and the risk of distortion, caused by the fragmentation of information offered by the internet, demand a continuous monitoring work by historians.

13.4.3 The Incorporation of Scientific Data in History Research

History research has fully incorporated scientific data in historical data since the middle of the twentieth century. On closer inspection, the previously mentioned interdisciplinary approach at the University of Göttingen prepared natural conditions for this integration. G. Droysen, who studied at the University of Berlin, which was closely related to the School of Göttingen [25], has a wide view of the areas with which history studies have concerned [46]. In the handouts of the lectures that Droysen gave from 1857 to 1883, he writes: «Historical material is partly what is still immediately present, hailing from the times which we are seeking to understand (Remains), partly whatever ideas human beings have obtained of those times, and transmitted to be remembered (Sources), partly things wherein both these forms of material are combined (Monuments)». Among the "Remains", Droysen lists a wide range of items, such as «works whose form is due to human agency,—artistic, technical etc., as roads, plots of leveled ground, and the like» [29].

The use of scientific sources for history studies is fully realized in the middle of the twentieth century, even if the conditions for the introjection of scientific data into

Fig. 13.3 Portrait of Johann Gustav Droysen, by unknown author, published in "Imagines philologorum", Berlin/Leipzig, in 1911. *Source* https://commons.wikimedia.org/wiki/Category:Johann_Gustav_Droysen?uselang=it#/media/File:Johann_Gustav_Droysen_-_Imagines_philologorum.jpg

historical studies had existed since a century before. This is due to two circumstance, which developed only towards the 1950s: the growth of technologies that can be used in history, and the growth of interdisciplinary approach in humanities (Fig. 13.3).

13.4.4 The Application of New Technologies in History Research

Chemical and physical applications to archaeology have a well-established heritage. The employment of chemistry to the analysis of historical source dates from the end of the eighteenth century. M. H. Klaproth, one the founders of analytical chemistry, published a series of analysis results: the first series of ancient coin analyses in 1799; the results of the chemical investigation on an ancient mirror in 1801;

the compositions of some ancient bronze weapons and implements in 1809. The attention concerning wall painting pigments, glass and metal, found during archaeological excavation, has been increased since the beginning of the nineteenth century [47]. The discovery of X-Rays, by W. Röntgen in 1896, immediately had significant implications in archaeology: X-Rays were applied in the study of Egyptian mummies [48]. The science and technology developments during and after World War II had considerable implications for archaeology: in the mid-fifties, the word "archaeometry" was coined as «a formal field of study whose practitioners apply techniques and approaches from the physical, chemical, biological, and earth sciences and engineering to address archaeological questions and problems» [49].

In particular, Ancient DNA (aDNA) analysis—the study of DNA from archaeological or paleontological specimens—enables continuously expanding research. The introduction of the polymerase chain reaction (PCR) in the 1980s, and the profound impact of next-generation sequencing technologies (NGS) in the decades that followed, have allowed aDNA studies. Examination of DNA from archaeological sample has helped reconstruct the evolutionary history of the human species. The Neanderthal genome was initially sequenced by Professor Svante Pääbo—who was awarded the 2022 Nobel Prize in Medicine, "for his discoveries concerning the genomes of extinct hominins and human evolution". Recent research, that examine the genetic variants introduced into the ancestors of modern humans from inter-breeding with Neanderthals, shows how that inter-breeding impacted on genes that are important for the immune system, development, and metabolism [50].

Genetic analysis allows to deepen the origin of genetic characteristics of modern human, and identify new branches of the human family tree, such as the Denisovans, who emerged about 370,000 years ago during the Pleistocene Epoch in Eurasia, spreading throughout eastern and southern Asia and parts of Melanesia before disappearing sometime after about 30,000 years ago [51].

As an outcome of the wide range of scientific approaches currently relievable, current trends are interdisciplinary, and closely related with a broader description of the context of the objects under study. Time limit of archaeological research is also expanding to the most recent past, like to Industrial revolution, involving even contemporary history: since the '60s technological tools and scientific methods had integrated into historical research in increasingly pervasive ways [52].

13.4.5 The Growth of Interdisciplinary Approach in Humanities

At the same time as historical research was integrating the new technologies, it was also developing interdisciplinary interests to new advances. A key role is played by the *nouvelle histoire* mainstream, expressed within the Annales School—already mentioned in 3.7—which acknowledges the validity of a wide range of data in order to investigate human life. This school is influenced by structuralism. Structuralism conceives history as a set of structures—from beliefs, to economic practices—that function organically; each structure changes slowly and functions as a set in which elements are linked, matched and swapped; and all structures are related to each other [53]. Structural historiography uses several source materials, information and types of data—belonging to sociology, economics, psychology, linguistics, ethnology, sciences and the like—to form historical frameworks. F. Braudel, one of the leading exponents of *nouvelle histoire*, published "Civilization and Capitalism, 15th–18th Century" between 1967 and 1979. The first of the three volumes, "The Structures of Everyday Life", is the emblem and promoter of historical studies on material culture [54]. Material culture is constituted by the impact that material objects produced by humans—such as tools, machines, weapons—have on the intangible culture. This area of study involves the connections between scientific data and history. The Annales' interest in cultural anthropology is equally important for the relationship between sciences and history, because cultural anthropology also focuses on everyday life [55]. The study of everyday objects and works is a field of research in which science and history intersect. The Annales's historiography developed while the interest for the interdisciplinary studies grows in Europe. Environmental and ecological studies, anthropology, ethnography, ethnoscience, ethnobotany, and the like, were widely circulated in the second half of the twentieth century, also reaching a non-specialized audience through information channels.

Since the late twentieth century the intersection of historiography with other branches of knowledge has led to many trends, schools and specialised fields. This resulted in the multiplication of studies and publications, the difficulty of communication between the specialised branches of historiography, and the increase of neologisms characterizing each micro-domain [56]. Consequently, mastering interdisciplinary studies becomes complex, and educating historians who cultivate wide interests requires considerable efforts.

Interdisciplinary approaches, particularly the incorporation of scientific data in historical research, enrich the notion of historical contextualization. An example is offered by works focusing on the implications of the relationships between humans and the environment, and the consequences of human behaviour from a historical and economic perspective. For instance, P. Bevilacqua's studies describe the relationships between Italian environmental policies, the process of anthropization, and the traditional strategies for interacting with the territory [57]. Bevilacqua focuses on the transformation of productive ecosystems related to ancient cultivation, indigenous and not, like olive, almond tree, or mulberry. The dissemination of rice fields in Italy—since their introduction by Arabs in the fifteenth century, mainly in Sicily—follows economic strategies, transforms human settlements, and modifies traditional agricultural systems [57]. The rows of eucalyptus, a non-indigenous tree planted along the Italian coasts, testify to a State policy at the beginning of the twentieth century aimed at absorbing water, regenerating the air, and withstanding the corrosion of winds and saline. Public technicians and administrators were also taking measures against malaria and against hydrogeological instability by draining water end by planting trees in a strategically ways [58].

Other projects of historical research focus on the relationships between specialised productions linked to land resources, microeconomies, and human settlements. An example is the research about mountain economic systems in the European Alps fostered by the International Association for Alpine History (IAAH), and its journal "Histoire des Alpes-Storia delle Alpi-Geschichte der Alpen". The comparative study of the integrated peasant economy developed in the Alps—entailing different exploitation options for different available resources, related to the ecological characteristics of the territories—shows deep connsections between the knowledge of the territory and social organisations [59].

Several branches of scientific research enable a deepening of our knowledge of the material culture of a given human community. Material culture is one aspect of historical knowledge which can complement historical sources, especially regarding cultures not having written documentation. For these cultures, oral narration and material products are the main sources, rather than written documents. European rural culture was largely an unwritten culture until the mid-1900s. For example, Michele Mezzaroba's testimony enables us to observe how scientific data can enrich historical sources [60]. Mezzaroba was a political deportee in Mauthausen, during World War II. He was born in Frisanco, in the valley of Colvera, in Friuli Venezia Giulia, in northeaster Italy. Mezzaroba tells: «Tullio Tomasso—whom I grew with in Frisanco—sometimes went to work outside, engaged on train lain, in the Mauthausen camp. He brought some grass back. He collected grass that we all used to eat when we were kids, in Frisanco: it is called "radichesa", or "pestenàla". [...] We both went around a less visible corner, Tullio took out the grass, and we shared and ate» [60]. This testimony is consistent with exhaustive studies on the internment of political detainees in Nazi concentration camps, coming from Friuli Venezia Giulia. Both Mezzaroba and Tomasso were recorded in national and regional works about political deportees [61] [62]. Other testimonies convey the forms of cooperation which were sometimes formed between persons connected by the same origin, or family bonds, despite the precarious existence in the Nazi camps—like the friendship between P. Levi and A. Della Volta, described in "If this a man" [63]. The Mezzaroba's account is consistent with the scenery of Nazi labour camps and death camps, as reconstructed from existing documentation, evidences, and researches. This testimony provides further specific information. The edible herb, "radichesa", or "pestenàla", that Mezzaroba uses to define the type of herbs consumed both in Mauthausen and in Frisanco, is *Taraxacum officinale* [64]. *Taraxacum* was one of the over fifty wild herbs traditionally eaten in the pre-alpine valley of Frisanco, Val Colvera. Several ethnobotanical studies trace the use of edible wild herbs in the Alpine region to the distant past. These researches map out several places that have in common some historical and archaeological evidence of Celtic culture [65]. Mezzaroba and Tomasso carry with them the traditional know-how of the inhabitants of the Alpine valleys, even in a war that belongs to the full contemporaneity. That knowledge is key to understanding Mezzaroba's survival strategy. If we consider that this cultural tool (and others) was widespread in rural Europe in the 1900s, we can infer that it was part of survival strategies in the Nazis camps also by people from others communities. In the case described here, a scientific source, an ethnobotanical source in particular, also turns out to be a historical source.

The discussion in this section leads to the conclusion that scientific-mathematical data and scientific sources are important acquisitions for historical studies. These interdisciplinary methodologies enable complex historical frameworks to be traced and historical events to be contextualized in an increasingly deeper way.

Current historiography diversifies the methods of collecting and analysing historical sources, and includes those related to the scientific field. The diversification of methods involves, on the one hand, the growth of specialized studies of a sectorial character and, on the other hand, the necessity of interdisciplinary approaches, which enable historical research to construct complex contextualisation.

13.5 Historical Prospective and Science—The History of Science

13.5.1 Some Outlines of the Historical Development of the History of Science

The development of historiography—also prompted by the comparisons with the scientific research—has also changed the way we deal with the history of science, which is one of the branches of historiography.

The history of science is the retrospective view, the historical investigation, about the development of science over time. It is the way in which historiography views the historical developments of science and the results achieved by humankind in the present. As explained at the beginning of this chapter, the interweaving of past human events with more recent human events is one of the defining meanings of the word "history".

The inclusion of the developments of scientific ideas in science books is a centuries-old practice. In this category can be mentioned the books of arithmetic and geometry compiled by Italian mathematicians and architects in the fifteenth century, such as "Summa de arithmetica, geometria, proportioni et proportionalità" (1494), by Luca Pacioli. Pacioli exposed techniques and results known at the time and taught in Italian Universities and in the Abacus schools (Botteghe d'abaco). His book collects the practice of abacus' masters, the translations of Greek classic studies, Leonardo Pisano Fibonacci's works, and the inheritance of Arab mathematicians. Some scholars began to collect the history of mathematical knowledge. For example, Benedetto da Firenze's "Pratica d'Arismetrica" (1463) is a detailed treatise about the equations of degree higher than the second, but reducible to them. Benedetto collected numerous "reasons" from previous masters, and built the first history of medieval algebra in his book [66]. This kind of treatises remained a work tool used only by specialists for a long time.

The history of science was recognized as a specific field of study and institutionalized in the first three decades of the twentieth century [67].

This transition was prompted by scientist-historians inspired by Positivist thought. Some German re-edition of the classic scientific works reached a broad public, and contributed to spread interest in the history of science at the end of the eighteenth century. A. Comte attempted—unfortunately unsuccessful—to create a professorship in the history of science at the Collège de France in 1832. E. Littré, a Comte's student, continued this academic challenge. The undergraduate teaching of "General history of the sciences" was established in 1892. The congress of comparative history, organised during the Universal exhibition in Paris, in 1900, hosted a session on the history of science. That meeting was an important incentive to establish relationships among the historians of science [2].

G. Sarton, a positivist scientist, founded "Isis", an international review of the history and philosophy of science, in 1912 in Belgium. This review continued to be published in the United States and currently is the official journal of the History of Science Society. Sarton defined science as «the totality of positive knowledge» [68]. His purpose on compiling the history of science stemmed from the conviction that such a study was the key to put together the history of humanity in its fullest form. As Sarton wrote in "Introduction to the History of Science": «No history of civilization can be tolerably complete which does not give considerable space to the explanation of scientific progress» [69]. Sarton thought that the history of science was a complex system of historical knowledge. He wrote in 1916: «it is actually impossible to write the history of any important discovery without writing, willingly or not, a chapter of the history of science—I mean the history of science as a whole. How could we explain, for instance, the discovery of the circulation of the blood if we did not explain the evolution of anatomy, of comparative zoology, of general biology, of natural philosophy, of chemistry, of mechanics?» [70].

Some scholars at the time of Sarton didn't share these positions. For instance, the Italian philosopher G. Gentile formulated an opposite thought in 1919, like others historicist philosophers, namely that science cannot have history, and a unique history of science cannot exist. Indeed, Gentile argued that «It is evident that for science there is no alternative, it must exclude the concept of a unique history of science, for the very reason that science breaks up into sciences, each of which, in so far as it is science and not philosophy, is separated from the others and has therefore no essential relations with them». Gentile continues: «it is impossible even to conceive the history of science. There being a definitive reality to know, either we know it or we do not know it. [...] History ought to be the development of science, and science as such can have no development, because it presupposes a perfect truth which we cannot reach by degrees but to which we suddenly leap» [71].

The history of science has pointed out that science develops in quite different ways, compared to what Gentile claims. Scientific research develops by questioning the principles of previously accepted theories, correcting results, or delimiting their validity, establishing new relationships between one branch of science and another, making the field of observations clearer, and using the available technological and mathematical tools. L. Geymonat observes that two historiographical issues are still open, concerning the historical development of scientific knowledge: how scientific research develops, and what are the causes of the development of scientific knowledge [72] (Fig. 13.4).

Summa de Arithmetica geo metria. Proportioni: et proportionalita:

Continentia de tutta lopera:

Fig. 13.4 A page from the Luca Pacioli's book "Summa de arithmetica, geometria, proportioni, et proportionalità", published in 1523. This copy is held by the Stockholm University Library. *Source* https://it.wikipedia.org/wiki/Summa_de_arithmetica#/media/File:Titelbladet_till_%22Summa_de_arithmetica_...%22.jpg

13.5.2 How Scientific Knowledge Develops Through History

There are two hypotheses concerning how scientific knowledge develops over time. The first one asserts that scientific knowledge is constantly growing, and the discoveries of an era are linked to the previous researches. The second one claims that scientific knowledge is interspersed with breaking points: it is necessary to formulate new hypotheses when a theory cannot describe new experimental results or it fails to explain contradictions or gaps. [72]. P. Duhem was an authoritative exponent of the continuity hypothesis. He began the publication of "Le Système du monde; Histoire des doctrines cosmologiques, de Platon à Copernic" ("The World System; History of Cosmological Doctrines from Plato to Copernicus") in 1913. Duhem aimed at underlining the elements of continuity between medieval thought about nature and modern science in the seventeenth century. The continuity hypothesis was taken up in the 1950s, with different perspectives and focuses, by some Anglo-American historians, such as A.C. Crombie—for example in "Augustine to Galileo. The history of science a.d. 400–1650". The discontinuity hypothesis was supported by others scholars, such as G. Bachelard. He highlighted the moments of discontinuity and epistemological rupture that characterized the succession of scientific theories. His doctoral thesis focused on the developments of the nineteenth century technology, and highlighted how the improvement of measurement techniques reveals deep epistemological changes, far from indicating a process of linear and cumulative technical improvement [73].

13.5.3 Factors Influencing the Historical Development of Sciences

There are two schools of thought concerning the factors influencing the historical developments of sciences. The first one argues that the factors influencing the development of sciences consist in results, theories and methods, which are achieved in previous stages by science itself. The second one asserts that advances in scientific research are the result of social structure in which researchers operate and are employed [72].

The Congress of the History of Science held in Oxford in 1961 was animated by a debate on the respective role of external factors—sociological, economic, theological or ideological—and internal factors—the conceptual development of theories—in the evaluation of the origins of modern science. On that occasion A. Koyré and A. R. Hall emphasized the irrelevance of external factors, while Needham linked the development of scientific enterprises to cultural and social conditions related to a specific civilization. A significant speech was held by H. Guerlac, a historian of chemistry: he observed that making a distinction between ideas and experience, and treating ideas as if they had their own independent life, separate from reality, is wrong [2].

Since the mid-twentieth century, the history of science has increased attention to external factors involved in the development of scientific knowledge. This attitude is linked to the novelty that changed historiography, and has mentioned in previous sections, i.e. historiography in general, and the history of science in particular, has incorporated questions from the other fields of historical, sociological, and philosophical research. The history of science is currently a disciplinary sector characterised by methodological intertwining: the different conceptions of historical research are debated, and inputs from the other branches of research are integrated. The debate aims at suggesting new interpretations of the development of science.

13.5.4 Sociology and the History of Science

Sociology is an important branch of studies which integrates the history of science. The first studies of this type date back to the second half of the nineteenth century. Botanist A. de Candolle published "Histoire des sciences et des savants depuis deux siècles" (History of science and scholars for two centuries), in 1873. In this book, Candolle observed most of the protagonists of new sciences belong to the middle classes. The first large-scale, quantitative and qualitative, assessment of the scientific community is undertaken by the American sociologist R. K. Merton, in an essay entitled "Puritanism, pietism and science", in 1936, followed by "Science, technology and society in seventeenth century England", in 1938. Merton's essay is characterized by statistical analysis of scientists' biographies, aimed at deepening Weber's theses on the link between Protestantism and the birth of capitalist society. Merton suggested a causal connection between rationalism and empiricism, reinforced by puritan ethics, and the spread of scientific education in England in the seventeenth century [74]. "The structure of scientific revolutions", by T.S. Kuhn, physicist and historian, is an important work in which the history of science is intertwined with sociological studies. Kuhn focused on the dynamics of scientific growth, as opposed to the theses of logical empiricism. He drew a distinction between "normal" science, which characterizes the daily work of scientists who adhere to the same "paradigm", and the periods of crisis that lead to the affirmation of new paradigms. Each crisis-period is followed by a new phase of "normal" science. This cycle, according to Khun, assigns a great role to sociological factors: the choice of paradigm by scientists is based on social and psychological basis. Therefore, the reasons for scientific changes do not lie in abstract verification procedures, but in dynamics within scientific community. As Maza observes: «Around the same time that the historical discipline was turning to the "new social history" [...], Kuhn's book conveyed in lucid and persuasive form the message that professional structures and dynamics mattered more than individual genius for understanding scientific change» [75].

The field of research known as "Science and technology studies" develops from these concepts. These studies view science and technology from the point of view of their social construction: facts described by scientists are not natural, but social, and they result from professional rules, individual aspirations, social pressures, cultural prejudices and technical tools. For instance, T. P. Hughes, in "Networks of Power" (1983), argues that technological differences in Western nations are related to cultural rather than techno-scientific factors. He examines the role of users in technological evolution in subsequent researches [76].

13.5.5 Relationships Between State, Politics and Scientific Research in the History of Science

P. Corsi observes that the history of science focused on an additional aspect between the Second World War and the decades of Cold War: the relationships among State, politics and scientific research. This interest was closely related to the previous experience in warfare, during which the various sectors of research established close relationships with political power, big industry, and military apparatus. Arms race and the conquest of space involved the growth of investments in scientific and technological research, which reached significant percentage of national products and was linked to an increase in the number of personnel in charge of research. These circumstances contributed to the development of the studies about the relationships between science and State apparatuses during the second World War, and about the science in the post-war period. For examples, A.H. Dupree studied the non-linear history of the relationships between scientists and the federal government in the United States, in "Science in the Federal Government: a history of policies and activities to 1940", in 1957 [2, 77]. C. P. Snow examined the relationships between politics and military choices in the air-war between Great Britain and Germany during the second World War in "Science and government", in 1961 [2, 78].

Studies on the relationships between power and science are increasing, and currently cover a period extending from the scientific Revolution in the sixteenth century to the most recent times. For example, M. Biagioli studies «the relationship between mechanics and the Medici court, as a situation capable of accounting for the rise and fall of Galileo Galilei, a courtier among many, and like everyone subjected to the strict protocol and deference, which bound—and bind?—the patron to the client. In this perspective, the court becomes a sort of space of rules that determine behaviour and research. There was an intrinsic relationship between a court where the mechanics was (or would have been) one of the languages of competition for the visibility and affirmation/defence of one's status—both as a client and as a patron—and the specificity of Galilean mechanics» [79, 80].

13.5.6 *Impact of Scientific Innovations on the Social Perception*

The area of research that takes into account the consequences of scientific innovations on social perception is one of the integrated studies, which is undergoing a fast transformation. For instance, M. Di Giovanni studies the mobilization of scientists and technicians during the Fascist period in Italy. The Fascist government deployed scientist and technicians to military, strategic, and economic purpose with new institutional tools. This mobilization was capable of mitopoiesis: it attributed prescribed cultural and social values to real facts through a politics-guided narration of facts. The Regime's public policy financed forms of mass communication, such as exhibitions, which conveyed messages that combined the developments of theoretical research to the politics of technical application. The large amount of exhibitions and awards were mostly meant for Italian authoritative scientists and research institutes. Beside these, visibility was also given to minor characters, who exemplified the national inventive myth. Minor inventions were associated to "big science and great scientists" and to the industrial power. The mass media presentation of lesser renowned inventors communicated the mass mobilization and the power of the Nation. These myths were part of the Fascist ideology [81, 82].

The field of studies that focuses on the consequences of scientific innovations on society is quite wide. This field of studies borders with the study of the material culture. Scientific discoveries and their technological application can produce profound changes in human history. These changes involve society, daily life, and the people's mind-set. An example can be photography: photography comes from chemical experimentation in darkrooms, and was a tool used by the wealthier social classes until the end of the nineteenth century. Originally used especially for artistic purposes, photography became a tool for the dissemination of information to make travel books more attractive. Then, war photographs and portraiture became a part of a wider market that included a vast number of people from the middle class. During the twentieth century, photography was a form of art, and an integral tool of advertising, communication, and information. Photography was a mirror of everyday life, but also influenced daily routine by way of advertising images. The use of camera goes along with the development of the public language linked to images. In the twentieth century, both the public and private use of photography became increasingly massive, until it was used extensively through social media in the 21th century. The development of photography and photographic equipment has brought profound changes in society. For example, before the diffusion of photographic ateliers in the early twentieth century, less well-off families did not possess images of their ancestors. The dissemination of family photographs, later changed people's relationship with their family history, even in the lower social class. At the same time, photography changed historiography, because photos have become historical sources which require a specific historical analysis method (Fig. 13.5).

Fig. 13.5 An example of a family photography that portrays poor social class. Peasant family on the steps of the Temple of Vesta, in Rome, Italy. *Source* https://www.flickr.com/photos/hwmobs/8157536838

13.5.7 Interdisciplinary Approach to the History of Science

Historical research currently goes in two directions. The first is the specialization. In this field, the history of science deals with sectorial questions that require specific skills. Researchers who works in these specialised fields often have a scientific background, and are often themselves scientists. The second direction goes towards a widening contextualisation, involving many branches of study. These historical studies are increasingly open to interdisciplinarity, and the interests of developing scientific and technological research involve several areas of historiography.

This interdisciplinary approach implies a dialogue between several areas of historiography, sociology, economics, political science, the history of science, and other branches of knowledge. Some studies on the age of municipalities may provide an example of such interdisciplinary approaches. From the end of the twelfth century, European society and economy underwent a process of transformation, which continued until the middle of the sixteenth century. It began in Italy and spread in

Flanders, England, northern Germany, and southern France in the thirteenth century. A new form of social organization developed: civilization of municipalities. The development of a typical town of the late Middle Ages, or, better still, Renaissance and Humanism period, led to important scientific and technological innovations. New calculation systems were applied to both architecture and economics. Important technological innovations were applied in the textile sector, such as the wide loom. The invention of mechanical clock enabled accurate measurement of time. Scientific and technological innovations were closely linked to bourgeois society which developed during this period. E. Scerpanti and S. Zamagni observe: «The hero of humanism is an active subject, open to innovation, a lover of freedom who is proud of his civil virtues: in short a merchant-manufacturer. This new Hercules combined action and reflection, art and accountancy, religion and politics, theory and practice, in the immense effort to create the birth of political economy his own world and in that world, a space for his own freedom» [83]. The rise of the bourgeoisie, the new relationships between citizens and power, innovations in the economic system, the increase of population, the new mononuclear family model, and the growth in the percentage of schooling [84]: these are all factors that help to explain the history of scientific and economic developments of the early modern age in Italy and in Europe.

13.6 The Language of History, the Language of Science

13.6.1 Comparison Between the Communication of Facts in History and Science

Mammino L. writes that "narrative abilities imply the ability to talk about something in a complete way, providing all the relevant details in the proper places, organising the talk according to a logical thread and achieving communication, so that those who listen or read can understand all the information meant to be conveyed". The exposition of any discipline is based on language. History and science have thus an important feature in common: the exposition of concepts through language.

Beyond the narrative abilities, each discipline have their own specific features and sectorial language. Both science and history are cognitive systems, which have their own specific peculiarity. Formalised language is a distinctive and constitutive feature of specialised knowledge systems, but this does not disregard the common basis of narrative abilities.

13.6.2 The Narrative Character of Historical Studies

The historical narrative is frequently traced back to the poietic aspect of history. The description of events is an important element of history. In the root of the term "history"—as was pointed out at the beginning of the chapter—there is the meaning of "recount". Some historians and philosophers, reflecting on the methodology of historical studies, pointed out that history has, from its origins, a significant creative component. This issue is clearly explained by Droysen, who defined history as «the only science enjoying the ambiguous fortune of being required to be at the same time an art» [86].

Since the 1960s, interest in the topic of historical narrative has grown. Postmodern historiography theorized a gap between language and reality. Scholars who focused on the poietic feature of historical narrative had several points of view: from the position which rejected the classical notion of "truth" —arguing that there is no "correspondence" between language and reality—to the extreme position which theorized that there is no object of research outside the texts, as promoted, for instance, by J. Derrida. Historiographical streams that refused traditional epistemology considered as there precursors G. Vico, F. Nietzche, M. Hedegger, L. Wittgenstein, M. Foucault [87].

The link between reality, language and creativity become the object of at least two major questions in postmodern historiography. The first one is the relationship between historical object, historical narrative, and historical interpretation. R. Barthes suggested, as Nietzsche already did, that there is no story-telling without interpretation; this interpretation is inevitably ideological and the relationships between story and facts are rhetorical rather than descriptive [25]. Barthes questioned the objectivity of all the reconstructions of the past.

The second issue is the relationship between historiography and the proprieties of narrative language. This approach focused on the analogies of the historical account with literary narrative and contested specificity of the language of historiography. These scholars, inspired by Postmodernism, reduced the gap between historical narrative and literary fiction. J. Topolsky proposed an analysis of the historians' narrative tools, dividing them into an objective and informative level, and two other subjective levels—persuasive and ideological—linked to the personality of the historian [87].

According to other studies, the narrative property of historical account separates historiography from scientific research. For example, H. White (1973)—the main representative of the linguistic turn school—asserted that historian is not like a scientist who does laboratory researches and discovers the meaning of facts, the hidden reality; historian, like a novelist, uses the language and rhetorical conventions to create different combinations of facts, to «emplot» historical events. White affirmed that historians use figurative language and non-scientific discourse to configure history communication [76].

13.6.3 The Deductive Type of Writing—Common Writing Strategy Between the Different Fields of Research

During the period when the postmodern theories developed, the ways of communicating history were greatly diversified, and they are still diversifying. There are currently many forms of exposition of the historical researches: extremely popular forms; specialised and scientific approaches, such as those used in monographs and papers in history reviews; and several other options that combine scientific and narrative approaches. Some forms of technical and specialized writings of historical research utilise the deductive type of writing, thus approaching to the definition of scientific language. A number of linguists, such as Altieri Biagi (1990) [88], De Mauro (1994) [89], Cortelazzo (2004) [90], distinguish four features that regulate linguistic choices in scientific texts: precision, concatenation, condensation, preference for the passive and impersonal verbal forms. These characteristics are common to some forms of history's writing, such as those used in papers published in specialized journals.

All the forms of communication of history research share the logical-argumentative account.

Recent studies [85] highlight that the logic argumentation is a necessary component for the presentation of thoughts, transversal to many fields of knowledge. In this sense, history and science have in common communication skills.

13.6.4 Communication Between Disciplines

Currently, some branches of historical research are not clearly distinct from scientific research: for example, the application of archaeology to the study of the industrial age, or the studies of the influences of the natural environment on humankind.

Interdisciplinary approaches to historiography may lead to two different outcomes, which have significant consequences on the communication of knowledge. The first outcome is that historiography is divided into increasingly specialized domains, by adopting sectorial languages. In this case, the fields of study tend to separate, and the fields that are expressed with specialized language do not communicate easily. The lack of communication among specialist sectors can cause the difficulty in composing a comprehensive and detailed historical overview. The second outcome of the growing importance of interdisciplinary approaches to historiography is a deeper integration between areas of study. To achieve this result, the branches of study need to be able to communicate. Successful examples of interdisciplinary dialogue are based on the openness of research areas. As concern the integration between science and history, the culture of openness leads historians to open their own questions to the interaction with scientific answers; and scientists, interested in providing answers to historiographical questions, to open their specialised language and dialogue with historians.

13.7 The History of Science in History Teaching

13.7.1 About History Teaching

The previous sections have shown that history and science are distinct and different forms of human knowledge of reality, which have dialogued with each other through the ages.

It may be interesting to consider whether relationships between science and history currently exist beyond the interests of specialists in these fields. The teaching practices in public schools can offer a testing ground to examine there current presence and developments. Teaching aims at conveying contemporary knowledge to younger generations. Best teaching practices gather relevant recent information in different disciplines, and link different disciplines to each other to shape learner's acquisition of knowledge in a holistic way.

Best history teaching practices relate history to science. Thus, history teaching integrates all the meanings of the term "history"—"events", "narration", "research", and "link of the present with the past"—to foster students' deepen understanding of history. History teaching entails a variety of components aimed that training learners to ask questions about the past and the present: a selection of events, chosen for their role in historical changes; the explanation of connections between facts, to accustom learners to history concepts and language; learners' training to express and narrate history utilizing information, based on the investigation of historical sources. The guidelines of the Italian Ministry of Education, consistent with the guidelines provided by the European Union, express and encourage these characteristics of history teaching. Teachers aim at achieving these goals through history teaching; and learners are expected to be able to express the acquired knowledge, and utilized acquired skills and competences [91].

The next section considers two different views and approaches to the inclusion of history of science in history teaching: signs of history of science as a special component of history teaching, or history of science integrated in the overall history teaching. In particular, the next section considers teaching activities developed in the public schools.

13.7.2 The History of Science as a Special Part of History Teaching

The history of science is often considered as a special component of history teaching. In this perspective, selected aspects of the history of science are presented as focused-themes. Then, public schools often provide the special activities on the history of science. These activities may consist in multidisciplinary lessons, in which teachers of different subjects converge a programmed number of lessons-hours to therewith a preselected topic, and develop it with from the point of view to each discipline. In the Italian educational system, the Triennial Plan of Training Offer of each school reports these specific activities. Every school organizes these works independently. The following part considers two examples of focused activities, one of national interest, the other of local interest.

An example of activity that may concern topics of national interest is produced by a middle school in Miane (Treviso province, Veneto region, Italy) invited students to prepare scaled models of Leonardo Da Vinci's machines, during the technical education classes, and then organized an exhibition, since 2019 as celebration of the 500th anniversary of Da Vinci's death. The quality of exhibition was such it was then shown in several Italian cities [92].

An interesting example of activity having a local character is the project carried out in the 2013/2014 school year in a professional secondary school in Maniago (Pordenone province, Friuli Venezia Giulia region, Italy) and focused on the Vajont disaster, which occurred in 1963, in the Vajont mountain valley, which is located 40 km form Maniago. The Vajont dam was built at the feet of Mount Tóc, in the Erto-Casso municipality. A huge landslide broke off from Mount Tóc and plunged into the artificial lake created by the dam, causing a massive wave (48 million cubic meters) that hit nearby towns, razing the closest town (Longarone) to the ground, and partially destroying Erto and Casso, with over than 1900 dead and over than a thousand of missing people. This is one of the most serious human-induced natural disasters in Europe in the twentieth century. The disaster changed the morphology of the site and its human settlements. The construction of the Vajont dam had been part of the process of Italian industrialization in the twentieth century: the dam was meant to provide a wide area with electricity. The above mentioned activities in the Maniago School comprised a series of lessons in which the topic was presented by different teachers from the viewpoints of different disciplines. For example, physicists and engineers addressed the functioning of dams, comparing some types of dams built in Italy since mid-1800s, and retrieving specific information from academic publications. Some activities, carried out directly by students, involved the preparation of some small scale models of the Vajont dam in mechanical workshop, for students to have a complete understanding of the details, and the search for the historical documentation [93] from national and local sources, such as the photographs of the Vajont valley before and after the disaster, and articles that chronicled the dam's construction and the later tragedy.

In addition to school activities, other components considering the history of science as a special part of history teaching, are the thematic insights, through boxes, focussed on specific aspects of the history of science provided within manuals and school textbooks. These insights are more or less detailed, according to the level of instruction and type of secondary schools. For instance, the level at which the content is treated in books for technical high schools is different from that meant to scientific high schools, or different strategies are used for primary schools and secondary schools.

Some topics include boxes in history books whichever the type of school in the Italian educational system. For instance, chapter presenting the eighteenth century industrial revolution, or the eleventh century agricultural revolution, comprises boxes focusing on relevant aspects of the history of technologies and sciences in that period. In addiction, boxes in history books offered The insights on the history deal with the biography of authoritative scientists, who were innovators in their time. An example is Galileo Galilei who contributed to change philosophy and scientific thinking, as well as Italian language. His contributions to culture, thought, and knowledge, were so fundamental that his biography is included in language, history, astronomy, and physic text books, at least. The same can be said for William Harvey, Marie Curie, Werner Karl Heisenberg, Rita Levi Montalcini, and many other scientists, who have been central *fig*ures in the history of science and consequently also in the historiography (Fig. 13.6a, b).

13.7.3 The History of Science as an Integral Part of History Teaching

The relationship between history and the history of science is a complex and pervasive link between the two areas of knowledge. In the teaching of history, the history of science does not only provide sources or themes for insights, for a "special education": teaching experience reveals that students who have basic knowledge in mathematics and sciences clearly and rightly understand some history topics. For instance, the correct reading of statistical data, tables, and charts, allows students to understand the significant development of Italian literacy rates between 1861 and the First World War. To get an idea scientifically based about the consequences of malnutrition enables to figure out the relationships between malnutrition and child mortality in Italian rural villages before mid-1900's. When scientific cultural basis are unclear, or some scientific knowledge are missing, the understanding of historical dynamics by students is inaccurate and often incorrect.

The scientific background provides students for a deeper understanding of history. When the reading of scientific data reaches a sufficiently reliable level, students often reach the ability to contextualize historical events in a more appropriate way, to seize more broadly the relationships between humans and environment, and to deepen the historical dynamics. C. Pavone observes that broad historical contextualization «can contribute "to the understanding of a reality that was believed to be already known". [...] the context makes the location of events over time less arbitrary, it unmasks the anachronisms, prevents indulging in the temptation of omnia ubique and forces us to compare sources of a different nature» [94].

Besides that, oral and written exposition is positively influenced by having assimilated clear and articulate concepts. Students who possess a basic scientific culture, and who better understand some aspects of history, have a clearer and more correct exposition.

The relevance of scientific background in historical studies does not only concern the understanding of data. The practise in classify, analyse and relating data is a fundamental praxis to history studies. This practice is exercised in many scientific

(a) (b)

Fig. 13.6 The Vajont dam site before and after the October 1963 catastrophe. a (left): the photograph was taken in September 1963 and shows the dam and its artificial lake. **b** (right): photo taken on October 9, 1963, soon after the catastrophe. The basin closed by the still standing dam is filled with the debris of the landslide, which fell from Mount Tòc into the water reservoir. *Sources* https://upload.wikimedia.org/wikipedia/commons/b/b7/Invaso_Vajont_estate_1963.jpg

subjects, and is the basis of historical research. When history teachers work with sources materials, students are gradually accustomed to use an historical method. For example, when students collect oral sources, they are asked to explain when, how and why they have collected that specific source. Students are asked to descript the content of oral source, who is the producer of the source, what is the subject matter, and with which other data it can be related. Historical study can benefit from the students' accustomed approach to rigorous collection and description of scientific data. And, hopefully, there is a mutual exchange: the ability to study objectively, to describe and analyse historical sources, can facilitate the habit of objective description necessary to the study of scientific disciplines. Moreover, learning to select sources, based on reliability, is a necessary exercise in both science and history.

Another aspect of the relationships between history and science in the practice of history teaching, is that historical view contributes to the reflection about sciences by students. The errors of sciences, the accidental discoveries, and the incorrect dissemination of scientific contents, have an important place in history teaching, alongside the scientific and technical discoveries. For example, in a history course the Röntgen's researches and the accidental discovery of X-ray are set in a frame which consider the historical time, characterised by interweavings among the impulse given by the second Reich to universities, to scientific research, to industrialization and to military power. The Darwinian theory are treated broadly: the dissemination of the evolution theory is explained in parallel with the problem of de-contextualization of this theory. The Darwinian theory was revised and transposed into other disciplines in European culture, between the end of the 19th and the beginning of the twentieth century. These transitions between different cultural fields were not subjected to scientific criticism. Those steps transformed scientific knowledge into ideological content, and generated pseudo-scientific contents that spread among the European and non-European circles of scholars in the twentieth century.

History teaching addresses scientific discoveries in the framework of scientific development. And scientific development is linked to social, economic, cultural, and political context. Historical approach allow students to consider past acquisitions and beliefs about natural world—that converge in the history of sciences—in their historical frame. Students learn to approach the past not through today's culture and values, but through the study of the culture and values of epoch in which scientific acquisition was obtained, or in which beliefs about natural world was shared.

The relationship between history teaching and sciences in public school is full of implications. The integration of the teaching of history with scientific knowledge makes students more competent in historical studies. Historical studies allow students to think of science as a historical pathway; and to exercise fundamental skills at the base of both historical and scientific studies. In order to achieve this goal, it is necessary to be able to use a clear language, specific, scientific, but shareable. A language capable of linking the different modes of knowledge. The construction of interdisciplinary language finds a privileged time in the years of the education in public school.

13.8 Conclusions

Both historical research and scientific research aim at systematically observing and understanding the reality. Each of them studies some different aspects of reality: history focuses on acquiring new knowledge on the reality of humankind in the past and in the present; science deals with the acquisition of new knowledge on reality in which human lives, and on human being. History and science, as research activities related to aspects of reality, share some commonalities. Each of them uses a specific method of research; each of them is based on the analysis of data—which are called historical sources, in history studies, or the documents and materials of which historians use for structuring their researches; each of them uses communication skills. Some of these features have become intertwining over time. New acquisitions in different fields of knowledge have often reflected in other fields, since ancient times. Historical research and scientific research have undergone dialogical developments.

Different methods are required to gather and interpret different types of sources. This has prompted historiography toward a double outcome: on the one hand, toward specialised and increasingly sectorial works; from the other, towards an interdisciplinary approach, that has become fundamental in some fields of history research. As range and types of data can be regarded as historical sources and of the methods to analyze them increase, historians have the opportunity to reconstruct increasingly more complex historical contexts, using extremely varied information.

The transformation that accord within historiography are reflected to the history of science. At present, the history of science follows paths that need a deep integration between the specialized knowledge of scientific subjects and the new methods of historical research. Both historical research and the history of science tend towards increasing recognition of the complexity of the context involved. Currently, many studies on the history of science do not simply focus on the history of scientific and technological discoveries, but also on the outcome of the mistakes in sciences, the social impact of scientific development, the social, cultural, political, and economic context in which they occur, and the relationships between scientific innovation and power.

The developments of historiography entail an increasing need for dialogue between history and science. That need for dialogue brings up the question of language. Science and history must be able to communicate, to enable historical studies to integrate scientific acquisitions, and vice versa.

On the other hand, both history and science are characterized by their own specialised language, in their quality of research domains. As regards historical research, the combination of history and narrative belongs to the meaning of the word "history". Historiography is closely related to story-telling. In addition to the vocation to narrative, history has developed over time a variety of languages, both for dissemination and for specialised researches. Certain areas of historical writing have developed a language that approaches to the scientific language, in relation to the features of "precision, concatenation, condensation, preference of passive and impersonal verbal forms". Scientific language and the historical language cannot be neatly separated also because language is the common basis for the communication of all knowledge.

Teaching the history of science in public schools make clear contemporary developments in the relationships between science and history. Teaching is a good trial field to understand how knowledge is conveyed. Teaching the history of science, in the context of history teaching, is influenced by dichotomy between the sectorial approach and the interdisciplinary approach of historical research. On the one hand, the history of science is often proposed in the form of thematic insights in the history lessons. On the other hand, in teaching practice it may be observed that basic scientific culture is necessary for students to understand historical matter in a non-superficial way. History teaching often reveals that scientific culture is an element that needs to be integrated into historical discipline. Basic scientific culture is an essential prerequisite for students to have a fully understanding of history. Equally, during teaching practice, it is observed that the ability to understand the basis of historical research is required for a non-superficial understanding of the history of science. Beyond that, the practice of gathering, describing, analysing, integrating historic sources, and the practice of verifying the reliability and comparability of sources, exerted by students in history courses, is also useful in science courses. Finally, the historical approach to science requires students to consider scientific acquisitions and beliefs about natural world as a process that has developed over time. This process needs to be contextualised in the historical periods.

Integration between history and science is currently increasing. Scientific acquisitions are quickly integrated into historical studies. For example, the potential offered by personal computers is used for storing and searching data, or for applying computational linguistic to historical studies. Technological innovations provide tools for reconstructing material culture and for studying archaeological or anthropised sites, as for construct marine drones used for mapping submerged archaeological sites. The field of DNA study enable to reconstruct the evolutionary history of the human species as well the different environments of the planet.

The integrations between history research and science require that historians are open to the progresses in scientific disciplines, and that scientists are open to opportunities offered by interdisciplinary approaches. Communicating knowledge between different areas of research requires the mastering of narrative skills that can lead to cross-knowledge. The construction of the bases of knowledge is accompanied by exercises in the language of study. Public school is fundamental for practising communication skills and learning to integrate languages from different fields of knowledge. The dialogue between history and science has proved fruitful for historical studies in the past. It is currently crucial for the development of the fields of historical studies that require an interdisciplinary approach, and for the convergence between the skills of different disciplines.

References

1. Britannica.com, sub voce «history of science», https://www.britannica.com/science/history-of-science. Accessed 7 Sept 2021
2. P. Corsi, sub voce «Scienza». In Treccani Enciclopedia Treccani—V Appendice (Istituto della Enciclopedia Italiana, Roma, 1994), https://www.treccani.it/enciclopedia/scienza_%28Enciclopedia-Italiana%29/. Accessed 6 Sept 2021
3. Britannica.com, sub voce «history», https://www.britannica.com/topic/history, Accessed 8 Sept 2021
4. Treccani.it. Dizionario di filosofia, sub voce «storia» (2009), https://www.treccani.it/enciclopedia/storia_%28Dizionario-di-filosofia%29/. Accessed 10 Sept 2021
5. G. Devoto, G.C. Oli, Vocabolario della lingua italiana, sub voce «storia» (Le Monnier, Milano, 2012), p. 2805
6. K. Pomian, in Enciclopedia Einaudi sub voce «periodizzazione», vol. 10 (Einaudi, Torino, 1980), p. 627
7. F. Rendich, Dizionario etimologico comparato delle lingue classiche indoeuropee (Paolombi, Roma, 2010), p. XXIX, 407; cit. p. 433
8. R. Beekes, L. van Beek, sub voce ἵστωρ, Etymological Dictionary of Greek (Brill, Leiden, Boston, 2010)
9. Treccani.it, sub voce «storia», https://www.treccani.it/enciclopedia/storia/, Accessed 21 Sept 2021
10. Cf. sub voce sciō, M. de Vaan, *Etymological Dictionary of Latin and the other Italic Languages* (Brill, Leiden, Boston, 2008)
11. F. Rendich, L'origine delle lingue europee. Struttura e genesi della lingua madre del sanscrito, del greco e del latino (Palombi&Partner, Roma, 2005), p. 191
12. Britannica.com, sub voce «science», https://www.britannica.com/science/science. Accessed 5 Sept 2021
13. Treccani.it, sub voce «scienza», https://www.treccani.it/enciclopedia/scienza/. Accessed 5 September 2021
14. F.Bearzatto Lo sguardo femminile sulla Resistenza nel mondo rurale. Figure di donne nella guerra partigiana nel Comune di Frisanco, in Il difficile cammino della Resistenza di confine. Nuove prospettive di ricerca e fonti inedite per una storia della Resistenza nel Friuli Venezia Giulia, ed. by Anna Maria Vinci (Irsml FVG, Trieste, 2016)
15. Britannica.com, sub voce «epistemology», https://www.britannica.com/topic/epistemology. Accessed 5 Sept 2021
16. E. Garin, L'umanesimo italiano. Filosofia e vita civile nel Rinascimento (Laterza, Roma-Bari, 1986), pp. 9–10; 229–230
17. G. Galilei, Il Saggiatore, in Opere. Volume Primo, a cura di F. Brunetti (UTET, Torino, 1964), pp. 631–632
18. G. Mori, Cartesio (Carocci Editore, Roma, 2010), p.40
19. A. Koyré, Lezioni su Cartesio, (Tranchida Edizioni, Milano, 1994), p.58
20. C. Borghero, La certezza e la storia: cartesianesimo, pirronismo e conoscenza storica / di Carlo Borghero (Angeli, Milano, 1983)
21. P. Bayle, Dissertation contenant le Projet d'un Dictionnaire critique, in Dictionnaire Historique Et Critique, Paris: ed. Beuchot, vol. XV, p. 241, cit. in Cantelli G. (ed. by) (1976) Bayle. Dizionario storico-critico, Roma-Bari: Editori Laterza, p. XII (1820)
22. Cit. Hazard P., *First published in France as La Crise de la conscience européenne in 1935) The Crisis of the European Mind: 1680–1715* (New York review book, New York, 2013), p.30
23. F. Chabot, Lezioni di metodo storico (Editori Laterza, Roma-Bari, 2000), pp. 49–51
24. M. Marcialis, Reviewed Work: La certezza e la storia. Cartesianesimo, pirronismo e conoscenza storica by Carlo Borghero. Riv. Di Stor. Della Filos. (1984 -) **41**(2), 369–374 (1986)
25. G.P. Romagnani, Storia della storiografia. Dall'antichità a oggi (Carrocci ed., Roma, 2020), p. 31; 222; 167–168

26. Mary Jo Nye, (ed.), *The Cambridge History of Science* (The Cambridge History of Science) (Cambridge University Press, Cambridge, 2002), p. I; 28. https://doi.org/10.1017/CHOL9780521571999
27. Cfr. Comte, (written in 1925, cit. in Pietro Rossi, Positivismo e società industriale, Loescher Torino 1973), Casini Paolo, Benvenuti Mario, Ragione e storia. L'attività filosofica nella cultura delle società occidentali, vol. 3, p. 345–346
28. C.H. Carr, *(1961)) What is History?* (Penguin Books, cit, London, 1990), pp.56–57
29. J.G. Droysen, Histórica, *Outline of the Principles of History: Grundriss Der Historik*, trans. by E. Benjamin Andrews (Ginn & Company, Boston, 1897), p. 18; 10
30. Cit. F. Gilbert, The New Edition of Johann Gustav Droysen's Historik. J. Hist. Ideas. **44**(2), 327–336 (1983) (p. 330). https://doi.org/10.2307/2709143
31. Cit. F.W. Nietzsche, The use and abuse of history. Translated by Adrian Collins (Liberal Arts Press, New York, 1957 (1949)), pp. 37–38. Internet Archive https://archive.org/, urn:oclc:record:1036909872
32. Britannica.com, sub voce «Annales school», https://www.britannica.com/topic/Annales-school. Accessed 5 Sept 2021
33. M. Bloch, *The historian's craft* (Manchester University Press, Oxford, 1992), pp.14–15
34. Herodotus, with an English trans. by A.D. Godley, vol. II, book IV, 172, I (1928), p. 375
35. A. Momigliano, *The Classical Foundations of Modern Historiography* (University of California Press, Berkley and Los Angelese, London, 1990), pp.50–52
36. T. Porter, *The Rise of Statistical Thinking, 1820–1900* (Princeton University Press. Ebook, Princeton, New Jersey, 2020). https://doi.org/10.2307/j.ctvxcrz1v
37. Cit. K. Brush, The Cultural Historian Karl Lamprecht: Practitioner and Progenitor of Art History. Cent.L Eur. Hist. **26**(2), 139–164 (1993), http://www.jstor.org/stable/4546329. Accessed 5 Aug 2021, cit. p. 143
38. T. Barker, The Beginnings of the Economic History Society. Econ. Hist. Rev., **30**(1), new series, 1–19 (1977), p. 5. https://doi.org/10.2307/2595495
39. J. Sheehan, Conference on Quantitative Data in Western European History: A Report. Comput. Hum.Ities. **2**(3), 114–116 (1968), http://www.jstor.org/stable/30199253. Accessed 6 Aug 2021
40. L. Stone, The Revival of Narrative: Reflections on a New Old History. Past & Present. (85), 3–24 (1979), http://www.jstor.org/stable/650677. Accessed 6 Aug 2021
41. G. Levi, Il consumo a Venezia. Una fonte contabile, in Prima lezione di metodo storico, ed. by S. Luzzatto (Laterza, Roma-Bari, 2010), p. 54
42. E. Grendi, Microanalisi e storia sociale. Quaderni Storici **12**(35 (2)), 506–520 (1977)
43. Cit. L. Stone, Quantifying the Holocaust: Hyperintense kill rates during the Nazi genocide. Sci. Adv. **5**(1) (2019). https://doi.org/10.1126/sciadv.aau7292, p. 1; et seq.
44. C. Vetter, Introduzione. La Rivoluzione francese: evidenze lessicologiche, evidenze lessicometriche e interpretazioni storiografiche, in La felicità è un'idea nuova in Europa. Contributo al lessico della rivoluzione francese, Tomo II, ed. by C. Vetter, M. Marin (EUT Edizioni Università di Trieste, Trieste, 2013), pp. 13–33
45. F. Socrate, Sessantotto. Due generazioni (Editori Laterza, Bari-Roma, 2018), pp. XXI-XXIV
46. F. Gilbert, The new edition of Johann Gustav Droysen's Historik. J History Ideas. **44**(2), 327–336 (1983). https://doi.org/10.2307/2709143
47. J. Biederer, Archäologie und Chemie. Einblicke in die Vergangenheit (Staatliche Museen Preußischer Kulturbesitz, Berlin, 1987)
48. S. Curto, I raggi X al servizio dell'archeologia. Aegypt. **76**(1/2), 3–6 S. (1996), http://www.jstor.org/stable/41217180. Accessed 11 Aug 2021, p. 3
49. E.C. Wells, Archaeometry: Definition, in *Encyclopedia of Global Archaeology*, ed. by C. Smith (Springer, New York, 2014). https://doi.org/10.1007/978-1-4419-0465-2_360
50. M. Campbell, *Leveling the Field in Ancient DNA Research, The Scientific Observer*, issue 22, pp.12–13 (2023)
51. X. Wei, C.R. Robles, A. Pazokitoroudi, A. Ganna, A. Gusev, A. Durvasula, S. Gazal, P.R. Loh, D. Reich, S. Sankararaman, The lingering effects of Neanderthal introgression on human complex traits. Elife (2023), https://doi.org/10.7554/eLife.80757

52. D. Manacorda, Archeologia: Definizione, in *Treccani. Enciclopedia del Novecento II Supplemento* (Istituto della Enciclopedia Italiana, Roma, 1998), https://www.treccani.it/enciclopedia/archeologia_res-9fd4181d-87f0-11dc-8e9d-0016357eee51_%28Enciclopedia-del-Novecento%29/. Accessed 5 Sept 2021
53. F. Botturi, Strutturalismo e sapere storico. Riv. Di Filos. Neo-Scolastica **75**(4), 563–81 (1983)
54. P. Burke, Una rivoluzione storiografica. La scuola delle «Annales», 1929–1989 (Laterza, Roma-Bari, 1992), pp. 48–50
55. C. Wulf, Antropologia dell'uomo globale. Storia e concetti (Bollati Boringhieri, Torino, 2013), pp. 84–92
56. G. Galasso, *Storiografia e storici del Novecento* (Roma, Salerno Editrice, 2016), pp.27–40
57. P. Bevilacqua, Tra natura e storia. Ambiente, economie, risorse in Italia (Donzelli, Roma, 2000), pp. 263–217; 39–47
58. P. Bevilacqua, L'utilità della storia. Il passato e gli altri mondi possibili, Roma: Donzelli, ebbok (IV Locale e universale. 2. Il territorio: un libro aperto) (2007)
59. Pluriactivité, Économie et organisation du travail: Alpes et Apennins (XIIIe –XXe siècles), Histoire des Alpes / Association internationale pour l'histoire des Alpes (25/2020)
60. F. Bearzatto, Il sale sul taràssaco. Michele Mezzaroba, dall'infanzia friulana a Mauthausen, Portogruaro: Nuova Dimensione, (translated by the author) (2007), p. 150
61. B. Mantelli, N. Tranfaglia (ed. by), Il libro dei deportati, vol. I, I deportati politici, 1943–1945, Milano: Mursia, Tomo 2, p. 1410 (sub voce Michele Mezzaroba); Tomo 3, p. 2124 (sub voce Tullio Tomasso) (2009)
62. F. Fabbroni, La deportazione dal Friuli nei campi di sterminio nazisti, Udine: Istituto friulano per la storia del movimento di liberazione (1984)
63. P. Levi, *Se questo è un uomo* (F. De Silva, Torino, 1947)
64. S. Pignatti, Flora d'Italia, vol. III (Taraxacum officinale, Weber) (Edagricole, Bologna, 1997)
65. M.G. Paoletti, A.L. Dreon, G.G. Lorenzoni, Pistic, traditional food from Western Friuli, NE Italy. Econ. Bot. **49**(1), 26–30 (1995). https://doi.org/10.1007/BF02862273
66. P. Rossi, edited by Storia della scienza. La rivoluzione scientifica: dal Rinascimento a Newton vol. 1 (Istituto Geografico De Agostini, Torino, (2006 (1988))), pp. 60–63
67. J. Alexander, T. Liba, The Cambridge history of science, vol. I, p. XVII
68. G. Sarton, The History of Science and the New Humanism (Indiana University Press, Bloomington, 1962), p. 118, cit. in: Garfield E. The life and career of George Sarton: the father of the history of science, Journal of the history of the behavioral sciences, vol. 21, no. 2 (April 1985), doi.org/https://doi.org/10.1002/1520-6696(198504)21:2<107::AID-JHBS2300210202>3.0.CO;2-K p. 107
69. G. Sarton, Introduction to the History of Science, vol. I (published for the Carnegie Institution of Washington by Williams & Wilkins, Baltimore, 1927), p. 3
70. G. Sarton, The Monist **26**(3), 321–365 (1927)
71. G. Gentile, The theory of mind as pure act. Trans. by H. Wildon Carr, London: Macmillan and Company Limited, pp. 233–234 (1922(1919)), https://archive.org/details/theoryofmindaspu032443mbp. Accessed 5 Sept 2021
72. Geymonat L. (2006 (1985)) Lineamenti di filosofia della scienza, Torino: UTET Università, pp. 70–77; ibidem; 87 (cit., translated by the author)
73. S.W. Gaukroger, Bachelard and the problem of epistemological analysis. Stud. Hist. Philos. Sci. Part A **7**(3), 189–244 (1976). https://doi.org/10.1016/0039-3681(76)90006-6
74. A. Zocchi, Scienza e società: la rivincita di Robert K. Merton. Studi Di Sociologia **42**(2), 161–175 (2004)
75. S.C. Maza, Thinking about history (The University of Chicago Press, Chicago, 2017), ebook (3. The history of what?—III. Thomas Kuhn's scientific revolution)
76. C. Cfr. Sorba, F. Mazzini, La svolta culturale. Come è cambiata la pratica storiografica (Editori Laterza, Roma, 2021), pp. 38–30; 40; 154–155
77. A.H. Dupree, *Science in the Federal Government. A History of Policies and Activities to 1940* (Belknap Press of Harvard University, Cambridge, 1957)
78. C.P. Snow, *Science and Government* (Harvard University Press, Cambridge, 1961)

79. P. Corsi, La storia delle scienze e delle tecniche, 3. Nuove storie vecchi problemi, in Le radici storiche dell'Europa. L'età moderna, edited by M.A. Visceglia (Viella, Roma), ebook, trans. by the Author (2007)
80. M. Biagioli, *(1993) Galileo, courtier: the practice of science in the culture of absolutism* (The University of Chicago Press, Chicago; London, 1993)
81. M. Di Giovanni, Scienza e potenza. Miti della guerra moderna, istituzioni scientifiche e politica di massa nell'Italia fascista. 1935–1945, (Zamorani, Torino, 2005)
82. M. Di Giovanni, Scienze, tecnica e potenza militare: istituzioni e ideologia tra fascismo e seconda guerra mondiale, AperTO—Archivio Istituzionale Open Access dell'Università di Torino—Centro studi piemontesi (2011). http://hdl.handle.net/2318/95291, since 2016-10-19T12:03:47Z, Accessed 5 Sept 2021
83. E. Scerpanti, S. Zamagni, *An Outline of the History of Economic Thought* (Oxford University Press, Oxford, 2005), pp.23–24
84. V. Zamagni, Una historia económica. De la Edad Media a la crisis del euro, Barcelona: Editorial Planeta, ebook, (Capítulo 7. Tecnología, cambios socioeconómicos y formas empresariales) (2016)
85. L. Mammino, Language mastery, narrative abilities and oral expression abilities in chemistry learning. Importance Mother-Tongue Tradit. Narrat., S. Afr. **35**(1), 19–27 (2015). https://doi.org/10.1080/02572117.2015.1056459
86. Cit. in Evans R. J., *In Defence of History, New York* (W.W. Norton & Company, London, 1997), p.23
87. T.J. Righini, Narrare la storia. Nuovi principi di metodologia storica, p. 9; 24; 143–145 (Bruno Mondadori, Milano, 1997)
88. M.L. Altieri Biagi, L'avventura della mente. Studi sulla lingua scientifica (Morano, Napoli)
89. T. De Mauro, (edited by) Studi sul trattamento linguistico dell'informazione scientifica (Bulzoni, Roma, 1994)
90. M.A. Cortelazzo, La lingua delle scienze: appunti di un linguista, in Premio «Città di Monselice» per la traduzione letteraria e scientifica, 31–32–33, ed. by G. Peron (Il Poligrafo, Monselice, 2004), pp. 185–195.
91. INDIRE (National Institute of Documentation, Innovation and Educational Research) *High school national guidelines*: *Indicazioni nazionali per i licei*, https://www.indire.it/lucabas/lkmw_file/licei2010/indicazioni_nuovo_impaginato/_decreto_indicazioni_nazionali.pdf, and *Linee guida degli istituti tecnici,* https://www.indire.it/lucabas/lkmw_file/nuovi_tecnici/INDIC/_LINEE_GUIDA_TECNICI_.pdf, Accessed 10 Sept 2021
92. Cfr. http://www.macchinedileonardo.it/, Accessed 15 Sept 2021
93. Commissione parlamentare d'inchiesta sul disastro del Vajont: inventario e documenti, Senato della Repubblica Italiana, ed. by A. Lucidi (Rubbettino, Soveria Mannelli, 2003)
94. C. Pavone, Prima lezione di storia contemporanea (Editori Laterza, Roma-Bari, 2007) (and ivi. cit. D. Rodogno, Il nuovo ordine mediterraneo (Bollati Boringhieri, Torino, 2003)), p. 119

Chapter 14
Pedagogical Integration of the Humanities and Sciences

Francis M. Burns

Abstract This chapter examines the rationale for integrated pedagogy from a holistic perspective. Our modern world needs cooperation and collaboration to develop and use innovation. Innovation also entails cultural change. Many major issues span multiple disciplines requiring a variety of perspectives and epistemologies. As a result, universities play a central role in innovation development and cultural change. Disciplinary segregation impedes cooperation and collaboration through competition and division. The origin of disciplinary segregation is outlined through history and philosophy associated with the humanities and sciences. The implication of disciplinary segregation is examined using organizational silos. An organizational silo is any repository of knowledge that cannot be readily shared. Integration offers potential remedies by bridging silos between the sciences and humanities. After defining common types of integration, a short review of the latest literature is provided. Exemplary materials for practitioners are highlighted.

Keywords Art · Art education · Higher education · Humanistic education · Humanities · Interdisciplinary approach · Liberal arts · Sciences · STEM education

14.1 Introduction

Interdependence marks our modern world. International commerce moves food and products around the globe. Canadian wheat is shipped to Italy. American cotton is shipped to the Far East. Computers are assembled in one country using parts manufactured in other countries and then shipped to yet another country. Technology has improved the lives of millions of people. It has also increased human misery through its use in war and environmental degradation. Our population has increased to the point that humanity can affect major ecosystems, such as the polar ice caps and tropical forests. These problems have international components. Nations

F. M. Burns (✉)
Division of Math, Science and Professional Studies, University of South Carolina Salkehatchie, Walterboro, SC, USA
e-mail: fmburns@mailbox.sc.edu

L. Mammino (ed.), *Insights into the Relationships between Humanities and Sciences*,
https://doi.org/10.1007/978-3-031-97900-2_14

participate in international bodies, such as the United Nations, and international agreements, such as the Montreal Protocol on Substances That Deplete the Ozone Layer. Collaboration and cooperation start within nations. Universities play an important role in fostering collaboration and cooperation within and between societies. American universities serve three social functions. First, these institutions shape the future through educating students. These students become tomorrow's professionals, politicians, and leaders. Second, universities solve problems and answer questions through scholarly research. Finally, most universities serve their local communities through outreach, such as collaborating with cities and nongovernmental organization. Universities support international activities, too. The University of South Carolina maintains Global Carolina, which facilitates international activities [1]. This university is not unique with respect to its support for international activities. The European Union strongly promotes international activities for students and young people through Erasmus + . The program supports the educational, professional, and personal development of people in education, training, youth, and sport. In doing so, the program contributes to sustainable growth, quality jobs and social cohesion within the European Union. European universities are important collaborators in efforts to implement the program's objectives [2]. Many students travel to other countries for educational and cultural purposes. Faculty travel to collaborate with international colleagues and attend conferences. Universities develop best practices, then widely disseminate. Even though universities need to foster cooperation with societies, these institutions also mirror the societies that they serve.

Universities are complex organizations characterized by both collaboration and competition. Individuals, departments, and colleges work together. They also compete for recognition and political power. Credentialed and earnest people objecting to favored ideas or beliefs can be bullied relentlessly or even lose their position. Institutional policies could emphasize cooperation instead of rivalry. Academic departments could support interdisciplinary activities. Universities can function as role models for interdependence to students.

Many of tomorrow's leaders receive a university education. Students' education typically focuses on a single discipline, such as engineering or history. They may or may not have additional courses in the humanities and sciences. Students' educational programs can be as fragmented as society. Literature and history courses tend to pay little attention to the sciences. Courses in the sciences rarely address large questions, such as ethics and values. Leaders need knowledge and skills from the humanities and sciences to function effectively. Some skills are broader than a single discipline, such cooperation and collaboration. Cooperation and collaboration are better learned as a student.

This chapter examines *integration* as a potent tool for educating students and solving problems. Science and humanities have evolved into two largely independent domains. Disciplinary segregation started during the Scientific Revolution, but it affects modern society. Many problems transcend single disciplines or a domain. For example, obesity is a global problem. The World Health Organization has acknowledged that obesity is a global health crisis [3]. Although obesity is a medical condition, it has social implications because obese individuals face stigmatization through popular culture. Social stigmatization can negatively affect obese individuals' employment, medical treatment, social interactions, and mental health. Effective solutions require destigmatizing obesity even as the root causes are addressed. This approach requires multi-disciplinary efforts [4]. Research has made considerable progress in understanding the causes of obesity [5], but considerable work remains that requires collaboration, cooperation, and information sharing. Substance abuse is another global problem requiring contributions from every discipline. There is an array of associated outcomes, such as increased health care, losses due to increased crime and traffic accidents, and reduced employment productivity. Based upon the meta-analysis of Manthey and others, alcohol abuse costs 2.6% of the GDP in the following countries: Australia, Canada, European Union, Russian Federation, South Africa, South Korea, Thailand, and the United States [6]. Another metanalytic study determined that tobacco addiction costs 1.8% of global GDP [7]. The United Nations published a comprehensive report on drug use and crime. Thirty-one million people suffered from drug use disorders in 2016. Their drug use was sufficiently harmful to require treatment. Approximately 450,000 people died directly or indirectly from drug abuse in 2015 [8]. Science has unraveled many of the chemical and biological aspects of addiction. The interactions of drugs of dependence and the nerve receptors have been delineated. Different categories of drugs interact with different sets of receptors. Several processes are possible points of attack by drugs, including depolarization of the receptor membrane upon nerve stimulation, link between depolarization and neurochemical transmitter liberation, or the uptake of the transmitter upon its release [9]. Similar gains have been made by the social sciences. Psychologists have incorporated neurobiology into addiction models and treatment protocols [10]. Despite these efforts, substance abuse continued to plague humanity. The problem transcends the sciences. Collaboration, cooperation, and information sharing must become existential societal skills because today's problems require them. Organizational silos tend to inhibit these skills. requires the contributions of every discipline. Integration can bridge the two domains. After establishing the rationale for integration within the liberal arts, this chapter reviews some of the latest publications. The chapter concludes by making the case that integration should be a priority for higher education institutions.

This chapter needs to define some terms. Humans and humanity refer to our species, *Homo sapiens*. The sciences and humanities are less obvious. The United States government funds the Educational Resources Information Center (ERIC), which defines descriptors for indexing purposes. ERIC has indexed 1.6 million citations, which is freely available to the world [11]. ERIC does not define the sciences, but it includes information science, natural sciences, social sciences, and space sciences as narrower terms. Applied sciences, medicine, and mathematics are not indexed as sciences. ERIC does not define humanities, but includes fine arts, history, literature, and philosophy as narrower terms. For purposes of this chapter, the science domain or "sciences" consists of science, technology, engineering, mathematics, and medicine. The humanities domain or "humanities" consist of languages, history, philosophy, performance arts, visual arts, architecture, and design. Both lists are not exhaustive. Scientists are practitioners of the sciences. For purposes of this chapter, scientists include chemists, physicists, biologists, engineers, computer scientists, mathematicians, and medical professionals. What term can be used for practitioners of humanities? "humanities scholars" seems too narrow by implicating excluding writers and poets, religious leaders, and artists. For purposes of this chapter, humanists are practitioners of the humanities. Humanists include artists, educators, writers, historians, religious leaders, philosophers, and politicians. Neither list of practitioners is exhaustive. ERIC defines humanism as philosophy that asserts the dignity and worth of man. ERIC defines humanism narrowly. This chapter defines humanist narrowly, too.

14.2 Rationale for the Integration of the Humanities and Sciences

14.2.1 Background Considerations

The current national and global problems besetting humanity have precedents. The Fertile Crescent served as one of the cradles of civilization. This region invented agriculture during the Bronze Age. As the climate changed and became more arid and famine threatened. Irrigation was invented to solve drought. Ruth DeFries addressed these events and more in her book, *The big ratchet: How humanity thrives in the face of natural crisis*. She placed major technological events within the context of social evolution. When a human discovered a novel technology or idea giving that individual an advantage, then she or he probably shared the discovery with kin, and eventually it spread to neighboring communities. Perhaps the individual developed a better weapon, tamed an animal, grew an edible plant, or discovered a practice that improved crop yields. With increased resources, success increased or "ratcheted up" the number and distribution of our species. Innovations do not produce infinite resources, though. Side effects resulted from the use of an innovative technology or idea, such as excessive population growth, environmental degradation, and economic displacement. Time and time again, prospects of doom appeared, such as famine, disease, war, and so on. DeFries labeled these events as "hatchets." When ecological, social, economic, or political structures start to fail, societies are at risk. Time and time again, a clever individual or group managed to discover solutions. DeFries labeled these solutions as "pivots." She framed the growth of humanity as a cyclic process. Three words describe DeFries' social theory: "Ratchet, hatchet, pivot." Success increased the human population, which increased the gravity of future hatchets, such as pandemics, climate change, and social unrest [12]. Her book provided a model for successful pivots, but today's pivots need to be global, even if they initially start small and locally.

How does humanity pivot to solve major catastrophes, such as pandemics and climate change? Science and technology can provide solutions, but science and technology cannot implement the solutions. Vaccines have been developed for a variety of infectious diseases, but vaccines are only useful if they are used. Vaccine hesitancy involves various factors including social norms, religion, and communication [13]. Science and technology cannot address many of these factors. Climate change also demonstrates that science and technology are insufficient for today's global problems. Many scientific questions about climate change were answered in the 2014 report by the Intergovernmental Panel on Climate Change (IPCC), except for the magnitude of the change, the magnitude of risks and effects upon humanity, and the timeline of the change. In addition to summarizing the latest climate science and assessing future risks and impacts, the IPCC report also proposed adaptation and mitigation strategies [14]. Little progress has been made since the report was published in 2014. The culture needs to change to improve humanity's adaptability. Today's hatchets have become global in scale. Pivots are no longer easily implemented; otherwise, climate change would have already been solved. The science and technology associated with climate change are well developed and continuously improving, but greenhouse gases continue to increase in concentration. Several vaccines have been developed for the COVID-19 virus, but vaccine hesitancy has left millions of people unprotected. Scientists cannot convince large segments of society to adopt their recommendations.

Cultural change has become a prerequisite for successful pivots. Science and technology no longer have sufficient power to implement pivots, and thus drive social change. Social change needs to be in progress prior to adopting the necessary scientific and technical solutions to make pivots. This is almost the reverse of DeFries' social theory. Humans, working together across geographical and political boundaries, will set cultural change in motion. They will demand greater agency in global solutions having environmental, economic, and socio-political implications. Successful pivots must incorporate many perspectives and approaches to produce effect solutions to systematic problems.

14.2.2 The Humanities and Sciences

Suppose we separated sciences and humanities, including ethics such that there can be no dialogue between the two domains. Would the solutions provided by science be capable of being culturally implemented and ethically sustainable? Probably not. Scientists have frequently overlooked human aspects of their work, but humanists have an important role speaking to scientific matters. Mary Shelley wrote *Frankenstein; or, The Modern Prometheus* in 1818. Prior to writing the novel, Shelley listened to conversations about Galvanism. Galvanism attempted to explain the role of electricity in animating frog legs or possibly a corpse. Without considering the moral consequences, Dr. Frankenstein "created" a new being in his laboratory. After coming to terms with his mistake, the doctor unsuccessfully tried to destroy his creation [15].

This ethical lapse has been repeated many times. Scientists often served as innovators during war. The chemical industry dominated pre-World War I Germany, producing large amounts of chlorine, phosgene, and other toxic gases for industrial applications. Germany's industry had well-developed research and development programs, too. Professor Fritz Haber was famous for developing agricultural chemicals. He also directed all of Germany's poison gas research and greatly enhanced their effectiveness. As a result, chlorine gas, mustard gas and other chemicals killed and maimed soldiers during trench warfare in World War I [16]. Environmentalism became a significant force during the 1960s due to the publication of Rachel Carson's *Silent Spring*. The book decried the use of DDT and other chlorinated pesticides, herbicides, radiation, and other man-made artifacts. Carson mixed scientific facts with lyrical prose. It became her third bestselling book [17]. *Silent Spring* made an immediate impact upon the social and political landscape that still reverberates. Carson attacked the integrity and moral leadership of the scientific enterprise. She questioned its pivots. She argued that the world was headed towards destruction. Carson framed the pesticide debate as an ongoing public debate. In doing so, she raised controversies and uncertainties about science. A political movement sprang up because of the public debates between the chemical industry and Carson [18]. This political movement raised many questions about the use of technology, including pesticides, nuclear power, as well as logging and mining practices. Science cannot be separated from the technology that is based upon it. Scientists bear responsibility for the applications arising from their research. The physicist Freeman Dyson wrote, "It makes no sense to me to separate science from technology, technology from ethics or ethics from religion [19]. Today's scientists are not Frankenstein clones, but it is difficult to consider ethical and social implications of one's work as one pursues it. Once scientific findings are published, there is no retraction. When scientific applications are developed, then they will spread throughout societies.

Science tends to remain separated from the studies of humanities and ethics also because humanists have frequently disparaged scientists and engineers, but scientists have improved life for humans through many innovations like the sanitization of drinking water and the discovery of electricity for lighting and power. Leon Wieseltier was the literary editor for the *New Republic*. He spoke at a commencement ceremony at Brandeis University in 2013. Wieseltier addressed the graduates as "fellow humanists." He stated that society needed the humanities more than ever because science and technology have enslaved humans. Science needs to stay within its boundaries: the study of nature. The humanities should be used to study humans. He concluded his speech by stating that "people pursuing the humanities are members of *the resistance* [20]. A 2002 research study examined the effect of mass media on public perceptions of scientists. Although newspapers and scientific television programs tend to promote science, general television viewing seems to "cultivate" reservations about science and technology [21]. Many factors affect the credibility of science, including the topics discussed. Many individuals like to cite "science" when it agrees with them. Humanists can wield enormous power within society. Perhaps humanists have lower average salaries upon graduation compared to scientists, but humanists have a greater influence through media, education, politics, and business. Instead of maligning science and technology, humanists need to educate and guide society.

Scientists and humanists need to work together. Scientists need to focus on solving scientific and technological problems, but they also need to maintain an ethical framework. Scientists recognize this issue. Professional organizations often possess codes of conduct. The American Chemical Society (ACS) requires ethics as part of undergraduate chemistry education [22]. Scientists need to avoid Dr. Frankenstein's mistake. Before seeking to create something new, they need to consider whether it is wise and ethical to do so. Similarly, humanists need to consider science and technology as more than gadgets and medicines. Science has developed cognitive tools to understand the world and then verify our understanding. These tools include skepticism, open debate, formal precision, and empirical tests [23]. Humanists need play an advisory role for scientific research and the use of technology like institutional review boards (IRBs) [24] or the advisory process for environmental impacts [25]. The scientific and humanist domains need to collaborate to solve problems, including educational ones. Both domains possess unique knowledge and skills meriting similar recognition. Humanists and scientists need to consider the state of the intellectual world.

C.P. Snow worked as a physical chemist and writer. He observed that the humanists and scientists exist as two separate cultures. The two domains operate independently of each other. Scientists have little use for literature and the arts. Writers, artists, and other intellectuals look upon science and technology with revulsion. Smugness justifies individuals' self-impoverishment. Snow focused upon educational reform as a remedy [26]. Leslie Real is a biologist studying infectious diseases. He described the differences between scientists and humanists with respect to collaboration. Collaboration between scientists frequently facilitates discovery and dissemination, whereas humanists tend to work alone [27]. Real's observations identified a fundamental difference. Multiauthor scientific articles are commonplace. In contrast, sole authorships are common in the humanities [28]. There are important questions to consider as one attempts to bridge the divide. When did intellectuals separate into two domains: science and humanities? Snow asserted that the division occurred during the twentieth century. He commented that very little of modern science has been assimilated into modern art. What pushed the two domains into nonoverlapping spheres of intellectual pursuits? Snow thought that humanists were natural luddites. They abhor change, which occurs regularly in science. These are important questions to consider as one attempts to bridge the divide. Sustainable remedies need adequate answers. These answers need not call for the abolition of academic disciplines, but collaboration needs a more congenial environment.

The divide between scientists and humanists dates before the twentieth century. It started before the Scientific Revolution (1543–1727). Medieval universities' curriculum focused on classical education: *Trivium* (grammar, rhetoric, and logic) and *Quadrivium* (music, arithmetic, geometry, and astronomy). Due to ties to the medieval church, universities were frequently unfriendly places for natural philosophers and alchemists [29]. While Renaissance humanists looked to Greeks and Romans for answers, natural philosophers looked to the natural world for answers. Humanists played an important role in the emergence of modern science because they retranslated texts written by Aristotle, Euclid, Pythagoras, and others. These ideas provided a starting point for new investigations. Some ideas were unsuitable, such as cosmology and alchemy. Some ideas were useful, such as mathematics and mathematical physics. Humanism provided a mixed blessing to natural philosophy [30]. Ultimately, natural philosophers rejected religious and scholastic authority even as deism maintained its credibility within natural philosophy. Nicholas Copernicus studied mathematics and astronomy. Copernicus wrote his manuscript for his heliocentric theory but delayed its publication due to his concern about its political and ecclesiastical reception [31]. Galileo Galilei wrote his tome favoring the Copernican theory, but the Inquisition banned the book's sale, and then tried and convicted Galileo for heresy [32].

Conflicts between the humanities and sciences grew during the Scientific Revolution. Francis Bacon developed methods of induction and founded the philosophy of science. He wrote *New Atlantis*, which was a utopian fantasy about a science-based society [33]. The dichotomy between the humanities and sciences grew during the seventeenth century. Four scholars exemplify this discord: Johannes Kepler, René Descartes, Blaise Pascal, and Isaac Newton. Each scholar pursued interests in the sciences (astronomy, physics, or mathematics) and humanities (philosophy, chronology, or theology). Kepler followed his passion for astronomy but used recorded astronomical events to date Jesus' birth. Pascal prepared a list of two domains. One domain consisted of history, geography, law, languages, and theology, which obtained knowledge by reading books. The other domain consisted of geometry, arithmetic, music, physics, medicine, and architecture, which obtained knowledge through experimentation. Isaac Newton discovered calculus and the Newtonian laws of physics. He also studied chronology and theology. René Descartes invented analytic geometry and cartesian systems. Starting with his fundamental concept, *I think, therefore I am*, Descartes built a rational philosophy for explaining the nature of everything within the natural world [34, 35]. The philosopher John Locke disagreed with Descartes assertions: "And if men had eyes like microscopes, though they might see further into the texture of bodies they would lose all perspective and never see their way to the 'market and exchange.'" Locke argued that humanity was rational and capable of free choice, but humanity was not capable of understanding all of God's will or workings [36]. Both Bacon and Descartes promoted the use of reason and logic at the expense of other human traits such as creativity, memory, and imagination. Opponents labeled their philosophies and similar ones as "scientism," which became derogatory in the twentieth century [37].

The Age of Enlightenment (1685–1815) was pivotal for Eurocentric culture and modern science. Advancements encouraged specialization within the humanities and sciences. Universities continued to promote classical education at the expense of science education and research. As a result, scientific academies and societies were established: the Royal Society of London, the Paris Academy of Sciences, and the Berlin Academy of Sciences. As the sciences were separating from the humanities, modern scientific disciplines began appearing. Mathematics became the method of solving problems by reducing them to equations. Mathematics was commonly applied to astronomy, physics and engineering problems. Experimental physics investigated the nature of heat, light, electricity, and magnetism. Chemistry emerged as an independent discipline with the published works of Antoine Lavoisier. Industrial processes needed large quantities of acids, alkalis, and organic chemicals. Industrial demand spurred chemical research. New discoveries produced new applications, such as vaccines and hot air balloons. Many of these discoveries would later support the Industrial Revolution. Scientific discoveries deeply affected philosophy, too. Philosophy was considered an umbrella for *moral sciences* even as economics, political science, and sociology separated from philosophy. David Hume wrote essays and books related to history and philosophy. Hume critiqued theories of causality and contributed ideas supporting the philosophy of science. He emphasized the empirical and probabilistic nature of the physical world. Hume's historical writings emphasized that history led to an understanding of human nature instead of an understanding of God's will. The German philosopher Immanuel Kant established modern philosophy. He categorized knowledge. Kant's works examined empiricism, moral philosophy, and aesthetic value judgments [38]. The humanities were also affected by the burgeoning scientific discoveries, which legitimized new ideas and literature.

The Industrial Revolution (1760–1880) brought science and technology to the forefront. Science was beginning to describe the natural world in novel and powerful ways, such as chemical bonds, biological evolution, geologic time and fossils, and electricity/magnetism. Joseph Priestley published *The History and Present State of Electricity*, which increased the importance of electricity. He compiled prior electricity research done by Benjamin Franklin, Charles François de Cisternay du Fay and other natural scientists. He also reported his own electrical experiments [39]. James Hutton published *Theory of the Earth*, which introduced the concept of geologic time. Although Hutton's work did not receive the recognition that Charles Darwin's did, it contradicted the biblical story of creation and laid the foundation for modern geology [40]. John Dalton introduced atomic theory, which contradicted Aristotle's ideas regarding matter [41]. The metric system was developed and adopted by the French Republic [42]. Samuel Morse invented Morse Code, which started the telecommunication revolution. The telegraph also was the first application using electricity. This created a need to generate electricity. James Watt received several patents for inventions that were necessary for the development of steam engines [43]. Steam-powered manufacturing and transportation transformed England and then Europe and North America. In addition to enabling mass production and transportation of consumer goods, industrialization intensified the effects of warfare [44]. Profound societal changes occurred. The revolution started in Britain, moved to Europe, and then to the United States and Canada. The world started to shrink due to new forms of manufacturing and trade, transportation, and communication.

Professionalism drove many changes during the Industrial Revolution. Professional engineers supported the mechanization of economic activities through their inventiveness. Engineers were almost absent from patent applications prior to 1760 but accounted for thirty percent of such applications by 1860 [45]. Professionalism also served as a potent force in dividing the liberal arts and science. Prior to the

Scientific Revolution, society was stratified. The elite (clergy and nobility) monopolized knowledge and power through politics and universities. Trade guilds also monopolized knowledge and skills. The pre-industrial professions were limited to lawyers, physicians, and architects. These professionals consisted of practitioners and specialists. Practitioners acquired their knowledge through apprenticeships. Specialists attended universities, which taught the liberal arts. They served the elite: clergy and nobility. The Industrial Revolution fostered the rise of professionalism. Specialization occurred in middle class professions, such as doctors, pharmacists, lawyers, architects, engineers, teachers, and librarians. These professions developed exclusive membership based upon their members' completion of certified curricula and examinations. The liberal arts were considered extraneous to professional education [46]. The ACS was established in New York City in 1876. The Society established minimum educational and training criteria to gain full membership [47]. Professionalism propelled the sciences by developing scientific epistemologies and methods. Professionalism also accelerated the division between the sciences and humanities. University education split into liberal arts and technical curricula.

The Industrial Revolution created still more divisions. It encouraged materialism and general belief in the power of science. Scientific discoveries and world events challenged world and philosophical views [48]. Darwin published *On the Origin of Species by Means of Natural Selection* and *The Descent of Man.* Darwin's publications spawned Social Darwinism. Herbert Spencer coined the term, "survival of the fittest." Social Darwinism was used to legitimize brutal political policies and economic practices maintaining class and racial hierarchies. Social Darwinism has come to be equated with racism, Nazism, and eugenics [49]. Auguste Comte formulated positivism from David Hume's empiricism and skepticism. Comte rejected theology as a source of wisdom. Comte claimed that the only valid data are acquired through the senses. Nothing was transcendent. Nothing metaphysical could have any claim to validity [50]. Religion and romanticism acted as counter forces to science and its derivative philosophies. Religion survived but ceased to be politically sanctioned. Religion remained important for some groups, but religious affiliation decreased. Methodism became established among British workers and formed a vital part of working-class life. Many American immigrants brought Catholicism with them [48]. Romanticism served as a secular religion for many humanists. Many artists and writers turned against the ugliness of the industrial setting and rationality of science. For example, a group of poets and artists felt that Newton had destroyed the mystery and beauty of the rainbow with his analysis of sunlight using a prism. They would raise their glasses and say, "to the confusion of Isaac Newton." These artists and writers expressed antipathy towards scientists, who threatened to destroy simple appreciation of nature. Humanists were also possibly envious of the scientists' ability to acquire financial support for research and development [51]. Charles L. Dodgson, aka Lewis Carroll, wrote many popular children's books, such as *Alice in Wonderland*. He complained about displacement of classical education by scientific education [52]. Scientists quoted poets in their textbooks and articles for the public. Many popular writers explored the implications of scientific discoveries [53]. Jules Verne wrote *Vingt mille lieues sous les mers* (*Twenty Thousand Leagues*

Under the Sea) in 1870 [54] and *L'île mystérieuse* (The Mysterious Island) in 1875 [55]. Verne's science-fiction novels used science and technology as it was known at the time, such as hot air balloons. Nineteenth century writers reflected the conflicted roles of science within society. The Industrial Revolution was a time of excitement due to political, economic and social changes. These changes produced conflicts.

The second half of the nineteenth century through the twenty-first century followed patterns established during the Scientific and Industrial Revolutions. By the end of the nineteenth century, disciplinary segregation had occurred. Most chemists did not interact with historians, except for academic governance. Literature professors did not pursue projects with mathematicians. Industrialization of agriculture, manufacturing, and shared transportation continued to spread throughout the globe, ratcheting up the human population. Societal hatchets appeared in the form of world wars, environmental crises, and global inequalities. Human societies continued to pivot and adapt to new situations. Science continued to make significant discoveries and inventions. Karl Popper and Thomas Kuhn made major contributions to science philosophy. Popper pioneered the concept of falsifiability, which formed the foundation for modern science [56]. Thomas Kuhn coined the term *paradigm shift* to describe major breakthroughs in science, such as elucidating the three-dimensional structure of DNA [57]. Science continued to enjoy considerable support from universities, governments, businesses, and the public. President Lincoln signed the Morrill Act in 1863 establishing land-grant universities, which emphasized the applied sciences of agriculture and engineering [58]. The United States made a major investment in science during World War II. The Manhattan Project involved multiple universities, nuclear scientists, engineers, thousands of people, and nearly two billion dollars. The remarkable effort produced nuclear weapons. President Truman ultimately decided to bomb Hiroshima and Nagasaki, which ended World War II and planted the seeds for the Cold War [59]. During the Cold War, the United States government established the National Science Foundation (NSF) for national security. The NSF received 8.3 billion dollars during 2020 [60]. Government support for the humanities is not commensurate. The United States also established the National Endowment for the Humanities (NEH), but the NEH received 0.10 billion dollars during 2020 [61]. This funding difference has not promoted harmony and cooperation between the sciences and humanities. This discrepancy between the two domains also manifests itself in architecture, building, and infrastructure usage. As higher education moved into the twenty-first century, universities have undergone a reversal from the medieval period. Science occupies prominent places at many universities. Scientists are clustered in scientific laboratories. Common spaces are shared for meetings and work. In contrast, humanists do not commonly use infrastructure in a similar manner. The library or art studios may serve as common workspaces, but otherwise there is little shared space [62]. Science education and scholarship should be highly valued. Arts and humanities education and scholarship also merit significant support. Funding inequities need to be addressed within higher education, government, and the public.

14.2.3 Integration Bridges Organizational Silos

The sciences and humanities need each other. Sciences need the humanities for effective integration of technology within society for humanity's welfare. The humanities benefit from scientific discoveries by expanding the horizons of possibilities. Collaboration requires extra effort because the sciences and humanities have maintained disciplinary segregation. This situation was highlighted by the National Academies of Sciences, Engineering and Medicine (NASEM) which examined the integration of the humanities and arts with the sciences, technology, and medicine. The comprehensive study highlighted disciplinary segregation as a major impediment to integrated pedagogy. For example, tenure and promotion tend to be based upon scholarship, teaching, and service within a department or college. As a result, collaborative activities are frequently viewed a distraction [63]. Segregation can constrain collaboration and problem solving by creating organizational silos. Diversity [64] and divergent thinking [65, 66] can enhance creativity and problem solving. Organizational silos are repositories of knowledge that cannot be readily shared like granary silos [67]. These silos are a feature of complex organizations, including universities. Integrated pedagogy offers powerful tools in bridging silos producing innovative solutions spanning disciplines. Cohen and Lloyd used the term "transdiscipline" to describe collaborations that have application across many disciplines and across disciplinary silos. Education and research within a transdiscipline share a common goal seeking to solve particular problems. Transdisciplines do not eliminate disciplines but bridge them [68]. Every academic discipline has developed a valuable set of disciplinary knowledge and skills. Effective collaborations do not require common epistemologies. Humanity needs all academic disciplines to contribute towards effective and sustainable solutions to a host of major issues, such as climate change, income inequality, adequate healthcare, and mass migrations.

14.3 Integration of the Humanities and Sciences or the Liberal Arts

Prior to discussing integration, an issue needs to be raised, which was noted by the NASEM study. There is an overabundance of acronyms derived from STEM (Science, Technology, Engineering, and Mathematics). The NASEM study used STEMM [STEM + Medicine] as recommended by the NSF. STEAM [STEM + Art] has been popularized by art education professionals, but this does not explicitly include humanities. As a result, SHTEAM has been coined to combine science, humanities, technology, engineering, arts, and mathematics. Others prefer the older term "liberal education" [69]. The ERIC database uses "liberal arts" to index citations with the sciences and humanities. STREAM [STEAM + Reading] has been recently proposed to promote reading and writing [70] HD-STEM has been used to promote humanities [71]. For purposes of this chapter, the science domain or "science" includes science, technology, engineering, mathematics, and medicine. The humanities domain or "humanities" includes languages, history, philosophy, performance arts, visual arts, architecture, and design. Both lists are not exhaustive of disciplines. A key aspect of DeFries social theory is the dissemination of innovation. The careless use of nonstandard acronyms and terms impedes the adoption and advancement of integration, which is also a nonstandard term.

14.3.1 The Different Forms of Integration: Multidisciplinary, Interdisciplinary, and Transdisciplinary

Integration within the classroom involves more than one discipline. For example, chemistry and physics are commonly integrated within engineering or material science courses. Integration can occur within a single assignment within a course, an entire course, an entire curriculum, or outside of a curriculum (extracurricular). In addition, different levels of sophistication can characterize integration pedagogy. This range is frequently characterized by the choice of terms: "multidisciplinary," "interdisciplinary," or "transdisciplinary" (Fig. 14.1). Several articles have defined these categories within a pedagogical context [63, 72–74]. Multidisciplinary curricula use the concepts and epistemology of specific disciplines. The individual instructors teach in parallel or independently. For example, Cronin and Hawthorne developed activities in which pre-nursing students wrote poetry within an introductory nursing course. Cronin was a nursing faculty member. Hawthorne was a tutor in academic skills and English. The poetry activity was a discrete session within the course [75]. Interdisciplinary teaching also requires instructors to address a common theme using two or more disciplines. Instructors need to directly interact from their disciplinary perspectives, working together to address the theme or issue. For example, *War, Machine, Culture, and Society: History and Engineering in the Second World War* was team-taught with an engineer and a historian teaching side-by-side. Students learned how engineering can change physical landscape during war and peace. They also learned how society grappled with difficult engineering decisions, such as the impact of scientific and engineering discoveries. Students examined the design and manufacture of war technologies as well as the ecological costs of war [71]. Finally, transdisciplinary courses possess the greatest level of sophistication. Biostatistical courses utilize the methods and epistemologies from biology and statistics because relevant problems overlap the two disciplines. Real-world problems or themes drive the construction of meaning within transdisciplinary curricula. The boundaries between the conventional disciplines dissolve within the context of the course [76]. The present chapter focuses upon integration involving at least one science discipline and one humanities discipline.

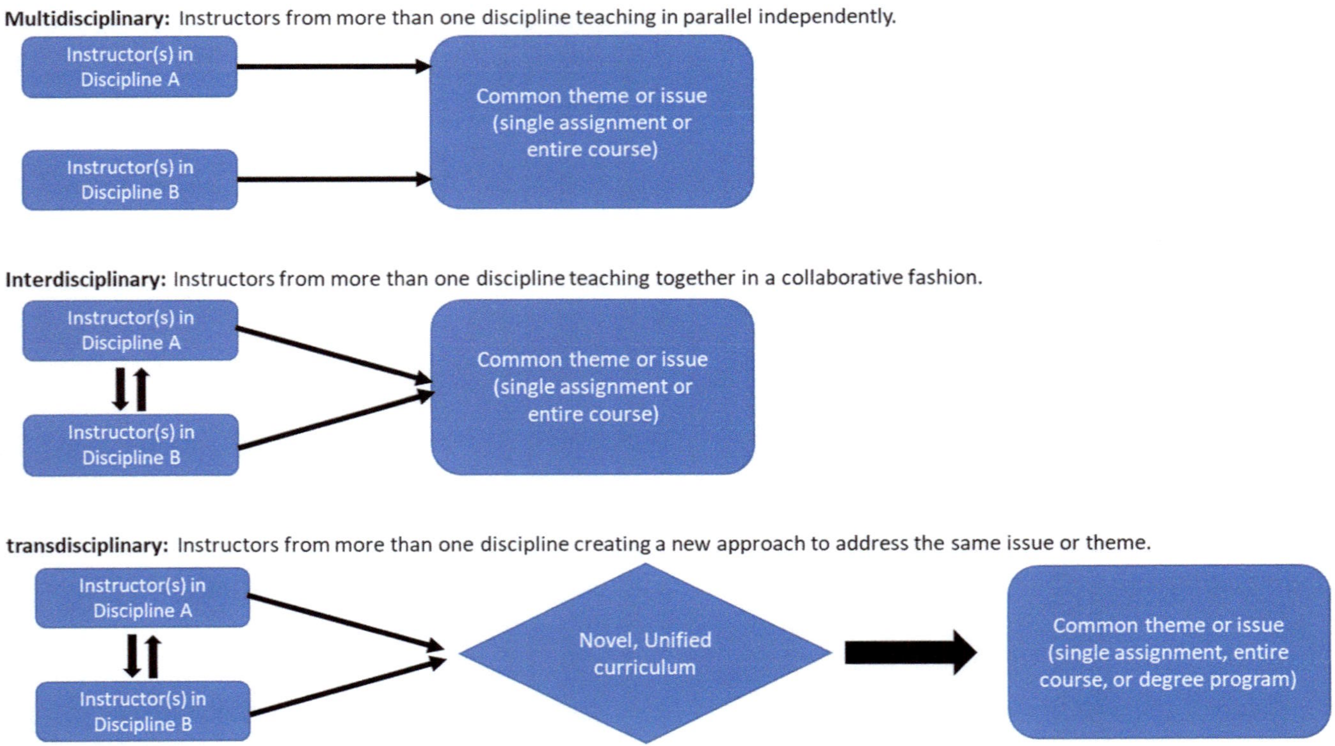

Fig. 14.1 Visual representation of integrated education: multidisciplinary, interdisciplinary, and transdisciplinary. The figure is based upon Begg and Vaughan [77]

14.3.2 The Direction of Integration: Path Dependence

There are two common approaches integrating the humanities and sciences. Science educators reach out to humanities educators to establish collaborations. Alternatively, humanities educators reach out to science educators for collaborations. These approaches are not path independent. How collaborations are started and managed will affect outcomes. Two examples illustrate this point. The Rhode Island School of Design (RISD) has been lauded for its efforts to integrate education by the NASEM study [78]. RISD degrees are awarded in Fine Arts and Design, such as printmaking and architecture. Liberal Arts faculty teach biology, history, English, and many other foundational courses. The institution was an early proponent of STEAM education, which integrates STEM education with Art education [79]. RISD's Nature Lab received a large grant from the NSF to support bio-design education. Art and design students will address social and environmental issues through their disciplines. Biology provides students with a scientific perspective. The University of San Diego's School of Engineering has developed a new program called "Integrated Engineering," This program seeks to prepare students to address societal and environmental challenges. This new program was also funded by a grant from the NSF [72, 80]. The "Engineering Peace" course illustrates the school's approach. An engineering faculty member, a sociology faculty member, and a humanities faculty member (Peace Studies) developed an engineering design module in which students develop peaceful applications for aerial drones [72]. Engineering students are applying their discipline even as they learn about ethic consequences. Both approaches have merit. Humanities students need to integrate the sciences within their disciplines. Science students need to integrate the humanities within their disciplines. Efforts to integrate multiple disciplines within teaching and scholarship need to be redoubled.

The NASEM study considered the question, "Does the direction of the integration matter?" The study concluded that direction matters only when an integrative course is directed towards a specific discipline either through availability, advertising, or learning outcomes. The supporting evidence seems to be contrary to the conclusion. Assessment methods used in science courses are often different from ones used in humanities courses. This difference makes direct comparisons of learning outcomes difficult. It also illustrates that integration is not path independent. As a result, a direct answer to the posed question is difficult. The study provided an indirect answer, though. It provided prodigious evidence supporting the integration of the humanities into science programs. In contrast, the study reported that "The evidence base for the impact on students of courses and programs that integrate STEM knowledge and pedagogy into the arts and humanities is extremely limited, particularly in the peer-reviewed literature [81]. The study's authors found interesting reports related to improving scientific literacy in humanities students, these reports are few in comparison to reports addressing humanistic literacy of science students. As a result, direction of integration does matter. The reasons underlying this difference appear to be unknown. Perhaps science and technology programs have ample financial support due to their political and economic importance. Perhaps resources affect integration. Philosophical and disciplinary differences may play a role, too. One review emphasized the value of adding chemistry to all disciplines, in the sciences and humanities alike. The value of adding humanities disciplines to chemistry or other science disciplines was not discussed to any degree [73]. Scholars and practitioners need to consider the overarching perspective of their respective courses. Does a prospective course function as introduction to biology in which literature and philosophy are integrated? This perspective would direct the selection of content and learning activities. Alternatively, a prospective course may serve as a capstone for a small group of science and humanities students. The instructor may be a historian directing students in disciplinary collaborations prior to finishing their academic program. She would probably approach the course from a humanist perspective instead of a scientific one.

Concerted efforts are underway within the United States to integrate the humanities and sciences within higher education. K-12 schools and community organizations have also collaborated. These efforts have been broad-based: academic institutions, nongovernmental organizations, and the federal government. The NEH funded *Connections*, which supports innovative curricula integrating a minimum of one discipline in the humanities and with one discipline outside of the humanities [82]. The NEH supported the development of a dual degree program for German and Engineering at Auburn University. German and engineering faculty developed four interdisciplinary German courses and transformed an introductory engineering course to include a cross-cultural component [83]. The National Endowment for the Arts (NEA) funded individual STEAM (STEM + Art) proposals, such as a grant to the Fashion Institute of Technology promoting sustainable fashion design [84]. The NEH and NEA cosponsored the NASEM study. The study's committee produced a comprehensive report, including findings, conclusions, and recommendations based on information gathered by the committee [78]. The NASEM website maintains a database of integrated courses taught within the United States [85]. Individual practitioners have designed, tested, and reported their experiences. Peer reviewed literature provides an abundant resource of different models of integration. Authors have published recent reviews associated with the integration of the humanities and sciences [74, 86–88]. Institutions and individuals are developing theoretical foundation, as well as the pedagogical and assessment methods, needed for disciplinary integration. The potential for integrating courses is limitless. For example, one could consider the complementary skills and knowledge provided by a chemist, mathematician, and writer team-teaching a course. Students could investigate the differences between different epistemologies. Students may benefit more from two professors with different perspectives on the topic of the class. This combination might create greater and more sophisticated understanding of the topic. It would require the professors to be open-minded and humble.

14.4 Summary of Recent Publications

The NASEM study finished their review in 2018. As a result, I searched for articles published after this comprehensive review. This search produced thirteen peer-reviewed articles deemed appropriate for this chapter. The selected articles are either reviews or case studies examining and documenting integrated pedagogy. Most of the case studies consisted of either individual assignments or the entire course within higher education. Most of the articles are written from a scientific perspective. The humanities (e.g., creative writing or visual arts) are brought into the existing structure. Efforts were made to find integration from a humanities perspective, but only two peer-reviewed articles were found [71, 89]. Pedagogical articles generally fit within the three categories of integration: multidisciplinary, interdisciplinary, or transdisciplinary.

14.4.1 Multidisciplinary Pedagogy

Multidisciplinary pedagogy consists of two or more individuals teaching ideas and skills based upon their knowledge and skills. Traditional liberal arts programs are multidisciplinary because students complete an array of courses spanning the sciences and humanities. Students are supposed to integrate disciplines within their own heads. Integration is implicit (Fig. 14.1). Three reports used an interdisciplinary approach. Sara Lindsay published her keynote address to the American Microscopy Society. She highlighted the synergy between biology and art. Students would examine diatoms under light microscopes, sketch them, and then discuss their results [90]. Both courses were introductory courses for nursing. Camille Cronin and Caroline Hawthorne sought to improve pre-nursing students' communication skills, ability to process difficult emotions, or improve learning of nursing content. They used poetry as an instructional aid for beginning nursing students. The authors stated that students needed to develop writing and reflection skills for their profession. Cronin and Hawthorne analyzed the student work using qualitative analysis characterizing the work in terms of writing skills (e.g., repetition) and emotional content (e.g., job satisfaction) [75]. Heather Lewis described the use of haikus in her pathophysiology course for pre-nursing students. The poetry exercises were treated as learning strategy by the instructor. Assessment was limited to a short student survey. Students were generally positive about haikus in pathophysiology, except only 30% of the students were positive about "I was able to use the haikus from class as a study tool [91]. These reports used writing poetry or sketching diatoms as multidisciplinary activities because the activities remained discrete and independent within the course. One can consider multidisciplinary pedagogy to be the least integrated of the three categories. It has the longest history of the three categories. Finally, multidisciplinary methods are the easiest to implement because courses and degree programs could be sliced into segments, then divided among disciplines. Students would need to integrate the courses through mental processes.

14.4.2 *Interdisciplinary Pedagogy*

Compared to multidisciplinary courses, interdisciplinary courses have greater integration between the involved disciplines. The different concepts, epistemologies, and skills interact to solve problems, such as communicating science to different audiences [92] Integration is explicit because the disciplinary experts actively seek to meld the disciplines (Fig. 14.1). Three reports described interdisciplinary courses. These courses integrated knowledge and methods from different disciplines, which combines perspectives. Arias-Alfonso and Franco employed the interdisciplinary model. They developed pedagogical activities combining physics, mathematics and visual art. Mathematics was viewed from an artistic perspective. The authors used art as a metacognitive tool [93]. Carrell, Keaty and Wong described a course taught by an engineer and historian. Two professors approached World War II from historical and technical perspectives [71]. Watkins and Tehrani described two interdisciplinary courses taught within an honors program. One course approached literature from a scientific perspective. The other course approached science from a humanities perspective [89]. Two reports approached the humanities from a scientific perspective. One report approached the sciences from a humanities perspective. These reports details efforts to explicitly integrate disciplines such that students could appreciate the different epistemologies and methods.

14.4.3 *Transdisciplinary Pedagogy*

Transdisciplinary pedagogy is the most sophisticated category. The approach seeks to explicitly integrate disciplinary knowledge and skills. Transdisciplinary pedagogy builds new frameworks that transcends disciplinary perspectives. Four reports of transdisciplinary instructional methods were selected. Hammons and colleagues provided an interesting example of transdisciplinary education. They developed a flipped hybrid course addressing obesity. Experts in twelve disciplines contributed micro-lectures as video, which were viewed by students prior to face-to-face lectures. Face-to-face lectures started with questions and answers but spent most of the time on transdisciplinary group work [94]. Rodier and colleagues developed a course for engineering and art students. They designed small-scale prototypes of interactive art installations to communicate key environmental issues. The course was based on several theoretical models, such as on the Stanford Design Thinking 5 phase framework: empathize, define, ideate, prototype, and test. The course had progressed to the prototype phase, but the COVID-19 pandemic interrupted its completion. Outcome assessment was incomplete. Engineering and art students combined their knowledge and skills to produce something outside of their disciplines [74]. Hoople, Choi-Fitzpatrick and Reddy reported a *Drones for Peace* course involving engineering and peace studies students. Students developed several deliverables: videotaped brainstorming ideas for peaceful applications for aerial drones, built an open-source drone, and then developed an entrepreneurial pitch. Student written reflections were a key component [72]. Oehlberg and colleagues reported innovative Human-Centered Design (HCD) courses. A diverse group of students enrolled in courses: engineering, environmental design, and liberal studies. Weekly topics of discussion included failure and ambiguity, sustainability, and humanitarian design. Students also completed short design projects. These courses did not adopt a humanist or scientific perspective but managed to be agnostic with respect to the two domains [95]. These reports illustrate that transdisciplinary methods explicitly integrate disciplines like interdisciplinary methods, which aids student learning. In addition, transdisciplinary methods bridge disciplinary segregation and organizational silos. Innovation is an attribute of transdisciplinary approaches due to the hybridization of disciplinary epistemologies and skills. Transdisciplinary teaching can serve as an effective tool for pivoting. Based upon these reports, the theoretical foundations have been developed for pedagogy facilitating innovative activities and curricula.

14.4.4 Learning Outcomes Assessment

Learning outcomes and outcome assessment remained a weakness for some peer-reviewed publications. The NASEM study stated teaching methods and learning outcomes are often constructed, tested with students, but not evaluated in a systematic fashion. Creativity and innovation can defy evaluation because benchmarks and rubrics may be ill-suited. The sciences and humanities often have different learning outcomes and use different methods. Assessment instruments and evaluation protocols are needed to establish best practices.

Two reports clearly stated key learning outcomes and assessment methods. Rodier and colleagues performed a systematic review of the STEAM literature then applied these results to a course at the University of Ottawa. This course was entitled *STEAM Creation*: twelve studio-art students and twelve engineering students designed interactive art installations focusing on specific environmental issues. The instructors matched activities to learning outcomes. They also identified a theoretical framework for assessment. Unfortunately, the pandemic disrupted the authors' instructional and assessment plans. The authors reported what collected to disseminate their findings [74]. Watkins and Tehrani used Science, Technology, and Society (STS) pedagogy as a model for their work. The professors focused on expanding students' understanding of the historical and sociocultural aspects of science and technology. Watkins assessed students' achievement of this outcome by evaluating creative writing assignments. He determined that ten out of seventeen students expanded their understanding of social aspects of science. Tehrani engaged her students in a thought experiment examining digital immortality. Students wrote an essay exploring "mind uploading." Mind uploading is speculative, but a day may come in which a person's mind could copied and digitized. This would be a form of immortality. She assessed her students' essays and determined that twelve out of eighteen students developed insights into immortality and its implications [89]. These reports provided rich details of their work from which practitioners could glean ideas to develop best practices.

14.4.5 Exemplary Materials for Practitioners

Two groups provided exemplary materials for practitioners. The reports provided extensive literature, theory, methods, and assessment. In addition, the curricula are currently available for practitioners and scholars to download or purchase. Hoople, Reddy, and Choi-Fitzpatrick integrated engineering and peace studies using drones as a "boundary object." Boundary objects are concepts or objects possessing different definitions for different people in different contexts. Aerial drones are commonly used for photography and recreation. The devices have also been equipped with missiles and bombs, then used to kill people. The authors taught a course in which engineering and peace studies students collaborated to develop peaceful applications for the technology. Students recognized the impact of engineering on important economic, environmental, and societal issues. The authors assessed this outcome using qualitative and quantitative methods. Students had positive gains with respect to the impact of engineering on society. They also reported positive outcomes for collaborating with students with a different mindset [72]. Hoople and Choi-Fitzpatrick later developed a pedagogical method to engage faculty to create a more interdisciplinary learning experience [96]. Hammonsa and colleagues developed an interdisciplinary flipped course addressing obesity. Obesity is multifaceted and requires many perspectives to treat effectively. Multiple experts contributed to the curriculum. A survey assessing the interdisciplinary nature of the course was administered at the beginning and end of the course. In addition, the evaluator held focus groups with students. The course was well received by the students, who demonstrated improved competence in interdisciplinary activities [94].

14.5 Conclusions

Interdependence has been a hallmark of human societies. Technological and cultural changes have occurred since the dawn of civilization. Ruth DeFries proposed a model of cultural change in which innovations gave humans temporary advantages, then new problems threatened, and humans adapted again. Societies' pivots required clever individuals, but cooperation and collaboration were needed as well. Suppose James Watt invented parts of the steam engine but did not share his inventions with others.

The Industrial Revolution might have been slowed or even stopped. Humans have relied on others for millennia. DeFries did not discuss the requisites for innovations or pivots. The implementation of these innovations ratcheted up the humans' success until new problems appear. These problems, such as famine, acted as hatchets. Today one observes many hatchets affecting societies: economic and cultural inequities, climate change, environmental degradation, and political instability. Innovative practices are being developed, tested, and then disseminated. Problems have grown to such a scale that individual countries cannot solve them; today's hatchets threatening to destroy human societies. A new social paradigm has appeared. Pivots used to drive cultural change. Steam engines transformed agrarian economies into industrial economies, which transformed cultures. Today's cultures are more resistant to change. Potential technologies can slow or reverse rising greenhouse gas concentration. Vaccines have become more potent tools in preventing diseases. And yet societies resist their adoption. Cultural change has become a prerequisite for pivots. Interdependence has become an existential reality. Scientists and other agents are sources of technological innovation but cannot drive cultural change. Humanists can drive cultural change but are unable to solve scientific or technical problems. Sustainable solutions require scientific and humanist collaborations. Universities serve as incubators for tomorrow's leaders. These institutions play critical roles in today's pivots. Collaborations require complex skills, which are best learned as a student. Integrative pedagogy is predisposed towards cooperation and collaboration because they are requisites for successful courses.

Although universities are the incubators of tomorrow's leaders, integrative education is not the nature of higher education. Most students complete multidisciplinary programs, which consist of checkoff lists. These lists comprise two domains: sciences and humanities. A student's program emphasizes one discipline, such as history, chemistry, or business. Language and literature, mathematics, and science are common courses for all students. Each course uses epistemology, pedagogical methods, and assessment appropriate for the corresponding discipline. Integration is implicit. Students are expected to find connections between their courses and to transfer learning from one discipline to another. A chemistry instructor using reflective journals in an introductory course can expect student resistance. A historian emphasizing science can expect resistance. One study determined that students had difficulty transferring concepts from calculus to physics even though the students recognized that the concepts were interrelated. The connections needed to be explicit for students [97]. Cooperation and collaboration need to be explicit for students, too. Integrated education needs to be explicit.

Universities suffer from organizational silos. Hierarchical divisions tend to divide the organization and personnel into units, such as colleges and academic departments. Colleges consist of similar disciplines, such as chemistry, physics, geology, and biology. Instructors tend to work within departments or colleges. The NASEM study highlighted disciplinary segregation as a major impediment to course integration. Many academic programs must meet accreditation requirements. The exchange of knowledge, skills, and personnel can be constrained due to politics. Traditional practices afford considerable power to academic departments: hiring (and sometimes firing) of faculty, tenure and promotion, course assignments, budgetary support, and even determining whether an individual has the appropriate credentials to teach specific courses. Tenure, promotion, and grants often depend upon publications within one's academic discipline. Interdisciplinary courses exist outside of the department structure, which can discourage faculty participation and make basic administration tasks difficult. If universities are to lead their respective countries and societies towards a more collaborative future, then the institutions need to either remove disciplinary silos or establish bridges between them.

Disciplinary segregation developed over centuries. Medieval universities taught Classical Education, which was a unified curriculum. Geometry and physics were minor components of Classical Education. The Scientific and Industrial Revolutions brought science to the forefront of society. Scientists achieved many breakthroughs, such as atomic theory and laws of motion. Innovations in engineering and medicine, such as steam engines and general anesthesia, have transformed humanity. Science changed universities. Some institutions, such as the Massachusetts Institute of Technology, focus heavily upon the sciences. What about the humanities? There are fewer higher education institutions focusing on the humanities. Humanities have carved out niches for their academic departments. Philosophers, writers, and other humanists maintained a complicated relationship with science. Some drew upon science for inspiration. Rene Descartes and August Comte based their philosophies upon science and mathematics. Other humanists viewed science with scorn or concern. Mary Shelley raised questions about scientific conduct. Leon Wieseltier decried the enslavement of the world by science and technology, but reality contradicts him. Humanists retain considerable cultural power within society and higher education. The They are experts in cultural change. Writers, journalists, filmmakers, philosophers, politicians, and religious leaders are major influencers in society. Scientists, engineers, mathematicians, and medical professors are sources of innovation. Neither scientists nor humanists have primacy in solving humanity's problems. Everyone's ideas and efforts are needed. Effective solutions to sustainability issues exist outside of academic disciplines. Cooperation and collaboration are needed to identify, develop, and then implement these solutions.

A small set of recent publications was reviewed. Each of the articles offers practitioners and scholars ideas for future endeavors. An important issue was identified during literature searches. STEM has been recognized as an acronym for Science, Technology, Engineering, and Mathematics. A search of databases for "STEM education" quickly identifies an enormous list of potential sources. Further refinement of search terms can quickly provide a comprehensive set of literature for a particular aspect, such as STEM education outreach. Integration has too many acronyms proposed by scholars and practitioners, including STEMM, STEAM, HD-STEM, H-STEAM, and STREAM. Perhaps authors want to highlight the innovative nature of their work, but they risk anonymity due to poor dissemination. ERIC does not use any variation of "transdiscipline" as a descriptor. "Integration" is not used, either. Only "Interdisciplinary Approach" is used. It can be difficult for practitioners and scholars to find the latest articles related to integration. A standard set of acronyms, descriptors, and definitions needs to be developed and disseminated.

It is all well and good to espouse collaboration, cooperation, and integration, but difficult work remains. Disciplines have been building epistemological systems for many years. Higher education has encouraged disciplinary segregation though organizational and academic policies. As a result, integration activities require extra effort and entail extra risk. Successful activities occur with regularity in research, teaching and policy. Hope is generated by results. Integration of the humanities and sciences is an imperative for sustainable solutions for many environmental and social problems. Ultimately, we need results and solutions—not just words or hope.

Acknowledgements I would like to acknowledge my colleagues, friends and family who critiqued my chapter: Carolyn Burns, Jim Burns, John Catalano, David Dangerfield, Chris Gecik, Bryan Love, Liliana Mammino, Sarah Miller, and Leslie Scott. They generously shared their time and ideas with me.

I would like to thank Liliana Mammino for our communications and discussions over the years. Her words required me to clarify my thinking. Her actions inspired my actions. I am grateful to have been invited to ACRICE 2017 held at the University of Venda. I am grateful for invitation to contribute to this book.

Annotated Bibliography

- A compilation of peer-reviewed literature associated with the integration of the humanities and sciences.

Arias-Alfonso, A. F.; Franco, C. A. The creative act in the dialogue between art and mathematics. *Mathematics* **2021**, *9*, 1517-1542.

Arias-Alfonso and Franco reported their Interdisciplinary efforts to incorporate the visual arts within secondary mathematics. Arias-Alfonso was a humanities faculty member. Franco was an engineering faculty member. Students used simply apparatus to make drawings: pendulums, Chladni plates, and spinning motor. The authors listed no explicit outcomes. Arias-Alfonso and Franco stated that students need to be able to construct new knowledge using both art and mathematics. The authors interviewed six of the participants as a focus group. The study used interesting interdisciplinary activities involving mathematics and the visual arts, but it was more of a pilot study.

Carrell, J.; Keaty, H.; Wong, A. Humanities-Driven STEM—Using History as a Foundation for STEM Education in Honors. *Honors in Practice* **2020**, *16*, 53–69.

A pilot Humanities-Driven STEM (HDSTEM) course offered by Texas Tech University's Honors College was described: *War, Machine, Culture, and Society: History and Engineering in the Second World War*. The interdisciplinary course was team-taught by an engineer and historian. The authors outlined its curriculum, outcomes, and evaluation. Both STEM and non-STEM students enrolled in the course. Course evaluations, interviews, and surveys showed a positive response by students. Although the authors listed a variety of artifacts used to evaluate course outcomes, only a few sample artifacts were provided. The report provides interesting teaching methods for integrating humanities and sciences. Carrell and colleagues also provided a summary of best practices learned from their experience.

Cronin, C.; Hawthorne, C. Poetry in motion' a place in the classroom: Using poetry to develop writing confidence and reflective skills. *Nurse Education Today* **2019**, *76*, 73–77.

Cronin and Hawthorne reported a multidisciplinary application of poetry to nursing education. Cornin was a healthcare faculty member. Hawthorne worked as a tutor for English and academic skills. Healthcare professionals need effective communication and reflection skills for processing emotions and work experiences. Explicit learning outcomes were not stated. Poetry was treated as a multidisciplinary activity within the nursing program. Students were introduced to poetry in a writing session. They wrote poems in reaction to poems and photographs presented describing the illness and death of a young man. Subsequently, the authors analyzed the student work using thematic analysis looking at emotional content. In addition, student reactions were assessed before and after the writing session. The authors provided qualitative evidence that students demonstrated effective communication and self-reflection skills. Cronin and Hawthorne provided an interesting pilot study integrating poetry and healthcare, which assessed student work in a novel fashion.

Hammons, A. J.; Fiese, B.; Koester, B.; Garcia, G. L.; Parker, L.; Teegarden, D. Increasing undergraduate interdisciplinary exposure through an interdisciplinary web-based video series. *Innovations in Education and Teaching International* **2019**, *57*, 317–327.

The authors developed a transdisciplinary curriculum for obesity education extending from the cellular to social levels. All of the authors are scientists, such as statistics or nutrition. The course involved experts from a range of disciplines, including mass media. The pedagogy utilized flipped classroom methods. Main concepts were presented through web-based microlectures prior to face-to-face meetings. Class time was mostly used for solving problem-based activities. Twenty-six experts from twelve disciplines contributed videotaped lectures and corresponding in-class activities. Student outcomes focused on the differences between disciplinary and interdisciplinary perspectives. Quantitative and qualitative methods were used to assess student learning. The pre-/post-course surveys were collected to measure student perspectives on interdisciplinary activities. Open-ended survey questions and focus groups also gathered qualitative information. This information was used to identify themes. For example, students reported that exploring obesity from an interdisciplinary lens was insightful in addressing this issue. Students gained insights into the nature of interdisciplinary problem-solving. The evidence-based curriculum is posted online: https://stemedhub.org/groups/toprs. The curriculum offers a template for other transdisciplinary topics, such as criminal justice, environmental science and sustainability education.

Hardy, J. G.; Sdepanian, S.; Stowell, A. F.; Aljohani, A. D.; Allen, M. J.; Anwar, A.; Barton, D.; Baum, J. V.; Bird, D.; Blaney, A. Potential for chemistry in multidisciplinary, interdisciplinary, and transdisciplinary teaching activities in higher education. *J. Chem. Ed.* **2021**, *98*, 1124–1145.

The authors prepared a comprehensive review that focused on chemistry's potential for multidisciplinary, interdisciplinary, and transdisciplinary teaching (MITT). The article lists more than 300 citations for additional information. All of the authors were scientists with the majority working as chemists. Hardy and colleagues provided a detailed examination of the practical and theoretical issues associated with MITT. The review emphasized the value of chemistry education using MITT with other science and technology disciplines. The authors mentioned the value of chemistry education to non-STEMM disciplines: arts, humanities and business. Analysis of MITT courses and programs showed that student outcomes, retention, and success improved. The authors emphasized the value of adding chemistry to all disciplines, such as public administration and fashion design. The value of these disciplines to chemistry was not discussed. As a result, the authors provided a useful report for integrative education literature, but it needs corresponding reports from other disciplines. Individuals seeking to integrate chemistry would find this article to be useful.

Hoople, G.; Choi-Fitzpatrick, A.; Reddy, E. Drones for Good: Interdisciplinary Project-Based Learning Between Engineering and Peace Studies. *International Journal of Engineering Education* **2019**, *35*: 1378–1391.

Hoople, Choi-Fitzpatrick, and Reddy reported their efforts to develop a curriculum for transdisciplinary education, which is an amalgam of two disciplines. Hoople is an engineering faculty member. Choi-Fitzpatrick is a political sociology faculty member. Reddy is also a social scientist. The authors developed a course involving both engineering and peace study students. The students work collaboratively using project-based learning to develop peaceful applications for drone technology, such as wildlife conservation. After describing their rationale and background literature, the authors described the course: learning outcomes, activities supporting the outcomes, and then assessment. Student outcomes were divided into three categories corresponding to three segments of the course. The first set of outcomes addressed the processes associated with any interdisciplinary project. A second set of outcomes focused on the specific interdisciplinary project (drones). The final set of outcomes addressed student reflection skills. Classroom activities and projects were designed to "operationalize" these learning outcomes. The authors collected considerable evidence during the course: pre-/post-course surveys, essays, focus groups, reflections, and student projects. Both qualitative and quantitative methods were used to assess the outcomes, which were lumped together into three categories. Results from both qualitative analytic methods and statistical methods indicated that the students achieved the outcomes for the first two categories. Results were ambiguous for the third category, which focused on self-reflection. The article provided an excellent overview of the design and implementation of courses integrating multiple disciplines

Hoople, G. D.; Choi-Fitzpatrick, A. *Drones for Good: How to Bring Sociotechnical Thinking into the Classroom*; Morgan & Claypool: San Rafael, CA, 2020.

Hoople and Choi-Fitzpatrick provided a comprehensive curriculum for developing transdisciplinary courses. Although the authors intend for their book to serve as a detailed blueprint for teaching their course previously described (Hoople et al, 2019), they organized the book so that other instructors could implement their methodology. The book examined many aspects commonly associated with the humanities, such as social change, empathy, and ethics. In addition, the authors provided their course's learning outcomes, lesson plans, and assessment methods and philosophy. This book provided excellent material for developing transdisciplinary courses.

Lewis, H. A. Haiku to enhance student learning: Experiences from a pathophysiology classroom. *Nurse Education Today* **2018**, *60*, 98–100.

Lewis briefly described a set of poetry assignments used in nursing pathophysiology course. The author is a nursing faculty member. Students are introduced to an alternative learning strategy: constructing haikus. She used a multidisciplinary model in which poetry was a separate activity from the pathophysiological material. Students create haikus based upon key concepts related to fluids and blood electrolytes. Student outcome focused on improved content mastery. The author did not report any attempt to assess content mastery compared to haiku quality. She surveyed students to assess their opinions regarding the use of haiku as a learning tool. Lewis' work illustrates that a single instructor can incorporate other disciplines within a course, which can be considerably easier than team-teaching a course.

Lindsay, S. M. Integrating microscopy, art, and humanities to power STEAM learning in biology. *Invertebrate Biology* **2021**, *140*, e12327.

A marine biologist, Lindsay summarized her keynote address to the American Microscopical Society in 2020. She describes a case study integrating biology, visual arts, and creative writing using an interdisciplinary model. A set of activities span the course in which students examine marine algae and invertebrates using stereomicroscopes or compound microscopes. Students would write down their observations, draw sketches, and then use analogies to describe the animals. Student outcomes centered upon the development of skills, such as measuring and observing. Qualitative methods were assessed to assess student performance. The author also cited a variety of other activities integrating the sciences, humanities and arts, such as art gallery shows and film festivals. Although Lindsay did not provide any explicit learning outcomes for the highlighted activities, she provided a pallet of ideas and resources for interdisciplinary and transdisciplinary education.

National Academies of Sciences Engineering and Medicine (2018). *The integration of the humanities and arts with sciences, engineering, and medicine in higher education: Branches from the same tree*; The National Academies Press**:** Washington, DC.

This landmark study summarized the work of a large committee of spanning higher education. The committee investigated the tensions between disciplinary education and integrative education or "liberal education." They noted that disciplinary fragmentation had occurred within the American higher education even as employers wanted graduates to have broad "liberal arts" skills, such as critical thinking and communication skills and the capacity for lifelong learning. The committee also examined issues associated with diversity and equity, such as reasons discouraging minorities and females from entering or finishing science programs. Although a comprehensive review of published literature provided a plethora of information, the authors noted that assessing student learning outcomes was a daunting task due to the diversity of educational institutions, students, academic disciplines, and assessment methods. The committee reported considerable evidence related to the development of integrated education. The authors concluded that considerable effort needs to be exerted to develop and disseminate pedagogy for integrated education. They also concluded that the assessment of this pedagogy needs development. Individuals interested in developing or evaluating integrated courses will find this report to be quite useful. Although the report is less than four years old, it has already been cited more in more than 180 publications in Google Scholar.

Oehlberg, L.; Leighton, I.; Agogino, A.; Hartmann, B. Teaching human-centered design innovation across engineering, humanities and social sciences. *International Journal of Engineering Education* **2012**, *28*, 484.

All of the authors are either engineering faculty members or engineering students. The authors describe a pair of courses for human-centered design (HCD). HCD seeks to solve problems using knowledge and skills from a range of disciplines, including biology, engineering, English, arts, rhetoric, and business. The transdisciplinary approach is open-ended with respect to the number and type of disciplines involved. Two HCD courses were reported. The student-initiated design course sought to recruit and train potential club members. This course was taught by students with faculty oversight. The faculty course was required for a human-centered design (HCD) certificate. Student outcomes for both courses were modest and focused on introducing students to HCD and increasing students' experience in multidisciplinary projects. The authors provided participation demographics and anecdotes. Although HCD offers promising ideas for integration, more curriculum development and assessment were needed.

Perignat, E.; Katz-Buonincontro, J. STEAM in practice and research: An integrative literature review. *Thinking Skills and Creativity* **2019**, *31*, 31–43.

Perignat and Katz-Buonincontro reviewed 44 articles that integrated the sciences and arts. While the scientific disciplines were consistently identified as STEM, Art had no broad agreement. The authors identified three main categories of Art: Arts, any non-STEM discipline, or project-based learning, problem-based learning, technology-based learning, or making. Learning outcomes was an issue in the STEAM articles. The development of creativity was frequently reported by STEAM literature as a student outcome, but no articles report creativity as a measured outcome. In fact, 24 articles failed to describe arts education learning outcomes. Some of the STEAM articles prioritized on STEM education at the expense of the arts. The authors pointed out the original idea of integrating the Arts into the STEM placed arts education as secondary to STEM education. Practitioners, researchers and scholars need to develop a common definition for the arts in STEAM. Pedagogy integrating the arts into STEM needs development. Assessment methods need development, too. Individuals interested in integrating the arts and sciences would find this article useful.

Rodier, C.; Galaleldin, M.; Boudreau, J.; Anis, H.; Peyton, *L. STEAM–Arts Integration Frameworks for Transdisciplinarity*, Proceedings of the 2021 Canadian Engineering Education Association (CEEA-ACEG21) Conference, 2021.

Rodier and colleagues developed a transdisciplinary course in which visual art and engineering students address environmental problems using studio art and engineering design practices. Rodier is an artist. Galaleldin is an engineer. Boudreau is an education faculty member. Anis and Peyton are engineering faculty members. The authors approached this goal in a systematic fashion. First, the authors completed a systematic literature review of STEAM education, conceptual frameworks of creativity, and frameworks for combining art with STEM. Their literature review produced several notable items. First, they compiled key terms, such as "creativity" and "transdisciplinary approach," articles, and then various definitions used in the literature. Second, the authors distilled major rationales for STEAM education: (1) a pedagogy addressing the need to increase student interest and skills in STEM; (2) improving workforce skills, such as creativity, communication, and teamwork; and (3) dissolving discipline boundaries to solve complex problems. Third, Rodier and other provided short summaries of models describing creativity and art combined with STEM (STEAM). After the extensive literature review, the authors described University of Ottawa's STEAM course, which was based upon the Stanford Design Thinking 5 phase model: empathize, define, ideate, prototype, and then test. Art and engineering students worked collaboratively to address a specific environmental issue. Learning outcomes were not specifically listed by the authors, but they inferred the following outcomes in the text: (1) increase diversity and inclusivity within design; (2) learn twenty-first century skills; and (3) solve problems by dissolving discipline boundaries. Unfortunately, the course was disrupted by the pandemic, forcing the authors to convert the course to a fully online. This changed the final deliverables of the student groups. The authors did not provide much evidence from student surveys (taken during the course) or assessing the deliverables. This article provides readers with an excellent method for integrating humanities and sciences, but the results from evaluation methods were scant.

Watkins, A.; Tehrani, Z. Brave New Worlds: Transcending the Humanities/STEM Divide through Creative Writing. *Honors in Practice* **2020**, *16*, 29–51.

Watkins and Tehrani described two interdisciplinary courses using creative writing. Most of the students enrolled in the two courses were science students. Watkins has a doctorate in English literature and teaches composition. He developed a course entitled: Human Redesign, which examined the relationship between literature and science during the 19th century. They read selected reading: Mary Shelley's *Frankenstein*, George Eliot's *The Lifted Veil*, and Robert Louis Stevenson's *The Strange Case of Dr. Jekyll and Mr. Hyde.* Students learned that 19th century authors examined new concepts about being human with respect to 19th century discoveries. Watkins wanted his science students to experience the power of stories through creative writing. Tehrani earned her PhD in cell and molecular biology and teaches molecular biology. She developed an honors seminar titled "Immortality," which looked scientific basis of aging and extending life to extreme lengths. The social implications of extreme longevity were examined, too. The seminar also examined other possible forms of immortality, such as digital immortality though mind uploading. "Mind uploading is the process of constructing a one-to-one model of every neural connection in the entire brain on software such that it behaves essentially the same way as the original brain." Students watched futuristic films using mind uploading. Finally, students engaged in a thought experiment in which they wrote short story examining one idea related to mind uploading. Both courses used creative writing and science in a transdisciplinary fashion.

Students approached scientific concepts from a humanistic perspective and humanistic inquiry can provide creative and critical thinking skills.

References

1. University of South Carolina, *Global Carolina*. https://sc.edu/about/offices_and_divisions/global_carolina/about/index.php. Accessed 20 Aug 2022
2. European Commission, *Erasmus+ Programme Guide: Version 1*, 24 November 2021, https://erasmus-plus.ec.europa.eu/sites/default/files/2021-11/2022-erasmusplus-programme-guide.pdf. Accessed 21 Aug 2022
3. W. James, WHO recognition of the global obesity epidemic. Int. J. Obes. **32**(Suppl 7), S120–S126 (2008)
4. L. MacLean, N. Edwards, M. Garrard, N. Sims-Jones, K. Clinton, L. Ashley, Obesity, stigma and public health planning. Health Promot. Int. **24**(1), 88–93 (2009)
5. M. Geserick, M. Vogel, R. Gausche, T. Lipek, U. Spielau, E. Keller, R. Pfäffle, W. Kiess, A. Körner, Acceleration of BMI in early childhood and risk of sustained obesity. N. Engl. J. Med. **379**(14), 1303–1312 (2018)
6. J. Manthey, S.A. Hassan, S. Carr, C. Kilian, S. Kuitunen-Paul, J. Rehm, What are the economic costs to society attributable to alcohol use? A systematic review and modelling study. Pharmacoeconomics **39**, 809–822 (2021)
7. M. Goodchild, N. Nargis, E.T. d'Espaignet, Global economic cost of smoking-attributable diseases. Tob. Control. **27**(1), 58–64 (2018)
8. Executive Summary: Conclusions and Policty Implications, in *World Drug Report 2018* (United Nations publication, Sales No. E.18.XI.9), https://www.unodc.org/wdr2018/prelaunch/WDR18_Booklet_1_EXSUM.pdf. Accessed 21 Aug 2022
9. S. Ehrenpreis, D.N. Teller, Interaction of drugs of dependence with receptors, in *Chemical & biological aspects of drug dependence*. ed. by S.J. Mulé, H. Brill (CRC Press, Boca Raton, 2019), pp.177–217
10. A. Bechara, K.C. Berridge, W.K. Bickel, J.A. Morón, S.B. Williams, J.S. Stein, A neurobehavioral approach to addiction: implications for the opioid epidemic and the psychology of addiction. Psychol.Sci. Publ. Int. **20**(2), 96–127 (2019)
11. ERIC: Educational Resources Information Center, U.S. Department of Education Website, https://ies.ed.gov/ncee/projects/eric.asp. Accessed 17 Jun 2022
12. R. DeFries, *The big ratchet: How humanity thrives in the face of natural crisis* (Basic Books, New York, 2014)
13. E. Dubé, C. Laberge, M. Guay, P. Bramadat, R. Roy, J.A. Bettinger, Vaccine hesitancy: An overview. Hum. Vaccines Immunother. **9**(8), 1763–1773 (2013)
14. IPCC, Climate change 2014: Synthesis Report, in *Contribution of Working groups I, II and III to the fifth assessment report of the intergovernmental panel on climate change*. ed. by R.K. Pachauri, L.A. Meyer (Geneva, Switzerland, 2014)
15. M. Shelley, *Frankenstein, OR* (The Modern Prometheus; Henry Colburn and Richard Bentley, London, 1831)
16. K. Coleman, *A history of chemical warfare* (Palgrave Macmillan, New York, 2005)
17. R. Carson, *Silent spring* (Houghton Mifflin Company, Boston, 2002)
18. M.K. Murphy, Scientific argument without a scientific consensus: Rachel Carson's rhetorical strategies in the silent spring debates. Argum. Advocacy **55**, 194–210 (2019)
19. F. Dyson, *Disturbing the university* (Basic Books, New York, 1979), p.5
20. L. Wieseltier, Perhaps culture is now the counterculture: A defense of the humanities. *New Repub* **28** (2013), https://newrepublic.com/article/113299/leon-wieseltier-commencement-speech-brandeis-university-2013

21. M.C. Nisbet, D.A. Scheufele, J. Shanahan, P. Moy, D. Brossard, B.V. Lewenstein, Knowledge, reservations, or promise?: A media effects model for public perceptions of science and technology. Commun. Res. **29**, 584–608 (2002)
22. Committee on Professional Training, *ACS guidelines and evaluation procedures for bachelor's degree programs* (American Chemical Society, Washington, 2015)
23. S. Pinker, Science Is Not Your Enemy: An impassioned plea to neglected novelists, embattled professors, and tenure-less historians. *New Repub* **6** (2013), https://newrepublic.com/article/114127/science-not-enemy-humanities
24. S. Grant, K.E. Bouskill, Opinion: Why institutional review boards should have a role in the open science movement. Proc. Natl. Acad. Sci. U. S. A. **116**, 21336–21338 (2019)
25. R.J. Burdge, R.A. Robertson, Social impact assessment and the public involvement process. Environ. Impact Assess. Rev. **10**, 81–90 (1990)
26. C.P. Snow, *The two cultures and the scientific revolution* (Cambridge University Press, New York, NY, 1961)
27. L.A. Real, Collaboration in the sciences and the humanities: A comparative phenomenology. Arts Hum.Ities High. Educ **11**, 250–261 (2012)
28. B. Macfarlane, E. Devine, T. Drake, A. Gilbert, M. Robinson, I. White, *Co-authorship in the humanities and social sciences* (White Paper for Taylor & Francis Group, London, 2017)
29. S. Falk, *The light ages: The surprising story of medieval science* (WW Norton & Company, New York, 2020)
30. H.F. Cohen, *The scientific revolution: A historiographical inquiry* (University of Chicago Press, Chicago, 1994)
31. S. Rabin, Nicolaus Copernicus, 2004 (Revised in 2019). The stanford encyclopedia of philosophy website, https://plato.stanford.edu/archives/fall2019/entries/copernicus/. Accessed 3 Mar 2022
32. W. Bixby, Galileo's fight to reconcile science and faith, in *The scientific revolution*. ed. by M. Young (Greenhaven Press, Farmington Hills, Michigan, 2006), pp.73–82
33. E.O. Wlison, Francis bacon charts the course of science, in *The scientific revolution*. ed. by M. Young (Greenhaven Press, Farmington Hills, Michigan, 2006), pp.104–111
34. H.F. Cohen, The natural sciences and the humanities in the seventeenth century, in *The making of the humanities*, vol. III, ed. by R. Bod, J. Maat, T. Weststeijn (Amsterdam University Press, Amsterdam, 2014), pp.43–52
35. R. Descartes, *Discours de la méthode*, trans. by J. Veitch, Released by Project Gutenberg, 2008 (Bobbs Merrill, New York, 1950)
36. G. Parry, *John locke* (Routledge, New York, 2004)
37. T. Sorell, *Scientism: Philosophy and the infatuation with science* (Routledge, New York, 1991)
38. J. Hoyrup, The Enlightenment, in *Human sciences: Reappraising the humanities through history and philosophy* (SUNY Press, Albany, 2000), pp. 127–163
39. Priestley, J, *The history and present state of electricity, with original experiments, by Joseph Priestley*, England, C. Bathurst, T. Lowndes [etc.] (London, 1775), https://hdl.handle.net/2027/mdp.39015073731369
40. J. Hutton, *Theory of the earth, with proofs and illustrations*, Messrs Cadell, junior, Davies, W. Creech (London, 1795), https://hdl.handle.net/2027/uc1.31822011183134 and https://hdl.handle.net/2027/uc1.31822000672394
41. J. Dalton, *A new system of chemical philosophy* (Bickerstaff, Strand, London, 1827), https://hdl.handle.net/2027/chi.57141596
42. J.H. Williams, Defining and Measuring Nature, in *The make of all things* (Morgan & Claypool, San Rafael, CA, 2014), pp. 3–3.
43. J. Wolfe, *The industrial revolution: Steam and steel* (The Rosen Publishing Group, New York, NY, 2015)
44. P. Deane, *The first industrial revolution* (Cambridge University Press, London, England, 1965)
45. W.W. Hanlon, *The rise of the engineer: Inventing the professional inventor during the industrial revolution*. Working Paper for National Bureau of Economic Research (Cambridge, MA, 2022), http://www.nber.org/papers/w29751.

46. M.S. Larson, *The rise of professionalism: A sociological analysis* (University of California Press, Berkeley, CA, 1977)
47. ACS History, The American Chemical society website, https://www.acs.org/content/acs/en/about/history.html. Accessed 10 Feb 2022
48. P.N. Stearns, The Social Impact of the Industrial Revolution, in *The industrial revolution in world history*; (Routledge, New York, 2018)
49. L. Helene, Chapter 2: social Darwinism: A brief outline of Social Darwinism and its influence on 19th Century Britain and elsewhere. J. Psychohist. **44**, 154–161 (2016)
50. A. Comte, *A general view of positivism* (Reeves & Turner, England, 1880), https://hdl.handle.net/2027/hvd.hn1eej.
51. C. Bohen, The fear and distrust of science in historical perspective: Some first thoughts. in *Fear of science, trust in science: Conditions for change in the climate of opinion*, ed. by A.S. Markovits, K.W. Deutsc (Oelgeschlager, Gunn & Hain, Cambridge, MA, 1980), pp. 29–58
52. S.D. Collingwood, *The Life and Letters of Lewis Carroll (Rev. C.L. Dodgson)* (Century Co, New York, 1899), pp. 187–191, https://hdl.handle.net/2027/uva.x004340084
53. L. Otis, *Literature and science in the nineteenth century: An anthology* (Oxford University Press, New York, NY, 2009)
54. Verne, J. *Twenty thousand leagues under the sea* (First Avenue Editions, Minneapolis, 2015)
55. J. Verne, *The mysterious island* (Open Road Media Integrated Media New York, NY, 2015)
56. K. Popper, *The logic of scientific discovery* (Taylor & Francis, London, England, 2005)
57. T.S. Kuhn, *The structure of scientific revolutions* (University of Chicago Press, Chicago, IL, 1970)
58. D.M. Fogel, E. Malson-Huddle, *Precipice or crossroads? Where America's great public universities stand and where they are going midway through their second century* (SUNY Press, Albany, 2012)
59. B.C. Reed, *Atomic bomb: The story of the Manhattan project. How nuclear physics became a global geopolitical game-changer* (Morgan & Claypool Publishers, San Rafael, CA, 2015)
60. Final Action Completed on Appropriations for FY20, National science foundation website, https://www.nsf.gov/about/congress/117/highlights/cu20.jsp. Accessed 13 May 2022
61. NEH Appropriations History, National endowment for the humanities website, https://www.neh.gov/neh-appropriations-history. Accessed 13 May 2022
62. U. Pawlicka-Deger, Place matters: Thinking about spaces for humanities practices. Arts Hum.Ities High. Educ. **20**, 320–338 (2021)
63. National Academies of Sciences Engineering and Medicine, *The integration of the humanities and arts with sciences, engineering, and medicine in higher education: Branches from the same tree* (The National Academies Press, Washington, DC, 2018), pp.63–64
64. T.R. Kurtzberg, Feeling creative, being creative: An empirical study of diversity and creativity in teams. Creat. Res. J. **17**, 51–65 (2005)
65. D. Hocevar, Intelligence, divergent thinking, and creativity. Intell. **4**, 25–40 (1980)
66. Z. Yang, I.W. Hung, Creative thinking facilitates perspective taking. J. Pers. Soc. Psychol. **120**, 278–299 (2020)
67. J. Cromity, U. De Stricker, Silo persistence: It's not the technology, it's the culture! New Rev. Inf. Netw. **16**, 167–184 (2011)
68. E.B. Cohen, S.J. Lloyd, Disciplinary evolution and the rise of the transdiscipline. Informing Sci.: Int. J. Emerg. Transdiscipl. **17**, 189–215 (2014)
69. National Academies of Sciences Engineering and Medicine, *The integration of the humanities and arts with sciences, engineering, and medicine in higher education: Branches from the same tree* (The National Academies Press, Washington, DC, 2018), pp.23–24
70. D. Sharma, B. Eng, A. Sharma, STEAM and sustainability: Lessons from the fourth industrial revolution, in *Handbook of Research on Integrating ICTs in STEAM Education* (IGI Global 2022), pp. 390–407
71. J. Carrell, H. Keaty, A. Wong, Humanities-driven STEM—Using history as a foundation for STEM education in honors. Honor. Pract **16**, 53–69 (2020)

72. G. Hoople, A. Choi-Fitzpatrick, E. Reddy, Drones for good: Interdisciplinary project-based learning between engineering and peace studies. Int. J. Eng. Educ. **35**, 1378–1391 (2019)
73. J.G. Hardy, S. Sdepanian, A.F. Stowell, A.D. Aljohani, M.J. Allen, A. Anwar, D. Barton, J.V. Baum, D. Bird, A. Blaney, Potential for chemistry in multidisciplinary, interdisciplinary, and transdisciplinary teaching activities in higher education. J. Chem. Ed. **98**, 1124–1145 (2021)
74. C. Rodier, M. Galaleldin, J. Boudreau, H. Anis, L. Peyton, *STEAM–Arts Integration Frameworks for Transdisciplinarity*, Proceedings of the 2021 Canadian Engineering Education Association (CEEA-ACEG21) Conference, 2021
75. C. Cronin, C. Hawthorne, Poetry in motion' a place in the classroom: Using poetry to develop writing confidence and reflective skills. Nurse Educ. Today **76**, 73–77 (2019)
76. C.D. Trott, T.L. Even, S.M. Frame, Merging the arts and sciences for collaborative sustainability action: A methodological framework. Sustain. Sci. **15**, 1067–1085 (2020)
77. M.D. Begg, R.D. Vaughan, Are biostatistics students prepared to succeed in the era of interdisciplinary science? (And how will we know?). Am. Stat. **65**, 71–79 (2011)
78. National Academies of Sciences Engineering and Medicine, *The integration of the humanities and arts with sciences, engineering, and medicine in higher education: Branches from the same tree* (The National Academies Press, Washington, DC, 2018)
79. B. Allina, The development of STEAM educational policy to promote student creativity and social empowerment. Arts Educ. Policy Rev. **119**, 77–87 (2018)
80. Changemaking Engineers in Action, Shiley-Marcos school of engineering website, https://www.sandiego.edu/engineering/student-innovation/changemaking-engineers.php. Accessed 15 May 2022
81. National Academies of Sciences Engineering and Medicine, *The integration of the humanities and arts with sciences, engineering, and medicine in higher education: Branches from the same tree* (The National Academies Press, Washington, DC, 2018), p.122
82. Humanities Connections, National endowment for the humanities website, https://www.neh.gov/grants/education/humanities-connections. Accessed 16 May 2022
83. Building Lasting Bridges: German and Engineering. National Endowment for the Humanities (Washington, DC, 2020), https://www.neh.gov/sites/default/files/inline-files/Auburn_Sample.pdf
84. NEA Grant Funds Sustainability STEAM Education, Fashion Institute of Technology (FIT) (New York, 2021)
85. Integrative Courses Compendium, National Academies of Sciences Engineering and Medicine Website. https://app.smartsheet.com/b/publish?EQBCT=ea0a8d78cd4948e798fcf3fbb2d4382d. Accessed 21 July 2021
86. C. Yanniris, Education for sustainability, peace, and global citizenship: An integrative approach. Educ. Sci. **11**, 430 (2021)
87. E. Perignat, J. Katz-Buonincontro, STEAM in practice and research: An integrative literature review. Think. Ski. Creat. **31**, 31–43 (2019)
88. S.R. Januchowski-Hartley, N. Sopinka, B.G. Merkle, C. Lux, A. Zivian, P. Goff, S. Oester, Poetry as a creative practice to enhance engagement and learning in conservation science. Biosci. **68**, 905–911 (2018)
89. A. Watkins, Z. Tehrani, Brave new worlds: Transcending the Humanities/STEM divide through creative writing. Honor. Pract. **16**, 29–51 (2020)
90. S.M. Lindsay, Integrating microscopy, art, and humanities to power STEAM learning in biology. Invertebr. Biol. **140**, e12327 (2021)
91. H.A. Lewis, Haiku to enhance student learning: Experiences from a pathophysiology classroom. Nurse Educ. Today **60**, 98–100 (2018)
92. C.M. Gillen, A.J. Kerkhoff, D.H. Lynn, H.G. McFarlane, A.J. Niemiec, S.C. Petersen, A.D. Reach, Does creative writing improve scientific writing and learning? FASEB J. **34**(S1), 1–1 (2020)
93. A.F. Arias-Alfonso, C.A. Franco, The creative act in the dialogue between art and mathematics. Mathematics **9**, 1517–1542 (2021)

94. A.J. Hammons, B. Fiese, B. Koester, G.L. Garcia, L. Parker, D. Teegarden, Increasing undergraduate interdisciplinary exposure through an interdisciplinary web-based video series. Innov. Educ. Teach. Int. **57**, 317–327 (2019)
95. L. Oehlberg, I. Leighton, A. Agogino, B. Hartmann, Teaching human-centered design innovation across engineering, humanities and social sciences. Int. J. Eng. Educ. **28**, 484 (2012)
96. G.D. Hoople, Choi-Fitzpatrick, A. *Drones for Good: How to Bring Sociotechnical Thinking into the Classroom* (Morgan & Claypool, San Rafael, CA, 2020)
97. L. Cui, N.S. Rebello, P.R. Fletcher, A.G. Bennett, Transfer of learning from college calculus to physics courses, in *Proceedings of the Annual Meeting of the National Association for Research in Science Teaching* (San Francisco, CA, 2006), 3–6

Chapter 15
Teaching Normative Literacy in STEM Classrooms: The Essential Role of Storytelling, Fiction, and Critical Thinking

Jan Mehlich

Abstract This chapter argues for the usefulness and efficiency of responsible research & innovation (RRI) courses in tertiary STEM education that put a focus on normative argumentation in academic, corporate and public innovation discourses. These skills of the future generation of scientific experts and innovation drivers, here framed as normative literacy, are expected to play a crucial role for sustainable development and innovation with beneficial impact on society, economy, and environment. A course that has been conceptualised, designed, and offered to chemistry students by the author is described and its outcomes are assessed. It is built largely on conceptual frameworks and models from normative and social disciplines and familiarises the students with the modes and cognitive attitudes of these fields. According to student feedback, they assess their ability to engage in normative discussions of their research work and their confidence collaborating with a wide range of actors on normative decisions in the context of research and innovation as significantly increased. This result is taken as a strong argument in this chapter for spending greater efforts on more interdisciplinary education of STEM students who will benefit greatly from their encounter with normative sciences and the humanities in their later jobs as innovators and responsible actors in the context of scientific and technological progress.

Keywords Normative literacy · Interdisciplinarity · Education for responsibility · Responsible research and innovation (RRI)

J. Mehlich (✉)
Center for Life Ethics, Rheinische Friedrich-Wilhelms-Universität Bonn, Bonn, Germany
e-mail: jmehlich@uni-bonn.de

L. Mammino (ed.), *Insights into the Relationships between Humanities and Sciences*,
https://doi.org/10.1007/978-3-031-97900-2_15

15.1 Introduction

Imagine innovation processes that involve, amongst other contributions, the expertise of scientists, researchers, engineers, and technical designers: a chemist working in the research and development (R&D) department of a company that produces sustainable agriculture products; a computer scientist developing diagnostic AI software for a medical technology company; a robotics engineer designing an exoskeleton that may help paralysed people walk again. In all cases, these technical experts make choices among several possible options concerning the direction the innovation takes. Some of these choices are assessed with scientific, technical, or numerical means: life cycle assessments, risk assessments, technical feasibility, economic competitiveness, existing expert knowledge. Yet, all the choices have elements that cannot be settled with scientific-technical facts, numbers, or advanced experimentation. What something *is good for*, what is *desirable* or *preferable*, what serves a particular social or ethical *value* (for example, health, safety, integrity, justice, freedom, sustainability, etc.), and why one value is *prioritised* over another, are *normative* judgments that are an inherent and inevitable part of design decisions in socio-techno-scientific innovation processes [1–3]. In terms of the examples above, the chemist might notice that her definition of sustainability differs drastically from the management's idea of it; the computer scientist's AI configuration breakthrough is rejected because of data privacy regulations in medical diagnostics contexts; or the roboticist learns that his safety-by-design concept ignores the mental well-being of the wearers of exoskeleton protheses when not considering aesthetic design features such as the invisibility of joints, bolts, and screws.

Science, technology, engineering, and mathematics (STEM) faculties at universities and colleges educate their students in the state-of-the-art of their academic and applied discipline, of course. Yet, in their later careers, as scientists, engineers, innovators, developers, etc., one of their key responsibilities is making the *right* choices and decisions in the context of their research, innovation, or regulation activity [4]. The competence to fully survey and judge what is *right* is not acquired through the successful study of physics, chemistry, biology, math, informatics, engineering, and related subfields [5]. At the same time, as this chapter will illustrate, such judgments can neither be delegated entirely to other stakeholders of the innovation process, nor can they be made with mere common sense or intuition in a satisfying, plausible, and scrutiny-withstanding manner. As participants of an interdisciplinary innovation team—it is assumed, here, that all innovation processes including academic research, corporate R&D, and techno-scientific governance are teamwork—, scientific-technical experts contribute factual knowledge and technical competence concerning the issue at hand. This input, however, is always set into perspective of goals, purposes, ends, or values [6]. Experiences show that innovation efforts have a higher chance of *value co-creation* (for example, functionality, utility, profitability, sustainability, personal and social integrity, ethical acceptability) when the contributing experts exhibit a strong normative judgment competence [7].

The idea of scientific and technical experts with the ability to see the larger picture of their work and to make competent informed decisions has been framed and communicated by many approaches such as post-normal science and technology (PNST) [8], *X*-by-design (with *X* being safety, ethics, sustainability, or other norms or values considered important) [9, 10], open innovation [11], responsible research & innovation (RRI) [12], or constructive technology assessment [13]. Some of these are rather theoretical, others focus on pragmatic-practical aspects. All share the overarching goal of guiding technological progress into a direction that is beneficial for society and environment. While the concepts are around for decades, the educatory efforts to train innovators (here: scientists, engineers, tech designers) in RRI, ethical judgment, or interdisciplinary discourse performance are still in their infancy and seldom leave their own disciplinary boundaries (for example, courses on PNST in sociology departments, philosophy and ethics of science in philosophy departments, or RRI courses in technology assessment institutes) [14].

In the following, it will be argued that normative-ethical competences can and should be trained during the STEM education at universities. Whereas STEM students learn *scientific information literacy* in their main curricular courses, they learn *normative literacy* in courses on science, technology, innovation, and engineering (STIE) ethics that cover aspects of scientific integrity and societal impact of innovation. After the elaboration of a clear definition of normative literacy, the chapter will explain the role of the Humanities for teaching it. A course that has been conceptualised and taught based on these considerations is, then, described and evaluated in order to show that normative literacy is teachable, useful, and effective in view of the goals of RRI.

15.2 What is *Normative Literacy*?

Normative literacy is the competence to make plausible and scrutiny-withstanding judgments concerning evaluative aspects of a situation, development, process, decision, or action (performed or intended). Such judgments assess *what ought*: a norm, moral, law, rule, preference, priority, or maxim. Knowledge of *what is*—facts, scientific insights, descriptions—is insufficient for making conclusions on *what should* [4]. In many contexts, we use common sense to make normative judgments, for example the culturally embedded and manifested moral commitments that members of a society share. The more complex the context of a judgment, for example the choice between several alternative designs of a technical innovation, the more sophisticated the demands on the normative judgment competence of the decision-maker [15].

Normative is, in several respects, the complement to *factual*. Factual knowledge is validated by scientific means along the lines of some concept of truth while correct normative judgments—the term *knowledge* seems inappropriate here—are plausible logic deductions from some kind of basic conviction, commitment, or belief that may be rooted in non-naturalistic realms such as culture, society, history, or humanity.

Factual discourses may be settled by agreement on methodologies of verification, by setting-into-perspective with other relevant knowledge, or by studying broader and deeper; normative discourses start from convictions and settle with conclusions that are both validated by mutual agreement, democratic methods, or pragmatic reason. Yet, both share similarities: Both the acquisition of specific or complex knowledge and the elaboration of normative insights benefit from respective expertise. Both are cognitively—that means, with rational reason—accessible and assessable. It is a misconception that normative—for example, ethical and moral—aspects of an argument or a decision are somehow random, a matter of personal taste or opinion, mere intuition or feeling, or intangible for rational assessment [16].

What is required from a decision-maker being competent in normative judgment? The following non-exhaustive list explains important elements of normative literacy:

- *Awareness of normativity as such:*

For many decision-makers in innovation contexts, the difference between facts and norms is not clear. One of the first steps towards normative literacy is, thus, the awareness that norms (including morals, ethical judgments, laws, perhaps aesthetic judgments) have characteristics that are different from descriptive facts [17]. For example, the decision to equip a novel diagnostic imaging device with data encryption software for the safety and privacy of patients' health information is not justified by the fact that a majority of citizen values privacy (as, for example, surveyed in a sociological poll), but by the innovators' assessment of the social and ethical implications of how the new device changes medical practices in the future. The descriptive knowledge about norms and values is only a small part of the argument. More importantly, proponents of particular norms and values or their prioritisation need to provide plausible arguments (rather than facts) that explain and justify their judgments. Innovators who know that prescriptive conclusions (what *should*) need the input from facts (what *is*) and norms (what *ought*), and that both need to be circumstantiated and justified, will engage in innovation discourses with a different attitude.

- *Knowledge of values, their justification, and their relations ('orientational' knowledge):*

Scientific researchers and engineers often assess the product of their activities (knowledge, artefacts, processes) in terms of utility (for example, gain or improvement of a function, enabling of procedures and follow-up applications, risks), and sometimes, especially in corporate R&D, in terms of profitability (economic competitiveness, marketability, revenue streams, etc.). Common values that guide strategic and conceptual choices are safety, sustainability, health, or wealth. In science-intense innovation fields with high expected social and/or environmental impact such as nano- and biotechnology, digitalisation including AI and internet of things, medical technology, cognitive sciences, etc., a much wider spectrum of norms and values is affected by the innovation choices [18, 19]. These include privacy, autonomy, freedom, justice, equity, and, in environmental contexts, biodiversity, recoverability and health of the eco-systems, or local and global climate justice, amongst others.

It is impossible to provide a complete list, here, because the norms and values vary with the type and field of innovation that is assessed. Orientations for engineers and innovators have been proposed, defended, and communicated in the past decades, for example the influential *value octogon* of the German Association of Engineers (Verband Deutscher Ingenieure, VDI) [20]. It is exactly the normative literacy skill that enables the innovator to construe and comprehend the norms and values that are relevant for the respective context.

More important than the norms and values themselves are the strategies with which they are justified and weighed against each other, if necessary. For example, a safe design of a service robot might stand in conflict with its profitability; or its function (say, a voice command system that requires constant online connection) raises legal or ethical concerns about privacy issues. It would surely be asked too much from scientific-technical experts to provide sophisticated reasons for priorities in the way that ethical experts could. Rather, the normative literacy is manifested in the competence to relate value considerations to different design alternatives (of artefacts, of research plans, of solutions for problems) and inform the choice with that *orientational knowledge* [21].

- *A holistic stance and vision:*

Normative literacy requires from the decision-maker to step outside of disciplinary and other intellectual boundaries and make mental space for a broad perspective and pluralistic vision of possible futures. What might sound a bit cryptic is, after all, a down-to-earth pragmatic attitude. The following example illustrates the difference between an innovator with a narrow view and one with a holistic vision (taken from [22]): For a pill-sized lab-on-a-chip system that travels through a patient's digestive system as a colon cancer screening tool, a researcher has a choice between an implemented RFID device to transmit positive tumour detections to an external device, or a dye capsule that bursts in case of a positive detection and allows for visual detection in the toilet. With a purely scientific-technical approach, the researcher would choose after assessing the occurrence of false positives and false negatives which is considered a scientific method of risk assessment. Yet, a broader perspective reveals that the design choice has significant implications for the mass-production and dissemination of the pill, the patients' acceptance of the screening method, the doctor-patient relationship, and the assessment of the device's legal classification concerning patient data treatment. Normative literacy, here construed as an overview of norm-related implications of design decisions, supports the researcher in making choices that result in the realisation of the envisioned benefits of the undertaken innovation efforts. This is in line with the image of normative literacy as a prism that refracts the components of a conflict into a spectrum of views and options, contrasted with the mistaken view of normative literacy as a lens that focuses on the one right solution for a problem (Fig. 15.1).

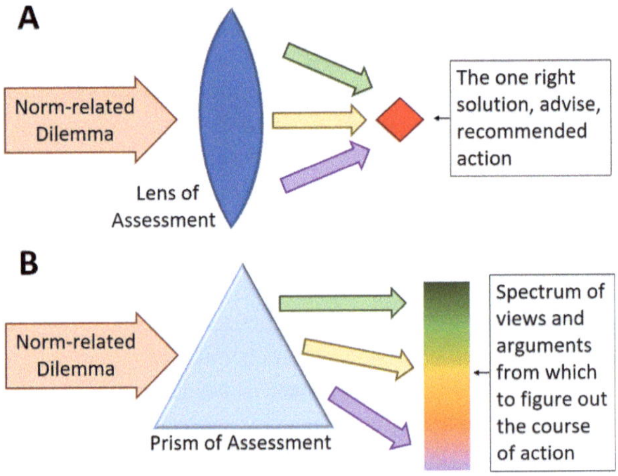

Fig. 15.1 A: Normative literacy as a focussing lens. B: Normative literacy as a refracting prism

- *Critical thinking:*

Critical thinking is the competence to comprehend, analyse, and refine own and other people's arguments and viewpoints including the cognitive processes that led to these statements [23]. A critical thinker is trying to fulfil standards of adequacy and accuracy appropriate to the thought process, for example avoidance of biases and fallacies, logic coherence and consistency, devotion to truth (however conceived), and careful goal-directedness. This attitude is of utmost importance especially in the context of normative reflection exactly *because* orientational knowledge is non-textbook knowledge and cannot be validated in the same way as scientific insights.

- *Discourse skill:*

Discourse performance is the applied communicative aspect of critical thinking. Decisions in innovation but also in any other context are best when they went through a scrutinization procedure. This scrutiny is only possible with communication between the participants of the innovation process. In the same way as knowledge reporting is an element of information literacy, normative discourse and argumentation are elements of normative literacy. Skills, in this context, range from language proficiency (speaking/listening in oral communication, reading/writing in written exchange) to logic and reason to interpersonal and intercultural competence (*knowing when to say what and how*). Convincing others of one's own convictions is one goal of discourse, of course; yet, admitting that others have the more plausible arguments and changing one's own stance is another important discourse skill.

The character of normative insights as explained above—situation- and context-dependent, often new and unprecedented in innovation contexts, participative, democratic, deliberative—requires reflection and decision-making strategies that are different from technical assessments. It is impossible to apply simple or even complex schemes and protocols to solve a normative issue. At the same time, norm-related choices should not be random, the personal taste of the innovator, or—at worst—ignored or underestimated. If we take RRI approaches serious, the medical technologist must consider the relation between her design choices and normative dimensions such as data safety, privacy, gain of life quality, gain of functionality, or impact on the health care system. The sustainable chemist must be able to align research directions with plausible and holistic conceptions of sustainability as a value framework. The AI programmer must be able to explain to his company's executive board how the software of the self-driving vehicle fulfils safety requirements and other legal regulations.

It is important to note, though, that the scientific-technical experts in innovation processes are not required to be, at the same time, social or ethical experts [24]. Innovation teams are, usually, interdisciplinary [25]. Normative choices are made after deliberative discourses that take place formally or informally throughout the process. Often, for example in EU-funded research consortia or in *open innovation* processes, the innovators are accompanied and supported by social scientists, legal experts, or ethicists who contribute input concerning the *ethical, legal, and social implications* (ELSI) of the envisioned innovation [26]. In the past decades, this mode of *post-normal science* (and innovation) has brought about effective and practicable ways of advancing innovation processes into directions that deserve the labels beneficial, responsible, or sustainable [27]. Yet, many of these approaches play out their potential only under cooperation with scientific and technical experts that have a basic understanding of the normative dimensions of their activities [28]. A constructive discourse on innovations and their social and environmental impact needs scientists, researchers, engineers, and designers with normative literacy [29].

15.3 The Role of the Humanities for Normative Literacy Training

Normative choices in the context of research, innovation and design are best facilitated with competences that are trained with methods from the Humanities disciplines (literature, history, philosophy and applied ethics, arts). Among these are the interpretation of case studies including the transfer of the insights to other situations, the extraction of normative implications from fictional settings and application of those to real-world contexts, discourse performance trained via storytelling and role-playing, and critical thinking including argumentation strategies and the assessment of own and other's viewpoints and commitments.

As outlined above, normative literacy is the visionary anticipation of the ethical, legal and norm-related implications of one's own or others' decisions and actions. This vision requires more than mere knowledge of values and norms. It is framed as a practical competence of reasoning, scrutinising, validating, and communicating normative judgments. Training it, thus, must include practical exercises in scenario analysis, normative judgment, argumentation, communication, and discourse. Moreover, understanding interrelations of meaning ascriptions in a larger picture (see above) is a matter of analysis and interpretation of statements, claims, expressions, and imagery. These are the domains of the Humanities. Cases and scenarios employed in normative literacy classes are chosen and discussed in view of their potential to foster the students' ability to recognise the relevant normative implications, make valid judgments, and transfer the acquired insights onto the real-life cases of their intra-disciplinary activities. The following list presents some examples:

- Classical literature: Famous stories, novels, plays, or legends are well-suited to illustrate the consequences of scientific progress and technological development, and the role of scientists, engineers, and other developers in these processes. *Pandora's Box*, *The Flight of Icarus*, *Prometheus*, or Mary Shelley's *Frankenstein* are powerful depictions of the interplay between intentions, goals, biases, decisions, and actions in the context of science and technology. Ought we do what we can? Are good intentions enough to justify a decision, even if they are limited or mistaken? Are there research or innovation directions we better stay away from? Who is responsible for unintended and unanticipated effects of techno-scientific progress?
- Modern literature: Science-fiction and technology-related dramas and thrillers, often well-known to the students, put novel or futuristic technologies into the context of society, governance, or individual fate. For example, Michael Crichton's *Prey* is an efficient trigger of a discussion about the chances and risks of nanotechnology, or Daniel Suarez's *Daemon* and *Freedom* duology surveys a wide array of questions arising from progress in the fields of computer science, cognitive sciences, artificial intelligence, digitalisation, and industry 4.0.
- Movies: The implications of scientific and technological innovation are effectively illustrated in popular movies, too [30]. Examples include the *Terminator* series (robotics, corporate power), *Minority Report* (artificial intelligence), *Gattaca* (biotechnology, genetics), but also classical movies like *The Man in the White Suit* (1951, innovation impact and resistance of other stakeholders) or *Westworld* (1974, robotics).
- Historical and contemporary cases: Applicable insights can also be gained from the study and interpretation of real cases in the history of science and technology development. A look at the Manhattan Project, for example, reveals important lessons concerning the responsibility of scientists and the non-neutrality of all scientific and technological activity. Other famous cases are the development of Agent Orange (unintended use of chemicals), the Tuskegee experiments (medical studies with humans), the Baltimore case (scientific integrity), or the long-term implications of Fritz Haber's and Carl Bosch's discovery of the nitrogen fixation process.
- Discourse and speech situations: There are many documented cases of scientists and engineers in historically important discourse situations, such as the silenced engineer in the space shuttle *Challenger* accident, climate scientists at UN panels, or confrontations between animal rights activists and researchers doing animal

experiments. The analysis of such communicative processes can help students practice their own discourse skills and give them the confidence to engage in similar discourses in their later career.

The main goal of discussing these cases is the facilitation of students' hermeneutic competences [31]. Hermeneutics concerns the interpretation and *viabilisation* (to make something *viable* = to unleash something's potential to have an effect, application, or realisation) of texts (spoken or written), media, historical incidents, etc. In view of the intended training goal, it would be insufficient to merely present the stories and cases and teach their quintessence as normative knowledge. Students need to engage with the content in a way that the interconnections with their own present or future work become clear, for example the attitude with which to discuss with a superior or a non-expert stakeholder, the incorporation of a value proposition into an innovative design, or the defence of a guideline of good scientific or engineering practice. The elicitation and comprehension of meaning from stories, case studies, thought experiments and historical narratives is best facilitated by group discussions, creative exercises such as drawing or text production, role-playing, and practical discourse scenarios, all guided and observed by a competent instructor who gives feedback and helps the students out of dead-ends and loops.

STEM students tend to believe that a conclusion or solution requires the identification of a validated and true statement. They may feel unsatisfied or confused when a case discussion does not yield a generalisable and universalisable 'end point' or a tangible strategy that can be applied like a tool to other cases. Indeed, many normative discourses remain open-ended in the sense that solutions and agreements may have to be scrutinised and re-assessed again in the future. A course with elements of normative literacy should familiarise these analytic-empirically minded learners with the methods of hermeneutics as cognitive and rational processes that, despite being soft or non-axiomatic, rely on reason and are far from random or a matter of personal taste. Practical discourse exercises with science-, technology-, or innovation-related cases can illustrate this difference effectively. The Humanities, in this context, provide orientations and guidance for students in their assessment of own and other's arguments, standards for differentiating an argument from a mere opinion, and the conceptual framework for normative judgments.

15.4 Putting It to the Test: A Course on Normative Literacy for Chemistry Students

The above considerations found their way into the syllabus design of a course *Good Chemistry—Methodological, Ethical, and Social Dimensions* that the author conceptualised and realised as a *massive open online course* (MOOC) on behalf of the European Chemical Society (EuChemS). It comprises of 16 classes that each contain of 1 h video lecture, exercises, group discussions, assignments, further study material (texts, websites, videos, etc.), and quizzes to assess the learning progress, exploiting the technical possibilities of the e-learning platform *Moodle*. A related textbook provides additional in-depth information and exercise opportunities [32]. The completion of the entire course earns the student 2 ECTS credits. The course is offered since fall 2018 to all chemistry students in countries whose national chemical societies are EuChemS members [33].

The course aims to equip the attendees with competences and skills in basic research methodology and its philosophical foundations on the one hand, and in overseeing, understanding, evaluating, and assessing contemporary ethical and social issues arising from scientific and technological (S&T) activity and progress on the other hand. The course is designed and planned in particular for chemistry students and their related fields, requiring no background knowledge in philosophy, ethics, sociology or history of science. The course content is strongly related to the students' daily research activity: Science conduct, logic and theory of science, experimentation, writing publications, dealing with uncertainty, social impact of scientific and innovation activity. The overall objective of this course is to contribute to a more complete education of young researchers and scientists as important enactors of progress and influential decision-makers in the future. It shall provide them with normative literacy as the skill to reflect on and deal with the major contemporary challenges in society and environment so that their choices and actions contribute efficiently to the larger goal *responsible research & innovation*. These skills shall be applied in all kinds of chemical professions and not just in academic environments at the student level. The course content, thus, covers three important domains of chemical expertise and activity: Academic science, private-sector industrial R&D and innovation, and public service.

As the course title suggests, the course is divided roughly into three parts (see Table 15.1): Chemistry is good—or: done well—on the *methodological level* when the scientist understands and correctly applies the practical and theoretical competences of scientific agency [34]. A chemist is good in terms of *research ethics* (internal responsibility) when he or she complies with the guidelines of good professional conduct (as in, amongst many other works, [35]). On the level of external responsibility, chemistry is good when its outcome is beneficial for *society and environment* [36]. While the first part is mostly informed by science theory, epistemology, the philosophy of science, and the respective practical methodologies of scientific agency, and the second part is a matter of compliance with moral common-sense commitments, the third part requires ethical reasoning and argumentation skills and an awareness of the societally embedded and justified role of scientific expertise. On a side note, the combination of these three topics in one course also follows a strategic rationale: Promoters of such courses usually face reluctance of STEM faculties to provide space in the tightly organised curriculum of their programs. A comprehensive syllabus addressing all three obviously relevant fields of responsible research in one course increases the acceptance significantly.

All classes contain of exercises and case explorations that require the students to make value- and norm-related judgments. The strategies for inferring and validating these judgments are informed by principles of applied ethics [37], discourse conduct [15], history of science [38], and sociology [39]. It proved useful to let students

Table 15.1 Overview of the content of the 16 classes of the MOOC *Good Chemistry* as offered on the EuChemS e-learning platform

Part	Class #	Class title	Content and key themes
	1	Introduction	What is professionalism? When is chemistry (a) good? Interplay of facts and norms. Discourse skills. Rationale and structure of the course
Methodology	2	Science Theory	Epistemology, construction of scientific knowledge, power and limits of scientific methods
	3	Scientific Method(s)	Scientific research, elements of every scientific research project, the Scientific Knowledge Acquisition Web
	4	Scientific Reasoning	Logic, biases and fallacies, heuristic and methodological analysis, statistical analysis, frequentism vs. Bayesianism
Good scientific practice	5	Scientific Misconduct	Virtues of scientific practice, the ethos of science, fabrication and falsification of data, reasons and prevention of misconduct
	6	Scientific Publishing	Authorship, citation practices, peer review, impact factors, predatory journals, publishing controversial research
	7	Scientific Teamwork	Chemistry as a network activity, multi-, trans- and interdisciplinary collaboration, mentorship,
	8	Conflicts of Interest, Academic Freedom	Chemistry in academia and in the private sector, intellectual property, policy and governance impact on science
	9	Animal Experiments	Ethical perspectives, facing activism, regulations, 3R
Social and environmental impact	10	Science and Values	Chemistry as social sphere; Chemistry as science and/or technology driver; Neutrality theses; Dual use; Social constructivist views
	11	Sustainability	Concept and definition of sustainability; Sustainability as a normative framework for S&T discourse; sustainability as value co-creation
	12	Responsibility	Four dimensions of responsibility (who is held responsible, by whom, for what, in view of what rules or knowledge?); types of responsibility (legal, social, organisational, moral)
	13	Risk, Uncertainty, Precaution	Definitions of risk & uncertainty; Risk assessment; Risk management; Risk discourse types; Precautionary principles

(continued)

Table 15.1 (continued)

Part	Class #	Class title	Content and key themes
	14	Science and Technology Governance	The role of scientific expertise in S&T policy and governance; technology assessment; EU frameworks (ELSI, RRI, 3O); knowledge reporting
	15	Public Communication of Science	Science and mass media; Scientists and science journalism; Public participation in S&T discourse;
	16	Course Summary	Taking responsibility = taking the right action; Good chemistry as a discourse skill

engage in a warm-up example case, explore their state-of-the-art knowledge and skills, proceed to knowledge and practical approaches for solving comparable issues, and finally test their progress with a comprehensive final case that, by comparison with their performance in the introductory case, illustrates the acquired skill. This problem-based learning concept is in line with an inductive teaching approach. It should be noted that the goal of the lectures and course activities is the development of normative literacy as a practical skill and as an aware mindset of the scientific expert. In terms of Bloom's taxonomy, this requires reaching at least level 5 (creation). The conceptual foundations of the course content—philosophy of science, applied ethics, and RRI—are not taught theoretically as such, but are manifested in concrete exercises and real-world situations of science and innovation activity.

The following overview lists the most prominent exercises that were employed in the three course sections and conducted in the sense of the abovementioned considerations concerning hermeneutics, that means, with case analysis, interpretation, meaning extraction, and transfer of the quintessence to contemporary issues of chemistry and its output's impacts (real cases marked with *).

- *Methodological dimension of good chemistry*: Alchemy vs. Chemistry, Wilhelm Ostwald's monism*, Trofim Lysenko's biochemistry dogma*, Thomas Kuhn's scientific revolutions*, alloy knowledge of bronze age smith vs. modern chemistry student, defence of umbrellaology, Edward Calabrese's hormesis claims*.
- *Research ethics dimension*: The Baltimore case (O'Toole vs. Imanishi-Kari)*, Baconian Oath (a possible ethos for scientists), the Bengü Sezen case*, Millikan's oil drop experiment*, publication of horsepox virus fabrication*, animal experimentation opponents' arguments.
- *Social and environmental impact*: The Manhattan project*, Arthur Galston's discovery of 'Agent Orange'*, research on explosives and chemical warfare agents*, Mengzi's sustainability definition (classical Chinese philosophy)*, chemical leasing as sustainable business model*, REACH's sustainability approach*, Frans van Anraat's complicity in war crime*, the UN convention

on persistent organic pollutants*, nanoparticles in sunscreens*, scientific policy-advise on plastic pollution, various examples of poor science journalism*, various examples of expert-layman consultations.

After each semester, student feedback is collected in a survey. Attendees completing the course are requested to evaluate the course with a questionnaire that leaves a lot of space for free-text comments. Valuable insights on the effectiveness of the course approach and content can be extracted from this feedback. In the three years that this course has been offered, students rated the learning opportunities and the practical relevance outstandingly positive. At that stage of their career, most are not aware of the requirements on innovation discourse in academia, industry, and S&T governance. Those who have experiences with interdisciplinary collaboration, for example in EU-funded research consortia with attached ELSI work packages, report that the competences taught in this course significantly helped them understand the approaches of such efforts and gain the confidence to contribute actively to them. Class 11 on sustainability received the highest score for popularity ("*The class I liked the most is class No.___*"). A student commented aptly, "*I finally know how to argue with my Sustainable Chemistry professor that sustainability is more than environmental friendliness*".

Does the course change anything in the attendees' attitude and behaviour? This question is notoriously difficult to assess. Empirical research indicates that classes on scientific integrity cannot decrease the scientists' susceptibility for scientific misconduct [40]. On the other side, learning discourse skills and interdisciplinary competences in the context of scientific and technological innovation—elements of RRI—does, according to the students' feedback, induce a change in their attitude and professional mentality. At the current stage—three years after its start—it is not impossible to measure long-term effects of this course's efforts, for example whether S&T-related innovations that are instantiated with the contribution from scientific and technical experts that attended this course, indeed, are more sustainable or shift dual use potentials to the benefit side (a method that may be employed for this purpose in future assessments is described in [41]).

Qualitatively, a learning effect can be observed from students' contributions on the e-learning platform of the course. For example, comparisons between posts to an introductory case and discussions of a concluding case at the end of a class, or between the initial self-assessment and the final quiz, are very enlightening. In the following, two examples of learning progress and skill formation indications are provided.

The first example is a risk assessment and related design choice scenario that is based on a real case, the development of a lab-on-a-chip for colon cancer screening (taken from [22], also discussed in [42]). This case was employed prominently in class 12 (*Risk, Uncertainty, and Precaution*). At the beginning of the class, students were confronted with the situation of making a choice between two options for indicating a detection of DNA strands associated with the presence of a cancer tumour. One is the trigger of a RFID signal, the other the release of a strong blue dye from a bursting capsule for visual detection in the toilet. A representative student reply, posted on the online platform, reads as follows:

> "I would investigate whether the blue dye, the capsule material, or the components of the RFID chip, are harmless in terms of toxicity and effect on the human body. A second consideration is how reliable the two options are. The one with fewer errors should be chosen. Third, it may be considered how difficult it would be to build the components into the lab-on-a-chip, how it would affect the size, or how expensive the device would become."

This reply illustrates the student's scientific-technical thinking. It is all about toxicology, technicality, feasibility, and utility. In the following lectures, exercises, reading assignments, and other activities, the students learn about value assessments, multi-stakeholder perspectives, ELSI research, and ethics-by-design approaches in R&D projects. When the above case is revisited at the end of the unit, the same student wrote the following:

> "I think, a good decision can only be made after thinking about all the people who have to do with the device in the end. Maybe, it affects the mass production of the device, whether it can be offered in pharmacies or only in doctor offices, or the communication between patients and doctors. The RFID chip solution produces a lot of data, so it must be checked if it is in accordance with data protection regulations. On the other side, the blue dye solution might be a challenge for elderly patients who feel insecure when they have to observe the results by themselves without the help of a doctor or nurse. These risks may even be more important for some people than toxicity or error rates. Therefore, it would be good to ask the users of the device about their needs."

Obviously, this student tries consciously to incorporate the learned course content into the reply, such as the multi-stakeholder view or the parlance of user needs. Yet, without having discussed these details in the lectures, the student successfully identifies a broad range of stakeholders and their possible challenges (production, dissemination, communication, regulation, even psychological impact). This indicates that the student changed the mode of assessment from a protocol approach (the systematic study of material safety and error rates) to a narrative approach (thinking through application scenarios, anticipating demands and preferences, identifying ambiguities and value clashes). Interestingly, the student leaves the question which of the two options is the better one open, perhaps knowing that it is not a decision the researcher would have to make alone but with a team of assessors.

The second example is taken from an activity in class 13 (*Science and Technology Governance*) in which the role of scientific policy-advise is discussed. Students are asked to prepare a short presentation about the relevant chemistry of plastics, their recycling, and their effect on the environment, to be presented at a panel that is seeking solutions for plastic pollution of the oceans. While some students remarked that it would be impossible since they are not experts on that particular matter and could not oversee what is relevant, others prepared highly specific overviews of polymerisation chemistry, industrial scale plastic production and recycling, and of use patterns or environmental impact assessment. When being asked what solution for the issue at hand (marine pollution), most students did not give a clear answer but referred to the just delivered 'facts', occasionally accompanied with a remark that 'the panel' may use these scientific insights for finding a solution, implying that it is not the scientific expert's job to decide on that. During the class, students got to know Pielke's typology of scientific knowledge communication [28]. He distinguishes four types:

- The pure knowledge exponent who presents in detail all the scientific knowledge underlying the issue at hand but remains neutral on the solution;
- The advocate who believes to have identified the one best solution, explains that in scientific detail, and tries to convince the panel to do exactly that;
- The arbiter who asks the panel what the priorities are (for example, economic feasibility, or environmental risks, or social justice), and—dependent on the reply—proposes a solution that is best suited to achieve the envisioned goal;
- The honest broker who—like the arbiter—inquires about the promoted values, but questions the claims in a constructive discourse in which factual insights and normative positions are brought together.

Students realised that most of their initial approaches would be close to the *pure knowledge exponent*, and that this input would not bring the panel any further. Equipped with Pielke's typology, they are asked to rethink their presentation and try again. In the online course, that could be recorded videos of short presentations; in a face-to-face class, this could be done in semi-prepared role-playing form. Experiences from the course described above show that students clearly shift their perspectives. Presentations became much less scientific-technical, and much more goal-oriented in terms of identifying a solution for a problem that requires both a lot of factual knowledge and a clarification on normative value propositions. Students, obviously, felt more comfortable with reacting flexibly on normative input from the inquiring panel. The former conference-style impersonal attitude in presenting knowledge was substituted by an imagined dialogue-style communication (*"What will they need to know from me? What do I need to know about them in order to choose the right breadth and depth of knowledge?"*). For example, students chose phrases such as *"If you endorse economic feasibility as the most important aspect, the problem is not a chemical one, but one of logistics and externalities. If you are seeking for the most environmentally sustainable solution, a specific plastic polymer digesting microorganism seems promising. But beware of the risks and the potential backfires that threaten the success of the intervention."* Here, the playful practice of discourse, the imagination of a possible narrative, and the forced expansion of their perspective towards a multi-stakeholder setting, help students to understand that there is no sharp distinction between the factual and the normative elements of a problem at hand, but that the two always must be thought, discussed, and acted upon together and in dependence of each other. Students who express this understanding in their different presentations before and after this course have acquired normative literacy.

15.5 Conclusion

Responsible researchers and innovators as envisioned by the RRI framework make the *right* choices in their research and innovation projects. What is *right* cannot be determined solely on the grounds of scientific or technical assessments but include the elicitation, deliberation, and realisation of ethical, legal, and social implications that go along with the intended real-world impact of the development product (knowledge, processes, artefacts, etc.). This judgment competence has been described in this essay as *normative literacy* since it concerns the discourse of norm- and value-related factors that are different from mere factual knowledge. While more and more STEM faculties at colleges and university incorporate research ethics and good scientific practice courses into their curricula, normative literacy as a RRI skill that addresses the societal and environmental impact of scientific, technological, and innovative activities is still underrepresented in these courses. This chapter described and evaluated a novel approach to teaching normative literacy in a course on ethical and social dimensions of chemistry. It employed content—case studies, exercises, discourse training, etc.—that features elements characteristic for the humanities disciplines, such as interpretation of fiction with subsequent transfer to personal real-life situations, effective discourse performance, critical thinking, and ethical vision assessment. Attendees' feedback and analysis of the students' contributions to the online course reveal that the course participants develop not only their awareness of normative dimensions of their scientific activities and their practical skill of reflecting, communicating, and defending normative arguments, but also their self-confidence when actually engaging in discussions and confrontations on normative issues of their work.

The most important rationale for reaching out to the Humanities for the training of RRI competences is the non-predictability of the kind and scope of normative concerns in the context of scientific progress and innovation. There is no standard procedure according to the book (in German *Schema F*) that future scientists and innovators could learn and apply like a tool to resolve ethical and social questions of their projects. Instead, a confrontation with hermeneutic techniques (interpreting fictional or real cases, extracting their meaning and quintessence, transferring the insights to new cases) and discourse approaches (critical thinking, argument analysis, performative aspects) prepares the students for a professional life in which they operate as responsible drivers of sustainable and beneficial scientific and technological progress.

References

1. W.J. Gonzalez (ed.), *New Perspectives on Technology, Values, and Ethics. Theoretical and Practical* (Springer, Heidelberg, 2015)
2. B.-J. Koops, I. Oosterlaken, H. Romijn, Swierstra Tsjalling; van den Hoven, Jeroen, (eds.), *Responsible Innovation 2. Concepts, Approaches, and Applications* (Springer, Cham, 2015)

3. L. Ursin, P.-E. Movik, A. Alvarez, The normative dimensions of new technologies. Nord. J. Appl. Ethics **9**(1), 1–4 (2015)
4. E. Kurz-Milcke, G. Gigerenzer (eds.), *Experts in Science and Society* (Kluwer, New York, 2004)
5. M. Vogt, C. Weber, The role of universities in a sustainable society. Why value-free research is neither possible nor desirable. Sustainability **12**(7), 2811 (2020). https://doi.org/10.3390/su12072811
6. J. Mehlich, Is, ought, should - the role of scientists in the discourse on ethical and social implications of science and technology. Palgrave Commun. **3**, 17006 (2017)
7. R. Owen, J. Bessant, M. Heintz (eds.), *Responsible Innovation. Managing the responsible emergence of science and innovation in society* (Wiley, Chichester, 2013)
8. R. Strand, Post-normal Science, in *Routledge Handbook of Ecological Economics*. ed. by C.L. Spash (Routledge, London, 2017)
9. V. Nurock, R. Chatila, M.-H. Parizeau, What does "ethical by design" mean?, in *Reflections on Artificial Intelligence for Humanity*. ed. by B. Braunschweig, M. Ghallab (Springer, Cham, 2021)
10. B. Ibiricu, M.L. van der Made, Ethics by design: a code of ethics for the digital age. Rec. Manag. J. **30**(3), 395–414 (2020)
11. E. Commission, Open Innovation, Open Science, Open to the World - a vision for. Europe (2016). https://doi.org/10.2777/552370
12. R. Gianni, J. Pearson, B. Reber (eds.), *Responsible Research and Innovation. From Concepts to Practices* (Routledge, Abingdon, 2018)
13. A. Grunwald, The objects of technology assessment. Hermeneutic extension of consequentialist reasoning. J. Respons. Innov. (2019) https://doi.org/10.1080/23299460.2019.1647086.
14. F. Ferri et al. (eds.), *Responsible Research and Innovation Actions in Science Education, Gender and Ethics. Cases and Experiences* (Springer, Cham, 2018)
15. D.L. Zeidler (ed.), *The Role of Moral Reasoning on Socioscientific Issues and Discourse in Science Education* (Kluwer, Dordrecht, 2003)
16. A. Soeteman, Normative Judgements, in *Logic in Law*, pp.23–46. (Springer, Dordrecht, 1989). https://doi.org/10.1007/978-94-015-7821-9_2.
17. G. Valkenburg, A. Mamidipudi, P. Pandey, W.E. Bijker, Responsible innovation as empowering ways of knowing. J. Respons. Innov. **7**(1), 6–25 (2020). https://doi.org/10.1080/23299460.2019.1647087
18. J. van den Hoven, P.E. Vermaas, I. van de Poel (eds.), *Handbook of Ethics, Values, and Technological Design. Sources, Theory, Values and Application Domains* (Springer, Dordrecht, 2015)
19. K.C. Elliott, *A Tapestry of Values. An Introduction to Values in Science* (Oxford University Press, New York, 2017)
20. VDI, *VDI Richtlinie 3780, Technikbewertung—Begriffe und Grundlagen (Technology Evaluation—Terminology and Fundamentals)* (VDI, Düsseldorf, 1991)
21. J. Hahn et al., Technology assessment for a changing world. TATuP (J. Technol. Assess. Theory Pract.) **29**(3), 74–75 (2020)
22. F. Lucivero, *Ethical Assessment of Emerging Technologies* (Springer, Heidelberg, 2016)
23. D. Hitchcock, Critical Thinking, in edited by E.N. Zalta, *The Stanford Encyclopedia of Philosophy* (Fall 2020 Edition) (2020). URL = <https://plato.stanford.edu/archives/fall2020/entries/critical-thinking/>.
24. M. Cotton, *Ethics and Technology Assessment: A participatory approach* (Springer, Berlin, 2014)
25. A. Barry, G. Born (eds.), *Interdisciplinarity - Reconfigurations of the Social and Natural Sciences* (Routledge, Abingdon, 2013)
26. A. Hullmann, *European Activities in the Field of Ethical, Legal and Social Aspects (ELSA) and Governance of Nanotechnology* (European Commission, DG Research, EU, 2008)
27. K. Konrad et al. (eds.), *Shaping Emerging Technologies: Governance, Innovation, Discourse* (IOS Press, Netherlands, 2013)

28. R.A. Pielke, *The Honest Broker: Making Sense of Science in Policy and Politics* (Cambridge Univ. Press, Cambridge, 2007)
29. F. Sabirova et al., Professional Competences in STEM Education. Int. J. Emerg. Technol. Learn. (iJET) **15**(14), 179–193 (2020)
30. W. Teays, *Seeing the Light. Exploring Ethics Through Movies* (Wiley-Blackwell, Chichester, 2012)
31. A. Grunwald, *The Hermeneutic Side of Responsible Research and Innovation* (ISTE, London, 2016)
32. J. Mehlich, *Good Chemistry. Methodological, Ethical, and Social Dimensions* (Royal Chemical Society, London, 2021)
33. https://elearning-euchems.eu
34. P. Pruzan, *Research Methodology. The Aims, Practices and Ethics of Science* (Springer, Cham, 2016)
35. R. Iphofen (ed.), *Handbook of Research Ethics and Scientific Integrity* (Springer, Cham, 2020)
36. M. Develaki, Social and ethical dimension of the natural sciences, complex problems of the age, interdisciplinarity, and the contribution of education. Sci. Educ. **17**(8), 873–888 (2008). https://doi.org/10.1007/s11191-007-9077-7
37. S.O. Hansson (ed.), *The Ethics of Technology. Methods and Approaches* (Rowman & Littlefield Intl, London, 2017)
38. B. Bensaude-Vincent, J. Simon, *Chemistry, the impure science*, 2nd edn. (Imperial College Press, London, 2012)
39. U. Felt, R. Fouché, C. Miller, L. Smith-Doerr (eds.), *Handbook of Science and Technology Studies*, 4th edn. (The MIT Press, Cambridge, 2017)
40. M. Mumford, Assessing the effectiveness of responsible conduct of research training: key findings and viable procedures, in edited by National Academies of Sciences, Engineering, and Medicine. Fostering Integrity in Research (The National Academies Press, Washington, 2017).
41. M. Heras, I. Ruiz-Mallén, Responsible research and innovation indicators for science education assessment: how to measure the impact? Int. J. Sci. Educ. **39**(18), 2482–2507 (2017). https://doi.org/10.1080/09500693.2017.1392643
42. J. Mehlich, Chemistry and dual use: from scientific integrity to social responsibility. Helv. Chim. Acta **101**, e1800098 (2018). https://doi.org/10.1002/hlca.201800098

Chapter 16
The Humanities in Science Teacher Education: A Perspective from South Africa

J. D. Bradley

Abstract Science teachers' education has a fundamental influence on the quality of science education at all levels of instruction. This chapter focuses on the benefits of infusing a humanities perspective into the education of science teachers for sustainable development. Although Education for Sustainable Development (ESD) was initiated 30 years ago, there has been little evidence of reaction from the national science curricula of South Africa. It is suggested that science teacher education curricula should change appropriately in advance of school curricula changes which must be made eventually. This requires careful development of parts of the existing curricula for prospective science teachers to include content, skills and attitudes suited to ESD. This would lead to a more humanistic science teacher education.

16.1 Introduction

In South Africa most science teachers qualify through a 4—year tertiary education leading to a Bachelor of Education (BEd) degree. School teaching does not attract high-school matriculants with good marks on the whole, and so the courses they follow for the BEd degree tend to focus quite narrowly on preparing them to teach their chosen subjects as required by the current national curricula. This policy is understandable for short-term purposes, but it inevitably makes the qualifying students vulnerable to future changes. More widely, it inhibits curriculum change.

School teachers have a key role in preparing young people for the future. As regards this role in science education, there has long been tension between preparing for more advanced education in the subject and preparing for other, or more general, directions. The place of the humanities in this connection has also been a long-standing issue. At the present time, we need to re-examine the role of school science teachers with a particular urgency because the future now includes sustainability

J. D. Bradley (✉)
Division of Science Education, School of Education, University of the Witwatersrand, Johannesburg, South Africa
e-mail: R.bradley@lse.ac.uk

L. Mammino (ed.), *Insights into the Relationships between Humanities and Sciences*,
https://doi.org/10.1007/978-3-031-97900-2_16

issues which require specific attention, such as climate change, which affects us all and to which the next sections will give particular attention.

Preparing for climate change is not a local, national project but a global one, as is illustrated by the activities of the UN and its agencies. (eg Agenda 21 (UNCED 1992) [1]). A particular concern is the evidence of public mistrust of scientists' advice in respect of climate change—a mistrust also encountered for other issues, including COVID mitigation measures. Thus science teachers need to understand not only the science of climate change and its mitigation, but also the scope and limitations of global actions requiring broad public support. Science teachers need informed awareness of contemporary issues in which science and technology are relevant, if learners are to perceive their science teachers as relevant.

Against this background, this chapter interrogates the situation of present and future science teachers, using experiences derived from South Africa. It is argued that science teacher education should:

(1) change in anticipation of school curriculum changes;
(2) provide a more humanistic science education.

16.2 Sustainable Development Goals vs Disciplinary Demands of Science Curricula

Deeper study of how to deal with the threats of climate change shows that there are many aspects to consider. The description 'sustainable development' widens the scope of thinking in this regard. The 17 UN Sustainable Development Goals (SDGs) summarise the international consensus covering social, economic, cultural and material needs all of which interact with each other [2]. Education for Sustainable Development (ESD) was initiated 30 years ago as part of Agenda 21, but the impact in many school systems (including South Africa's) is weak. We may doubt if many current curricula achieve the intended 'ethical and environmental awareness, values and attitudes, skills and behaviour consistent with sustainable development and for effective public participation in decision-making.' Indeed we do not see 'the dynamics of both the physical/biological and socio-economic environment, and human (which may include spiritual) development… integrated in all disciplines…' [1].

The apparent failure to embrace the ideas quoted above, may be understood when we compare the titles of the 17 SDGs with the titles of the school subjects that prospective school teachers are prepared to teach. There is little or no correspondence. This is related to the framework of established disciplines that dominates the traditional repertoire of school subjects. The disciplines learned at school are together supposed to provide the basic content and skill tools with which to deal with the future needs of young learners. This is how they have always been envisaged in school curriculum policy. So the lack of overlap between SDGs and school subjects is to be expected. However the problem remains; the future needs now include how to comprehend climate change and how to achieve sustainable development.

As an example of how the traditions of disciplinary education should be re-examined, we may refer again to the interactions between the SDGs, the evident relevance of more than one discipline to a particular SDG and hence the need for multidisciplinary and interdisciplinary strategies in addressing a particular goal. Do school curricula and subjects provide exposure to such strategies? We may doubt this is the case for most countries. Focusing now on the current science curricula for South Africa for example [3, 4] we may find encouragement amongst the General Aims in that learners should be able to:

- *Use science and technology effectively and critically showing responsibility towards the environment and the health of others;*
- *Demonstrate an understanding of the world as a set of related systems by recognising that problem-solving contexts do not exist in isolation.*

However, the more detailed elaborations of subject curricula and their assessment generally fail to recognise these aims. This implies that the average teacher will devote little or no time to achieving them, focusing instead on the realities of matriculation examinations and departmental assessments. The weight of educational interventions is towards improving the pass rate by focusing on a relatively small range of standard types of questions. Furthermore, interdisciplinary links are very hard to find. Overall, the impression is that what matters is passing an exam where limited demands of the school subject/discipline are tested. The two General Aims quoted are wishful thinking, and the reality is that probably the majority of science teachers in South Africa are not equipped to realise them.

Given the key role of teachers in preparing young people and the evidence of global changes already under way, it is important to ask if teachers are ready to be agents of educational changes that will surely come. Probably many are not, and this is not entirely their fault. They too are products of an educational system with a national curriculum usually structured into disciplinary subjects with rather narrow objectives.

Science teachers might feel that the laws of science are universal so they need not pay much attention to talk of climate change. Climate is for the geography teacher to deal with, they may say. And if national examinations continue in traditional ways, teachers are constrained to prepare their learners for them, however progressive the teacher may be. Indeed where the passing of matriculation exams is seen as the ticket to future employment and escape from poverty, all efforts in secondary school science teaching and learning are aimed at achieving this. So the traditional disciplinary emphases, supposedly needed for further study of the subject, remain.

In view of these considerations it seems probable that there will be big challenges to successful implementation of a new curriculum which reflects the ambitions expressed 30 years ago [1]. The capacity for curriculum design may exist within the Department of Basic Education, but the capacity for design and implementation of new curriculum resources, in-service training and support is very unlikely to be adequate. All the evidence of school curriculum changes during the past 25 years points to this problem. And the more substantial the changes in curricula, the more the challenge is like dealing with a culture change [5]. With climate change increasingly evident each year, we need to speed up curriculum change.

16.3 A Role for Science Teacher Education in Catalysing School Science Curriculum Change

Whilst existing school teachers are trapped in simple conformity with national directives and assessments the same is not necessarily the case for prospective teachers (which in South Africa means students studying for a BEd degree). It is of course true that such students must be equipped to teach the existing science curricula, but there is no specific restriction on what else may be taught and learned. Hence new teachers might be graduated not only with sufficient skills and content knowledge to be competent in meeting traditional needs, but with sufficient awareness and experience of new, future needs. Such teachers would be able to support curriculum changes in the future, not least by assisting older colleagues to adapt. In other words the culture change likely to be faced by existing teachers in the future could be facilitated by new graduates who have already been through it! They would have been through it, because they succeeded at school by learning traditional disciplines but have been challenged to reflect on the implications of climate change and the SDGs for these disciplines during their BEd studies.

The strategy proposed then is to teach BEd students the subject content required by the current curriculum but in ways that reflect the future needs—as already to some extent reflected in the General Aims.

In the following two sections this strategy is exemplified in relation to the South African school curriculum. The first of these sections discusses STS (Science, Technology and Society), a science curriculum development with a substantial following in many parts of the World, including to some degree in South Africa. The second section reflects the author's experiences in teaching a Bachelor of Education Honours (BEdHons) course entitled 'Acids through the Ages' for a number of years.

16.3.1 Science, Technology and Society (STS)

Interest in teaching science in context has a relatively long history. Amongst a number of notable school curricula in which this aim is manifested are the Salter's Science (UK, [6]), Chemistry in the Community (USA, [7]) and Chemie in Kontext (Germany, [8]) curricula. These have been implemented and researched widely and enthusiasm for them continues. Climate change and sustainable development were not significant features of these past initiatives originally but they would be readily adopted in newer versions.

In South Africa, the STS curriculum concept was first given a place in the secondary science curriculum in the form of Optional Topics. Most of these focused on a specific product of the mining and mineral processing industry (e.g., copper, manganese, coal, chromium, gold) and covered a range of aspects, as implied by science, technology and society. With the support of the Chamber of Mines of South Africa, very substantial teacher resources were developed, which ranged from practical activities to economic data [9, 10].

STS became a compulsory feature within a completely changed secondary school Physical Sciences curriculum (NCS) in 2005 as a strand (officially termed a knowledge area) running through grades 10–12. It was called Chemical Systems, anticipating the present interest in systems-thinking as part of ESD [11, 12]. The strand included topics such as the hydrosphere, the atmosphere and the lithosphere as well as chemical industry. These topics not only fit under the banner of STS, but already invite interest in climate change. Indeed one of the subsections was devoted to global warming and climate change. The lithosphere topic had an emphasis on mining and mineral processing, a subject of particular relevance in South Africa and most suitable for some exploration of the impact of such activities on the environment.

In the present national curriculum (CAPS) for Physical Sciences [3] the time allocated to Chemical Systems in the NCS curriculum has been reduced whilst that allocated to traditional disciplinary topics has been increased. Despite this, the General Aims of the CAPS curriculum still include the two interdisciplinary and sustainability aims highlighted earlier. These aims are appropriate in relation to the development of national curricula reflecting education for sustainable development. Designing courses and topics for initial science teacher preparation which emphasise these aims would seem fully justified as responding to national policy. Although this is feasible it may require culture change for some BEd course lecturers. Clearly this challenge has to be met if the aim of preparing new science teachers for change is to be initiated. BEd course lecturers are of course not alone, and there are initiatives in many countries and from many organisations, including UNESCO [13]. In South Africa, Tsakeni [14] has proposed an ESD conceptual framework with five pillars, namely social, economic, environmental, citizen skills and pedagogies for ESD, to help analyse the potential of the Chemical Systems knowledge area over grades 10–12. Science teacher educators designing courses for their students can draw inspiration from these local and international sources, and integrate relevant aspects into the preparation of new science teachers for the South African curriculum.

16.3.2 Acids Through the Ages

The concept of acids has a history extending over more than 2 000 years [15]. It features in all school science curricula and must be one of the universally-recognised science concepts even amongst those who did not receive secondary education. The science curricula of different countries follow rather similar patterns of development of the concept. This pattern, presented in 12 years of school education, has some similarities to how scientists developed the concept through 20 centuries. Recognising this provides an opportunity to discover how science and scientists develop, making mistakes along the way, and correcting them eventually. It also reveals how developments in one area (e.g. physics) influence those in another (e.g. chemistry), and the potential for progress that lies in interdisciplinary work. Without formality, the nature of science and human endeavour can be revealed.

In order to substantiate such claims it is necessary to make a quick journey through the curricular development of the acid concept observing how it reflects the history. So we begin in grade 7 (no science in grades 4–6) [4] with the sour taste characteristic of acids, known to Romans as 'acidum'. In grade 7 you are in the same state of ignorance of chemistry as a Roman at the time of Christ. Indeed, although acids were known, chemistry was not. In those days there were philosophers who thought about the nature of matter, but they had no way of explaining why some substances were acids and others not. Some philosophers thought that all of the huge variety of materials in their environment were composed of a small number of 'elements'. Mixing these together in different proportions gave the variety of materials. They speculated on what these might be (fire, earth, air, water were popular suggestions), intending to reflect some of the different qualities or properties of materials (perhaps representing energy, solid, gas and liquid respectively). Other philosophers thought rather of atoms, particles of a material that had the properties of the material. For example gold was made of gold-coloured atoms.

But, like today's young learners, philosophers of long ago had no way of proving their ideas, and several centuries passed in which observations could be made to enrich the store of knowledge about acids but without understanding. Some of this store of knowledge is laid before learners in grades 8 and 9 but mixed with the concepts of elements (and compounds) and atoms. The effect is to miss the opportunity to recognise the revolution of our understanding of materials (including acids) that the new conception of elements and atoms triggered. Lavoisier (following Boyle) made it possible to decide by experiment (not by verbal persuasion) which substances are elements and which are compounds in 1787. Thereafter, within 25 years, Dalton linked the other ancient idea of atoms with that of elements, but showed that atoms were not just tiny bits of the substance itself. He even determined atomic weights for the atoms of elements!

Lavoisier has often been called the father of chemistry because his breakthrough may be seen as when the science of chemistry began [16]. The element symbols and the naming of compounds (introduced in grade 8) started with him too. Yet this revolution is not identified in the school science curriculum. Dalton too is unidentified even though the grade 8 curriculum goes as far as describing the sub atomic particles.

Lavoisier was able to apply the old idea of elements explaining the properties of substances in explicit ways that were not possible before. He proposed that oxygen was the acidifying element, meaning that when oxygen reacted with another element the compound produced was acidic. Based upon this idea he gave oxygen its name—which is based upon ancient Greek and means acid-forming. The idea was later proven incorrect, but the name has stuck. Ironically acid rain, ocean acidification, and acid-mine-drainage (AMD) today illustrate that Lavoisier was not entirely wrong!

Although the curriculum scrambles the sequence of history, the bits and pieces of scientific history are there like jigsaw pieces that can be fitted together as and when there is opportunity. By the end of grade 9 the acid jigsaw is far from complete. Further pieces come to hand through grades 10–12, but it is not appropriate here to follow the story further, except to note the impact of Volta (a contemporary of Lavoisier and Dalton). This is not so much for his chemical work, but for making available current electricity from his pile (battery of cells). Chemists seized upon this new invention to electrolyse water (for centuries considered an element), to isolate new elements from compounds that had resisted Lavoisier's attempts to break them down, and to develop a whole new awareness of ions (introduced in grade 10). Berzelius, electrolysing aqueous salt solutions, suggested that the acid and base combination to form salts (known for centuries and represented symbolically as $A + B = AB$) was reversed in the electrolysis and that perhaps acids and bases had opposite charges. Thus the time-honoured acid–base reaction called neutralisation was also a neutralisation of charges. This understanding was only partly correct, but it was revived by Arrhenius 50 years later as a combination of hydrogen ions (from the acid) and hydroxide ions (from the base).

What then is the purpose of considering acids through the ages? The school science curriculum prepares young people for the future, not the past. So it may be argued there should be a strict focus on the disciplinary content most likely to be useful for them. Yet it must be accepted that learning will need to start with gathering some knowledge which is largely free of abstraction. Learners may work towards abstraction to a greater or lesser degree, adding to their store of important general knowledge as they do so. Concepts develop from primitive ones (e.g. acids and bases are opposites, a concept reflected in Arrhenius' hydrogen and hydroxide ion theory). To understand how this happens is to understand how science develops. And understanding this contributes to understanding how technology and society develop (see below).

Preparing new science teachers for sustainable development needs this understanding. Hence when teaching student teachers about acids as part of courses dealing with the school subjects Natural Sciences and Physical Sciences, the important breakthroughs of the past should be emphasised.

16.3.3 Aims of Increasing Attention on Historical, Social and Economic Aspects of Science

Teaching about the interaction between science, technology and society should develop explicit awareness of the role of science in our present-day lives. The awareness should be sufficiently substantive that a student would be able to describe and exemplify this to others. Similarly, following the development of even a single science concept over decades or centuries can develop awareness of the human endeavour to understand the world. (see also [17]) Strengthening these awarenesses provides a basis for mature consideration of how sustainable development could be achieved in principle. Future science teachers need this maturity.

16.3.4 How Did We Get Here? What Are the Implications for Science Teacher Education?

History on a global scale provides insights that can help us now and in the future. Learning from experience and learning from our mistakes are familiar reflections. This is surely needed in moving from unsustainable development to sustainable development. The presentation of historic data by Matlin [18] points the way. The two graphs showing global GDP and population changes over many centuries surely provoke the question 'How did we get here?'. Another graph links the GDP growth with developments in chemistry and the chemical and allied industries. They show our great success (by these measures) in the last two or three centuries: they show how, through science and technology, society benefitted in terms of wealth (GDP) and health (population). Newton (seventeenth century), followed by Lavoisier and Dalton (eighteenth century), enabled scientists and entrepreneurs to make discoveries and applications thereof in a continuing cycle of progress and development—STS in action, at its best! Yet the message today is that however good it has been, it is no longer sufficient. New understandings and considerations are required as we enter the Anthropocene Epoch [19]. We now require, what some call, "STSE" (E = Environment) in our evolving and sustainable development.

None of this kind of perspective is provided for in our school science curricula. However, it would be perfectly feasible to include it in a BEd curriculum for prospective school science teachers.

At the University of the Witwatersrand, Johannesburg, there are signs suggesting that this may not only be feasible but encouraged. New students entering Wits University participate in an induction programme before the start of the academic programme. This year (2022) the programme included for the first time a short course with the title 'Climate Change and Me'. This suggests that the university aimed to highlight the challenge we all face but also that it is something that should engage each student. It is not a problem for somebody else to solve. There are profound implications in this message as far as prospective science teachers are concerned. Present science teacher education curricula (as noted earlier) tend to respond to the needs of students to get a job by preparing, what we may call, classroom technicians. This serves a purpose, but what the future requires is a more humanistic science education.

Humanistic science education emphasises 'the scientific enterprise as a human endeavour grounded in its own culture, values, ideologies, knowledge and processes, as well as its myriad of interactions with society' [20]. 'Its central focus is to provide students the context they need to make connections with other aspects of their lives' [21].

In the great majority of South African schools (including secondary schools), teachers are expected to be competent to teach more than one school subject. This reality is catered for in the BEd curriculum by providing for two teaching subjects. It is an open question as to whether or not there are opportunities for interaction between the subjects concerned. The point is that science teachers prepared through a 4-year BEd degree have a somewhat broader educational preparation than a teacher prepared to specialise in a single subject. A recent comparison between South African and German science teacher preparation reveals this clearly [22]. Disciplinarity is less emphatic in the South African approach. This might make it easier for a science teacher to engage in 'promoting sustainable development and improving the capacity of the people to address environment and development issues'. Education is critical for 'achieving environmental and ethical awareness and attitudes, skills and behaviour consistent with sustainable development and for effective public participation in decision-making' (UNCED 1992) [1]. Science teachers need to be prepared accordingly and a more humanistic science teacher education curriculum would appear to be a way forward.

References

1. UNCED, Agenda 21 (1992), http://www.un.org/esa/dsd/agenda21/
2. UN, Transforming our world: the 2030 agenda for sustainable development (2015), http://sustainabledevelopment.un.org/post2015/transformingourworld/
3. Department of Basic Education Curriculum and Assessment Policy Statement (CAPS): Physical Sciences. Further Education and Training Phase Grade 10–12. (Government Printing Works, Pretoria, 2011)
4. Department of Basic Education Curriculum and Assessment Policy Statement (CAPS): Natural Sciences, General Education and Training Phase Grades 7–9. (Government Printing Works, Pretoria, 2011)
5. J.M. Rogan, D.J. Grayson, Towards a theory of curriculum implementation with particular reference to science education in developing countries. Int. J. Sci. Educ. **25**, 1171–1204 (2003)
6. B. Campbell, J. Lazonby, R. Millar, P. Nicolson, J. Ramsden, D. Waddington, Science: The Salter's approach—a case study of the process of large-scale curriculum development. Sci. Educ. **78**, 415–447 (1994)
7. A.T. Schwartz, Contextualised chemistry education: the American experience. Int. J. Sci. Educ. **28**(9), 977–998 (2006)
8. I. Parchmann, C. Gräsel, A. Baer, B. Ralle, R. Demuth, 'Chemie im Kontext': A symbiotic implementation of a context-based teaching and learning approach. Int. J. Sci. Educ. **28**(9), 1041–1062 (2006)
9. J.D. Bradley, K. McCarogher, 'PHOENIX—From Ashes to Flight in South Africa' *in* Five Case Studies in Education-Industry Partnerships. York: Chemical Industry Information Centre, pp 53–69 (1995)
10. J.D. Bradley, M. Brand, M. Demmer, M. Judd, Copper—an optional topic resource package for the South African physical science syllabus. Johannesburg: Chamber of Mines of South Africa
11. P.G. Mahaffy, S.A. Matlin, J.M. Whalen, T.A. Holme, Integrating the molecular basis of sustainability into General Chemistry through systems thinking. J. Chem. Educ. **96**(12), 2730–2741 (2019)

12. A.C. Eaton, S. Delaney, M. Schultz, Situating sustainable development within secondary chemistry education via systems thinking. J. Chem. Educ. **96**(12), 2968–2974 (2019)
13. UNESCO, Teaching and learning for a sustainable future—a multimedia teacher education programme (2005), http://www.unesco.org/education/dsf/index.htm
14. M. Tsakeni, Opportunities for teaching sustainable development through the chemistry component of CAPS Physical Sciences. African J. Res. Mathem. Sci. Technol. Educ. **22**(1), 125–136 (2018)
15. D. Kolb, Acids and bases. J. Chem. Educ. **55**, 459–464 (1978)
16. D. McKie, Antoine Lavoisier: scientist, economist, social reformer. (Republished by Da Capo Press, 1990, New York: H. Schuman, 1952)
17. I. Galili, A. Hazan, The effect of a history-based course on optics on students' views about science. Chapter 16, pp 229–254 *in* Science Education and Culture, Bevilacqua, F., Gianetto, E., and Matthews, M.R. eds. (Kluwer Academic, Dordrecht, 2001)
18. S. Matlin, *Chemistry Education for the 21st Century*. Paper presented at the CHEMRAWN-IOCD Meeting "Education in Chemistry: Scoping the Future", Namur, 13–15 January (2014)
19. J.D. Bradley, Microscience in the IYPT and the Anthropocence Epoch. African Journal of Chemical Education **9**(3), 4–17 (2019)
20. L. Klopfer, G. Aikenhead, Humanistic science education: a history of science and other relevant contexts. Sci. Educ. **106**(3), 490–504 (2022)
21. S.A. Southerland, J. Settlage, Humanistic science education: updating our collective consciousness and offering fresh consideration. Sci. Educ. **106**(3), 479–481 (2022)
22. J.-P. Burde, J.D. Bradley, P. Moodie, Teaching and learning about electricity in two different curricular contexts. Chapter 10 pp 149–161 *in* Teacher Education in Germany and South Africa, Hofmann, E., Mavhunga, E., Bohl, T., and Maringe, F. eds. (Tübingen University Press, Tübingen, 2022)

Chapter 17
The Motivations for the Merger of the International Council for Science and the International Social Science Council

Natalia P. Tarasova and Anna A. Dodonova

Abstract Today science plays an important role in many sectors and spheres of human life. The progress in science serves as one of the main indicators of the development of a society, as well as that of the individual states.

If analysing all scientific trends, two global scientific currents can be seen—natural science and social science.

This division of the sciences historically led to the formation of two global international councils in the twentieth century.

- The International Council for Science (ICS), founded in 1931 as the International Council of Scientific Unions (ICSU).
- The International Social Science Council (ISSC), founded in 1952.

The world has changed significantly over the past 100 years. The global problems of humankind are so complex that for any science domain it is simply impossible to solve them in isolation. High speed of decision-making, evaluation of social and physical consequences, and a systematic and comprehensive approach to information processing are important features of solving global problems. These trends initiated the process of merging of the two major international scientific unions, ICS and ISSC. Over the decade 2005–2015, ICS and ISSC worked increasingly closely together. The International Science Council (ISC) was created in 2018 from a merger of the two international councils. This merger is the reflection of the objective need for the world science community to unite for the sake of solving global problems. This chapter focuses on the motivations for the merger of the two councils, and on the main aspects of the process.

Keywords The International Council for Science (ICS) · The International Council of Scientific Unions (ICSU) · The International Social Science Council (ISSC) · The International Science Council (ISC) · Merger

N. P. Tarasova (✉) · A. A. Dodonova
D. Mendeleev University of Chemical Technology of Russia, Moscow, Russia
e-mail: tarasova.n.p@muctr.ru

L. Mammino (ed.), *Insights into the Relationships between Humanities and Sciences*,
https://doi.org/10.1007/978-3-031-97900-2_17

17.1 Introduction

Developments in today's world force us to think about the evaluation and validity of the knowledge and information that we receive in our studies or in ordinary life. Sometimes we see how an area of science in which we are proficient is completely incomprehensible to others, and conversely, we cannot appreciate the complexity of the work of our colleagues in other fields.

The processes of unification and globalization in society have not passed over science. The interdisciplinary approach, the integration of scientific knowledge into related fields, and the adoption and dissemination of Sustainable Development Goals have led to a close intertwining of social and natural science research.

Achieving sustainable development in society is impossible without an integrated interdisciplinary approach to modern science. Usually, every scientist or researcher is a specialist in a rather narrow field of science. This is both good and bad. Good because he/she is a specialist who can solve any problem in his/her field. Bad, because the importance of this research for the society is not always obvious, until it becomes an object or phenomenon that helps people in their daily life or improves its quality. Today, the importance of a scientific discovery is defined not only by its usefulness, but also by its environmental friendliness, the minimization of resources and energy consumption in the process of its implementation, the cost of implementation, the attitude of the society to the implementation of this discovery, the absence of human rights' violation and the violation of the laws of the state where this discovery is being implemented.

Scientists should come together to solve global problems and not rely solely on the natural sciences or solely on the social sciences in meeting the global needs.

Recent decades have seen an increased demand from citizens, civic groups and non-governmental organisations for greater scrutiny of the evidence that underpins scientific conclusions. In some fields, there is growing participation by members of the public in research programmes, as so-called citizen scientists: blurring the divide between professional and amateur in new ways.

The merger of the two global scientific unions, The International Council for Science (ICS), and The International Social Science Council (ISSC), was the reflection of the objective need for the world science community to unite so that to provide an essential international capacity to meet the above-mentioned needs: to creatively draw on excellence across the full range of research disciplines, and support science to work with society in addressing critical global problems.

17.2 Modern Understanding of Science

The word "science" is used to mean the systematic organisation of knowledge that can be rationally explained and reliably applied. As in most languages except English, people use it to embrace all domains, including humanities and social sciences as well as the science, technology, engineering, and medical disciplines [1].

It is convenient to classify the sciences according to the "world" in which, i.e., in which area of knowledge, science "operates". Four such "worlds" can be distinguished: the world of ideas, the natural world, the cultural world, and the human world (vital, or practical). According to this criterion, the sciences are grouped into four classes: intellectualism, natural science, cultural studies, and praxeology.

Intellectualism as a subject uses the world of ideas, concepts of numbers, figures, and values. Such sciences include mathematics, philosophy, theology, etc.

Natural science as a class of sciences is fundamentally different from intellectualism. Its subject is nature, living and non-living. Natural science emerges in the process of humans' encounter with the surrounding reality. The basis of natural science is experience, which is gained through direct study of objects or phenomena. This experience cannot be gained only through reflection.

Culturology combines the social and historical sciences: sociology, history, ethnography, etc.

Praxeology unites sciences aimed at practical application; they are also called applied sciences. Applied physics, mathematics, chemistry, psychology, etc. apply their knowledge wherever possible. Praxeology also includes economics, pedagogy, political science, jurisprudence, and other sciences that implement generally accepted or meaningful values using scientific methods. Unlike natural science, praxeology is subjective—the practical application of knowledge can be contradictory. For example, chemical knowledge can be used to create modern medicines or, conversely, chemical weapons.

If analysing all scientific trends, two global scientific currents can be seen—natural science and social science.

This division of the sciences historically led to the formation of two global international councils in the middle of the twentieth century.

- The International Council for Science (ICS), founded in 1931 as the International Council of Scientific Unions (ICSU).
- The International Social Science Council (ISSC), founded in 1952.

17.3 Comprehensive Credible Scientific knowledge—the Key to Solving Global Problems

Knowledge derived from scientific research is a staple of human understanding and creativity. It is fundamental to the evidence that should inform societal decision-making and public policy. The importance of deliberative scientific understanding to society has never been greater, as humanity grapples with the problems of living sustainably and equitably on planet Earth [2].

It is vital therefore that science is safeguarded as a global public good. Its knowledge, data and expertise must be universally accessible and its benefits universally shared. A mutually supportive global community of science carries responsibility for this by ensuring inclusivity and equity, also in opportunities for scientific education and capacity development [3].

Scientists are increasingly expected not only to advance scientific understanding, but also to contribute solutions to pressing real-world problems, and to support transformative societal responses to them. As the range of global challenges embedded in the United Nations 2030 Agenda for Sustainable Development shows, these problems are often highly coupled and inherently complex [4].

The priorities and practices of science are determined both by the internal dynamics of scientific inquiry and by contemporary socio-political contexts. The pressure on science is to produce "actionable" knowledge that responds to the needs and expectations of society and that supports transformative societal responses to challenges of the present and foreseeable future [5]:

- *The imperative to help solve global problems.*

  Major contemporary challenges confronting humanity have global impacts that demand global responses that almost invariably require strong engagement from the world of science [6].

However, effective communication of science embodies a dilemma. A major principle of scientific enquiry is to "take nobody's word for it". Yet many areas of science demand levels of skill and understanding that are beyond the grasp of most people, including those of scientists working in other fields. A virologist is likely to have a poor understanding of cosmology, and vice versa. Most citizens have little alternative but to put their trust in what they can judge about scientific practice and standards, rather than in personal familiarity with the evidence. If democratic consent is to be gained for public policies that depend on difficult or uncertain science, the nature of that trust will depend to a significant extent on open and effective communication within expert scientific communities and their participation in public debate.

The achievements of science should be understood by everyone, but often society is not interested in basic research. They only become relevant when they become applied.

A realistic means of making data open to the wider public needs to ensure that the data that are most relevant to the public are accessible, intelligible, assessable, and usable for the likely purposes of non-specialists. The effort required to do this is far greater than making data available to fellow specialists and might require focused efforts to do so in the public interest or where there is strong interest in making use of research findings. However, open data is only part of the spectrum of public engagement with science. Communication of data is a necessary, though not a sufficient element of the wider project to make science a publicly robust enterprise.

The publicity of scientific knowledge is obvious and necessary. Society is aware of innovative changes in development, provided that these changes are massive and can be implemented not only in individual countries, but worldwide.

- *The need to defend the inherent value of scientific enquiry and interpretation.*

  A new digital world is providing unprecedented levels of global connectivity that has powerful implications for the relationships between citizens, the media, elected representatives, interest groups and experts, and more broadly, between science and society. The ubiquitous use of software tools and social media enable the democratisation of the processes whereby knowledge and information are generated and used. For science, this digital world offers great opportunities to reach new audiences. But it also drives a "post-expert" dynamic in which people regard access to information as obviating the need for scientific interpretation.

It enables the spread of misinformation and its growing use as an agent of political activism, strategy, and policymaking. The consequences of such "information accessibility" in the interpretation of scientific facts can be disastrous. Reduced trust in institutions, accusations of elitism, and broader trends towards populist politics all pose fundamental challenges to the value of deliberative scientific enquiry. Although scientists still enjoy high levels of public trust in many parts of the world, these developments change political dynamics in ways that make it harder for the scientific voice to be heard [7].

These were not new trends, but they were intensifying as the enduring consequences of ongoing changes that are technological, social, and cultural in nature. They created a context in which there was a distinctive need for international collective action to:

- *Enable science for the future by securing its practical relevance to complex global problems that no one country and no one discipline can address on its own.*

  This required strengthened international scientific collaboration that harnesses:

(1) scientific perspectives and expertise from all parts of the globe.
(2) a strengthened integration of knowledge through enhanced inter-disciplinary collaboration, involving the joint framing of issues and the collaborative design, execution, and application of research across different fields of science.

It required as well new ways of working in the trans-disciplinary mode of engaging decision-makers, policy-shapers, and practitioners, as well as actors from civil society and the private sector, as partners in the co-design and co-production of solutions-oriented knowledge [7].

- *Secure the future of science by advocating the inherent value and values of all science—from fundamental to stakeholder-engaged science—to society.*

This includes effectively communicating scientific knowledge in relation to a wide range of contemporary issues. It means promoting continued support for focused disciplinary enquiry and unfettered scientific curiosity. It also meant championing investment in scientific education, research, and development, particularly in the Least Developed Countries of the world [8].

Effective exchange of ideas, expertise and people between the public and private sectors is key to delivering value from research. The economic benefit and public interest in research should influence how and when data, information, and knowledge from publicly or privately funded research are made widely available.

These trends initiated the process of merging the two major international scientific unions, ICS [9] and ISSC [10].

17.4 The Motivation to Merge

The motivation to merge ISSC and ICS was based on the need to significantly strengthen the roles played by both organizations in response to insistent demands for science to address a series of major, often highly coupled challenges that have global impacts and are of unprecedented complexity. Many of these challenges cross physical–chemical, environmental, biological, and medical, socio-economic, political, and cultural domains and meeting them thus requires strong collaboration involving the full range of research disciplines. It also needs science to engage with a wide range of public and private stakeholders, including citizens, in the transdisciplinary creation of solutions-oriented knowledge for policy and practice.

Over the decade 2005–2015, ICS and ISSC worked increasingly closely together. As the result, they co-sponsored major international research programs (such as Future Earth [11] and the Integrated Research on Disaster Risk Programme [12]), partnered in delivering international capacity development initiatives (such as the Leading Integrated Research for Agenda 2030 in Africa grants and training scheme [13]), and collaborated in coordinating scientific input into the UN policy processes.

Apart from enabling more dynamic and responsive decision-making (through unified governance and management arrangements) and making more efficient use of available resources (for example, by reducing the costs of maintaining organizational structures and dual representation), the fundamental argument in favour of a merger was necessarily two-fold:

- The international setting within which science is done and the demands for it to contribute more directly to human wellbeing has changed in ways that needed greater integration of understanding from the different domains of science represented by the two Councils. This could be based on the joint framing of ideas and co-design of collaborative initiatives.
- The level of integration required was greater than could be achieved by two separate organizations working together. It depended on overcoming institutional silos and persistent barriers to effective inter-disciplinarity, for example, through regular opportunities for dialogue to understand fundamental differences between the natural and social sciences, and to build trust and confidence in each other through collaboration.

Given these features of today's global realities, the case for greater engagement by, and attention to, the social sciences is clear. Their knowledge is indispensable in the search for a clearer understanding of the causes and consequences of global environmental change, and for informing more effective, equitable and durable solutions to today's sustainability challenges.

So today we have a view of many global problems from two sides.

1. from the basic natural sciences and research,
2. from the side of society, economics, politics, and human perception of the problem itself.

The analysis highlighted that these two major challenges in the context of global science were likely to intensify rather than diminish [14]. They presented a distinctive need for a unified response from the global scientific community. Moreover, as the international scientific landscape has evolved, it has been occupied by many new bodies, thereby shrinking the traditional niche of ISSC and ICS. The proposed strategy identified a major area of unsatisfied need for action from the international scientific community. It was evident that the merged body would be uniquely qualified to respond.

The International Science Council (ISC) was created in 2018 from a merger of the International Council for Science [9] and the International Social Science Council [10].

The merger was intended to build a strong foundation for advancing science across the disciplines and in all parts of the world, and for protecting the Council's vital role in shaping humanity's future on planet Earth. The new organisation has drawn strength from its unique membership, which brings together scientific unions and associations, academies, and research councils.

The ICS main activities focused on its international programs; ISSU's priority was its world reports and fora. Both types of activity were retained in the new Council but reoriented to serve a set of highly focused global aims that would increase impact and reach. Checks and balances to safeguard equal voice and visibility of the natural and social sciences in meeting those aims have been addressed in proposals for the new Council's governance arrangements.

17.5 Global Challenges Require Unity

The world faces great challenges and society increasingly looks towards science to address them. This place demands on all fields of science in all parts of the world. It compels a global response, involving strengthened collaboration within the international scientific community and between it and the world of policy and business, civil society, and the public at large.

Global sustainability requires concerted action to protect the planet's bounty and, simultaneously, to safeguard social equity, human dignity, and well-being for all.

Today's global environmental problems can serve as a prime example. The inseparability of social and environmental systems and problems is more evident than ever.

Environmental problems cannot be separated from the other risks and crises that comprise current global realities. They are not disconnected challenges; they do not occur in discrete, autonomous systems rooted in the environment on the one hand, or in society on the other. Instead, they are part of a single complex system. Global environmental change is simultaneously an environmental and a social problem. Social science research helps us to comprehend the complex dynamics of 'social-ecological' or 'coupled human—natural' systems and can help explain how these systems unfold and interconnect across space, from the local to the global, and in time, from the past and present into the future.

Humans are living at a time when the Earth's land surface and climate, its elemental cycles, oceans, fresh water, ice, air and ecosystems, have all been altered fundamentally from the state they were in even just a few centuries ago. Scientists now know with great confidence that these changes are attributable primarily to human activity. Indeed, the 'Anthropocene' is increasingly regarded as a new geological era in Earth's history, one in which people take centre stage as the defining geological force.

This makes the causes, consequences, and responses to global environmental change fundamentally social in nature. Global environmental change is about humans changing global environments, and about humans, individually and collectively, shaping the direction of planetary and social evolution. The social sciences therefore have a vital role in enriching society's understanding of what it means to live—and maybe thrive—in the Anthropocene, and in raising awareness of the opportunities, accountabilities, and responsibilities this brings with it.

Today, environmental change research aims more than ever to integrate the social, natural, human, engineering, and health sciences. In this context, integration does not imply the loss of disciplinary strength or identity. On the contrary, it means being confident in one's disciplinary base and engaging with colleagues from other disciplines and fields in the joint, reciprocal framing of problems, and in the collaborative design, performance and application of research.

The call for more integrated science is dictated by the complexity of the environmental and sustainability challenges that society faces, and the inability of any single discipline or scientific domain to understand, let alone address, this complexity. Despite the progress made by many academic groups and scientific institutions across the world, the task of bringing the different sciences together in integrated global change research remains difficult. Much work remains to be done to clarify what integration means in practice, find effective ways of realizing it, and adjust institutional practices to support it.

No single discipline or scientific domain can understand, let alone address, the complex challenges involved in environmental change and sustainability. This is where an interdisciplinary approach to solving environmental problems becomes important.

In terms of acting in response to global environmental change, it is imperative to understand the need for a 'safe and just space' towards and within which pathways

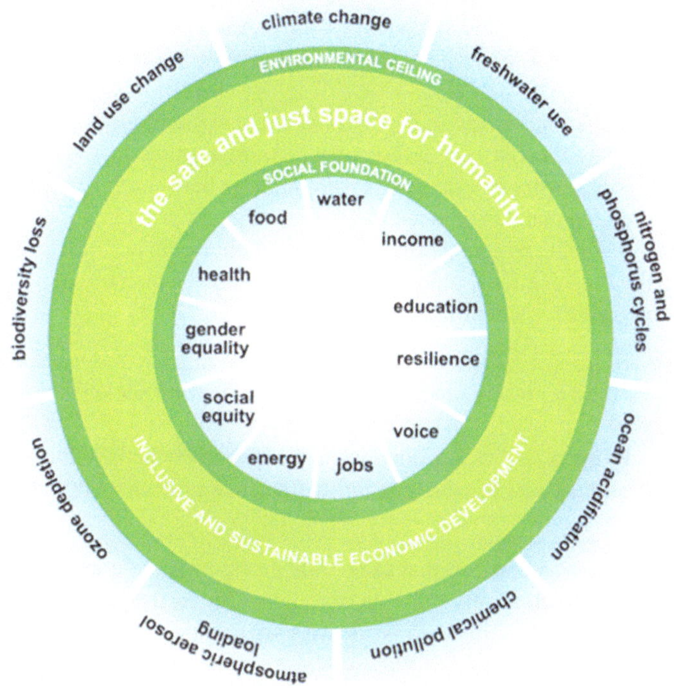

Fig. 17.1 Social and planetary boundaries [15]. *Source* K. Raworth (2012), "A safe and just space for humanity: Can we live within the doughnut?" discussion paper, Oxfam, Oxford, based on Rockstrom et al. (2009), "A safe operating space for humanity", Nature, No. 461, pp. 472–475

to sustainability must be steered. This space is defined by the planetary and social boundaries within which humanity can thrive without endangering the ecological resilience of the planet, or the well-being and security of its current and future inhabitants (Fig. 17.1).

17.6 Transforming Scientific Approaches

The world is changing every second and these changes require a new way of looking at many things. Science cannot stay the same. There are no mere national or international interests in scientific research, no mere humanities, or natural sciences. There are global problems that humanity must solve in the very near future. These problems hinder the achievement of sustainable development and endanger the lives of future generations and the existence of the planet.

One problem is the fragmentation of the scientific world and the inability, or sometimes unwillingness, to work together. Scientific advances have always been originally a national asset that has been shared over time. The time is ripe for a coming together of scientific knowledge and currents of knowledge to avert the crisis the world is about to experience. In whose hands is knowledge, in whose hands are the keys to success.

The international landscape of science has become increasingly competitive and fragmented during the last two decades. In this environment, ISSC and ICS were recognized for their work in stimulating collaboration between disciplines and countries in the creation and use of scientific knowledge and providing pathways for its international political and social influence. Their authority to perform these roles has come from the organizations' unique membership bases, which brought together the expertise and extensive scientific networks of

- the international scientific unions and associations that foster development of the disciplines they represent, and of the national scientific societies that adhere to them; and
- the national organizations—academies and research councils—that promote and support excellence and relevance of national science across a spectrum of scientific fields.

The merger was designed so that to provide an essential international capacity to meet these needs: to creatively draw on excellence across the full range of research disciplines, and support science to work with society in addressing critical global problems.

The natural sciences, and increasingly the social sciences, have played an important role in defining sustainability problems and environmental risks. Yet science alone cannot adequately define these problems or the solutions to them, partly because they mean different things to different people, and partly because science does not have universally accepted legitimacy for doing so. To make the knowledge claims underpinning environmental governance more salient and legitimate, social scientists have brought greater attention to the need for co-production of knowledge by scientists and the users of knowledge. The co-design and co-production of science and policy calls for new procedures.

The call on science to make a difference speaks to the social sciences no less than to the natural, physical, human, or engineering sciences. The concrete environmental challenges that societies face—water scarcity, the loss of biodiversity, the transition to a low-carbon society, food security, or greater preparedness for extreme events—are shared challenges, requiring joint scientific effort and priority setting.

The great achievements of science in recent centuries lie primarily in understanding uncoupled or weakly coupled systems. There is now a need for much deeper understanding of the nature and dynamics of complex systems through the integration of knowledge from disparate fields of the natural and social sciences. Exploiting the potential of "big data" and machine learning could contribute significantly to this end. But there is also a need for new ways of working, not only in the interdisciplinary mode, but also with approaches that integrate knowledge from academic and non-academic stakeholders in processes of mutual learning and problem-solving. Coupled with "open science", this "transdisciplinary" mode could be an effective means of improving the infusion of scientific knowledge into policy and practice [16].

Given that planet's systems are under rapidly growing and unsustainable pressures, and that human systems are inextricably linked to their fate, human security is clearly at stake. If societies are to maintain or establish such security, and successfully pursue together the larger quest for global sustainability, deep social transformation is needed. The social sciences are uniquely placed to clarify what this means, and what role science can play in finding solutions. Through engaged research, they can help society as a whole understand the changes required at individual, organizational and systemic levels, and how such changes could be realized in politically feasible and culturally acceptable ways.

New approaches to understanding transformative knowledge production emphasize the importance of open information and knowledge systems that facilitate collaborative learning and problem solving, around concrete challenges and in specific social-ecological contexts. In such systems, multiple sources of expertise are mobilized: scientists work with non-academic knowledge holders to co-design, co-produce and co-implement new knowledge, new priorities and mutual learning processes. In this way, open knowledge systems are arenas for the democratization of science, a process increasingly facilitated by cyberspace and new digital technologies.

A new digital world is providing unprecedented levels of global connectivity. This has powerful implications for the relationships between citizens, the media, elected representatives, interest groups and experts, and more broadly, between science and society. The ubiquitous use of software tools and social media means that the processes of generating and using knowledge and information are essentially democratized. For science, this digital world offers great opportunities to reach new audiences. But it also drives a "post-expert" dynamic in which people regard access to information as obviating the need for scientific interpretation. It enables the spread of misinformation and its growing use as an agent of political activism, strategy, and policymaking [17].

Growing concerns about the irreproducibility of otherwise highly regarded scientific work, reduced trust in institutions, accusations of elitism, and broader trends towards populist politics all pose fundamental challenges to the value of deliberative scientific enquiry based on verifiable facts. Although scientists still enjoy high levels of public trust in many regions of the world, these developments change political dynamics in ways that make it harder for scientific input to be heard [18, 19].

Open science, changing trends in education, information accessibility and the popularisation of fundamental knowledge are what will help humanity to maintain the sustainability of civilisation.

These are shifts towards much stronger forms of interdisciplinary collaboration between and across different fields of science in both the framing and execution of research; and towards the greater active engagement of scientists with a wide range of public and private stakeholders, including citizens, in the transdisciplinary co-production of knowledge for public policy and practical action.

The realization of such transdisciplinary, open knowledge processes involves significant challenges and opportunities for the scientific community, and for those responsible for organizing, funding, evaluating, and rewarding research. It calls for a fundamental transformation of the institutions and practices of science itself.

The social sciences have grown beyond traditional disciplinary boundaries in most developed countries. Interdisciplinary research is increasing across the social sciences and with the natural sciences and is encouraged by funding agencies. In Japan, for example, interdisciplinary research has been very much promoted since the triple Fukushima disaster, which cast doubt on natural science's capacity to anticipate or solve problems. Outside developed countries, however, interdisciplinary research is still rare.

17.7 A Unified, Global Voice for Science

Insights into the visions and conditions of change show that no single discipline or level of investigation can capture the complexity of how social change occurs. The story emerging from the contributions here is one of individuals richly and dynamically embedded in households, communities, sociotechnical systems, economies, and cultures. It goes a long way toward explaining the paradox of how the social drivers of global environmental change persist, or at least change only slowly, while environmental crises continue to unfold rapidly. Nonetheless, more research is needed on the power and embeddedness of individuals, and on the cross-scale connections between them in processes of social change.

The ISC provides a powerful institutional basis for addressing these needs at the international level. It has amplified a voice for science that can represent and support the full range of the scientific enterprise, and build the enhanced capacities required for it to address major global challenges in ways that neither the natural nor the social sciences can realize on their own.

In recent decades, the landscape of international representative bodies for science has become increasingly fragmented and by implication, competitive. Within this landscape, and in the broader context described above, there was an opportunity for the ISC to position itself as a unified global voice for science.

The ISC founding members and primary stakeholders were the former members of ISSC and ICS, including over 40 international scientific unions and associations, and over 140 national and regional organizations such as academies and research councils. These bodies represent approximately 70 per cent of the world's nations. They also encompass a substantial range of disciplines in the natural (including physical, mathematical, and life) and social (including behavioural and economic) sciences. In order to be a truly global voice for science, the ISC is seeking to expand its membership to include countries that have not been represented in ICS or ISSC, many of which are categorized as "least developed". It will also strengthen its standing as a unified voice for science by adding to its ranks unions and associations of key scientific or technological disciplines not represented by either organisation.

Most of the scientists believed that bringing these two domains of science together will create, for the first time, a unique opportunity for the co-framing of questions, co-creation of knowledge and co-framing and co-execution of responses to the major current and future challenges that confront humankind. The ISC actions during the Covid pandemic that humanity has been facing since 2019 confirmed the validity of this assumption [20].

The importance of scientific understanding to society has never been greater, as humanity grapples with the problems of living sustainably and equitably on planet Earth. The Council defends the inherent value and values of all science at a time when it has become harder for the scientific voice to be heard. It strengthens international, interdisciplinary collaboration and supports scientists to contribute solutions to complex and pressing matters of global public concern. It advises decision makers and practitioners on the use of science in achieving ambitious agendas such as the Sustainable Development Goals (SDGs) adopted by the UN in 2015 [21].

The vision of the ISC is to advance science as a global public good [22]. Scientific knowledge, data and expertise must be universally accessible and its benefits universally shared. The practice of science must be inclusive and equitable, also in opportunities for scientific education and capacity development.

The Council realizes its mission by convening the scientific expertise and resources needed to provide leadership in catalysing, incubating, and coordinating international action on issues of priority to the scientific community and the society of which it is a part. It directs its voice both "externally" on issues of major public relevance, and "internally," on matters that support effective scientific responses, particularly where new knowledge, capacities, resources or ways of working are needed. Its activities are delivered in collaboration with its members and key partners [23].

In the years since the merger of the ICS and ISSC to the present, the world has faced serious challenges, including numerous environmental crises, deepening social inequality and polarization, international armed conflicts and deteriorating human well-being. Problems often overlap and interact in unpredictable ways. They also influence and reflect problems in scientific systems around the world, including pressures on scientific freedom, tensions between different value and knowledge systems, the politicization of science, and declining trust in science as an institution—all of which make it difficult for science to make its best and necessary contribution to society.

ISC presents science to the world as a global public good, which means that scientific knowledge and practice should be universally regarded as a shared resource from which everyone should be able to benefit regardless of their location, gender, economic status, cultural background or other characteristics [24].

17.8 Conclusion

The achievement of sustainable development is impossible without scientific support, without a modern system of unification and dissemination of scientific knowledge, without public recognition of certain achievements of science. The merger of the International Council for Science and the International Social Science Council is an important step for the humankind towards solving global problems, achieving sustainable development, and preserving our planet. Humanity is gradually realizing that science and education are the basis for many changes, for the existence of future generations and for the sustainable development of planet Earth.

As for today, the ISC's unique global membership brings together 250 international scientific unions and associations, national and regional scientific organizations including academies and research councils, international federations and societies, and young academies and associations. The International Science Council actively works at the global level to catalyse and convene scientific expertise, advice and influence on issues of major concern to both science and society.

References

1. Violatti, Cristian. "Science". World History Encyclopedia. World History Encyclopedia, 28 May 2014 (2014). https://www.worldhistory.org/science/
2. Effective Integration of Science and Policy through Joint Fact Finding. Herman A. Karl, Lawrence E. Susskind, Katherine H. (Eds.), Wallace Environment: Science and Policy for Sustainable Development, Vol. 49, pp. 20–34 (2007). https://doi.org/10.3200/ENVT.49.1.20-34
3. Science without Borders, Richard Hodson. Nature **562**, S57 (2018). https://doi.org/10.1038/d41586-018-06970-5
4. Transforming our world: the 2030 Agenda for Sustainable Development. United Nations // Department of Economic and Social Affairs https://sdgs.un.org/2030agenda
5. Ethics in Scientific Research, An Examination of Ethical Principles and Emerging Topics by Cortney Weinbaum, Eric Landree, Marjory S. Blumenthal, Tepring Piquado, Carlos Ignacio Gutierrez Gaviria. Published by the RAND Corporation, Santa Monica, Calif (2019). doi.org/https://doi.org/10.7249/RR2912
6. Effectiveness of research and innovation management at policy and institutional levels in Cambodia, Malaysia, Thailand and Vietnam. https://www.oecd.org/sti/Effectiveness%20of%20research%20and%20innovation%20management%20at%20policy%20and%20institutional%20levels_Meek%20and%20Olsson.pdf
7. Knowledge Management and Institutional Accounting Functional Effectiveness: Evidence from Malaysia. Freida Ozavize Ayodele, Liu Yao, Hasnah Binti Haron and Eyasan Leslie Dabor. J. Inform. Knowl Manag. **18**(02), 1950017 (2019). https://doi.org/10.1142/S0219649219500175
8. More investment in Research and Development for better Education in the future? Rim Lahmandi-Ayed and Dhafer Malouche. ESSAI and MASE-ESSAI, University of Carthage. https://www.researchgate.net/publication/326570228_More_investment_in_Research_and_Development_for_better_Education_in_the_future
9. International Council for Science (ICSU). https://www.staff.city.ac.uk/tom.davies/ICSU.html
10. International Social Science Council (ISSC). https://www.worldsocialscience.org

11. Future Earth is a global network of scientists, researchers, and innovators collaborating for a more sustainable planet. https://futureearth.org
12. Global Platform for Disaster Risk Reduction. https://www.irdrinternational.org
13. Leading Integrated Research for Agenda 2030 in Africa. https://council.science/what-we-do/funding-programmes/lira2030/
14. N.P. Tarasova, A.S. Makarova, Assessment of the chemical pollution in the context of the planetary boundaries. Russian Chem. Bull. **65**(5), 1383–1394 (2016). http://www.kluweronline.com/issn/1066-5285 https://doi.org/10.1007/s11172-016-1467-z
15. K. Raworth, A safe and just space for humanity: can we live within the doughnut?" discussion paper, Oxfam, Oxford, based on Rockstrom et al., "A safe operating space for humanity", Nature, 2012 No. 461, pp. 472–475 (2009). https://www-cdn.oxfam.org/s3fs-public/file_attachments/dp-a-safe-and-just-space-for-humanity-130212-en_5.pdf
16. Reflecting on the geographer's contributions to the observation of climate change effects Soto, D., Thimonier-Rouzet, E., Carré, H., Langlois De Septenville, W., Renard, F.2016 Bulletin d'Association de Geographes Francais. **93**(3), c. 315–323. https://journals.openedition.org/bagf/1224
17. Sustainable development requires environmental justice. M.J. Ellerbrock, Intern. J. Sust. Devel. Plan. **13**(2), c. 208–214 (2018)
18. Will public trust in science survive the pandemic? Ryan Cross C&EN, **99**(3), 19–29 25 (2021). https://pubs.acs.org/doi/https://doi.org/10.1021/cen-09903-feature1
19. Science Is Political Cite this: Chem. Eng. News, **90**, 8, 50–51 Publication Date: February 19 (2012). https://doi.org/10.1021/cen-09008-books
20. The Territorial Impact of COVID-19: Managing the Crisis and Recovery across Levels of Government. 10 May 2021 (2021). OECD https://www.oecd.org/coronavirus/policy-responses/the-territorial-impact-of-covid-19-managing-the-crisis-and-recovery-across-levels-of-government-a2c6abaf/
21. Sustainable Development Goals: Why Do We Need Them? O. Mironenko, P. Lucas, N. Tarasova, J. Zlinsky. Social Evolution & History. Volume 14, Number 2. September 2015, pp. 176–190 (2015)
22. Scientific knowledge as a global public good: Contributions to innovation and the economy Dana Dalrymple The role of scientific and technical data and information in the public domain: proceedings of a symposium. National Academic Press, pp 35–49 (2003). https://www.ncbi.nlm.nih.gov/books/NBK221876/
23. International Science Council (ISC), Advancing science as a global public good—action plan 2019–2021. International Science Council, Paris (2019). https://doi.org/10.24948/2019.09
24. The ISC Draft Strategic Plan 2025–2028. https://council.science/about-us/strategic-plan/

Chapter 18
Choices, Motivations, Additional Reflections, and Take-Home Messages

Liliana Mammino

Abstract This chapter has a conclusive role for the discourse of the relationships between humanities and sciences developed throughout the previous chapters. It further emphasises the importance of identifying implementation routes for working conceptions of the relationships between humanities and sciences. It stresses the fuzzy nature of the boundaries of the two domains, resulting in the impossibility of giving definitions which can univocally indicate to which domain a given discipline or study area belongs in an exclusive manner. It provides summary overviews of topics which have not been devoted individual chapters because of contingent reasons, but which would deserve further elaboration. Finally, it outlines take-home messages from individual chapters and from the whole book, once more emphasising the working conceptions components and inferences.

Keywords Challenges for defining science and humanities · Ethics and science · Language and education · Take-home messages from this book · Working conceptions of the relationships between humanities and sciences

18.1 Introduction: The Role of This Chapter

This chapter has a conclusive role for the discourse of the relationships between humanities and sciences developed throughout the previous chapters, and performs a number of tasks, which can be summarised as follows:

- Reiterating and confirming the intended objectives and roles of the present book.
- Explaining the motivations of some relevant choices that have been made. This will also provide answers to some questions that might arise, above all in view of the novelty of the perspectives from which the relationships between humanities

L. Mammino (✉)
Faculty of Science, Engineering and Agriculture, University of Venda, Thohoyandou, South Africa
e-mail: mammino@univen.ac.za; sasdestria1@gmail.com

L. Mammino (ed.), *Insights into the Relationships between Humanities and Sciences*, https://doi.org/10.1007/978-3-031-97900-2_18

and sciences are viewed in this book. The explanations may be presented in different sections, for reasons of suitability with individual sections' contents.
- Providing summary overviews of topics that have not been included because the authors who had started writing about them did not maintain adequate confidence in the search for the needed interdisciplinary outlooks, or because no interested author was identified.
- Providing some reflections on the perspectives of analogous investigations within other cultures.
- Presenting take-home messages for individual chapters and for the entire book.

The chapter has been added after the realisation that working familiarity with cross-disciplinarity may not be so widely spread and, therefore, it may be useful to include some additional clarifications on the scope and objectives of the book, on the selection of individual topics and approaches, and on the envisaged outcomes. Reiterated emphasis on the fact that the book intends to foster working conceptions of the relationships between humanities and sciences can facilitate better understanding of its scope, and the summary presentation of take-home messages can make the 'working conceptions' concept more concrete.

For the sake of conciseness, the author of this chapter will be denoted with the LM acronym in the rest of the text.

18.2 The Objectives of the Present Book

This book is meant to foster working conceptions about the relationships between humanities and sciences, i.e., to foster cross-disciplinary attitudes as well as the outlooks needed to turn them into practice, including professional practice.

The *interdisciplinary* and *cross-disciplinary* terms are becoming increasingly fashionable, parallel to the realisation that the practices that they denote are becoming increasingly crucial to successfully tackle a variety of problems and pursue important objectives, including the pursuit of sustainable development. Specialists in areas that have for long operated separately find the need to converge on some issues and to integrate their contributions. For instance, there is increasing awareness of the benefits that experimental chemistry research can attain from synergies with theoretical/computational chemistry [1–4].

On the other hand, practices taking cross-disciplinarity between humanities and sciences into working account are largely still to be designed or internalised. For instance, LM published a book titled "language and science" in 1995 [5]; it turned out that bookshops' owners or managers did not know what to do with it, and often ended up not putting it on their shelves, because it did not respond to standard classifications by subjects: it could not fit in the humanities shelves because of the word "science" in its title, and it could not fit in the shelves devoted to the sciences because of the word "language" in its title (with the conjunction "and" putting the two areas on the same level); it would appear logical that a work whose content pertains to two areas should be classified within each of them, but the dominant compartmentalization of knowledge prevented it. Nearly 30 years later, LM has realised the frequent absence of protocols for a true multidisciplinary handling of multidisciplinary works. All this indicates that there is still a long way to go for the interdisciplinarity concept to be integrated into working options and practices.

In order to promote working conceptions of the relationships between humanities and sciences, this book highlights cases in which such relationships play important roles for the final results within each of the two domains. It thus shows how resources born and developed within one of the two domains can be fundamental for the other. The cases utilised as documentation and examples are derived from professional activities, written or media sources, and also everyday-life situations within professional contexts.

The plural form "insights" in the book's title clearly informs that the book considers selected aspects or components viewed as particularly relevant (it may be noted that language mastery is fundamental also to infer the purpose of a book from its title). The book is not a systematic treatise; the possibility of a systematic treatise may develop after many more investigations like the ones presented in this book are completed.

A number of chapters highlight how humanities-pertaining resources (first of all, language mastery and epistemology) are fundamental for the practice and development of sciences, and document how their absence or inadequacies jeopardise understanding and creativity within the sciences, or the efficiency of communication concerning science or education themes. On the one hand, this relates to the fact that nearly all the authors of these chapters are practising scientists and, therefore, they have direct experience of the roles of humanities-born skills in science practice and science education. A more important motivation than the authors' areas of expertise relates to the fact that some humanities areas risk becoming 'endangered species'. Official attention increasingly focuses on STEM (science, technology, engineering and mathematics), viewed as the motors and instruments of development. On the other hand, innovation requires creativity, and creativity sprouts more extensively and novelty-rich from a broad, fertile background. Creativity needs understanding, and the depth and completeness of understanding depend on language mastery and on adequate epistemological awareness. Language mastery, in turn, has its roots and growth within the humanities.

The consideration of specific themes is often accompanied by summary reviews of historical developments of the relationships between humanities and sciences relevant for the theme concerned. This is important because it is important to stimulate the awareness that the significance of the various relationships and connections between the two domains is not something appearing now 'out of the blue sky', but is something that has been object of major attention by scholars from different areas for many centuries and is, in a way, part of the intellectual and knowledge heritage from which modern research has developed. Like for many other aspects of human activities, the knowledge of the historical roots reinforces the foundation on which to build new advances, and provides motivations and criteria.

The book is meant for a broad range of readers—all those who may be interested in the considered themes and questions. In particular, it aims at stimulating reflections capable of having immediate impacts on the way in which:

- students perceive their subjects and reflect on their approaches to learning;
- teachers present the subjects to students and design options to facilitate understanding and to foster skills that will be useful throughout their professional careers;
- decision-makers—at all levels—pursue broad perspectives, promote interdisciplinarity, and refrain from decisions that might penalise one domain without the awareness that they would automatically penalise the other (in current times, the humanities are the domain more at risk of penalisation);
- anybody interested in how knowledge grows and is expressed is stimulated to reflect on the connections between the historical development of knowledge, and of the theories about it, and the practical aspects stemming from it and becoming part of intellectual and professional activities.

Besides the chapters specifically devoted to education, several other chapters include reflections about implementations within educational frameworks, because of the general importance of education for the future of all the areas.

18.3 The Definitions of Humanities and Sciences

Section 1.1 in Chapter 1 introduces the discourse on the relationships between humanities and sciences on the basis of the common understanding about them. On this basis, everybody would feel confident to ascribe physics, chemistry, mathematics or biology to the science domain and literature, art or philosophy to the humanities domain [6]. But there are disciplines that cannot be easily ascribed only to one domain, because they share characters from both of them. That section mentions the example of architecture as illustration. Various other disciplines have partial characters from both domains. For instance, history studies the development of humankind through all the times for which we can have records and, as such, pertains to the humanities domain. On the other hand, it does utilise tools developed by various sciences for its research, and it proceeds through the study of documents and artefacts, which can rightly be viewed as a scientific approach. The objection that could be put forward by those who consider the scientific method as something analogous to a cooking recipe is the absence of reproducibility, because historical events belong to the past and are not reproducible by their own nature; but this feature (absence of reproducibility) is also true for sciences like geology, palaeontology or astronomy.

All this indicates that the two domains do not have sharply defined (or crisp) boundaries. In more technical terms, one could say that their definitions would unavoidably be fuzzy, in a similar way in which mathematics speaks of 'fuzzy sets' [7]. The introduction to [7] explains that

> the classes of objects encountered in the real physical world do not have precisely defined criteria of membership. … Yet, the fact remains that such imprecisely defined "classes" play an important role in human thinking, particularly in the domains of pattern recognition, communication of information, and abstraction.

This fully applies to the humanities and sciences domains, and implies that each discipline could pertain for a certain percentage to one domain and for the complementary percentage to the other, and the proportions are different for different disciplines. For instance, physics, chemistry or mathematics will have a much higher proportion for the "pertaining to the sciences" classification, and a small proportion—if any—for pertaining to the humanities; on the other hand, the importance of language, epistemology and logic for their nature and development excludes the possibility of ascribing them as pertaining 100% to the sciences and 0% to the humanities. Architecture will likely have comparable proportions of 'belonging' to each domain. Identifying possible percentages would go beyond the scope of this book and would, in any case, be uninteresting. Identifying the interplay routes between the two domains would pose the challenges of identifying relationships between each pair of areas pertaining to one and the other domain, and utilising them to build an overall picture. This operation would show the complexity of the relationships between humanities and sciences. It would simultaneously highlight the richness and complexity of human intellectual activities, and the impossibility of compartmentalizing them into a series of *boxes* and sub-boxes [6] (the *boxes* image with reference to the compartmentalization of knowledge is effectively outlined in [8]).

The approach adopted in Sect. 1.1, Chapter 1, to introduce the discourse on the relationships humanities and sciences is meant to foster the perception that the boundaries of the two domains are non-crisp and to add it to the common perceptions of the two domains. It thus creates a perceptions-background which is fully adequate for satisfactory understanding of the other chapters, including all the details of the relationships and interplays of the humanities and sciences, which are explored in individual chapters. In other words, respecting the fuzzy character of the potential definitions of the two domains, and making it a component of the background information generated by the introductory chapter, does not create any obstacle to quality understanding, and has the advantage of not limiting the reader's perceptions with the introduction of artificial barriers in a work (like this book) that aims at overcoming barriers.

Many definitions of science have been proposed, in relation to specific philosophical views on its nature. The more comprehensive definitions (like the one reported in Sect. 8.2.3.1 in Chapter 8, viewing science as "*a generator of models and a flexible intermediary for our understanding and description of the reality in which we are inserted*" [9]) would require detailed explanations of the meaning of some of the terms and of the overall philosophical meaning. This further justifies the selected option of leaving the meaning and extent of the two domains to the sphere of common perceptions, and hinting about the challenges of defining them within sections providing some outlines of the relevant philosophical issues. The challenges associated with the fuzzy character of the hypothetical boundaries of the two domains are presented here because, at this stage, a reader has accumulated enough information to be in a position to grasp it.

Chapter 17 provides some formal definitions because it concerns an international body, and it is common practice that institutional bodies provide definitions. By the time a reader reaches that chapter, he/she will have developed sufficient personal opinions to be in a position to know the extent to which he/she agrees with any definition that might be reported.

18.4 The Chapters that Were not Written

The dichotomy between the two domains is already so largely established that viewing their relationships in various areas of human intellectual activities may pose challenges that may become paralysing. It has happened for a number of themes that had initially been envisaged to develop into chapters, but for which potential authors did not find pathways for the combination or integration of the two perspectives into some working outcomes, or for factual collaboration between authors from different domains, and the development of the chapters did not proceed beyond the initial stage. This further highlights the frequent lack of real possibility to 'talk to each other' between experts in different domains, stemming from not sufficiently 'knowing each other' (knowing the significance of each other's fields) and, consequently, not knowing *how* to talk to each other. Brief outlines (prepared by LM) are given here, for the sake of information completeness of the book; some of them can also be viewed as 'seeds' that might bloom into more complete and detailed works in future.

18.4.1 Ethics and Science

The term "science" in this section is used with a very general meaning; it often refers to all research areas, including the humanities ones, except when the discourse can be clearly identified as referred to the physical and natural sciences.

18.4.1.1 Ethics in Science Practice

Ethics is the part of philosophy which investigates the issue of moral 'right' and 'wrong' [1, 10]. The question has been investigated since ancient times and concerns all human activities: it is an area in which theoretical progress directly translates into practical behaviour recommendations. Within the current discussion, and notwithstanding the inherent simplification of the criterion, it can be taken that [11]

> Whatever promotes the most good for the highest number of people is morally right and ought to be pursued, whatever does the opposite is morally wrong and ought to be avoided.

In past centuries, ethical research could at least approximately keep the pace with the advances in other areas, and the new scientific knowledge that was progressively acquired did not go so deeply into the nature of matter and life as to provide the tools for possible large-scale destruction or large scale alterations of the nature of living forms. This has profoundly changed with the discoveries of the XX and XXI centuries. The knowledge about the nature of the atom has given humankind weapons capable of destroying all humans and most other forms of life on the planet. The knowledge of genetic mechanisms and other key aspects of the chemistry of living beings has given humans the possibility of interfering with the nature of life forms as they have resulted from millions years of evolution, and of creating new forms of life. These developments have been so fast and so deep-reaching that ethics research has not kept the pace. The concern about ethics research having remained behind with respect to the pace of scientific discoveries had already been expressed by Einstein. Albert Einstein and Bertrand Russell (both being simultaneously scientists and philosophers) were particularly concerned about the inherent unethical-ness and horror of the concept and existence of wars, and put great efforts in the search for effective ways for their prevention, also in view of the fact that the invention of the atomic bomb had introduced the risk of global destruction into human history. The gap between scientific discoveries and ethics has continued widening since then. Isaac Asimov's famous statement "*The saddest aspect of life right now is that science gathers knowledge faster than society gathers wisdom*" fully applies to the current situation.

The progress of science cannot be stopped: there will always be new discoveries following previous ones. The key ethical issues concern the application of new knowledge. The conflict between economic aspects (the possibility of profit from the application of new technologies based on new discoveries, suggesting fast implementations) and the normal caution which would recommend prior detailed investigation of possible mid-term and long-term impacts, has often been won by the former. In theory, the precautionary principle should constitute the safest guidance for the protection of human health and the environment [12–15]. The principle can be formulated as [16]

> a broad epistemological, philosophical and legal approach to innovations with potential for causing harm when extensive scientific knowledge on the matter is lacking. It emphasizes caution, pausing and review before leaping into new innovations that may prove disastrous

Decisions on large-scale applications of new discoveries should be guided by this principle. The complexity of the phenomena taking place in the environment, and the enormous complexity of those concerning living organisms, entail a huge number of possible impacts which can result from the introduction of new chemicals [3] or new organisms. The awareness of such complexity recommends consistency with the precautionary principle.

It also needs to be acknowledged that existing modelling tools do not enable fully certain predictions because they are not yet able to incorporate all the aspects of the complexity (many of which may not even have yet been fully recognised). Therefore, the precautionary principle would prevent interventions that might alter fundamental equilibria on the planet and would be irreversible. An example could be the hypothesis of ejecting large quantities of particulates into the upper atmosphere in order to partially shield sunlight and cool the planet. It is impossible to exactly predict the effects on a large scale, and even small errors in the prediction might engender catastrophic effects, including a new ice era, or impacts with in the plant ecosystems that could affect agriculture as well as the remaining natural areas, with heavy impacts both on human beings and on the other living species.

The open letter signed by many specialists and asking for a pause in the development of advanced AI system, to give time for a proper evaluation of the possible impacts and the preparation of suitable regulations (Sect. 6.3, Chapter 6) also responds to the precautionary principle. Issues like distinguishing between true and fabricated information is crucial for any application of the information provided by AI. Ensuring privacy to citizens (including ensuring that the information present in their computers is not picked up by a chatbot and added to its own storage) are fundamental components of human rights. It remains essential that humans are always fully aware and fully in control of whatever AI may develop.

18.4.1.2 Ethics in Science Education

Ethics needs to become an integral component of the training of scientists, not only in terms of ethical clearance forms when required by regulations, but in terms of thorough honesty with regard to all the aspects of research: integrity in all the research steps and in reporting results, and also systematic abstention from pursuing personal advantages from one's research-related leadership responsibilities (refraining from putting pressures on colleagues to include their names in publications to which they have not contributed, or from requesting that authors cite one's own publications when acting as reviewers or editors). It also needs to be an integral training component in terms of familiarising students with the philosophical background of ethical questions, so that they acquire the ability to relate them to the relevant aspects of the science in which they get trained. The moral imperative of avoiding causing damages should be part of the acquired awareness of anybody practising science, as well as anybody making decisions. Conversely, philosophy students should be given enough information on the complexity of the phenomena concerning living organisms, and those concerning the environment and the planetary equilibria, to be able to perform ethics research in a way that is functionally linked to the latest progress of scientific knowledge and the ensuing technological innovations. The design of teaching/learning materials suitable for these educational components is a challenge requiring the collaboration of scientists and philosophers. The habit of 'talking to each other' could be fostered through projects simultaneously involving science students and philosophy students, so that it gets sufficiently internalised before they start their professional careers.

At pre-university level, the incorporation of ethics components into science courses is an integral constituent of the preparation of responsible citizens. Responsible citizens need to have enough exposure to basic science to be able to make informed decisions when needed. Primary and secondary instruction needs to provide basic science knowledge and the basis of science-related ethics, to enable future citizens to make decisions with which they can be ethically comfortable (it is here taken for granted that uninformed decisions could not be considered ethically comfortable).

18.4.1.3 Ethics in Science Research and Practice, and Ideologies

The root word of ethics (ἔθος) had the meaning of custom, character and, by extension, behaviour. Therefore, fundamental attitudes that have been parts of science traditions and practices across centuries can rightfully be considered as integral components of the ethics of 'doing science' or 'being a scientist'. A tradition of paramount importance is the full independence of science, and all what is connected to science, from factors that tend to generate divisions among people. The absence-of-discrimination tradition was common in the ancient world. In post-classical Europe, the tradition goes back to the first European institution of higher learning (a university-type institution), the Salerno Medical School, which was already flourishing in the IX century; according to legends, it was founded by four scholars of different origins: the Latin Salerno, the Greek Pontus, the Jew Elino and the Arab Adela; it also enrolled women for all the levels of learning and specialisation. Such open-mindedness towards women did not transfer to universities after they were founded, and women remained excluded for long; the first time that a woman obtained a university degree was in 1678, when Elena Lucrezia Cornaro Piscopia graduated in philosophy at the University of Padua. On the other hand, scientists, artists and scholars did not practice discriminations on the basis of nationality, ethnic origin, and other criteria alien to intellectual activities. The idea that scholarship is incompatible with any form of discrimination has continued growing and being part of normal practices in the last centuries. It was infringed by some extremely dictatorial regimes (like the Nazis), but the general consensus in academic communities has been that scientists do not discriminate other scientists. On this basis, it becomes a moral duty for all scientists to make sure that no scientist is discriminated on the basis of their nationality, race, gender, religious or political belief, or any other factor that is alien to science. Individual scientists may campaign for their political, religious or other points of view in any other forum, but not use them as discriminatory criteria within the practice of science. Building the awareness of this moral duty needs to be part of education in any area. It can be substantiated by reading philosophical considerations and historical documents, and also recent documents such as [17].

The full independence of science is an objective that needs to be unremittingly pursued. It poses the challenges of looking into the relationships between science and ideologies. The term "ideology" can take an ample variety of meanings, stemming from its basic definition of "set of opinions or beliefs of a group or an individual" [18]. When considering the ethics of science practice, the meanings that are more relevant are those that refer to situations in which ideologies attempt to control science, or to incorrectly use science to justify their beliefs and actions; in such contexts, ideologies can be viewed as sets "of collectively held ideas about society, usually promoted in order to justify a certain type of political action" [19] and ideological thinking as something that "can be described as a style of thinking that is rigid in its adherence to a doctrine and resistance to evidence-based belief-updating (i.e., doctrinal) and favourably oriented toward an in-group and antagonistic to out-groups" [20]. Ideologies of this type may gain power either because they actually control the politics of a country, or because they have been absorbed by certain mobs up to the point of blurring rational thinking and behaviour. Probably the most famous historical case of an ideology wanting to control science, or what a scientist thinks, is the Galileo's case, where the ideology was the interpretation of the Catholic Church about the nature of the solar system and its assumption that it had the right to impose itself on scientists and philosophers.

There have been other prohibitions and impositions in the past: alchemy had been banned in a number of European countries; the idea that matter consists of particles was discouraged by the Catholic Church because Aristotle (whom they had adopted as the most truthful philosopher) had opposed it; the Nazi forced scientists to support their racist theories; Stalin forbade genetics research; and several others can be found through history. Examples of mobs becoming irrational on the basis of ideologies could be the murder of Hypathia, torn to pieces by a furious Christian mob because she was a woman-scientist and she was not a Christian, or the destruction of the Alexandria library—the most valuable and rich library of the ancient Mediterranean world.

Educating the young generation about the importance of maintaining the independence of science may consider two major categories: the verification of the scientific validity of the positions that an ideology attempts to impose on the one hand, and the misuse of true scientific information that ideologies may bend to their own belief patterns, on the one hand; and the position of scientists within a environment in which dictatorial-type ideologies have large control, on the other hand. Historical examples are the best for illustration purposes, because facts are established beyond the emotional involvements that may sometimes be associated with current ideologies. Once students learn to think independently, they will also be able, on their own (not as part of in-class exercises) to evaluate current ideologies in a rational way.

The form of racism considering the so-called Arian race as superior has engendered the worst horrors in XX century Europe. Some theories about the superiority of the German race, which appeared in the second half of the XIX century, had soon been discarded as pseudoscience by the majority of the European scientific community, because they were devoid of any experimental or rational basis. On the other hand, young Hitler was fascinated by them and used them as basis to develop his racist theories. He tasked some scientists to find proofs, including through expeditions to various parts of the planet, in search of affinities between the populations of Northern Europe and the populations of other areas with highly developed civilizations, and he used ideology-guided interpretations of the results to strengthen the credit given by the public to his ideology [21, 22]. At the same time, the scientists performing those studies and supporting those interpretations enjoyed credit and support. As it became known after the end of WWII, the Nazi's racial theories constituted the basis for the murder of many millions of people in concentration camps, and for additional horrors like the experiments that some Nazi doctors performed on the prisoners.

The other aspect concerns the ethical position of scientists in situations in which ideologies try to control their activities and interpretations. The theoretically ideal answer is self-evident: the only ethical position is that of opposing the dictates from the ideology. On the other hand, there are many context-related questions that may dim the clarity with which any person, including scientists, manages to understand all the aspects of the reality of a given ideology and its implementation; in addition, the weight of personal considerations, including one's safety or one's job, is unavoidably present. These aspects can be highlighted by recalling the situation of German physicists during the Nazi period [23]. At that time, German physics was the most advanced in the world, putting the basis of quantum mechanics and of the knowledge of the atom (the same knowledge that enabled the possibility of the atomic bomb). Some physicists (like Albert Einstein) openly opposed the regime and chose (or had) to emigrate; few agreed with the regime; many others tried to survive and continue their research as they could, under impositions such as the prohibition of using or mentioning the relativity theory because the regime viewed it as "Jewish science" (Einstein was a Jew). The final outcome (after WWII) was the dispersal of many of those physicists, with the strongest ones being claimed by the major Allied countries.

Another inference is important in terms of relationships between humanities and sciences: the need that experts in both domains join efforts to counteract impositions or diktats from ideologies, which entail inherent violations of the right to think independently. Such experts share the task of educating open-minded new generations, showing in practice that intellectual freedom is a value for both domains, and an essential requisite for their development, both for the specialised themes of each of them and for the growth of their common endeavours.

18.4.2 Medicine and the Humanities

Attempts to treat wounds and diseases are nearly as old as the human species. Their quality and efficacy has evolved parallel to the development of the various civilizations. Their progress required knowledge of anatomy and of the processes occurring within the body, and knowledge of the medicinal properties of materials available in a given context, with plants being the most widely diffuse. Papyri from ancient Egypt include descriptions of various recipes for the treatment of various ailments. In ancient Greece, the philosophers' views on matter influenced the models of the components of the human body. The four elements theory (earth, water, air and fire) for the non-living matter was paralleled by the four humours (fluids) for the human body: blood, yellow bile, black bile, and phlegm; then, a disease entailed an imbalance of the humours and medications were aimed at restoring the balance. Practical experience discovered a huge amount of remedies, mostly from plants; extensive information for ancient times is provided in Pliny's works [24].

In the XVI century, alchemists in Europe developed increasing interest in medications. Paracelsus asserted that the main purpose of alchemy was the discovery of medicines to treat diseases. The iatrochemists (XVI and XVII century) combined alchemy and medicine in an attempt to understand the chemistry of the human body and to search for medical remedies in their laboratories. The close association of medicine with the other sciences continued strengthening throughout the following centuries. The study of anatomy and of the processes occurring in the human body entailed overlap of medicine and biology. The contribution from chemistry grew into a new specialised area, biochemistry. The continuation of this association in our epoch is well known: the development of new drugs requires contributions from pharmacy, chemistry, biochemistry, and medicine.

It may also be interesting to note the continuity of an attitude tending to adopt the newest chemistry discoveries as the most promising remedies for as many diseases as possible. Like Paracelsus, fascinated by the properties of the newly discovered antimony, used to prescribe it for a number of diseases, medical doctors have prescribed newly discovered drugs much more frequently than warranted. After antibiotics were discovered, they were prescribed even for minor problems (like colds) for which they were not needed, with the outcome that a number of pathogens have developed resistance to them. Nowadays, the enthusiasm for the discovery of the structures of RNA and DNA, and the possibility of sequencing them, is prompting a tendency to explore expanding their use to the treatment of as many diseases as possible. Maybe the awareness of undesirable previous outcomes of this tendency to generalise the use of new remedies (from Paracelsus' extensive use of antimony to all the similar occurrences in the subsequent centuries) could provide cautionary guidelines and warn against hurried-up broadenings beyond specific and well-tested targets for which the use is surely beneficial.

The interfaces and overlaps of medicine with humanities areas are numerous and bring significant contributions to both. The history of medicine (like the history of any other science) is an overlap between history (a humanities-based area and approach) and the development of knowledge about the human organism, the nature of the various diseases, and the remedies to treat them. It also links to cultural aspects (many remedies were born as part of what is now termed *ethnomedicine*) and to the perceptions about the psychological features of diseases. And it can foster perceptions resulting in greater caution and greater specificity in the use of medications and medical procedures.

Ethics is crucial in medicine, because it deals with human beings. Medicine is the only science requiring an ethics-related oath from those who start practising it—a tradition that goes back to the IV century BC (Hippocrates oath). Familiarization with the philosophical bases of ethics provides the best route to engender an attitude of continuous enquiry towards ethical issues that relate to the treatment of patients.

Up to few decades ago, 'classical studies' (comprising Latin, ancient Greek and philosophy) were considered important for young people wishing to choose medicine for their future profession, and secondary schools like the Italian *Classical Lyceum* were the recommended option. These studies offered the practical advantage of familiarization with the root words from which medicine's technical terms are built. More important, they fostered an attitude of continuous attention to human aspects, which translated into care and empathy towards patients. The attention to a human being as a whole—as opposite to viewing the human body as an ensemble of 'pieces' that can be treated as if they were self-standing—enables safer and more effective treatment options and needs to be promoted.

The experience of dealing with the most difficult aspects of human beings' lives (disease, death) has prompted a number of medical doctors to write about these aspects in forms meant for a broad range of readers: collections of short stories, or entire novels, where the narration is at the overlap of medical knowledge, understanding of human nature, and literature. This phenomenon of doctors-writers is present in many contexts and corresponding languages: English, Italian, Spanish, Swedish, and many others. It integrates professional expertise and empathy with writing and communication abilities.

18.4.3 Architecture, Humanities and Sciences

Architecture belongs to both domains (Sect. 1.1, Chapter 1). It requires specialized knowledge from several sciences: physics, chemistry, materials sciences, engineering. It also requires various humanities-based components: the history, traditions and cultural roots of the place in which the new building/s will be located, and the features of its landscape, so that the structural shape and other aesthetical characteristics of the new building harmonize with the context. In this way, knowledge components from different areas of the two domains converge to generate new approaches, a new outlook, and new expertise [6].

The two historical milestone-treatises on architecture pertain to the history of science, of engineering, of the arts, and of literature. Towards the end of the first century BC, Marcus Vitruvius Pollio wrote a 10-books' work titled *De Architectura* (on architecture, [25]), in which he presents all the architecture-related knowledge of his time. The content comprises town planning, building materials (whose properties are analysed in terms of the four elements theory), various categories of buildings (temples, civil buildings, houses), decorations on walls and pavements, aqueducts, applications of sciences (geometry and measurements, astronomy) in architecture, and the construction and uses of machines for various purposes, including water pumping ones and military ones.

In 1570, in Venice, Andrea Palladio published *I quattro libri dell'architettura* (the four books of architecture), which influenced architectural styles in Europe for centuries and, later on, also in the USA. He describes architectural styles, all the components of a building, criteria for the selection of building materials, the construction of roads, bridges, squares, and public buildings like basilicas, selecting examples from different times (including ancient) and from various locations, including buildings that he had designed. He accompanies the descriptions with detailed drawings, starting a tradition that has become an essential component of architecture and engineering.

In modern times, architecture still continues being a convergence area of various science and humanities disciplines. Analysing these components, how they converge, and the outcomes of the convergence, can offer interesting pathways to overcome knowledge compartmentalisation both at educational level and within professional practices.

18.4.4 Music and Science

Music is an art that has accompanied mankind since its very origin. With time, different instruments were invented, capable of producing different types of sounds. Pythagoras was the first philosopher to relate music to mathematics (different pitches to different mathematical ratios), thus founding the theory of music. Music also relates to physics, as a musical instrument produces vibrations that we perceive as sounds; vibrations are described by physics using mathematical terms.

The fabrication of musical instruments also connects art and science. The characteristics of the instrument (shape, proportion of its parts) are determined by the music that one wants to obtain; so is also the choice of the material. Both criteria require practical knowledge of the scientific aspects of music. The hand-making of instruments became an art by itself, and some makers became very famous for their excellent mastery.

18.4.5 Astronomy and the Humanities

Light pollution was not present in ancient times, and many more stars than now could be seen with the naked eyes. Observing stars involved a variety of components: the recognition of patterns leading to the identification of groupings of stars (constellations) involved intuition, imagination and geometry; the perception of shapes for individual constellations involved imagination and led to the association of some of the shapes with myths; the recognition of the apparent motion of the sky during the 24 h of the day and the 365 days of the year involved systematicity in the observations and the ability to identify regularities (patterns) in the changes occurring during specific periods of time; geometry and mathematics brought considerable sophistication to astronomical knowledge.

The names of many constellations are associated with the perceived shapes; when these were associated with myths, the names reflect the mythological being (Sect. 7.4.3.3, Chapter 7). Some constellations were associated with different myths in different cultures. On the other hand, similarities in some perceptions are encountered across cultures; for instance, the planet Venus was named after the Latin goddess of beauty because of its beauty as a brightly shining object in the sky; its name in Tshivenda (*Naledzi masase*, meaning "beautiful star") responds to the same perception.

18.4.6 The History of Science in Science Teaching

The history of science is by itself an overlap between humanities and sciences because history is, by its nature, a humanities-pertaining area, and the history of science combines the historical approach (including its narrative character) with the information about the appearance of new science-related observations, interpretations and theories. To be complete, it also needs to include information about the views on the method that leads to scientific knowledge, i.e., the history of a crucial overlap between science and philosophy (Chapter 13).

The importance of incorporating the history of science into science education is widely acknowledged (e.g., [26–39]) and devoting more attention to it may be viewed as an emerging area of science education [34]. Some works [40] clearly recognise it as a humanities component in science education, which is apt to familiarise students with "the nature of science and scientists, cultural studies, and the multifarious interplay between science/scientists and society". However, its character of overlap between humanities and sciences is often missing within practical approaches.

Historical information by itself provides examples of the scientific approach in such a broad realisations' diversity as to give full evidence that it is not a rigid 'recipe'. The discussion of historical information can thus easily link with the philosophical aspects of the investigation methods in the sciences. This requires the recognition of the role that philosophy plays within the sciences. Without this recognition, the presentation of the history of science risks to turn into something totally different, as diagnosed in [41]:

> All of the chemistry textbooks deemphasize science as a way of thinking. Their authors do not stress the importance of how chemists discover ideas and experiments, the historical development of chemistry concepts, cause-and-effect relationships, evidence and proof, and self-examination of one's thinking in the pursuit of knowledge.

A typical example is offered by the presentation of the development of the particulate model of matter. Too often, it seems to start with Dalton's theory, sometimes with a quick mention of the fact that some ancient Greek philosophers hypothesised that matter is made of particles, but presenting it more as a lucky guesswork than as an interpretation of observed phenomena (the fact that substances can mix). All the debates on the nature of the particles of which matter consists, whether in ancient times or from alchemists and beyond, are ignored. This also prevents clear focus on the nature of Dalton's novel contribution, i.e., the identification of the mass of atoms as the main characteristic which relates them to the nature of different elements (an idea that was replaced by different terms only after the structure of the atom was understood). Neglecting the debates in the centuries prior to Dalton also ignores the long thread of views ascribing importance to the shapes of the particles—a thread that, from the uncertain and improvable initial steps of the past, is currently blooming through the studies of the geometries of molecules, the interactions within and between molecules [42] and the roles that shapes and shapes-complementarity may play for such interactions (a typical example being the investigation of how a biologically active molecule interacts with the active site of a protein).

Some aspects can be considered crucial for the presentation of the history of science within science courses, so that the presentation fosters the development of enquiring minds:

- outlining the enquiry nature of discoveries and proposed theories, the questions inherent in them and the search for answers;
- delineating the continuity of the search, by not presenting the history as spotlights on selected moments as if they were isolated in time, but highlighting the connections between prior and subsequent observations and interpretations;
- emphasising the value of each new step (observations, interpretations, hypotheses) in relation to their time and the insights or progress they brought, without criticising them in the light of modern views: they are part of a journey that has led to modern views, and need to be seen as such.

Acquiring the latter habit is actually crucial both for the history of science and for any other form of history. If we assume that we have reached a higher level of understanding and knowledge in our times, with respect to previous epochs, we also need to acknowledge that these levels have been achieved thanks to all the previous steps and the reflections that were developed to overcome each of them in due time. Alchemists assumed that the transmutation of lead into gold was the main objective of their research; now, we could dismiss this as naïve and dismiss the entire alchemy as something unrealistic, lacking foundations, responding to too primitive ways of thinking, too inferior to modern chemistry to be worthy of our attention; or we can consider all the contributions that alchemy has given to the birth of modern chemistry, first of all the design of laboratory instrumentation, techniques and procedures; only the latter option provides a constructive perception of historical development, which gives new significance to the perception of the development that we are witnessing in current chemistry. The historical development of a science, or of anything else pertaining to human society and knowledge, has not been linear: not every new proposition was better than the one preceding it, but every new proposition ended up being examined, analysed, and discarded when proven inadequate or totally incorrect; thus, the overall process resulted in some form of progress.

There is an ensemble of philosophy components associated with science in general (the question of the method to pursue the best approximations to knowledge, the issue of ethics, and others), and there are philosophical questions associated with each science (including the nature of a given science). These components can be considered from the point of view of their historical development, of their meaning now-a-days, of how they relate to practical aspects, including the understanding of the various concepts and the language we use to express them. It is another area that combines contributions from many types of expertise and whose familiarisation is beneficial to the development of an open and inquiring mind, which is essential in science practice.

Overall, acquiring the habit of seeing a science concept or model also through its historical perspective may enlarge the basis for creativity. To this purpose, it is important to include also information about recent history (e.g., [43]), because this can facilitate the understanding of the conceptual and practical tools currently in use, and also offer suggestions for further work (for instance, certain computational approaches that were too challenging to be convenient few decades ago may become promising if the recent enormous increase of computer power makes then affordable). It will also prevent attitudes that are too narrow to be compatible with a real understanding of how science works. For instance, a reviewer who states that a certain manuscript should not cite references prior to 2018 ignores the importance for a given research to be embedded in the frame of all what has been done previously, to ensure that no significant aspects are neglected; of course, not all the previous works can be cited, above all if there are hundreds of them, but the author will choose those relevant to the work in progress according to how they relate to it conceptually, not according only to their date. A research work without roots risks missing important components. Familiarisation with the history of chemistry also contributes to foster a sound research, research-writing and research-evaluating attitude.

18.4.7 Education—An Endeavour Across Domains

Education is a huge area comprising a high number of fields in relation to the different disciplines and specialisations, to the levels of instruction, to the general-type approaches and the approaches to the teaching of individual disciplines or topics within each discipline, to components such as assessment, and to the best approaches to educational research. Most of the previous chapters contain educational components, and some of them specifically focus on education (Chapters 14–16). The present section mostly relate to the nature of education as an endeavour that cannot be ascribed solely to the science domain or to the humanities domain, because it entails vast component from each of them. A quick historical excursus followed by some focused reflections provides a summary outline of this discourse.

Training the young ones in skills that are essential for survival is observed in a variety of animal species; ethologists study the approaches with which the adults carry on the training. In the human species, the transmission from the adults to the young ones has gradually evolved to incorporate all the already acquired knowledge, including the components that go beyond 'skills', and this enables the addition of new knowledge in each new generation. All cultures developed their forms of schools, and the teacher has become a revered figure in the community.

Socrates was one of the first scholars to investigate the issue of the best approaches to teaching. His maieutic method (etymologically meaning 'midwifery' method) entails the use of dialogues, where the teacher's questions and comments guide the pupils through subsequent logically connected reflections, until a comprehensive picture is formulated, or is 'given birth'. The method is still valid (e.g. [44]), and can be particularly effective for issues requiring students' active engagement, like the search for connections between what is learnt in a science class and everyday life.

Most of the ancient Greek philosophers had their own schools, where they engaged disciples in investigations on different questions. What we now classify as humanities or sciences were integrated not only in the search for answers, but also within education. Philosophers like Empedocles and Parmenides (515 – 460 BC) used the most appreciated literature form—poetry—to express and teach their views. Lucretius used poetry to explain the nature and details of Epicurus' theory on the particulate nature of matter; his book [45] is one of the greatest masterpieces of Latin literature and, simultaneously, the first book aimed at popularising a scientific theory. Poetry was used also to teach about agriculture works, like Hesiod's *Works and Days* (around 700 BC) or Virgil's *Georgics* (29 BC). The interest in education encompassed various areas. Rhetoric art (*ars oratoria*) was the key expertise of orators (the lawyers of those times); the study of geometry was considered an essential component of their training because its demonstration techniques develop the ability to argue in a logical way. The book *Institutio Oratoria*, by Marcus Fabius Quintilianus (≈95 DC), focused on the training of orators, but also expanded to analyse approaches to education in general.

With the birth of universities in the modern meaning of the term (the first one in Europe having been the University of Bologna, founded in 1088), education became increasingly connected with research, as scholars trained students. Other forms of transmission of knowledge and expertise, often combined with stimulation of the young persons' creativity, were realised in the artists' workshops or in the laboratories of the alchemists. As the Enlightenment ideas spread, they fostered the expansion of formal education to increasingly broader groups of youth. The *École Normale Supérieure* of Paris was established in 1794 by the Committee for Public Instruction, and was meant to prepare teachers for secondary schools, to ensure basically homogeneous education to the population of the entire country. Following a decree by Napoleon on "institutions for public instruction" (1810), an analogous institution (*Scuola Normale Superiore*) was established in Pisa (Italy). Both institutions have developed into excellence establishments and are widely known as such; it should also always be recalled that they were started with the objective of preparing high quality educators to promote high quality education in their countries.

The search for the best ways to realise education has gradually grown into a research area of its own, and faculties dedicated to it are currently present in most tertiary institutions across the world. As it happened for most areas, education has developed a range of specialisations (teaching methods, curriculum studies, assessment, and various others), which have often become compartmentalised. Its cross-disciplinary nature is often lost or not sufficiently acknowledged. On the other hand, as some of the previous sections have highlighted, education in one knowledge-area needs expertise also from other areas; for instance, humanities-based abilities like language mastery or epistemological awareness play crucial roles in science learning. Learning chemistry, or physics, or mathematics, cannot rely only on the content (which is in any case essential): teachers and learners need to master communication tools at a level that enables learners to understand the concepts embedded in the content. The content itself needs to be broad enough (including features like epistemological aspects, or history of the discipline, or connections with everyday life and with societal needs) to go beyond the transmission of a set of individual pieces of information and develop into a multisided comprehensive picture. This turns educational research into a type of research needing the convergence of many perspectives from both the humanities and the sciences, whether it focuses on the teaching of a science subject or a humanities subject.

The character and objectives of science education also depend on the understanding of its humanities components. As the general presence and role of humanities in education continue decreasing in recent times, the approaches to science education may change to directions that appear to be more and more dogmatically (or even fundamentalistically) empiricist. An example of this trend is the frequently declared shift of the emphasis from knowledge to skills, or from understanding to "being able to do things". It is easy to realize the connections between the decrease of the role of humanities in education and the decrease of the relevance given to understanding and to the acquisition of knowledge. Knowledge and understanding are holistic aspects pertaining to human intellectual activities, whichever area they are applied to. By their own nature, humanities are not apt to a "skills only" option, because their nature is that of knowledge and understanding. It is not easy to imagine, e.g., doing philosophy or literature critics on a "skills only" basis, without knowledge and without understanding. That the sciences have been regarded as more apt to a "skills only" option is by itself puzzling, unless one hypothesises factors like some misinterpretations of the roles of the practical components in the sciences. These components (laboratory activities, the use of algorithms to solve problems) are an integral part of science, but they are unable to have an independent existence, separated from the overall conceptual body of a science discipline. It is not possible to design a solution procedure without knowing the theoretical aspects on which the design needs to be based and which it needs to utilise, or to interpret an experiment without knowing the available theory related to its nature and scope. Skills without understanding and without knowledge are unable to produce anything beyond the repetition/reproduction of already encountered and memorised cases. Students' attitudes undergo deep changes. Instead of being able (or trying to become able) to discuss science concepts and solution procedures, they memorise what they call "formulas" (basic equations relating quantities) and they try to apply those formulas mechanically. They are often unable even to explain the meaning of the memorised formulas. In extreme cases, students have not been invited to analyse or learn definitions, because definitions pertain to 'knowledge'. Yet, definitions are also the references providing guidelines for the identification of the defined entities and for their utilisation (Sect. 7.3, Chapter 7, [46]). For instance, chemistry students trained on a 'skills only' perspective are not stimulated or required to try and understand the definition of *mole*. When asked to explain what we mean by *mole* they give answers like "it is something that we use when we make calculations on chemical reactions". But not knowing the meaning of the concept results in incorrect uses of the ‘formulas’ containing it, in confusions between the use of masses and the use of moles and, as final outcome, in serious mistakes in those same stoichiometric calculations for which students just know that the mole concept is meant to be used, without understanding its meaning and role. This is true also for technologies areas; for instance, a ‘skills only’ perspective for engineers may solely or dominantly focus on the skills relevant to prepare inputs for the standard engineering software and to “obey” the outputs; engineers trained in this way will never be able to address projects that are not completely standard, and might even lack the bases to identify errors in computer outputs, that might have results from possible imprecisions in the input and might jeopardise the quality or safety of the final product [3].

In broader terms, the “skills only” option results in complete loss of the science perspective. It transmits a narrow and distorted vision of the sciences. The body of science (concepts, interpretations, theories, open questions) remains unknown to students. The inquiring mind attitude is not given any chance to develop.

The “skills only” option is not compatible with the incorporation of humanities-born components that, by their own nature, stimulate conceptual reflection and are associated with it. The consideration of the method, of ethics, of historical development, is possible only if concepts are discussed, and such discussion is the foundation of understanding and knowledge. This highlights another role of the incorporation of humanities-born features into science course: that of preventing the implementation of options that would penalise the future of scientific preparation and, therefore, the future of science itself.

Finally, it may be important to highlight that the current fast progress in science and technologies makes a skills-only approach uneconomical. Required skills change fast in response to the appearance of new technologies. A young person who has been given adequate conceptual foundations will always be in a position to learn new skills, because of being in a position to understand their underpinnings. A young person who has only learnt a certain set of skills will be out of the job-market as soon as those skills become obsolete. This is likely one of the major reasons for which many experts envisage the future as a knowledge-based (not skills-based) society.

18.4.8 Emotional Aspects and Their Educational Relevance

Aspects like enthusiasm, the joy of learning, the feeling of beauty, pertain to the emotional domain and have fundamental roles in people’s behaviour, perceptions and choices. They also have a fundamental role in the practice of any intellectual activity (and not only): being a researcher without enthusiasm for what one is doing would be extremely boring. They also have a role in the effectiveness of teaching and learning processes and can impact on their outcomes up to the point of determining a student’s choice of future career (e.g., LM chose chemistry because her secondary school chemistry teacher was able to stimulate interest in all the aspects of chemistry and its interfaces with other areas).

It is not easy to stimulate enthusiasm in students. First of all, the teacher needs to have it to an extent that is both visible and communicable. It is more probable that teachers can have and communicate enthusiasm if they have that broad-perspective preparation and attitude that enables them to capture and present the links between one discourse and another and between the various disciplines, and make students perceive these links as if they were new, continuous discoveries. Young people minds (above all the very young ones) are not compartmentalised. Their curiosity embraces everything. The connections with everyday life stimulate enthusiasm because they are tantamount to re-discovering familiar facts, viewing them from new points of view and, thus, seeing more in them. The connections with other disciplines stimulate enthusiasm by highlighting the relevance and fundamental unity of the various human intellectual activities, by showing that courses are not isolated. Presenting new information in the course not as a list of facts to memorise, but as discoveries, with their experimental and/or method-related innovations, highlights the continuity of challenges that is inherent in science. Discussing ethical aspects makes them reflect on the ethical challenges posed to and by modern science. For instance, green chemistry is a scientific and technological challenge that can reconcile students with chemistry because of its ethical significance, countering the common public perception of chemistry as something producing and utilising only non-natural and polluting things [47].

The perception of beauty is in-born in human beings. We want our house, or the objects that we use, to be beautiful. Also other things, not always immediately tangible, stimulate perceptions of beauty. A broad-scope range education may enable students to recognise also the beauty of the products of pure thought, like the beauty of a mathematical demonstration or the beauty of logic. It is not unavoidably the privilege of mathematicians to be able to recognise it: others could be guided to it. Guiding students to the appreciation of the beauty of a logical demonstration or a fascinating theory would require collaboration between humanities teachers and science teachers at a considerable sophistication level. A commonly perceived difficulty derives from the lack of such appreciation from several humanities teachers, what, in turn, is often rooted either in the absence or poverty of basic familiarity with the sciences, or in memories of poor success in science courses during past instruction (where the non-succeeding was often the outcome of not having been sufficiently stimulated to consider the sciences as interesting and worth the challenge of trying and understand them). This problem will continue perpetuating through students' and teachers' generation overturns, unless some changes are introduced. The design of courses for humanities teachers, with the specific aim of highlighting the philosophical aspects of the science that is taught at pre-university levels, as well as the logic aspects and the motivational aspects, could be a viable option. These courses would not be science courses, but courses aimed at "knowing each other" between the two domains. Their contents and approaches constitute a whole area open to exploration, as it has not yet been tackled sufficiently.

The 'kingdom' of beauty in education pertains to the history-of-the-arts courses. These courses teach to recognise and appreciate the beauty created by human beings. Science and technology have provided the instruments for its creation (Chapter 10).

The collaboration between the art courses and the science courses can highlight aspects like the chemistry of painting and the broadening of colour-options made available by the discovery of new compounds, the chemistry of clay works, the role of geometry in the realisation of beautiful architecture works, and many others. In this way, the history-of-the-arts course would teach to recognise and appreciate beauty and also to appreciate the mastery of the artist in using available materials, instruments and technologies to create beauty. On the other hand, history-of-the-arts is often mentioned by inexperienced or scarcely informed public as one of the totally useless areas in education, with the implicit or open suggestion that it could be cancelled altogether; those who propose this are obviously not aware of how, for instance, design-related professions (which enjoy good recognition in the modern world) benefit enormously from the history of art.

Attention to emotional aspects in science education is currently growing (e.g., [48]).

18.5 Humanities and Sciences in Other Cultures

18.5.1 The Dichotomy Between Humanities and Sciences

The historical summaries included in the chapters of this book mostly refer to developments in the Mediterranean area and in Europe, from ancient times to modern days, because the classification of human intellectual activities into humanities and sciences appeared in Europe, and the dichotomy between them also appeared in Europe in recent centuries, and has been increasingly growing. Currently, both the classification and the dichotomy are widely spread, and most universities in the world have separate faculties for the two domains. The existence of specific faculties and departments is understandable and has functional roles in view of the ever-growing specialization within each disciplines, with the number of specialised sectors, as well as the specialisation depth, continuing growing. On the other hand, there is increasing need of ‘bridging the gaps’, or fostering the possibility and practice to ‘talk to each other’, in view of the benefits stemming from the synergies and of the damages engendered by inadequate knowledge of information and skills born within one domain and fundamental also for the other.

The types of knowledge and expertise that are now classified as ‘humanities’ and ‘sciences’ have been present in all cultures, mostly in an integrated form. While it would not be interesting or constructive to bring in artificial dichotomies where none existed, it can be interesting to investigate the various forms in which they were integrated in different cultures. Besides its general-character interests as important components of human development, the information from such studies is likely to bring novel insights and suggestions to the search for routes aimed at bridging the gap between the two domains.

No suggestions or recommendations can be outlined here concerning the history of humanities and sciences components in individual cultures, because doing so would require deep knowledge of those cultures. On the other hand, the outcomes would be interesting for everybody. It can be hoped that the present book might encourage new explorations in this regard and, possibly, also new collaborative studies.

Each culture contains some 'jewels' that may be valuable for the entire human community. The most suitable 'jewel' for the interactions among cultures, when features pertaining to individual cultures are concerned, is likely one that can be derived from Islamic teaching, and would be apt to become a guideline independently of one's religious or other convictions. There is an image according to which God has made human beings of different groups not for them to hate each other, but for them *to come to know each other*. This could be an ideal guideline whenever diversity is concerned, and could also prevent some unplanned or unrecognised blunders. It could also be an ideal guideline in all endeavours in which cultural heritages and knowledge are important. Keeping it in mind when investigating the roles and relationships between what we now call humanities and what we now call sciences in other cultures can favour a fully unbiased character of the research, thus enabling valuable contributions to the overall understanding of human intellectual activities.

The next paragraph provides a quick focus on Sub-Saharan Africa, not in terms of results about the presence and relationships of humanities and sciences in the traditional cultures (which still need a lot of investigation), but in terms of how synergies between humanities and sciences can play important roles for key issues that are currently the focuses of keen attention.

18.5.2 *A Quick Focus on Sub-Saharan Africa*

18.5.2.1 Humanities and Sciences in the Study of Indigenous Knowledge Systems

Indigenous Knowledge Systems (IKS) are the ensembles of knowledge that have developed autonomously within indigenous societies. They are objects of increasing attention, also in view of the contributions that they can bring to the pursuit of sustainability [49].

Components that would now be classified as pertaining to humanities and components that would now be classified as pertaining to sciences are fully integrated in traditional knowledge. For instance, traditional medicine involves 'scientific' type of knowledge for the knowledge of medicinal plants, how to prepare the medications, what is the right medication for a given disease (which also implies knowledge of the human body); it also involved interpretation aspects and ritualistic aspects that can be viewed as more closely linked to humanities.

The collaboration between experts in the humanities and experts in the sciences is important in IKS studies, to attain a more complete picture of individual areas of knowledge. For instance, indigenous chemistry practices are comparatively easy to identify for chemists who work on the basis of the definition of chemistry as the science of substances (e.g., [50, 51]). On the other hand, someone who is familiar with the history of chemistry may identify similarities across times and across cultures, thus broadening the overall picture. Someone who is familiar with collecting story-telling and folklore may identify perceptions related to substances and their transformations, which are interesting for a more complete picture of traditional chemistry. The potentialities of synergies between the two types of expertise are still largely to be explored.

18.5.2.2 Ensuring that African Languages Express Science Fully: An Integrated Enterprise Between Humanities and Sciences

The importance of language mastery for science learning and science creativity has been quickly outlined in Sect. 4.3.2 and is extensively documented in Chapters 4 and 5. Similarly, the role of the mother tongue for thought [52] and for clear and complete understanding in the sciences is recognised by many studies; an overview of the main concepts and related references is given in Chapter 4. Sub-Saharan Africa is the context where second language instruction is most extensively practiced, using the language of the former colonial power. Several studies consider the shift to mother tongue instruction to be fundamental for development, as the only option that can ensure the level of mass education that is necessary for development (earlier studies comprise [53–59]). The importance of mother tongue instruction is also recognised at policy-making level. The language policy framework in South Africa [60] is meant "*to provide a framework for the development and strengthening of indigenous languages as languages of scholarship, teaching and learning and communication at South African public higher education institutions, in particular, universities*". This marks a paradigm-shift with respect to previous practices in a number of Sub-Saharan Africa contexts, where mother tongue instruction has been limited to the early instruction stages, mostly not beyond primary school level, or even limited only to the very first years of primary school. Such a practice inherently fosters a second-rate perception about the mother tongue, thus prompting part of the public to feel that English-medium instruction throughout could better equip the young ones for their future work and professional life. On the other hand, not using the mother tongue as a medium of instruction entails not learning its theory adequately, and this deprives the young ones of a full development of their thinking abilities (including abstract thinking) and also of the possibility of mastering the second language to the level that is desirable or required in professional activities (good mastery of the mother tongue being an essential foundation for the good mastery of any second language [61, 62]).

For the mother tongue to be given the full respect that it deserves, it needs to be ready-for-use to express contents at any level, including the most sophisticated science, engineering and philosophical discourses. It is argued in this section that the integration of expertise from the humanities and the sciences is crucial to pursue this objective, and that they main focus needs to be the identification of the resources of indigenous languages to express all the details of concepts, including inherently epistemological ones. The next paragraphs briefly elaborate on this.

Two major language-related components can be identified in a scientific discourse: the terminology (technical terms denoting objects, phenomena and operations) and the selection of individual words and of the construction of sentences to express concepts and descriptions fully and rigorously (Chapter 7, [63]). When a language has not yet been extensively used to communicate science, terminology may not yet be adequately present; on the other hand, building it is not a formidable challenge, because it is an ensemble of names, and names can be chosen or formed according to the criteria that appear more convenient, such as importing the root words from another language and adapting the morphology to the indigenous language, or creating them de novo using root-words of that language [63]. These two options may be simultaneously adopted in any proportions that appear desirable, with the former option being preferable for terms that already use the same root-words in several languages (e.g., words like *atom* or *molecule*). This freedom in the choice of the most suitable options to build terminology is possible because technical terms are just names: they are not the knowledge carriers. The knowledge carrier is the entire sentence, and its meaning is determined by the common words linking the technical terms (Chapter 7, [63]). Therefore, the main task while building the possibility that a language can express science rigorously is the identification of the resources of that language, which can be used to express all the wanted details, including epistemological ones. This, in turn, implies the need to analyse each concept in detail in order to establish which language-resources are needed. In other words, the criterion is not that of translating words or sentences from English in some way, but that of analysing all the features involved in a certain concept and identifying the best way to express those features through the resources of the indigenous language.

A simple example may conveniently illustrate these ideas. In the sentence

1. The nucleus of the atom consists of protons and neutrons

The words "nucleus", "atom", "protons", and "neutrons" are technical terms, i.e., they are names of objects. The information is conveyed by the entire sentence, and is determined by the common words ("the", "of", "consists", "and") which connect the technical terms. Their roles can be analysed in the following way:

- Using the article "the" before "nucleus" is consistent with the fact that each atom has only one nucleus.
- The preposition "of" after "nucleus" identifies the 'owner' of the nucleus. In this case, both "of the atom" and "of an atom" would be correct in English.
- Using the verb "consists" informs that protons and neutrons are the only two types of particles present in the nucleus. It could be replaced by the verb "is made". The verb "contains" would not be appropriate, because it would allow the possibility of the presence of other types of particles.
- The preposition "of" after "consists" introduces the 'items' that make up the nucleus.

- The conjunction "and" between "protons" and "neutrons" stresses that the list of the particles is complete.
- The way in which the sentence is constructed (e.g., the word combination "*the nucleus of the atom*") informs that this statement has general validity, i.e., it is true for all the atoms.

These are the aspects that need to be conveyed through the expression in another language. The identification of the best options requires the collaboration of scientists and linguists: a scientist with adequate awareness of the details inherent in each concept, and a linguist with deep knowledge of the resources of the indigenous language concerned. For instance, statement *1* entails the convergence between chemistry knowledge and language knowledge: knowing that an atom contains only one nucleus pertains to chemistry knowledge, and knowing that this, in English, is communicated by the use of the article "the" before "nucleus" pertains to language knowledge; knowing that protons and neutrons are the only types of particles present in the nucleus pertains to chemistry knowledge, and knowing that this, in English, is communicated by the use of the verb "consists" pertains to language knowledge. How to express these features through the indigenous language pertains to the expertise of the linguist building the statement in the indigenous language by using its resources. For instance, a Venda colleague explained that Tshivenda has a linguistic tool to specify whether a certain statement has general validity or is true only for specific cases; that tool would likely be appropriate in the expression of statement *1* in Tshivenda.

Other simple examples can highlight a logical feature that is often missed by second language students, and that is conveyed in English by the use of the indefinite article "a":

2. The internal energy is a state function
3. KOH is a strong base
4. A chemical reaction is a process in which ....

The use of the indefinite article "a" before "state function" implies that the internal energy is not the only state function (there are others); its use before "strong base" implies that KOH is not the only existing strong base; its use before "process" implies that chemical reactions are not the only type of processes. In other words, the subjects of these three statements are elements in larger ensembles of elements pertaining to the same category (state functions, strong bases and processes respectively). Second language students often do not get this piece of information conveyed by the use of the article "a" and replace it by "the", thus building incorrect statements. The expression of these concepts through the mother tongue needs to make this aspect clear. In this way, the rigorous expression of a concept in the students' mother tongue becomes a mental reference facilitating a rigorous expression when the student uses a second language [61]. The mother tongue becomes a "mental home" [64] where concepts are clear in all their details. This would result in generalised better understanding of science concepts and also in more correct expression of those concepts through other languages.

An analysis of the errors in second-language students' works is capable of highlighting many details that students do not identify through the second language, and these are often details that need to be taken care of when building concepts' expressions through the mother tongue. Several of the examples analysed in Chapter 4 may also be viewed under this perspective.

It may be suitable to conclude this subsection with a comment from a lady who sells fruits and vegetables in the open market in Sibasa (Limpopo Province, South Africa). The issue of mother tongue instruction came up as a response to the question why countries in Sub-Saharan Africa have not yet managed to attain the type of development that countries like South Korea and other Asian economic giants have achieved, and noting that all those countries use the mother tongue as the language of instruction. The lady explained that "Our languages can express science, but they have not yet put the science into them". She meant that the indigenous languages have the resources to express science, but the technical terms have not yet been built. This concept summarises the guidelines for expressing sciences through the indigenous languages: building technical terms remembering that they are just names, and identifying the resources of each language to ensure rigorous expression of concepts and descriptions.

18.5.2.3 Interconnected Roles of Humanities and Sciences in the Design of Curricula Incorporating Africa-Related Contents

The incorporation of information pertaining to the students' culture has proven to enhance students' interest in the given subject, thus also facilitating understanding (e.g., [65, 66]). It also responds to the pedagogical importance of relating what is taught in the class to students' everyday-life experience ([51] and references therein).

In Sub-Saharan Africa, there is diffuse consensus on the importance of incorporating IKS into formal education ([67–72]; other references in [51]). The main challenge is the design of the optimal ways to realize it and to maximise the benefits for pupils and students. This requires collaboration between humanities and sciences for several reasons. The study of IKS requires this collaboration (Sect. 18.5.2.2), and IKS studies are the studies that provide the content materials. The design of pedagogical options is also a cross-domain issue (Sect. 18.4.7). The dichotomy between the two domains is not present in IKS, and this can be one of the values to be highlighted to students. Students may also be involved in IKS research [48, 51, 73], collecting information on the traditional practices with which they come in contact and being encouraged to appreciate the knowledge embedded in them (for instance, the knowledge about substances and their properties in traditional chemistry practices).

18.6 Take-Home Messages

This section highlights the main take-home messages, with particular emphasis on the components that more closely respond to the 'working conception' notion. Because of its nature, this section also constitutes (or replaces) the 'Discussion and Conclusions' section that customarily appears at the end of a chapter.

18.6.1 Take-Home Messages from Individual Chapters

The take-home messages from individual chapters are summarised in separate subsections for each chapter, for the sake of clarity. The take-home messages of the present chapter (18) are incorporated into the take-home messages of the whole book.

18.6.1.1 Chapter 1

Humanities and sciences were born together as integrated components of human intellectual activities, and remained closely linked up to 2–3 centuries ago. The current and increasingly growing dichotomy entails drawbacks for each of them. Each domain can benefit from components pertaining to the other domain and be enriched by them. In particular, science related activities, including science education, need adequate language mastery and at least basic epistemological awareness to be effective. The future of science depends substantially on the attention that will be given to these two humanities-based components. Exploring relationships between humanities and sciences, which can be infused into the ways of doing things, thus taking a working character, appears to be particularly relevant in the current situation, to counteract the dichotomy growth through the concrete benefits that bridging it can bring.

18.6.1.2 Chapter 2

It is from time to time necessary to re-evaluate the culture that shapes the beliefs and behaviour of society, and especially to reflect on the reigning ideas of a specific culture, including on what it considers to be the nature of "progress"—especially regarding the benefits and costs to society over the longer term. A new consciousness and appreciation of the potentially useful complementarities in the contributions of different scholastic traditions are needed. To discover new knowledge about the constituent elements, or workings, of matter and the living environment is necessary, but it is not sufficient to use this new knowledge wisely: a new way of thinking about the workings of complex living systems are also needed. Therefore, the contributions of both divergent and analytical thinkers, of humanists and scientists, of West and

East and of modern and traditional thinking should be considered—if only to break the mould of the culturally-shaped mind. To cross the cognitive divide between the humanities and the sciences, multidisciplinary problem-solving through dialogue, heuristic learning and soft systems thinking is proposed.

18.6.1.3 Chapter 3

Humans have been observing the physical world and trying to make sense of it for many millennia. Eventually, this activity was organized into a coherent practice known as Natural Philosophy. Humans have also observed one another and tried to understand human culture. One definition of The Humanities is the study of human creativity. A persistent notion in academic circles is that Natural Philosophy and The Humanities belong to two different domains and that there can be no useful dialogue between them. The nature of Natural Philosophy and of The Humanities has been examined through the consideration of examples and areas of complementarity between the two domains have been explored. The explorations demonstrate that both areas of human scholarly activity would benefit from considering the insights of the other approach.

18.6.1.4 Chapter 4

A sufficiently sophisticated level of language mastery is essential for science learning. Language mastery inadequacies severely impact on the clarity and completeness of conceptual understanding, and hamper the quality and correctness of the texts that students submit for assessment. The consideration of concrete examples referred to chemistry learning highlights a high variety of categories of language-related difficulties, and the modes in which they affect learning. Careful attention to rigour by the teacher, and the provision of expert guidance, are crucial for students to understand concepts and procedures and to be able to express the acquired knowledge. The integration of language training into chemistry courses (or other science courses) can offer the most permanent addressing option; the practical aspects of its implementation require intensive explorations, including at teachers' training level, and entails simultaneous and integrated humanities-based and science-based expertise.

18.6.1.5 Chapter 5

Engineering, as well as other implementation and production oriented areas going under the *technology* collective term, requires strong conceptual bases for the design and the implementation stages to have the required quality, leading to products where functionality and safety are both optimised. The acquisition of the needed knowledge largely depends on language mastery and associated logical abilities; its conceptual depth, and consequent suitability for the search of viable solutions for a broad variety of problems, requires basic familiarity with epistemological characters. The communication of instructions for practical implementations requires great precision in their expression, for them to be univocally understood and executed. Therefore, adequate mastery of language, adequate familiarity with logic, and basic familiarity with epistemology—all being humanities-based tools—are essential for both the design and the implementation stages, and need to be integral component of an engineer's training. A quick overview of the development—through history—of the investigation of the relationships between language and knowledge, and language and thought, provides the bases to better understand the motivations for the crucial roles of language in science and technology learning and practice.

18.6.1.6 Chapter 6

Computers provide invaluable aid for an increasingly high number of operations. However, this does not authorise human beings to decrease the amount and depth of knowledge that they need to acquire and apply: the extent and functionality of computer-provided aids depend on the extent and quality of the knowledge possessed by the humans using the computer. This is true also for text-writing. Computer-performed spelling correction, computer's suggestions for different wording, computer-performed translations into another language, as well as the indication about plagiarism and self-plagiarism detection, need careful verification by humans. This implies that good language mastery and adequate epistemological bases are fundamental for humans to be in a position to maximise the benefits of the information provided by the computer and simultaneously avoiding errors or loss of rigour. This remains true also with regard to the recently developed AI-based large language models, with the additional caution that the human in charge needs to verify the absence of fabricated references or other types of fabricated information. Since language mastery and epistemology familiarity are humanities-based abilities, optimal usage of the writing-related computer resources depends on the presence and adequacy of humanities-related expertise in the human's overall preparation.

18.6.1.7 Chapter 7

It is important to distinguish the roles of terminology and the roles of common language in scientific communication, with the latter being the one that realises communication, thus being the information carrier. Terminology is the ensemble of names, and related adjectives and verbs, denoting objects, qualities and processes that are of interest within a given discipline. The generation of these terms can be viewed as a long thread of interactions between the two domains, developing through a variety of aspects, in correspondence with the circumstances and contexts of each discovery, with the discoverers' impressions and perceptions, and with the richness of options made available by their deep humanities preparation. Presenting the origin of at least some of these terms within science education can stimulate the awareness that terminologies are ensemble of names (an important epistemological aspect), the perception of broader connections between the name and the time when it was introduced, and the perception that science practice is not isolated from culture.

18.6.1.8 Chapter 8

Epistemology is a cross area between humanities and sciences, focusing on the meaning of the knowledge humans attain, and analysing the details and power of the methods through which it is attained. While the epistemology of science has been the object of investigation since ancient times (when science was part of philosophy), and the practical implications of its subsequent advances have been implicitly incorporated into engineering design whenever expedient, specific attention to the epistemology of engineering is recent. Its development needs to give attention to the specificities of the connections of the major components of the epistemology of science with engineering practice, because the practical character of engineering determines the way in which it utilises theory, modelling and experiments in the design process—the component requiring theory-based investigations, complemented by simulations and experiments and, therefore, the component with more immediate links with epistemology. Works reporting the full bodies of projects which have utilised innovative design, including all the theoretical approaches and novelties, can extensively contribute to emphasise the value of epistemology in engineering practice. Building awareness of the epistemology of engineering in the training of engineers is bound to have relevant roles in their professional practices, above all when they encounter non-standard problems such as the design of non-standard structures, which entail the design of the relevant mathematical models taking into account all the details of the structure complexity; this, in turn, requires the integration of various types of knowledge, which is productively guided by epistemological awareness. The more complete will the body of 'epistemology of engineering' (currently in its construction stage) become, the more confident can engineers be on designing solutions, including those meant to address non-standard problems.

18.6.1.9 Chapter 9

The chapter presents 'multidisciplinarity at work', in a project in which different branches of engineering interact among themselves, with other sciences like geology and chemistry, with a discipline—architecture—that is by its nature a discipline where sciences and humanities converge, and with humanities-based disciplines such as history, literature and folklore. It is concluded that multidisciplinarity can take many forms, which have to be identified and tailored for each individual context or project. The best results are obtained when all possible contributions from different disciplines that may have some connections with a given project are explored, and the new information generated by their synergies is turned into operational approaches. The explorations need not to be restricted to only one domain (either science-based disciplines or humanities-based disciplines), as cross-domain synergies can bring new valuable perspectives.

18.6.1.10 Chapter 10

The progress of chemistry went hand in hand with new possibilities for materials to be used in the arts. This is particularly evident in the area of pigment production: new discoveries by chemists placed new artistic materials into the hands of artists and, at times, the search for pigments with specific shades of colours prompted new chemistry searches. The transfers between the two disciplines were so quick that one can see them as co-evolving as partners in progress.

18.6.1.11 Chapter 11

The dichotomy between art and science is only apparent, and the methodological analogy between artists and scientists remains as integral today as it was in the Renaissance. This can be highlighted by some aspects of chemistry.

Over time, scientists have learned how to synthesize new molecules not existing in nature, thus inventing new geometries and new functionalities. Similar to macroscopic-world buildings made of suitable materials, with construction order and mechanical strength that must meet the demands of its functionality, a molecule is characterized by its constituent atoms, by the succession and spatial order of chemical bonds and by the strength of the bonds themselves. From this perspective, chemistry is the equivalent of architecture in the atomic microcosm.

On the other hand, the arts could help inspire and contribute to the design of future technologies based on nanoscale materials, which hold great potential for improved energy conservation, conversion, and harvesting. Connections between visual and applied art are epitomized by the riveting tessellation of two dimensional materials, such as metal organic frameworks. The tremendous diversity of these atomically-thin 2D materials presents unique opportunities for creating new nanostructures through new vertical and horizontal stacking, with virtually limitless possibilities. Some forms of arts can inspire new patterns for these nanostructures.

18.6.1.12 Chapter 12

Through centuries, a number of scholars have had expertise both in the humanities and in the sciences, leaving impacts in both. In other cases, components from one area have been incorporated into the other with enriching outcomes. Synergies between poetry and science can yield results that are precious both for the messages conveyed by poetry and for the way in which expression through poetry may stimulate imagination in the sciences. The consideration of selected cases provides concrete illustrations. The educational relevance of attracting students' attention on these synergies also deserves adequate emphasis.

18.6.1.13 Chapter 13

The idea that sciences and humanities have nothing in common, either from the point of view of the methods they use or from the point of view of their investigation subjects, neglects all the mutual exchanges between scientific disciplines and humanities through many centuries and does not take into account the variety of current developments in both domains. In particular, history and scientific disciplines have been comparing, and at times sharing, research methods, analytical tools, and objects of study. The dialogue is still ongoing: history research incorporates tools designed within the sciences, and mutual influences open new pathways, including founding new branches of study, such as the ethnosciences. This motivates adequate presence of the history of science within history courses at all levels of instruction, as part of a route to building broad-basis background knowledge. The continuance of fruitful interchanges between history and the sciences relies on ample background knowledge, including the nurturing of language mastery as an essential tool for both learning and research, in any field of specialisation.

18.6.1.14 Chapter 14

Many current and emerging problems span humankind and higher education and require multiple perspectives. Disciplinary segregation impedes innovative solutions by isolating educators within organizational silos. Pedagogical integration provides collaborative processes for science and humanity educators seeking to address complex problems. Integration can vary from minimal (multidisciplinary) to moderate (interdisciplinary) to extensive (transdisciplinary). Multidisciplinary courses consist of two or more educators teaching from discrete perspectives. Students implicitly integrate different perspectives. Poetry writing was taught to pre-nursing students to improve their communication and emotional processing skills. Interdisciplinary courses explicitly integrate perspectives within the classroom. Hands-on activities combined mathematics, physics, and art to spark creativity within students. Transdisciplinary courses meld perspectives and new ideas and outlooks emerge. Engineering and peace study students worked together to develop drones that have a positive impact on society, such as examining palm trees for weevil infestations. Integration fosters innovation within higher education in which problems and solutions are viewed from holistic perspectives.

18.6.1.15 Chapter 15

Chemistry students prepare for a job life as innovators and contributors to techno-scientific advancements, which may have significant impact on the life-world of societies and on the planet as a whole. Because of this, they should not grow into scientific-technical innovation agents that shift responsibilities for the ethical dimensions of technological developments to other actors or that decide over these issues without adequate legitimation. The introduction of the normative literacy concept as a course component can foster a responsible-agency outlook for future researchers or technology developers. Thus equipped with normative literacy and discourse skills, acquired through the encounter with concepts and methods from the humanities, they can be expected to be able to make a difference in team dynamics and professional networks as repositories of the ethical and societal discourses that shape technologies' implementations.

18.6.1.16 Chapter 16

Science teachers' education has a fundamental influence on the quality of science education at all levels of instruction. Given the increasing importance of the education for sustainable development (ESD), science teacher education curricula should change appropriately in advance of school curricula changes, which must be made eventually. This requires careful development of parts of the existing curricula for prospective science teachers to include content, skills and attitudes suited to ESD. This would lead to an increase in the presence of human sciences components in science teachers' education curricula.

18.6.1.17 Chapter 17

The merger of the International Council for Science and the International Social Science Council into the International Science Council constitutes a significant indication of the importance that humanities and sciences work together for crucial issues concerning society and concerning the planet—first of all the pursuit of sustainable development. The identification and development of viable practical routes for 'working together' deserves to be the object of active research.

18.6.2 Take Home Messages from the Entire Book

Overall, this book stresses the importance of cross-disciplinary collaborations between humanities-based areas and science-based areas, the huge potentialities for mutual benefits and their relevance for education. It also encourages explorations for viable routes that can be used for implementations within concrete situations and for a variety of purposes; the examples outlined in the various chapters may thus serve as starting inspirations.

The main conceptual messages can be summarised as follows:

- General importance of cross-disciplinary collaborations;
- Importance of method-related (epistemological) awareness within each domain, and mutual enrichment from exchanges of views and expertise across domains;

- Importance of humanities-based components like language-mastery and epistemological awareness in science and engineering practice and education;
- Importance of science-derived tools for research within humanities-based areas, like history.

A variety of benefits, interests and reflections may ensue for different categories of readers. Some of them can be summarised as follows:

- Anybody interested in human intellectual activities: the book provides overviews of relationships between the two domains, comprising both working components and history reviews related to them and recalling their underpinnings;
- Teachers in all areas: ideas for the presentation of the relationships between the two domains to students. Ideas for possible students' projects (including the possibility of inviting students to 'invent' and design their own project after reading one of the chapters);
- Science and technology teachers: awareness of the roles of language mastery and epistemological bases in science teaching and learning, including concrete guidelines that can be derived from the numerous examples analysed in various chapters;
- Researchers in various areas: investigation of possible cross-disciplinary collaborations between the two domains, which may be relevant to their specific field (e.g., researchers in epistemology, researchers in education, specific science or engineering projects, etc.);
- Other professionals in various areas: the awareness of the benefits of cross-disciplinary collaborations between the two domains may foster new ideas relevant to their activities;
- Decision-makers: awareness of the importance of some humanities-based areas for the progress of science and technology. Necessity of avoiding downgrading areas that have essential roles in the progress of knowledge. Necessity of not decreasing the presence of humanities components in primary and secondary education.

In summary, this book presents selected synergies between humanities-based areas and science-based areas and outlines their roles and benefits, as well as viable routes for concrete implementations. The range of advantageous possibilities is vast, and the exploration potentialities are numerous. The examples presented in these chapters may also become seeds for the identification and development of many more synergy routes.

References

1. L. Mammino, Incorporating information on green chemistry into theoretical chemistry courses. Curr. Opin. Green Sustain. Chem. **13**, 76–80 (2018)
2. L. Mammino, Computational chemistry for green design in chemistry and pharmacy: building awareness in the classroom. Sust. Chem. Pharm. **18**, 100283 (2020)

3. L. Mammino, Computational chemistry and green chemistry: Familiarizing chemistry students with the modes and benefits of promising synergies. Sust. Chem. Pharm. **29**, 100743 (2022). https://doi.org/10.1016/j.scp.2022.100743
4. L. Mammino, Green chemistry and computational chemistry: a wealth of promising synergies. Sust. Chem. Pharm. **34**, 101151 (2023). https://doi.org/10.1016/j.scp.2023.101151
5. L. Mammino, *Il linguaggio e la scienza* (Societá Editrice Internazionale, Turin, 1995)
6. L. Mammino, Some general-character reflections on the relationships between humanities and sciences. Ometeca **IX**, 156−178 (2005)
7. A. Zadeh, Fuzzy sets. Inf. Control. **8**(3), 338–353 (1965). https://doi.org/10.1016/S0019-9958(65)90241-X
8. A. Bignoli, *Pensamientos de un ingeniero* (Ediciones Dunken, Buenos Aires, 1997)
9. C. Ranzoli, in *Dizionario di scienze filosofiche* (Hoepli, Milano, 1963)
10. J.C. Callahan, *Ethical Issues in Professional Life* (Oxford University Press, New York & Oxford, 1988), p. 7
11. J.B.R. Gaie, The Moral Basis of Green Chemistry, in *Green Chemistry in Africa*, eds. by P. Tundo, L. Mammino (IUPAC and INCA, Venice, 2002), pp. 16–30
12. M. Martuzzi, J.A. Tickner (eds.), *The Precautionary Principle Protecting Public Health, the Environment and the Future of Our Children* (WHO Europe, 2004)
13. D. Kriebel, J. Tickner, P. Epstein, J. Lemons, R. Levins, E.L. Loechler, M. Quinn, R. Rudel, T. Schettler, M. Stoto, The precautionary principle in environmental science. Environ. Health Perspect. **109**(9), 871–876 (2001). https://doi.org/10.1289/ehp.01109871
14. J. Peel, Precaution—A matter of principle, approach or process? Melb. J. Intern. Law **5**
15. A. Wallace Hayes, The precautionary principle. Arch. Ind. Hyg. Toxicol. **56**(2), 161–166 (2005)
16. Precautionary Principle. Wikipedia. Accessed 22 July 2022
17. ISC Position on Conference and Event Boycotts (issued May 2021), Freedom of association—International Science Council
18. https://www.vocabulary.com/dictionary/ideology#:~:text=An%20ideology%20is%20a%20set,not%20all%20%2Dism%20words%20are
19. I. Ijabs, Ideology and quality of life, in *Encyclopedia of Quality of Life and Well-Being Research*, ed. by A.C. Michalos (Springer, Dordrecht, 2014). https://doi.org/10.1007/978-94-007-0753-5_1368
20. L. Zmigrod, A psychology of ideology: unpacking the psychological structure of ideological thinking. Perspect. Psychol. Sci. **17**(4), 1072–1092 (2022). https://doi.org/10.1177/17456916211044140
21. J. Chapoutot, *Il nazismo e l'antichità* (Einaudi, Torino, 2017), pp.12–21
22. Alfonso Lucifredi, "L'invenzione della razza ariana. Storia della spedizione delle SS in Tibet, alla ricerca pseudoscientifica delle origini della razza pura", Il Tascabile, Treccani online, https://www.iltascabile.com/scienze/tibet-razza-ariana/, latest Accessed 29 June 2023
23. P. Ball, *Serving the Reich—The Struggle for the Soul of Physics Under Hitler* (University of Chicago Press , 2014). https://press.uchicago.edu/ucp/books/book/chicago/S/bo19196057.html
24. Pliny the Elder, *Naturalis Historia*; written across several years, until Plinius' death in 79
25. Vitruvius, *De Architectura* (Biblioteca Universale Rizzoli, Milan, 1988)
26. H. Siegel, Kuhn and Schwab on science texts and the goals of science education. Educ. Theory **28**, 302–309 (1978)
27. R.A. Duschl, Science education & philosophy of science twenty-five, years of mutually exclusive development. Sch. Sci. Math. **87**(7), 541–555 (1985)
28. R.W. Bybee, J.C. Powell, J.D. Ellis, J.R. Giese, L. Parisi, L. Singleton, Integrating the history and nature of science and technology in science and social studies curriculum. Sci. Educ. **75**(1), 143–155 (1991)
29. M. Monk, J. Osborne, Placing the history and philosophy of science on the curriculum: a model for the development of pedagogy. Sci. Educ. **81**, 405–424 (1997)
30. F. Seroglou, P. Koumaras, The contribution of the history of physics in physics education: a review. Sci. Educ. **10**, 153–172 (2001)

31. D. Koliopoulos, S. Dossis, E. Stamoulis, The use of history of science texts in teaching science: two cases of an innovative, constructivist approach. Sci. Educ. Rev. **6**(2), 44–56 (2007)
32. G. Gooday, J.M. Lynch, K.G. Wilson, C.K. Barsky, Does science education need the history of science? Isis **99**(2), 322–330 (2008). https://doi.org/10.1086/588690
33. M.R. Matthews, History, philosophy, and science teaching: the new engagement. Asia-Pacific Forum Sci. Learn. Teach. **10**(1), 1–14 (2009)
34. M. Niaz, L.A. Montes, Towards a history and philosophy of chemistry. Educ. Quím. 290–297, online 12/04/2012
35. M. Niaz, Review of Matthews, M.R., Science Teaching: The Contribution of History and Philosophy of Science (20th Anniversary Revised and Expanded Edition). Educ. Quím. **26**(2), 174–176 (2015)
36. F.H. Khazaei, B. Roucau, C.S. Kalman, Can learning about history of science and nature of science in a student-centred classroom change science students' conception of science? Creat. Educ. **9**(15) (2018). https://doi.org/10.4236/ce.2018.915194
37. R.M. Clary, J.H. Wandersee, The history of science in the science classroom: the past is the key to the future in science education. Earth Sci. Hist. **34** (2), 310–332 (2015). https://doi.org/10.17704/1944-6187-34.2.310
38. G.C. Powers, The history of chemistry in chemical education, *Isis.* **111** (3), 576–581 (2020)
39. Y. Yaru, X. Liang, Y. Yanan, H. Chenyang, G. Yingying, Chemistry history education infiltrating cultivation of scientific literacy. Front. Educ. Res. **3**(12), 1–6 (2020). https://doi.org/10.25236/FER.2020.031201
40. L.E. Klopfer, G.S. Aikenhead, Humanistic science education: the history of science and other relevant contexts. Sci. Educ. **106**(3), 490–504 (2022). https://doi.org/10.1002/sce.21700
41. E.L. Chiappetta, G.H. Sethna, D.A. Dillamn, A quantitative analysis of high school chemistry textbooks for scientific literacy themes and expository learning aids. J. Res. Sci. Teac. **28**, 939–951 (1991)
42. L. Mammino, Computational chemistry: studying the properties and behaviours of molecules, in *Green Chemistry and Computational Chemistry: Shared Lessons in Sustainability*, ed. by L. Mammino (Elsevier, 2021), pp. 1–39
43. L. Mammino, The recent history of theoretical chemistry presented from a method-related perspective. J. Molec. Struct. (Theochem) **769**(1–3), 61–68 (2006)
44. L. Mammino, The use of questions in the chemistry classroom: an interaction instrument with maieutic nature. Anu. Latinoam. Educ. Quím. **XXI**, 241–245 (2006)
45. Lucrezio, in *La natura delle cose*, ed. by G. Milanese (Mondadori, Milan, 1992)
46. L. Mammino, Il ruolo delle definizioni nell'insegnamento delle scienze. Sc. E Citta'. **1**, 25–37 (2000)
47. P. Tundo, T. Patti (eds.), *IUPAC Workshop on Green Chemistry Education Report* (Venice, (Italy), 2001), p. 19
48. V.M. Jiménez, L.J. Blanco Nieto, A.B. Borrachero Cortés, J.A. Cárdenas Lizarazo, *Las emociones en la enseñanza y el aprendizaje de las ciencias y las matemáticas*, (Spain), ISBN: 978-84-15090-10-6
49. S. Rist, F. Dahdouh-Guebas, Ethnosciences—a step towards the integration of scientific and indigenous forms of knowledge in the management of natural resources for the future. Environ. Dev. Sustain. **8**(4), 467–493 (2006). https://doi.org/10.1007/s10668-006-9050-7
50. L. Mammino, T. Tshiwawa, Chemistry practices in the Vhavenda indigenous society. Indilinga, African J. Indig. Knowl. Syst. **16**(2), 221–237 (2017)
51. L. Mammino, Indigenous knowledge and the roots of African chemistry: A wealth of potentialities for chemistry education, in *Problematising Local Indigenous Community Research—Afro-Sensed Perspectives*, eds. by. M.A. Masoga, A.L. Shokane, K.J. Gross. Book series: Knowledge Pathing: Multi-, Inter and Trans-Disciplining in Social Sciences Series, vol. 2. (AOSIS, South Africa, 2021), pp. 5–28
52. E. Tomaselli, Senza parole. L'impoverimento del linguaggio nell'era moderna. http://www.freakoutmagazine.it/09-03-2019/focus-on/101117/senza-parole./. Accessed 20 July 2022

53. K.K. Prah, *Mother Tongue for Scientific and Technological Development in Africa; German Foundation for International Development* (Bonn, Germany, 1993)
54. N.C. Nangu, How can we be expected to learn in science, in *Proceedings of the second annual meeting of the Southern African Association for Research in Mathematics and Science Education*, ed. by M.J. Glencross (University of DurbanWestville, Westville, 1994)
55. K.K. Prah, *African Languages for the Mass Education of Africans; German Foundation for International Development* (Bonn, Germany, 1995)
56. K.K. Prah, In Between Distinction and Extinction; K.K. Prah (Ed.), (Witswatersrand University Press, Johannesburg, South Africa, pp. 1–16 (1998)
57. K.K. Prah, in Rehabilitating African Languages, ed. by K.K. Prah (CASAS, Cape Town, South Africa, 2002), pp 1–6
58. Y.I. Rubanza, Competition through English. The Failure of Tanzania's Language Policy. in *Rehabilitating African Languages*, ed. by K.K. Prah (CASAS, Book series No. 18, 2002), 39–51
59. B. Brock-Utne, R.K. Hopson, (eds.), *Languages of Instruction for African Emancipation—Focus on Postcolonial Contexts and Consideration* (CASAS, Cape Town , 2005)
60. The language policy framework for public higher education institutions, Government Gazette, 30 October 2020, No. 43860 (2020)
61. L. Mammino, The mother tongue as a fundamental key to the mastering of chemistry language, in *Chemistry as a Second Language: Chemical Education in a Globalized Society*, ed. by C.F. Lovitt, P. Kelter (American Chemical Society, Washington, 2010), pp.7–42
62. M. Qorro, Language planning and policy in Tanzania: when practice does not make perfect, in *The Role of Language in Teaching and Learning Science and Mathematics*, eds. by Z. Desai, M. Qorro, B. Brock-Utne (African Minds, Somerset West, 2013), pp. 151–170
63. L. Mammino, *Terminology in Science and Technology—An Overview Through History and Options* (Ditlou Publishers, Thohoyandou, South Africa, 2006)
64. S. Seepe, A pedagogical justification for mother tongue instruction, in *The language of science*. ed. by S. Seepe, D. Dowling (Vivlia Publishers, South Africa, 2000), pp.40–51
65. B. Marasinghe, Ethnochemistry and ethnomedicine of ancient Papua New Guineans and their use in motivating secondary school children and university undergraduates in PNG. Univ J. Educ. Res. **4**(7), 1718–1720 (2016). https://doi.org/10.13189/ujer.2016.040726
66. A. Siwale, I.S. Singh, P. Hayumbu, Impact of ethnochemistry on learners' achievements and attitude towards experimental techniques. Intern. J. Res. Innov. Soc. Sci. (IJRIISS) **IV**(III), 534–542 (2020)
67. A. Breidlid, Culture, indigenous knowledge systems and sustainable development: a critical view of education in an African context. Int. J. Educ. Dev. **29**, 140–148 (2009)
68. H.O. Kaya, Y.N. Seleti, African indigenous knowledge systems and relevance of higher education in South Africa. Intern. Educ. J. Comp. Persp. **12**(1), 30–44 (2013)
69. M. Mawere, Indigenous knowledge and public education in Sub-Saharan Africa. Afr. Spectr. **2**, 57–71 (2015)
70. J.C. Ronoh, Indigenous knowledge in the school curriculum: teacher educator perceptions of place and position, M.Sc. thesis at the Nelson Mandela University, Port Elizabeth, South Africa, 2017
71. C. Ezeanya-Esiobu, *Indigenous Knowledge and Education in Africa* (Springer, Singapore, 2019)
72. D. van Kruger, J. De Beer, Infusing indigenous knowledge in the life sciences classroom: design principles for creating an ethnobiological matrix, in *8th Teaching & Education Conference*, Vienna, 17 September 2019. https://doi.org/10.20472/TEC.2019.008.014
73. L. Mammino, Chemical heritage in sub-Saharan Africa and its significance for chemical education. Southern African J. Folklore Stud. **33**(2), 147–158 (2020)
74. C. Horner, E. Westacott, *Thinking Through Philosophy: An Introduction* (Cambridge University Press, Cambridge, 2000), p. 119

MIX
Papier aus verantwortungsvollen Quellen
Paper from responsible sources
FSC® C105338

If you have any concerns about our products,
you can contact us on
ProductSafety@springernature.com

In case Publisher is established outside the EU,
the EU authorized representative is:
Springer Nature Customer Service Center GmbH
Europaplatz 3, 69115 Heidelberg, Germany

Printed by Libri Plureos GmbH
in Hamburg, Germany